Einstein Studies

Editors: Don Howard John Stachel

Published under the sponsorship of the Center for Einstein Studies, Boston University

Volume 1: Einstein and the History of General Relativity
Don Howard and John Stachel, editors

Volume 2: Conceptual Problems of Quantum Gravity
Abhay Ashtekar and John Stachel, editors

Volume 3: Studies in the History of General Relativity
Jean Eisenstaedt and A.J. Kox, editors

Volume 4: Recent Advances in General Relativity
Allen I. Janis and John R. Porter, editors

Volume 5: The Attraction of Gravitation: New Studies in the History of General Relativity
John Earman, Michel Janssen and John D. Norton, editors

Volume 6: Mach's Principle: From Newton's Bucket to Quantum Gravity
Julian B. Barbour and Herbert Pfister, editors

Volume 7: The Expanding Worlds of General Relativity
Hubert Goenner, Jürgen Renn, Jim Ritter, and Tilman Sauer, editors

Volume 8: Einstein: The Formative Years, 1879–1909
Don Howard and John Stachel, editors

Volume 9: Einstein from 'B' to 'Z'
John Stachel

Volume 10: Einstein Studies in Russia
Yuri Balashov and Vladimir Vizgin, editors

John Stachel

Einstein from 'B' to 'Z'

Birkhäuser
Boston • Basel • Berlin

John Stachel
Center for Einstein Studies
Boston University
Boston, MA 02215
U.S.A.

Library of Congress Cataloging-in-Publication Data

A CIP catalogue record for this book is available from the Library of Congress, Washington D.C., USA.

AMS Subject Classifications: 0079, 01-06, 83-03, 83-06, 83A06, 83A05, 83Cxx

Printed on acid-free paper.

The Einstein studies series is published under the sponsorship of the Center for Einstein Studies, Boston University.

Birkhäuser ®

ISBN 0-8176-4143-2 SPIN 19901785
ISBN 3-7643-4143-2

Typeset by TEXniques, Inc., Cambridge, MA.
Cover design by Mary Burgess, Cambridge, MA.
Cover sketch by Julius C. Turner, 1932.
Printed and bound by Hamilton Printing, Rensselaer, NY.
Printed in the United States of America.

9 8 7 6 5 4 3 2 1

Contents

Introduction

This volume assembles what I hope are my best articles and talks on Einstein prepared over the past quarter of a century. The title, *Einstein from 'B' to 'Z'*, is meant as a humorous reminder that the book makes no claim to offer a complete picture of Einstein — an account from 'A' to 'Z.' Of course, every human life is inexhaustible, and any account of it represent a selection from the available material and presentation of it from a standpoint adopted by the author. But, in spite of the enormous rivers of ink that have been devoted to the topic, I believe it is still too early to attempt a full picture of Einstein's entire life. The eight already-published volumes of *The Collected Papers of Albert Einstein* have revolutionized our picture of the young Einstein; but these volumes have only reached his fortieth year, and future volumes may be expected to present further surprises. Readers may consult the final essay in this volume, a review of of a recent biography of Einstein, for an idea of what happens when a premature synthesis is attempted.

Thus, the reader will not be surprised to learn that many of the essays in this volume are concerned with the first half of Einstein's life, the understanding of which—both of the man and of his work—has been considerably changed by new documentary evidence. From the mists of obscurity and myth, there starts to emerge the portrait of a human being, of his strengths and weaknesses, and of his often-contradictory strivings (for example, defiance of authority in physics coexisted with a longing for recognition of his work), who is a thousand times more interesting than the saintly figure of the legend. And there is also emerging a much clearer picture of the development of his ideas about relativity, both the special and the general theory, and about the quantum theory—his greatest contributions to modern physics.

An advantage of this collection of independently prepared essays is that each item may be read individually. The reader may dip into the book at his or her own pleasure—and I hope that every reader will find pleasure in at least some of the essays—with no sense of obligation to read the entire collection. On the other hand, I hope that at least the outlines of a picture of Einstein that differs in major respects from the conventional hagiographic accounts will emerge from reading all the essays. I find contemplation and critique of a living human being much more fascinating than worship of even the most beautiful plaster saint.

Rather than presenting them chronologically, the essays are grouped thematically. The first section, called "The Human Side," starts with "Einstein, the Man Beyond the Myths," and attempts to extract some features of the real Einstein from the cloud of myths that still surround him. Two encyclopedia articles on Einstein, one from an encyclopedia of physics and the other from an encyclopedia of Judaism, provide some of the factual information about Einstein that is assumed in other essays. The next two essays are concerned with the young Einstein, in particular, with the troubled relationship between him and his first wife. Again, there are a number of myths to be confronted, in particular the claim that she was responsible for a major portion of the work on the special theory of relativity published under his name. An attempt then follows to characterize "Einstein's Jewish Identity," how it developed and how it colored his approach to the Zionist movement. A short essay on "Einstein and Civil Liberty," prepared for a civil rights journal, pays tribute to the role Einstein played in the 1950s during the McCarthyite attack on basic American freedoms. The section ends with an essay that gathers together a number of Einstein's comments on the nature of creativity—scientific and artistic—which shed a good deal of light on his own creative process.

The second section, "Editing the Einstein Papers," presents a bit of the story of my work as founding editor of the *Collected Papers of Albert Einstein*. More importantly, it explains the principles of historical editing upon which that edition is based, and gives a few examples of the fruits of this method. Since I retired as editor some years ago, I hope it will not sound immodest if I assert that, in the future, all serious research about Einstein must be based on the use of this edition of his writings and correspondence.

The next four sections deal with various aspects of Einstein's scientific work. Two "Surveys of Einstein's Work" deal, respectively, with his work up to 1909, the year in which, at age thirty, he entered the academic world full time, and with an underlying dichotomy in Einstein's thought throughout his life: between his work on theories based on the space-time continuum and his feeling that an explanation of quantum phenomena might call for abandonment of that continuum. Then follow sections on "Special Relativity," "General Relativity," and "Quantum Theory," three areas of physics to which Einstein made fundamental contributions. Each section begins with some less technical surveys of the development of Einstein's work in this area and proceeds to more detailed—and hence sometimes more technical—consideration of some specialized topics. I have not tried to eliminate duplication in the content of these essays; hence, each one may be read independently of the others. The careful reader will also note that later essays occasionally express a somewhat different outlook on particular issues than earlier ones; such changes present the evolution, hopefully progressive, of my views in response to the discovery of new documents and/or a deeper understanding of already-known ones.

While I have not treated Einstein's important work on statistical physics in a special essay, readers interested in this topic will find a good deal about it in the essays on "Einstein's Light Quantum Hypothesis" and "Einstein and Bose."

The following section is on "Einstein and Others." A short piece discusses some of the parallels and the differences in the careers of the young Einstein and the young Newton, the physicist with whom Einstein has been most often compared. Five pieces follow elaborating Einstein's relationship with five of his contemporaries: Sir Arthur Eddington, Leopold Infeld, Cornelius Lanczos, Satyendra Nath Bose, and Wolfgang Pauli. These essays elucidate various aspects of Einstein's personality as it expressed itself in his personal contacts with five remarkable—and remarkably different—men, as well as in the significant intellectual exchanges between them.

The book concludes with reviews of two outstanding biographies of Einstein now available in English. In spite of some weaknesses in each, which I indicate, those wanting to delve further into Einstein's life and work can do no better today than to start with one or both of these books.

It was just a quarter of a century ago, in 1976, that I was called upon to become the founding editor of *The Collected Papers of Albert Einstein*. (Interested readers will find some of the ensuing story in the second.) This began my transformation from a theoretical physicist, working primarily on the theory of relativity and its philosophical implications, into an historical editor, concerned with the minutiae of Einstein's life and work and called upon to speak and write on this topic on many occasions. Since retiring as editor, I have continued my interest in Einstein and the many topics in theoretical physics, and the history and philosophy of science raised in the course of consideration of his career. Perhaps the most important lesson that I have learned from my own career is the vanity of such disciplinary boundaries: The unit of research should be the problem, and many of the problems that I have been led to consider have aspects that combine elements of what would ordinarily be classified as theoretical physics, history of science, and philosophy of science. Accordingly, I hope that readers from each of these disciplines, as well as many non-specialists interested in Einstein, will find something helpful and thought-provoking in this book.

John Stachel
Center for Einstein Studies
October, 2001

Part I

The Human Side

Albert Einstein: The Man Beyond the Myth

John Stachel

In the last analysis, fame is only the epitome of all the misunderstandings which gather about a new name.

RAINER MARIA RILKE

These words were first quoted with reference to Einstein in 1930 by his son-in-law Rudolf Kayser. Today, this comment seems even more appropriate. Few modern figures, and certainly no other scientists, have been the center of such an ever-growing list of myths and misconceptions. Great mythic figures of the past, such as Buddha, Lao Tse and Jesus, are known through the legends that have come down to us. But in the case of Einstein, we can watch the mythic process in operation.

Some Myths

Perhaps the most widespread myth about Einstein is that he was born old. His name calls to mind a white-haired, saintly figure, well advanced in years. It takes an effort to remember that he was born at a more tender age, passing through a childhood, adolescence and young manhood that were often quite stressful. He abhorred the drill-sergeant atmosphere in his Munich primary school and dropped out of high school to spend a half-year wandering about Italy. His father's repeated business failures left him unable to attend a university without financial help from relatives. His failure to get an academic job upon graduation—when every other member of his class did—left him on the brink of financial disaster. When he did secure a steady job, it was as a technical expert, third-class, in the Swiss Patent Office in Bern. Seven years later, he finally started an academic career, rising rapidly to a full professorship. In 1914, the year he started work at a full-time research position in Berlin, he had to deal with two major traumas: separation from his first wife, who left Berlin forever with their two sons, to whom he remained deeply attached; and the outbreak of World War I, when he found himself at the center of German militarism (which he depised) as a wave of national chauvinism

Bostonia Magazine
Vol. 56 pp. 8–17
February 1982

swept over most of his colleagues. And Einstein was just 35! There was a long and complex lifetime leading to the saintly, white-haired figure of the myth.

Associated with the myth of creation at age 70 goes another myth: "Einstein the beloved of all humanity." How could anyone help but love that saintly image? No one can ever be sure, but it is possible that Einstein was as much hated as he was loved during his lifetime. He was hated as a Jew, a pacifist, a democrat and civil libertarian, a radical and a socialist. He was hated not only during his years in Germany, where his life was threatened numerous times in the 1920s, but also in the United States, where his efforts on behalf of Jewish and other anti-fascist refugees earned him the hatred of "America Firsters" from his earliest days in this country. It is amusing—if you have a wry sense of humor—to compare the laudatory tributes to Einstein published in his centennial year of 1979 with what those same sources had to say about him during the early 1950s. Consider, for example, this *New York Times* critique. In June, 1953, after Einstein advised a teacher not to cooperate with the House Un-American Activities Committee, the *New York Times* chided: "To employ the unnatural and illegal forces of civil disobedience, as Professor Einstein advises, is, in this case, to replace one evil with another." In those days, Einstein stood like a bastion in defense of the Bill of Rights while so many other intellectuals were collapsing under the McCarthyite onslaught.

Turning to science, another myth is that Einstein created airtight, perfect physical theories that will endure forever. Could anyone ever challenge the theory of relativity? Of course, this is a total misconception of the nature of science, a misconception that Einstein never shared. He was quite aware that his way of trying to understand the physical universe had grown out of a long history of previous attempts. He had to criticize these attempts in order to go beyond them. But, as early as 1917, soon after he had perfected the general theory of relativity, he noted that just as his theory had gone beyond Newton's, sooner or later someone would have to go beyond his theories. He conceived the evolution of physics as a constant attempt to solve new problems using new clues, comparable to a detective novel; except that in science we can never turn to the last page and find out "who did it."

Another misconception is that Einstein was completely impractical, a purely abstract thinker. That is perhaps the opposite of the case. Einstein grasped the world in concrete images. His problem was to translate those images into words and equations that could be shared with other people.

"The words or the language, as they are written or spoken, do not seem to play any role in my mechanism of thought." Einstein wrote. "The physical entities which seem to serve as elements in thought are certain signs and more or less clear images which can be 'voluntarily' reproduced and combined."

"The above-mentioned elements are, in my case, of visual and some of muscular type. Conventional words or other signs have to be sought for laboriously only in a secondary stage, when the mentioned associative play is sufficiently established and can be reproduced at will."

Einstein first demonstrated his independent streak to professors at the Swiss Technical University. Later, this trait, in addition to his being a Jew, probably was responsible for the fact that he was not offered an assistantship at the University, although his student colleagues received offers. All other academic avenues were also closed to him. Through the kindness of one of his school friends, he was offered a position with the Patent Office in 1902. By the time he left that position, he had published over two dozen papers, at least four of which were profoundly significant.

Although one might imagine that a genius like Einstein would have hated devoting himself to the daily drudgery of the Patent Office routine for seven years, that was not the case. Einstein actually looked back on it as one of the happiest periods of his life: "The work on satisfactory formulation of technical patents was a true blessing for me. It compelled me to be many-sided in thought, and also offered important stimulation for thought about physics. Following a practical profession is a blessing for people of my type. Because the academic career puts a young person in a sort of compulsory situation to produce scientific papers in impressive quantity, a temptation to superficiality arises that only strong characters are able to resist."

Einstein maintained his interest in inventions throughout his life. He was promoted before he left the Patent Office, and was called in as a patent expert in various legal cases long after he had won the Nobel Prize for physics in 1922. In addition, he held numerous patents on inventions made alone or jointly with colleagues. It is true that financial matters were of relatively little importance to him; so people who equate practicality with solvency might well consider him impractical.

His Scientific Work

Einstein once listed three key questions he had posed to himself in the course of his study of physics. The pursuit of these questions had motivated his "proper life's work." The first question was: "How does the representation of a light ray depend on the state of motion of the coordinate system with respect to which it is referred?"

Einstein started to think about this problem at age 16. He began to wonder what would happen if you were to chase a light ray, running faster and faster. Could you catch up with it, and if you did, what would it look like? Many a precocious adolescent might think about such a question for five or 10 minutes, or even half an hour. Einstein thought about this question, and the even deeper problems that grew out of it, for 10 years. Looking back, he saw in this question the beginning of his work on the special theory of relativity, completed in 1905 at the age of 26, in which the speed of light is the ultimate, unattainable speed. Einstein was capable of thinking about questions like that, developing and elaborating them for a decade and then coming up with a profound new physical theory to answer them. His tremendous tenacity was combined with the ability to focus it on questions that not only troubled him, but were really vital for the advance of physics.

The second key question he mentions is: "What is the basis for the equality of the inertial and gravitational mass of bodies?"

It had been well known for 300 years, since Galileo and his famous Leaning Tower experiment, that all bodies fall with the same acceleration when dropped. Einstein saw this not as a simple fact, but as a major puzzle. He started thinking about this problem shortly after he completed the special theory of relativity. He continued thinking about it, and the other problems that grew out of it, for about eight years until he had developed the general theory of relativity. This was Einstein's theory of gravitation—still the best theory of gravitation we have. One cannot say that Einstein's theory is the final, ultimate theory, but so far no one has succeeded in probing more deeply into the nature of gravitation. There have been many attempts at such alternate theories, but Einstein's has withstood all tests.

The third question characterizing his most important scientific work was: "Can the gravitational field and the electromagnetic field be theoretically grasped in a unified manner?"

Einstein started to think about this question between 1916 and 1917, shortly after he developed the general theory of relativity, and continued to think about it until the end of his life. It seemed artificial that the two basic natural forces then known, gravitation and electromagnetism, should be explained by two quite distinct theories. This question occupied his mind for almost 40 years, but he never arrived at any solution that was completely satisfactory to himself, let alone to the rest of the physics community. Yet for Einstein, success was *not* the most important thing. He had come upon a deep problem he felt required an answer. So he felt justified in devoting the major part of his efforts to looking for that answer. Even if he did not succeed, he was sure there was an answer. For a long time, the search for such a unified theory of all the basic physical forces was out of fashion among physicists. But in recent years, it has come back into fashion, although in ways rather different from those Einstein worked on (several new forces have been discovered, for example).

Einstein concluded that: "These three questions characterize my own proper life's work. Whatever else I occupied my mind with was more occasional work... and is related to the current problems of physics."

"Whatever else" includes all his work on quantum theory and statistical mechanics. The significance of this work has been characterized as follows: If you ask most physicists today who is the greatest physicist of our century, they would name Einstein for his work in relativity theory. If you ask them who is the second-greatest physicist of our century, quite a reasonable answer would be Einstein for his work outside of relativity.

Einstein was motivated to find a solution to what he considered the really important questions because he felt a "logically simple interpretation of empirically known connections existed. This state of mind," he explained, "is comparable to the state of mind of a person who wants to solve a riddle or chess problem about which he is indeed convinced the solution certainly exists because the person who made up the riddle possesses it."

There have been many myths about Einstein's religious views. For him, cosmic religion was the guarantee that there is a solution to the puzzle, that the universe is lawful. This feeling of certainty that the universe is rational is what enabled Einstein to work for 10 years on special relativity, almost a decade on general relativity and then to work the rest of his life on the unified field theory problem. If he could not find the solution to some problem, it was not the fault of the universe!

Einstein's religiosity thus had little to do with religion in the conventional sense. He considered "ethics to be an exclusively human concern with no superhuman authority behind it." This did not mean he undervalued morality. As he noted, "... people of our type see in morality a purely human matter, albeit the most important in the human sphere."

Einstein was a good teacher, although he did tire of teaching the same basic conventional material over and over. But this is not uncommon even among good pedagogues. It is clear from accounts by former students that he was not always the conventional "Herr Professor" of the German academic system, but this was certainly not held against him by his students. It is also clear from these accounts that he loved to expound his ideas and that he was very successful at it.

"Einstein's delivery in his lectures was quite unrhetorical, anything but brilliant. With expressively opened eyes, the chalk in his slightly raised right hand, he stood at the board, often looking off into the distance. He spoke rather softly. He was at that moment no more, but also no less, than that which he was thinking about out loud, like a spotlight guided from without, which in vivid perspective lit up every new portion of a landscape, itself remaining modestly in the darkness and background..."

In 1914, after a conventional academic career of only five years, Einstein was called to Berlin to fill a post especially tailored to his talents. He could devote as much time to research as he wanted; he had the right to teach at the University of Berlin if he desired, but no fixed teaching obligations. He gave several courses of lectures at Berlin and elsewhere before he left Europe in 1933. Thereafter, he occasionally gave individual lectures at Princeton or elsewhere in the United States.

His university experiences left him rather cynical, yet not unappreciative: "In truth the university taken as a whole is a machine with a very poor efficiency, and yet irreplaceable, and indeed also not essentially capable of improvement. Here the public must assume the standpoint that the biblical God took toward Sodom and Gomorrah: for the sake of a few, the whole effort must be made. And it is worth it."

His Social Views

There are countless myths regarding Einstein's social and political views. In addition to being considered an "ancient sage," Einstein was thought of as the bleeding heart, the naive sufferer for all humanity, who would endorse any cause if it were presented to him in sufficiently pathetic terms. This myth completely misses a central element in Einstein's emotional makeup that he once described: "My passionate sense of social justice and social responsibility has always contrasted oddly with my pronounced lack of need for direct contact with other human beings and human communities. I am truly a 'lone traveler' and have never belonged to my country, my home, my friends or even my immediate family with my whole heart; in the face of all these ties, I have never lost a sense of distance and a need for solitude, feelings which increase with the years."

Einstein built his "inner equilibrium" upon the foundation of his scientific quest rather than upon personal relationships.

Although in later years he spent a large part of his time responding to the many requests for his help in various personal and scientific matters, as well as social and political causes, he maintained a certain inner distance. He also chose his issues carefully. He was well aware that his fame gave him a vast audience for his views; but he also realized that he had to use this instrument sparingly if it was to remain effective. So he refused many times to speak out on an issue. If he disagreed with the cause, he naturally refused; but he also refused to endorse appeals for causes that he favored if he was skeptical about the good faith of the sponsors, or if he thought his voice would not add significantly to the impact of what had already been said.

There is no evidence that Einstein was politically active before World War I broke out. It was the impact of that war, particularly seeing how most intellectuals fell victims to chauvinistic passions, that stirred him into action. His first political moves were in connection with the peace movement in Germany. He was still a Swiss citizen, but even so, such actions were rather risky in wartime Germany. His name appears on a list of notable pacifists prepared for the Berlin police chief. This was the beginning of a lifelong association with peace movements.

His pacifism sometimes gave rise to intense controversy. At the beginning of the 1930s, Einstein urged young men to refuse military service, provoking the hostility of nationalists in many countries, including the United States. After Hitler came to power, he felt this was no longer a viable tactic and urged the rearmament of the democratic states in the face of the fascist peril. This provoked cries of betrayal from many erstwhile pacifist colleagues. Einstein felt that changed circumstances called for changed actions: since advocacy of disarmament and refusal of service were impossible in fascist states, their advocacy elsewhere played into the hands of the dictators. He never regarded his actions as a renunciation of pacifism.

In this connection one meets another myth: Einstein, the father of the atomic bomb. Everyone knows Einstein showed that $E = mc^2$, even if he or she does not know exactly what E, m, c or squared mean; and everyone knows this has something to do with the A-bomb. Actually, the work leading to the discovery of nuclear fission and production of the A-bomb could have taken place just as well even if Einstein had never derived that formula from his special theory of relativity. After all, fires were lit long before a theory of chemical combustion existed. But Einstein sent a letter to Roosevelt that triggered the production of the A-bomb, didn't he? Einstein certainly sent the letter, but the role it played in the development of the American A-bomb project has often been greatly exaggerated. Einstein never worked in nuclear physics and played no other role in the Manhattan Project.

"My part in producing the atomic bomb consisted of a single act: I signed a letter to President Roosevelt pressing the need for experiments on a large scale in order to explore the possibilities for the production of an atomic bomb. I was fully aware of the terrible danger to mankind in case this attempt succeeded. But the likelihood that the Germans were working on the same problem with the chance of succeeding, forced me to this step."

"I could do nothing else, although I have always been a convinced pacifist. To my mind, to kill in war is not a whit better than to commit ordinary murder. As long, however, as nations are not resolved to abolish war through common actions and to solve their conflicts and protect their interests by peaceful decisions on a legal basis, they feel compelled to prepare for war. They feel obliged to prepare all possible means, even the most detestable ones, so as not to be left behind in the general armament race."

"This road [of an armaments race] necessarily leads to war, a war which under the present conditions means universal destruction. [He wrote this at the time the H-bomb was being developed.] Under these circumstances the fight against *means* has no chance of success. Only the radical abolition of wars and of the threat of war can help. That is what one has to work for. One has to be resolved not to let himself be forced to actions that run counter to this goal. This is a severe demand on an individual who is conscious of his dependence on society. But it is not an impossible demand."

Einstein was deeply concerned about the social and moral responsibility of the scientific community to try to end the arms race.

"We scientists whose tragic destiny it has been to help make the methods of annihilation ever more gruesome and more effective must consider it our solemn and transcendent duty to do all in our power to prevent these weapons from being used for the brutal purpose for which they were invented. What task could possibly be more important for us? What social aim could be closer to our hearts?"

His Jewish Identity

It was also after World War I that Einstein first started to be concerned with his identity as a Jew. He had been brought up in a rather secular Jewish home and had never identified closely with the Jewish community. It was witnessing the post-war growth of anti-Semitism, particularly as it flared up in Germany, that led him to identify with that community and decide to support the work of the Zionist movement. Some people have seen a paradox here. Einstein always pronounced himself, and indeed was, a confirmed internationalist: How could he reconcile internationalism with his activities on behalf of the Jewish people and their attempts to build a homeland in Palestine? He answered that question in 1926:

"Generally speaking, it does not accord with my ideal that communities bound together by the bond of race or tradition should make special efforts to cultivate and emphasize their separateness. Insofar, however, as a given community is attacked as such, it is bound to defend itself as such in order that its individual members may be able to maintain their material and spiritual interests. Corporate action is needed to save the individual from those spiritual dangers that isolation necessarily entails. Whoever understands this clearly must approve of united action by all Jews for a corporate purpose, be he ever so unsympathetic in principle to nationalism."

Along with his support of the Zionist movement went a deep awareness of the problem of the existence of two peoples in Palestine. He was quite active in trying to find some means of cooperation, some modus vivendi, for the Jewish and Arab communities in Palestine. Insofar as he could, to the end of his life he worked for a common Jewish–Arab dialogue and mutual understanding.

Einstein's concerns with economic and social inequalities growing out of the organization of our society ultimately led him to the conclusion that only a socialist reorganization could provide the framework for a solution to such problems.

> "...the essence of the crisis of our time... concerns the relationship of the individual to society. The individual has become more conscious than ever of his dependence upon society. But he does not experience this dependence as a positive asset, as an organic tie, as a protective force, but rather as a threat to his natural rights, or even to his existence. Moreover his position in society is such that the egotistical drives of his makeup are constantly being accentuated, while his social drives, which are by nature weaker, progessively deteriorate. All human beings, whatever their position in society, are suffering from this process of deterioration. Unknowingly prisoners of their own egotism, they feel insecure, lonely, and deprived of the naive, simple and unsophisticated enjoyment of life. Man can find meaning in life, short and perilous as it is, only through devoting himself to society."

And he pinpointed the cause of this crisis:

> "The economic anarchy of capitalist society as it exists today is, in my opinion, the real source of the evil. We see before us a huge community of producers, the members of which are unceasingly striving to deprive each other of the fruits of their collective labor—not by force, but on the whole by faithful compliance with legally established rules."

He was not naive, however, about what constituted socialism. He certainly did not think that just having a planned economy would solve all problems, being well aware of what had happened in the Soviet Union.

> "Nevertheless, it is necessary to remember that a planned economy is not yet socialism. A planned economy as such may be accompanied by the complete enslavement of the individual. The achievement of socialism requires the solution of some extremely difficult socio-political problems: how is it possible, in view of the far reaching centralization of political and economic power, to prevent bureaucracy from becoming all-powerful and overweening? How can the rights of the individual be protected and therewith a democratic counterweight to the power of bureaucracy be assured?"

Einstein's views on such issues as education, war and peace, discrimination, planning, and freedom are far from irrelevant to our current situation. In many ways he is still our contemporary, whom we would be well advised to praise less and read more.

His Fame

Fame came to Einstein in November, 1919, as a result of the publicity given to the English eclipse-expeditions, which confirmed his prediction of the so-called deflection of starlight by the sun's gravitational field. Einstein could never understand why the theory of relativity, with its elaborate conceptual but minute empirical refinements of the Newtonian theory of gravitation, gripped the popular imagination so vividly and for such a long period of time. Why did the results of the eclipse expedition have such an extraordinary impact on the general public? Why did Einstein's person become a center of attraction, gravitational or otherwise, for so many stories, legends and myths? Many reasons have been offered but I feel none of them has provided an adequate explanation.

When he was 70, Einstein summed up his feelings about fame in a letter to a relative: "It is a curious thing, to see how one appears from the perspective of others. It was my fate that my accomplishments had been overvalued beyond all bounds for incomprehensible reasons. Humanity needs a few romantic idols as spots of light in the drab field of earthly existence. I have been turned into such a spot of light. The particular choice of person is inexplicable and unimportant."

Considering some of the individuals idolized in the 20th Century, I find it comforting that a man like Einstein has become an enduring symbol of human aspirations. Referring to his latest attempt at a unified field theory, he continued:

"After years and years of ceaseless efforts I have rounded off the relativity theory. But on account of enormous mathematical difficulties, I am not myself in a position to judge if the completion of the theory has hit the correct mark. The contemporary generation of my colleagues does not believe in it, so that I stand rather isolated in my viewpoint on fundamentals without being certain that I have taken the right path. Anyway, the arguments are good and the remainder of my days is well employed, even if my powers bear no relation to the magnitude of the task..."

He closed the letter with this advice: "Be glad of your calm days and do not let yourself be deceived by the appearance of external circumstances. Take pleasure that only a few care about you and, believe me, it has its good side. Better an understanding spectator than an electrically-illuminated actor."

Albert Einstein*

John Stachel

Einstein's ancestors had long lived in small south German towns. The well-to-do family of his mother Pauline was in the wholesale grain trade, and his father Hermann was a small businessman. Like many German Jews of their generation, his parents never denied their origins but were nonobservant and culturally quite assimilated. Always independent minded and rather a "loner," young Albert was close to his sister Maja. A period of childhood religiosity ended at twelve when popular scientific literature made him a free thinker, but a feeling of wonder at the harmony of the universe never left him.

In 1880 the family moved to Munich, the largest south German city, where Hermann and his brother Jakob, a trained engineer, started one of the city's first electrotechnical firms. Raised in a technological milieu, Albert was originally destined to take over the family business, which at first flourished. In 1894 competition from larger German firms led the brothers to relocate to northern Italy, where further business reverses soon led to the breakup of the partnership. Helped occasionally by Albert, Hermann's small, debt-ridden business ended with his death.

Einstein attended primary and secondary school in Munich. He found much of the curriculum and above all the instructional methods distasteful, later comparing most of his teachers to drill sergeants. His slow but thorough and methodical approach to the subjects in which he was interested earned him good but not outstanding grades. Encouraged by Uncle Jakob, he developed a bent for mathematics, especially geometry and calculus, mainly through self-study; a family friend stimulated a precocious interest in the natural sciences and philosophy. When the family left Munich in 1894, Albert stayed to finish school, but soon left for

*born Ulm, Germany, March 14, 1879; died Princeton, New Jersey, April 18, 1955.

John S. Rigden, ed.
Macmillan Encyclopedia of Physics
Vol. 2, pp. 393–397

Italy after clashing with some of his teachers. He continued studying on his own, hoping to be admitted to the Poly (Swiss Federal Polytechnical School, a technical university) in Zurich, which recognized his talent but recommended he finish secondary school at the nearby Aarau Cantonal School. Its more liberal style of teaching and excellent scientific facilities soon changed his attitude towards schooling and made apparent his talent. In 1896 he entered the Poly as a physics student, earning generally good grades; but his independent attitude did not ingratiate him with his teachers. Happily working in the newly-equipped physics laboratories, he remedied the dearth of advanced physics courses by self-study of theoretical physics, often joined by fellow physics student Mileva Marić, whom he married in 1903.

Unable to find a university or secondary-school position in physics after graduation in 1900, he worked at a variety of temporary jobs until hired as an Examiner at the Swiss Patent Office. While there (1902–1909) he completed studies of the special theory of relativity, Brownian motion, and molecular dimensions (the topic for which he received his doctorate), all in 1905; published his first papers on quantum theory and started work on the general theory of relativity.

Growing recognition of his accomplishments by the physics community soon followed. He was appointed associate professor of physics at the University of Zurich in 1909 and two years later became a full professor in Prague (then part of the Austro-Hungarian Empire), but returned to Zurich—now at the Poly—in 1912. He was appointed to the Prussian Academy of Sciences in 1914, a full-time research position in Berlin, where he completed work on the general theory of relativity (1915). Separated from Marić in 1914, he married his cousin Elsa Einstein in 1919.

Einstein remained in Berlin throughout World War I, where his pacifist views set him against the mainstream of academic jingoism; and during the Weimar Republic, when his democratic views made him a hero to defenders of the republic and a target of the growing fascist movement. With Hitler's advent to power in 1933, Einstein left Germany for good and settled in Princeton, New Jersey, at the newly-established Institute for Advanced Studies, where he continued to live and work for the rest of his life.

Einstein regarded the special and general theory of relativity and the search for a unified field theory as the central thread of his life's work. The special theory grew out of the problem of reconciling Newtonian mechanics, which implies the equality of all inertial (i.e., nonaccelerated) frames-of-reference (principle of relativity), with Maxwell's theory of electromagnetism, which was taken to imply the existence of only one frame (the "ether frame"), in which the speed of light is constant. For the classical law of addition of velocities implies that, with respect to any frame moving through the ether, the velocity of light should depend on that frame's velocity. But all attempts to detect such a variation in the speed of light with respect to Earth as it moves around the Sun had failed. A number of prominent scientists worked on this problem, but Einstein was the first to see clearly that the way out was to give up the classical law of addition of velocities by replacing the Newtonian concept of absolute time that is the same for all inertial frames, by that of a time that is relative to each inertial frame. The special theory preserves the equality of all inertial frames, but postulates that the speed of light is absolute, that is, the same in all inertial frames. The failure to detect a variation in the speed of light thus becomes evidence supporting the new relativistic kinematics. Such counterintuitive but well-established effects as time dilatation and the twin paradox provide further supporting evidence. Physical theories, such as Newtonian mechanics, had to be reviewed and modified to ensure compatibility with special relativistic kinematics. Many surprising consequences followed from this review, notably the blending of the previously separate laws of mass and energy conservation into a single law of conservation of mass-energy (often known as the equivalence of mass and energy). The entire modern theory of elementary particles is built on the foundations of the special theory, as is the operation of high-energy particle accelerators.

In 1907 Einstein came to the conclusion that a modification of the theory was needed to incorporate gravitation because of a unique feature, already stressed by Galileo: Regardless of their mass, all bodies fall with the same acceleration in a gravitational field. Einstein realized this feature casts doubt on the privileged role of inertial frames because it implies that no mechanical experiment can distinguish between an inertial frame of reference with a uniform downward-acting gravitational field, and an accelerated frame-of-reference with no gravitational field, whose upward acceleration is numerically equal to that produced by the gravitational field in the first case. Einstein assumed that there is a complete equivalence between inertial frames with a gravitational field and accelerated frames without a gravitational field, and he made this principle of equivalence the foundation of an eight-year-long search for a relativistic theory of gravitation. In the resulting general theory of relativity, completed in 1915, all frames of reference are equally acceptible. Gravitation is an effect of matter on the structure of space and time, not a force pulling objects off their straight line (inertial) paths in a flat spacetime but a warping of spacetime in which objects attempt to follow the straightest possible paths. The same mathematical object that describes the structure of spacetime, the metric tensor, also characterizes the gravitational field; the very structure of spacetime is now a dynamical field. With the new theory, Einstein was able to explain the hitherto anomalous portion of the precession of Mercury's perihelion and suggested a number of astronomical tests of his theory, such as the gravitational red shift of stellar spectra, the apparent deflection of light rays passing near the sun, and the focusing effect that a concentrated massive object would have on light (gravitational lensing). All of these predictions have been confirmed with increasing accuracy by recent optical and radio wave observations. General relativistic corrections have proved important in the theory of such super-massive objects as neutron stars, and the theory also predicts novel phenomena such as black holes and gravitational waves, which have become the object of recent intense theoretical and observational study.

Starting in the 1920s, Einstein became increasingly absorbed by the search for a unified theory of the electromagnetic and gravitational fields. Always fascinated by the discovery of conceptual unity behind apparently different phenomena (Maxwell's unification of electricity, magnetism, and optics provided the outstanding example) and having developed a field theory of gravitation, Einstein felt that some generalization of the metric tensor should encompass the electromagnetic field. In the case of gravitation, the principle of equivalence provided a physical clue that led rather directly to the metric tensor, but Einstein never found a similar physical clue for its generalization, so he continued to explore a large number of mathematical possibilities on the basis of their formal simplicity until the end of his life, without ever really convincing himself or others that he was on the right track.

A major motivation for his decades-long search was the hope that such a unified field theory might have a discrete set of nonsingular solutions, thereby explaining the all-pervasive quantum effects he had been exploring since the turn of the century. Max Planck introduced the quantum of action, but Einstein first took the idea seriously enough to suggest in 1905 that electromagnetic radiation might consist of discrete quanta of energy. He was able thereby to offer simple, quantitative explanations of a number of puzzling phenomena involving the exchange of energy between matter and radiation, notably the photoelectric effect, mentioned in his 1921 Nobel Prize citation. Although Einstein continued to develop his concept of the quantum of radiation into that of a full-fledged particle, later named the photon, carrying momentum as well as energy, the idea was not taken seriously by most physicists—including Planck and Niels Bohr—until 1923, when the Compton effect turned the tide. In 1907 Einstein took Planck's idea of quantized material oscillators and developed it into the first quantum theory of the solid state, thereby providing an explanation for the anomalous low temperature specific heats of solids. This work, successfully tested almost immediately, was instrumental in making the study of quantum effects a central concern of the physics community. In 1924 Einstein made a major contribution to the development of quantum statistics by showing that the recent derivation of the black-body radiation spectrum by Satyendranath Bose, who treated the radiation as a gas of light quanta (photons), was tacitly based on a method of counting particle configurations that differed from the classical one used by Boltzmann. The resulting Bose–Einstein statistics, as it came to be called, was later shown to hold for all particles with integral spins. By applying Bose's method to a gas of material particles, Einstein showed that it would undergo condensation at a certain temperature, thus providing the first theoretical model of a phase transition.

However, when the new quantum mechanics began to explain a number of quantum phenomena from 1925 on, Einstein found himself out of sympathy with the basic approach of the theory. At first he tried to find flaws in it but soon acknowledged that, within its theoretical framework and when given a statistical interpretation, quantum mechanics is the best explanation that can be given for these phenomena. What he continued to challenge was the theory's alleged completeness—the claim that the theory gave the most complete possible characterization of the state of an individual system—and the assertion that no other theoretical framework could be devised that would avoid what he regarded as objectionable features of quantum mechanics: The introduction of probability as an irreducible feature of reality and the continued entanglement of two quantum systems once they have interacted—no matter how far apart they may subsequently move. He continued to hope that a suitable classical unified field theory, which by its nature would avoid these features, could explain quantum phenomena, a hope shared by few physicists today.

The discovery of the weak and strong nuclear forces made obsolete Einstein's original program of unification confined to gravitation and electromagnetism. On the other hand, it has made the idea of a unification of these four fundamental interactions more attractive. Major successes have been achieved in the unification of the electromagnetic and weak interactions, and then the electroweak and strong forces, although these unifications differ from Einstein's attempts in that they are based on quantum mechanics. But the general theory of relativity has so far resisted all attempts at conventional quantization, let alone its unification with the other fields. It is possible that Einstein was right to the extent that the unique features of gravitation—its character as a spacetime structure rather than a force—may require modifications of the quantum-mechanical formalism as well as of general relativity before any unification is possible.

While we have concentrated on those aspects of Einstein's work that go beyond classical physics, he was also a master of the latter, and developed many new applications of its methods. His explanation of Brownian motion and his method of estimating the size of molecules in a solution, both published in 1905, as well as his many studies of fluctuation phenomena over the years, provide outstanding examples.

After the successful testing of Einstein's prediction of the apparent deflection of light rays by two English solar eclipse expeditions in 1919, Einstein's name became well known to the nonscientific public. Indeed, he became the first scientific "superstar," often mobbed during his public appearances; as a consequence these became rarer and rarer over the years, especially after his move to the United States. While regarding his notoriety as a personal burden, it offered him a means of disseminating his views on a number of important political and social questions: whatever the great Einstein said was news. Increasingly identifying with the Jewish people as they became ever more frequent targets of anti-Semitic propaganda and physical attacks, first in Weimar Germany and especially after Hitler took power in 1933, he supported the Zionist ideal of a Jewish homeland in Palestine as a way of building up Jewish pride and self-confidence in the face of these attacks, and then as a place of refuge for Jews forced to flee Europe. After the Holocaust, he supported the establishment of Israel to ensure a haven for the remnants of European Jewry.

A convinced antimilitarist, he was first impelled to political action by his opposition to World War I. After the war, he supported the pacifist movement, advocating refusal of military service. When Hitler came to power, Einstein felt that pacifist tactics were powerless against fascism's ruthless threat to peace and democracy, and he advocated rearmament to deter and ultimately defeat aggression in World War II. When the development of nuclear weapons threatened the destruction of humanity, he advocated a world government as the only way to overcome national enmities and ensure disarmament. Ironically, he is often credited with—or blamed for–playing a major role in the development of the American atomic bomb, although his role was confined to alerting the American government in 1939 to the danger of Germany's doing so.

The economic chaos in Germany in the 1920s, followed by the worldwide economic crisis and collapse in the 1930s, convinced Einstein that the capitalist economic system needed drastic change, and he began to advocate a socialist reorganization of the economy. Well aware of the dictatorial features of the Soviet model, he stressed the need to preserve democratic political rights under socialism.

As may be imagined, Einstein's political and social views were not universally shared, and he became the object of intense personal attacks, often anti-Semitic in nature. He was denounced as unpatriotic for his stand against nationalism and war, and as a "red" for his social and economic views. This was true not only in Germany but also in the United States, especially during the "cold war" period, when he blamed the United States government for a large share of the rising tensions with Russia and urged resistance to all attempts at governmental inquisitions into individual beliefs. His defense of civil liberties in his adopted homeland was an inspiration to many during the McCarthy years.

References

Einstein, A. (1961). *Relativity: The Special and the General Theory*, 15th ed. New York: Crown.

—— [1954] (1993). *Ideas and Opinions*. New York: Modern Library, New York.

Einstein, A. and Infeld, L. (1938). *The Evolution of Physics: From Early Concepts to Relativity and Quanta*. New York: Simon & Schuster.

Pais, A. (1982). *"Subtle is the Lord...": The Science and the Life of Albert Einstein*. Oxford, England: Oxford University Press.

Stachel, J., ed. (1987). *The Collected Papers of Albert Einstein*. Princeton, NJ: Princeton University Press, vols. 1–8, to date.

Albert Einstein (1879–1955)

John Stachel

The most renowned physicist of the twentieth century. The child of non-religious German–Jewish parents, he was born in Ulm, but spent his early years (1881–95) in Munich. A perod of intense religiosity as a child was followed by a free-thinking adolescence, and belief in an impersonal, Spinozistic cosmic reason, manifested in nature, as an adult. He attended primary school, where he first encountered anti-Semitism among his fellow pupils, and *Gymnasium* in Munich. After his parents moved to Italy, he completed his secondary education in Switzerland, attended the Swiss Federal Institute of Technology (ETH) from 1896 to 1900, and became a Swiss citizen in 1901. Unable to obtain an academic position, he worked in the Swiss Patent Office from 1902 to 1909. In 1903 he married Mileva Marić, a fellow ETH physics student of Serbian descent, with whom he had a daughter and two sons.

His scientific work, notably on the special theory of relativity, the quantum hypothesis, and Brownian motion, brought him increasing prominence after 1905 in the physics community, of which he became an acknowledged leader by the end of the decade. His work in physics is characterized by concern for fundamental problems and profound conceptual innovations. In 1909 he got his first academic position at the University of Zurich, moving to the German University in Prague (1911–12), then to the ETH in Zurich (1912–14). In 1914 he accepted a specially-created research position in Berlin, then the world center of theoretical physics, where he stayed until 1933. He separated from his wife immediately after moving to Berlin, and in 1919 married his cousin Elsa Einstein.

He worked on a relativistic theory of gravitation for eight years, publishing his general theory of relativity in 1915, which introduced further changes in current concepts of space, time, force, and matter. In 1919 verification of the pre-

Glenda Abramson, ed.
The Blackwell Companion to Jewish Culture
pp. 198–199

dicted gravitational deflection of light rays by two British solar eclipse expeditions brought him international celebrity. He won the 1921 Nobel Prize in physics. Recruited to the Zionist cause, he accompanied Chaim Weizmann on a trip to the USA in 1921 to help raise money for the Hebrew University, and in 1923 spoke at the ceremony of its inauguration in Jerusalem. He refused to join the Jewish religious community, but identified himself with innumerable Jewish causes, Zionist and non-Zionist, for the rest of his life. He favored a binational state in Palestine before the Second World War, but after the Holocaust hailed the establishment of the State of Israel, while advocating full equality for its Arab citizens and reconciliation with the Arab states. Active in the pacifist movement in Germany during the First World War, he was one of its leading world figures during the 1920s, and was generally identified with the internationalist left. An anti-relativity campaign, often with overtly nationalistic and anti-Semitic overtones, developed in post-war Germany; at moments of political tension there were threats against Einstein's life. Visiting the USA when Hitler came to power, he resigned his position in Berlin and settled permanently in Princeton, New Jersey, taking US citizenship in 1940, and working at the Institute for Advanced Study until his death.

The grave deterioration in the international situation in the 1930s led Einstein to devote more time to politics. Holding pacifism useless as a tactic against Fascism, he called for rearmament and alliance of the anti-Fascist states in order to resist aggression, and supported the allied efforts in the Second World War. He aided numerous refugees, Jewish and non-Jewish, forced to flee Europe. Frequently credited—or blamed—for his role in suggesting the US atomic bomb project, after the war he warned incessantly against the dangers of nuclear weapons, favoring universal disarmament under the aegis of a world government. He staunchly defended civil liberties in the USA during the anti-communist witch-hunts accompanying the cold war. While condemning Soviet totalitarianism, he advocated a socialist economic order that embraced individual liberties. Offered the presidency of Israel in 1952, he refused on grounds of age and lack of aptitude. Working on his scientific ideas and political projects, he awaited the impending end calmly. In death, as in the later years of his life, he has become a mythic figure, whose popular image often has little to do with the complexities of the real human being.

The Young Einstein: Poetry and Truth

John Stachel

My title is meant to recall that of Goethe's autobiography, "Dichtung und Wahrheit." "Wahrheit" means "truth;" "Dichtung" can be translated as "poetry." But it can also be translated as "imagination" or even "fiction," and debate has long raged among Goethe scholars as to the exact ratio of "Dichtung" to "Wahrheit" in the master's memoirs.

Einstein also wrote two (much briefer) autobiographical memoirs ("Autobiographical Notes," and "Autobiographical Sketch"), about which one could raise similar questions. But I want to raise the question of "Dichtung" and "Wahrheit" in a larger sense. As a result of the recent discovery of a number of new documents about the early years of Albert Einstein—let us define these as the years from his birth in 1879 until 1905, his *annus mirabilis*—these years have begun to emerge from the considerable obscurity in which he and most of his biographers had left them. A stream of recent publications—which threatens to become a flood—attempt to paint a more vivid picture of "The Young Einstein," his family, his friends, and his fiancée. The question I want to discuss is: What is the ratio of "Dichtung" to "Wahrheit" in some of this recent literature?

After he became world famous in 1919, a number of myths started to accrete around Einstein, a process that by no means stopped with his death. Perhaps the most prevalent of these myths is that he was born old. When his name is mentioned, how many can resist the reflex impulse to picture the ancient sage, surrounded by his nimbus of gray hair, his eyes reflecting all the pain he had seen in his life? Insofar as his childhood and youth were discussed at all, they were also mythologized by projecting back in time a youthful version of the elderly plaster saint. The prevalence of such myths only hinders the study of the truly interesting

Talk delivered at the AAAS Session on "The Young Einstein"
New Orleans, February 18, 1990

questions about Einstein's early years, questions that are raised most acutely by some of the newly available documentary material.

With the help of such material, perhaps we can now begin to escape from the pervasive myths of saint and sage. If we do, will it be only to fall victim to a new set of myths? I shall try to address this question, taking as my texts the writings of some of my fellow speakers this afternoon. If this is not very polite, at least I cannot be accused of attacking them behind their backs.

I

Lewis Pyenson has done a large amount of imaginative and important scholarly work on *The Young Einstein*, the title of one of his books (my citations will be from it), uncovering a large amount of hitherto unknown information, and helping to lay to rest some of the old myths. Yet even he has not been immune from the mythologizing urge. Let me give some examples.

Pyenson is at pains to correct what he calls "the misperceived legacy of the Luitpold Gymnasium" (p. 1), the Munich academic secondary school that Einstein attended from 1889 to 1894. He wants to prove that Einstein was offered an excellent education in mathematics and physics at the Luitpold, a school at the forefront of the struggle then going on for reform of German science education. He emphases that the physics text used at the Luitpold took "issue with the dominant view of the time in Prussia, where physics was taught as a branch of mathematics" and presented physics as "'a science of experience' which had to be taught by appeal to intuition." Pyenson asserts that Einstein was exposed to the beneficent influence of this physics instruction "during his last year and a half there" under the tutelage of "Gottlieb Effert and Joseph Ducrue" (p. 3). The fact is that Einstein took physics at the Luitpold (with Ducrue—he never had Effert as a teacher) during his last term there. Since he left abruptly in the middle of that term, never to return (I shall return to the subject of his departure in a moment), no matter how good or bad the physics education at the Luitpold may have been, Einstein was exposed to only three months worth of it at a time of personal crisis, when he was presumably not very receptive to any instruction.

Having disposed of physics, Pyenson turns to mathematics at the Luitpold. "Einstein was privileged from 1889 to 1891 to have Adolf Sickenberger as a mathematics teacher," which is important for Pyenson's case since Sickenberger was "a vocal partisan of school reform" (p. 4). The fact is that Einstein *never* had Sickenberger as a teacher at the Luitpold. Pyenson states that Einstein used Sickenberger's mathematics text "throughout [his] five and a half years at the Luitpold Gymnasium" (p. 4), which is almost correct—except for the fact that Einstein spent almost *six* and a half years at the Luitpold. [All facts about the Luitpold cited here are taken from *The Collected Papers of Albert Einstein*, Vol. 1, *The Early Years*, where the original sources are given.]

Pyenson is at pains to project the image of Einstein the adult "loner" back into his youth. Indeed, "Einspänner—"loner"—is the title of one of the essays in *The Young Einstein*. He comments on Einstein's dramatic mid-semester departure from the Luitpold, which he quit in order to rejoin his parents, who had moved to Italy for business reasons: "[Einstein] knew that without the final certificate called the *Abitur* he ... placed himself outside the intellectual life of a society that valued culture and formal education highly. He wanted none of it" (p. 6).

This sounds very romantic indeed, but let us see what his sister (Maja Winteler-Einstein) tells us about the move in her biographical memoir of Einstein. She reports that their parents "were very upset over his arbitrary behavior, but he ... reassured them about his future by assuring them in the most definite terms that he would study by himself until fall in order to prepare for the entrance examination for the Zurich Polytechnical School" (vol. 1, pp. lxii–lxiv). Einstein himself confirms this intent in an "Autobiographical Sketch" [*Autobiographische Skizze*] written in the year of his death. He took the examination, doing very well in physics and mathematics, which he had indeed studied on his own (his copy of the physics text he studied is still in existence); but he failed to convince the authorities at the Polytechnical School (hereafter referred to as "the Poly") that he should be admitted without completing his secondary education (in addition to problems with the exams in other subjects, Einstein was still two years below the minimum entrance age).

They advised him to finish up in the technical section of the Cantonal School in Aarau, graduation from which guaranteed admission to the Zurich Poly without further examination. Writing about his Aarau schooling, Pyenson states: "The grades he received for industry and mastery of his course material were uniformly poor. For the final quarter, ending in April 1896, he showed little improvement... In arithmetic and algebra he received the lowest grade, one out of six. ... After Einstein registered for the final half-year at Aarau, his performance in the exact sciences improved dramatically. During the first quarter of the new year—May and June of 1896—he received a six in arithmetic and algebra, and in physics a six for industry and a five for mastery of the material" (p. 12). Elsewhere he states: "As a last-year pupil in the cantonal school at Aarau, his grades improved at the time that a splendid new physical laboratory opened there" (p. 52).

If correct, these facts would provide a sad commentary on Einstein's reaction to the splendid educational opportunities offered him at the Luitpold that Pyenson had previously depicted, and a dramatic proof of the beneficent influence of the Aarau milieu on Einstein's grades—another point that Pyenson wants to make. But closer examination shows that what changed dramatically during Einstein's final half-year in Aarau was not Einstein, but the grading system at the school: the ordering of grades from highest to lowest was reversed (see Vol. 1, p. 14, note [3]). His grades at Aarau were consistently high in the sciences.

One of the greatest debts that Einstein scholars owe to Pyenson is that he called attention to the significance of his family's involvement in the electrotechnological business—then on the high tech frontier—during the years when Einstein was growing up. In seeking to define the nature of the influence of this electrotechnological milieu on Einstein, however, Pyenson goes overboard at one point. He draws attention to the patent for an electric meter held by Einstein's uncle Jakob—the technical director of the family firm–and Sebastian Kornprobst, an employee. After a description of the meter, Pyenson waxes lyrical:

> Two clocks keeping different time, each in what might be called its own frame of reference, lay at the center of the Einstein–Kornprobst meter. The idea of synchronising and comparing clocks in moving frames of reference plays such an important role in the special theory of relativity that these patents fairly call out for further comment (p. 40).

After recalling some well-known but somewhat later "intimations of relativity," he goes on:

> One can also imagine Jakob spreading out his patent plans on the kitchen table and explaining them to his eleven-year-old nephew Albert, and in this way locate the germ of relativity at an even earlier date (pp. 40–41).

Again, this vignette would be enchanting if Pyenson had accurately described the meter—but he has not. The device involves a pendulum, which Pyenson in his description insists on calling a "pendulum clock," the function of which he has misunderstood. To quote him:

> Meters of this kind employed a pendulum clock with the pendulum designed as a shunt coil.... The rate of the shunt coil pendulum clock would obviously depend on the current, and its changing rate could be compared and counted relative to a standard clock.

The "pendulum clock" and the "standard clock" mentioned here are the "two clocks" so lyrically invoked above. If we examine the Einstein–Kornprobst patent (they also took out a patent in the United States, so I quote that patent, which is a translation of the German original cited by Pyenson) [See Figure 1]:

> The apparatus consists of a revolving body R movable about a shaft A and put into *uniform* rotation (my emphasis) by a mechanical or electric power motor, such as an electric pendulum. A shaft a is arranged parallel to one of the longitudinal sides s of the revolving body, the said shaft being adjustable in bearings l l and carrying a frictional disk V. This disk V is rotated by frictional contact from the revolving body and the rotation of the disk is more rapid the more the disk approaches the base B of the body R. The shaft a carries a worm or toothed wheel q by means of which the revolutions are registered by a counting device (US Patent No. 437,754, dated October 7, 1890).

In other words, the pendulum (call it a pendulum clock if you like) is used precisely to maintain a *steady*, *uniform* rotation of R. A contemporary description of the device, cited by Pyenson, uses the word "gleichbleibend"—constant, steady, unwavering—to describe the rotation produced by the pendulum [C. L. Imhoff, "Elektricitätszähler der Firma J. Einstein & Cie., München. System Kornprobst." *Elektrotechnische Zeitschrift* **22** (1891), 278–279]. Any change in the rate of the pendulum would be disastrous for the operation of the meter. So there is no "changing rate" of the "pendulum clock" to be "compared and counted relative to" a non-existent "standard clock" in the meter.

None of what I said is meant in any way to deny that his upbringing in a household involved so deeply in electrotechnology had a profound influence on the young Einstein. It is just meant to suggest that we employ less "Dichtung" and more "Wahrheit" in seeking to define that influence.

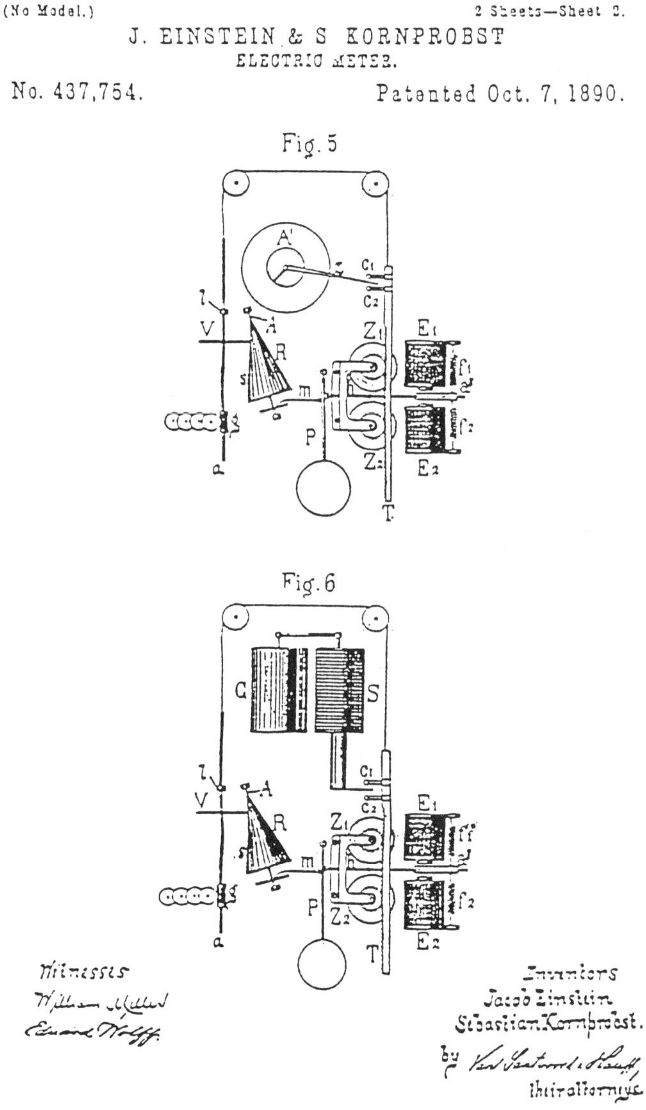
(No Model.)
2 Sheets—Sheet 2.
J. EINSTEIN & S KORNPROBST
ELECTRIC METER.
No. 437,754.
Patented Oct. 7, 1890.
Fig. 5
Fig. 6
Witnesses
Inventors

Figure 1:

II

Lew Pyenson, in my opinion, is a gifted and imaginative scholar, still in search of the real Einstein, who is constantly tripped up by his own imaginativeness combined with carelessness in checking information. Others, it seems to me, are more concerned with creating new myths. Since the detonation of the atomic bomb, the myth of the benevolent, saintly Einstein has had to contend with the counter-myth of the malevolent Einstein, arch-symbol of the evil that modern science has let loose upon the world.

Recently, there have been attempts to project this image of the malevolent Einstein back into his youth. Let us turn, for example, to Evan Harris Walker. His sole previous work on Einstein, as far as I know, is a letter published in *Physics Today* (February 1989) under the title (not of his choice, he assures me) "Did Einstein Espouse his Spouse's Ideas?" (the letter was published together with a reply by me). His professional affiliation is listed in the letter as "Walker Cancer Research Institute, Edgewood, Maryland," although I understand he is a physicist working at the U.S. Army Ballistic Research Laboratory at Aberdeen Proving Grounds. According to the *Journal of the National Cancer Institute*, "The mission of the institute is to test and develop" a form of cancer therapy based on "Walker's TDTA theory" ("chemically triggered time delay toxin activation chemotherapy"). The article includes critical comments on Walker's approach by a researcher at the National Cancer Institute.

I know nothing about cancer therapy, but if I had to judge Walker solely on the basis of his letter on Einstein, I would have to conclude that he is a fantasist, who judges reality on the basis of his own desires. However good-hearted he may be, such desires are no substitute for some knowledge of the facts of Einstein's life and that of Mileva Marić, his fiancée and first wife. In order to make a heroine of Marić, Walker is willing to make a monster of Einstein.

I do not want for a moment to suggest that it is *a priori* impossible that Einstein *was* a monster. I only want to emphasize that, if he was, this can only be established on the basis of some serious evidence. Yet, without providing any such evidence, Walker is willing to insinuate that: 1) Einstein destroyed letters by Marić indicating her contributions to their scientific work, 2) she wrote the doctoral thesis published by Einstein, 3) he took credit for her basic ideas that were "the turning points of relativity theory," and 4) he gave Marić his Nobel prize money to induce her to keep quiet about her role in developing the theory. Pretty good pickings from a one-page letter!

The level of Walker's logic may be assessed by his use of a quotation from one of my articles. After citing a sentence from a letter to Marić by Einstein while they were both students at the Poly, I wrote that "this comment raises the intriguing question of the nature of Marić's role in their collaboration." Walker says: "For me this statement is the proof of something I had suspected ever since Ronald Clark's *Einstein: The Life and Times* ... appeared in 1971." If someone interprets raising a question as providing proof of something, it is hard to know what would constitute a disproof for him. If Walker has looked at the "Introduction" to Vol. 1

of the *Einstein Papers*, he must be aware of the summary sentence there on this topic:

> Although the possibility that she played a more significant part cannot be excluded, the available evidence suggests that [Marić's] role was that of a sounding board for Einstein's ideas, a role also played on occasion by his friends Michele Besso and Conrad Habicht (pp. xxxix–xl).

Walker's expression of his desires is straightforward:

> I cannot help but see Mileva and Albert Einstein working as a team, hoping together to achieve the kind of recognition that had come to Pierre and Marie Curie.

He quotes another sentence from my article:

> [Einstein's] letters to [Marić] contain references to joint study of books, requests for her to look up data, and one or two other mentions of joint work; but these letters give no indication of any ideas she contributed to their work.

Undeterred by this lack of evidence, he goes on:

> Yet only *ten* letters from Mileva to Albert Einstein from this period have been found. One may wonder if there were other letters that for some reason were not so carefully kept.

The implication is clear: The evidence must have been destroyed. Why? Because this belief makes it possible for Walker to continue seeing Mileva and Albert "working as a team." In fact, Walker has intuited the division of labor within this team:

> I cannot help but feel [again that phrase–Dr. Walker really must learn to keep those feelings under control] that... those most basic capricious [sic] ideas that were the turning points of relativity theory came from Mileva, while the mathematics and proofs came largely from Albert.

Parenthetically, let me note that you can have it either way you like: Go with Dr. Walker, and Mileva gave him the important ideas, while Albert just filled in the mathematical details; go with Dr. Troemel-Ploetz (see Section III), and Mileva did the crucial mathematics. What do the details matter, as long as Einstein was a plagiarist who published the whole thing under his name?

We shall soon return to Mileva the mathematician, but let us get back to Dr. Walker's visions. Suffice it to say that, apart from one reference to "our work on the problem of relative motion" (Einstein to Marić, March 1901), there is not a shred of evidence known to me indicating that Marić worked on the subject. Of course, I cannot assert categorically that she did not; I merely ask that we weigh the probabilities in the light of all the available evidence.

Let us look, for example, at a similar case, in which we happen to have a little more evidence. This is the case of Einstein's first published paper, on capillarity. Here, too, Albert wrote to Mileva: "I have given him [who is not important] a copy of our paper" (letter of May ? 1901, Doc. 107, Vol. 1, p. 300). This seems at least as equally clear-cut a case, and even more impressive since it refers to an already-published paper rather than to some ideas on relative motion that were still nebulous in 1901 and only bore fruit in radically altered form four years later.

But in this case, we happen to have two other pieces of evidence that bear on the origins of the paper. Six months earlier, Einstein had written to Marić:

> The results on capillarity that I recently discovered in Zurich, seem in spite of their simplicity to be completely new. When we get to Zurich, let's try to collect empirical data on the question with the help of Kleiner [Professor of Physics at the University of Zurich]. If a law of nature results, then we'll send it to Wiedemann's Annals [of Physics, a journal] (letter of 3 October 1900, Doc. 79, Vol. 1, p. 267).

Here, Einstein clearly refers to *his* results, although he contemplates *their* doing various things with these ideas of his. And two months later, Mileva wrote to Helene Savić, a close friend and confidante:

> Albert has written a paper on physics, that will soon probably be published in the Annals of Physics. You can picture to yourself how proud I am of my dear treasure. It is in fact no every-day piece of work, but very significant, on the theory of fluids. We have sent it privately to Boltzmann, and really would like to know what he thinks of it, we hope that he will write to us.

She here asserts that *he* wrote the paper, taking no credit for herself under circumstances where this would have been quite natural and in no way endangered his job prospects (it has been alleged that she agreed not to sign their supposedly joint papers because this would have been detrimental to his search for a job).

If we did not have these additional letters, but only Albert's reference to "our paper," it would be natural to conjecture that it was based on joint ideas and jointly written. In this case, we just happen by luck to have additional evidence that contradicts this conjecture. I suggest that this example shows the need for extreme caution in drawing conclusions for the other case (relative motion) from Einstein's lone comment. I remind you that in this country our lives are ruled by a system of law, according to which the accused is innocent until proven guilty. I would hate to think that—with whatever noble motives—the history of science is going to be rewritten on the basis of the principle that one is guilty until proven innocent.

Einstein's preserved letters to Marić from 1899–1901 are studded with passages containing ideas that, with hindsight, we can see constitute the beginnings of work on the theory of relativity. [I have translated and discussed these passages in an article in *Physics Today*: "Einstein and Ether Drift Experiments," May 1987, pp. 45–47]; [See page 171, this volume.] None of Marić's preserved letters—some of which are direct replies to his—contains anything on this subject, or on *any* of the many topics in physics about which he wrote to her, nor indeed do they include any ideas of hers on physics (the one reference to physics in them recounts a lecture she heard). True, only ten of her letters to him are preserved, compared to fourty-one of his to her; but you could not pick ten of his that are so devoid of discussion of physics. Indeed, at times one might be tempted to accuse him of callousness in continuing to discuss physics at times of great personal stress in both their lives (she became pregnant and gave birth to their daughter out of wedlock during this period); but the fact remains that you can't get away from physics in his letters, while there is none in hers. This *proves* nothing, of course, and there is always the possibility that her masterpieces have all been lost; but what are the probabilities? If we are to grant *all* possibilities equal rights, then we would have to entertain the following possibility: Marić was totally without scientific talent: Einstein was able to carry her through all her course work, and wrote her final diploma thesis for her. The reason why she failed the final diploma examinations, not once but twice (1900 and 1901), is that he was unable to take them for her. I don't for a moment believe this fantasy of mine; but the point is that there is just as much—or rather, just as little—evidence for it as for Dr. Walker's effusions.

III

Marić's failure to graduate from the Poly brings us to Dr. Troemel-Ploetz's evocation of "Mileva Marić, the Woman Who Did Einstein's Mathematics." As I write this, I know nothing of Dr. Troemel-Ploetz's views beyond the title of her talk, since she has published nothing on the topic. A newspaper article, which she kindly sent me, mentions "her work on language and psychotherapy" [Lancaster, Pa. *Intelligencer Journal* for Monday 12 September 1988]. So all I can do is to discuss the only objective evidence that we have for Marić's mathematical talents, her grades at the Poly. During the first year there, she and Einstein took several mathematics courses that were graded. Her grades are never higher than his, and are lower in several of them. He got a total of 32 1/2 out of a possible 42; she got a total of 30. Marić twice failed the final diploma examination. We can compare her grades the first time (1900) with Einstein's, since they were the only two physics students to take the exam that year. In theoretical physics he got 5 out of a possible 5, she got 4 1/2; in practical physics they both got 5; in astronomy he got 5, she got 4. His diploma work (an independent research project) got 4 1/2, hers got 4. So far they are rather closely matched, her grades being slightly lower than his. The one big divergence in their grades is in mathematics: he got 5 1/2 in the theory of functions, she got 2 1/2 (see Doc. 67, Vol. 1, p. 247, for these grades). Nor was her low grade in mathematics a one-time fluke; when she retook the examination the following year she got a grade of 3 1/2 in the theory of functions, while the two other physics students who took the exam each got 5. One might be tempted to speculate that she was the sort of person who has difficulty in taking examinations even if well-versed in a subject; but her creditable grades in other subjects argue against the existence of any such examination phobia.

Incidentally, if Marić *had* written Einstein's Ph.D. thesis, as Walker suggests (without a shred of evidence, I emphasize again), we would have to convict *her* of a serious mathematical error in that thesis, for which we now hold *Einstein* responsible. [For details on this error, see *The Collected Papers of Albert Einstein*, Vol. 2, Doc. 15 and the headnote preceding it on Einstein's doctoral thesis.]

Marić's failure to graduate from the Poly has sometimes been attributed to her not taking the final examination,[1] which is false, or even to a school policy at that time of not granting diplomas to women, which is equally incorrect.[2] Indeed, while Marić was there, two women graduates were hired as Assistants by other Sections of the school. I cite this obviously low number only to prove that it was possible for women graduates to become Assistants at the Poly, since it seems clear from some of the letters that Marić had hopes in that direction. While I do not want to minimize sexism at the school—the subject of women at the Poly has not yet been studied in sufficient detail to permit comment on this question[3]—it is a fact that during the last quarter of the nineteenth century many women attended the Poly precisely because a number of other technical schools and universities were still closed to them (e.g., in Vienna).

I have compiled the following statistics for Section VI, the section that trained teachers of mathematics and science and that Marić and Einstein attended [figures from the Poly school programs]. The years listed include those Marić and Einstein attended (1896–1900), as well as the ones just before and after (she also attended the summer semester 1901 as a repeat student as part of her second attempt to graduate):

Female and Male Students in Section VIa

YEAR	VIA F	VIA M	VIB F	VIB M	TOTAL F	TOTAL M
1895/96	3	7	5	17	8	24
1896/97	4	16	5	17	9	33
1897/98	1	20	5	22	6	42
1898/99	1	20	2	23	3	43
1899/1900	1	23	6	23	7	46
1900/1901	1	26	7	22	8	48

Source: *Programm der eidgenössischen polytechnischen Schule* (Zürich: Orell Füssli, 1895–1901)

Thus, during this period about one-sixth of the students in Section VI were women; although Marić was the only woman in VIA during several of her years there, there were always several other women in Section VIB, some of whom did graduate. So it would be hard to attribute Marić's difficulties in graduating to her supposedly-unique position as a woman in Section VI.

All this evidence is not offered as *proof* that Marić was no mathematical genius. The point is that there is *no* evidence that she *was* particularly gifted mathematically, while there *is* some evidence that she was *not*. It is no service to her memory, to an understanding of her difficult situation, or more generally to an understanding of the real problems confronting women trying to start a career in science at the turn of the century, to make exaggerated and unsubstantiated claims about Marić's abilities.

IV

Perhaps the most serious consequence of claims for Marić's scientific abilities that appear to be highly exaggerated on the basis of all available evidence is the disservice they do to Marić's legitimate claims to our attention. Two questions come to my mind at once in this connection:

1) The question of what became of women of average talent in the sciences at the turn of the century—indeed what becomes today of such women who are not

Marie Curies or Sonia Kovalevskas? Marić's life story could provide material for an excellent case study of the fate of such a woman.

2) The question of the nature and evolution of her relationship with Einstein gives Marić a more unique claim to attention. Einstein's letters to her, published in Vol. 1 of the *Einstein Papers*, manifest a deep emotional dependence on her that seems later to have turned into an emotional coldness and even hostility on his part that led to, and continued beyond, their separation and divorce. To understand the reasons for the evolution of their relationship requires the intuition and tact of a master psychobiographer, rather than the facile skills of an advocate determined to win a case at all costs.

In the interests of "truth in advertising," I feel obliged to tell you that Lew Pyenson recently wrote an extremely critical review of *The Collected Papers of Albert Einstein*, Vol. 1, *The Early Years* [I can provide copies of his review and of the editors' reply]. His review is entitled "Just the facts," which is pretty clearly intended as a pejorative comment on the pedestrian spirit of the editorial work on that volume. I am well aware of the inevitable ambiguities involved in trying to draw a hard and fast line between "objective" facts and such "subjective" things as comments, opinions, interpretations, etc. Nevertheless, as a goal, I find nothing objectionable in the statement that, in Vol. 1, the editors were trying to establish the basis for a factually accurate account of Einstein's early years. I hope that the examples given in this paper may help to persuade you that the goal of establishing a base line of "just the facts," from which one may then take off for more imaginative—but well-grounded—speculations about the young Einstein, is not such a bad goal.

Preliminary Comments on the AAAS Papers (March 5, 1990)

A reply to sensational claims is always difficult. A German statesman of the nineteenth century put the problem well: "Denials never have the fascination and the impact of false reports." All that one can do is to weigh the evidence put forward in support of such a claim in the light of all the available evidence bearing on the question. This does not result in high drama, but perhaps in the long run it may do some good.

I do not yet have copies of Walker's or Troemel-Ploetz's papers; so all I can do is to add some preliminary comments based on some hasty notes on their talks that I took at the AAAS meeting.

Troemel-Ploetz states that it is common knowledge in Zurich that Marić collaborated with Einstein, solved Einstein's mathematical problems, etc. She also cites from the biography of Marić by Desanka Trubhović-Gjurić (*Im Schatten Albert Einsteins/Das tragische Leben der Mileva Einstein-Marić*) a number of comments by relatives and friends of the Marić family indicating that it was well-known in her home town that she did made important scientific contributions. Many things are well known in this world: For example, it was well known in Germany in the 1920's and 1930's that the Jews were the cause of all the people's misfortunes. And, as one of James Thurber's "Fables for our Times" points out, it is well known that rabbits, by stamping on the ground with their hind feet, cause earthquakes. The question is not whether some assertion is common knowledge, but what is the evidence that it is true? I shall go into their claims in detail when I have Walker's and Troemel-Ploetz's papers, but here I must emphasize that bare assertions, particularly by interested parties, do not constitute proof of such assertions, even when these assertions are repeated in print, even in a book.

Walker makes much of a claim by Trbuhović-Gjurić that Abram Joffe, the Russian physicist, saw the manuscript of the 1905 relativity paper before it was published, and that it had both Einstein's and Marić's names on it. Dr. Robert Schulmann, an editor of the Einstein Papers, interviewed Dr. Trbuhović-Djurić several years ago (she has since died), and asked her about the evidence for this claim. She cited the published memoirs of Joffe as her source. However, in his memoirs, Joffe makes *no* such claim. Indeed, apart from the fact that it is not made by Joffe, the claim is suspect—to put it politely—on a number of grounds. First of all, Joffe is supposed to have seen the manuscript because he was then working with Wilhelm Roentgen, a member of the Curatorium [Board of Editors] of the *Annalen der Physik*, the journal to which the paper was submitted. But Roentgen was an experimentalist, and there is no reason why a purely theoretical paper should have been submitted to him for review when two members of the Curatorium, Paul Drude (the editor) and Max Planck, were both outstanding theorists quite capable of evaluating the paper—indeed, both obviously did read it since they soon made reference to it in their own published writings. Second, if both names were on the paper when it was submitted to the *Annalen*, who removed Marić's name? The *Annalen* had no policy forbidding publication of articles by women! Third, if Roentgen read the paper even before it was published in 1905, why did he wait until 1906 to ask Einstein for a reprint of it? I think Trbuhović-Djurić's claim must be rejected on the grounds of both lack of evidence and of inherent implausibility.

I could give numerous other examples of assertions about Einstein that have been repeated from source to source but proved to be totally unsubstantiated when subject to close examination (for some examples, see Schulmann and Stachel, *Comments on Pyenson's Review of "The Collected Papers of Albert Einstein, Vol. 1, The Early Years,"* available from me on request).

Troemel-Ploetz claims that discrimination was responsible for Marić's failure to graduate from the Swiss Federal Polytechnical School, or Poly. She claims that both Einstein and Marić failed the final examinations for the diploma, but that Einstein was allowed to graduate while Marić was not. This claim is based on an unsubstantiated assertion that a grade of "5" (out of "6") was needed to pass at that time. I have searched the regulations of the Poly in vain for any such rule. The report of the grades by the head of Section VIA of the Poly (Vol. 1, p. 247) includes no statement that Einstein had failed the examination. Since Einstein and Marić were the only two students taking the examination in physics that year, all we can do is compare their grades, as I did in my paper, and note that Einstein's average was 4.91 while Marić's was 4.00. If there was discrimination against

Marić in grading their examinations, it must have been differential discrimination since it was her grade in mathematics (theory of functions) that was primarily responsible for pulling her average down significantly below Einstein's. But, as indicated in my paper, women had been graduating from the Poly for decades by the time Marić took the exams (twice) and failed, and continued to do so after she left. So the fact that she was a woman cannot by itself account for her failure. Any charge of discrimination will have to answer the question: What did the examiners have against Marić *specifically*?

Both Walker and Troemel-Ploetz suggest in no uncertain terms that, at the very least, Marić and Einstein collaborated on the special theory of relativity, if the basic ideas were not actually hers alone. The only piece of contemporary evidence offered in support of this claim is Einstein's statement in a letter to Marić:

> How happy and proud I will be when the two of us together will have brought our work on relative motion to a successful conclusion! (letter of 27 March 1901, Vol. 1, p. 282, translation from Stachel, *Physics Today*, May 1987, p. 46). [See this volume, p. 171].

First of all, let us look at the context in which this statement occurs. It does not occur in a scientific context, but in the course of Einstein's attempt to reassure Marić of his love in a letter written from his home in Milan during a separation occasioned by his attempt to find a job—a separation about which she was not very happy:

> Right now Michele [Besso] is staying in Trieste at his parents with his wife and child and only returns here [Milan] in about 10 days. You need have no fear that I will say a word to him or anyone else about you. You are and will remain a holy shrine to me into which no one may enter; I also know that of all people you love me most deeply and understand me best. I also assure you that no one here either dares to or wants to say anything bad about you. How happy and proud I will be when the two of us together will have brought our work on relative motion to a successful conclusion! When I look at other people, then I truly realize what you are! (27 March 1901, Vol. 1, p. 282).

Einstein is clearly a young man, deeply in love, who has relaxed his own ego boundaries to include Marić within them—a not uncommon phenomenon among lovers–although he maintained a wary and perhaps even hostile sense of separation between the two of them and the rest of the world. That a lover in such a situation should over-value the accomplishments of the beloved, and identify the achievements of one member of the couple as achievements of the pair is also not uncommon. (See the addendum at the end of this paper.)

Bearing this possibility in mind, let us examine Einstein's and Marić's other comments in their correspondence that bear on the topics that led in 1905 to the paper that Einstein called "On The Electrodynamics of Moving Bodies," the paper that constitutes the foundation of what is now called the special theory of relativity. The very term "electrodynamics of moving bodies" is used in the first letter by Einstein that touches on the subject, his letter of 10 August 1899. In this remarkable letter, he comments at some length on the topic, his remarks taking up about one-half of the letter:

> I am more and more convinced that the electrodynamics of moving bodies, as currently presented, is not correct, and that it should be possible to present it in a simpler way. The introduction of the term "ether" into theories of electricity leads to the notion of a medium of whose motion one can speak without, I believe, being able to associate any physical meaning with such a statement. I believe that electrical forces can only be directly defined for empty space, as Hertz also emphasizes. Further electrical currents are to be considered not as "the vanishing of electrical polarization in time," but as motion of true electric charges, the physical reality of which seems to be proven by the electrochemical equivalent. Mathematically, they are then always to be considered in the form $\frac{\partial X}{\partial x} + . +$. Electrodynamics would then be the theory of the motions of moving electricities and magnetisms in empty space: Which of the two conceptions must be selected must be decided by the radiation experiments.

This case is one of the few in which we happen to have Marić's reply to Einstein's letter (actually to this and Einstein's previous letter). What does she say in response to his comments on electrodynamics? Nothing! She discusses her happiness on getting his letters, she approves of his statement that he is not studying too much, she sends greetings to Einstein's mother and sister (with whom he was vacationing), she worries about her impending examinations. Near the end of the letter, she writes: "In truth, you aren't letting anyone read my letters, you must promise me that" (Vol. 1, p. 229), suggesting that she shares his feeling it is the two of them against the world (see his letter cited above). But there is not a word about *any* scientific topic in her letter, let alone a response to Einstein's lengthy discussion of his ideas about the electrodynamics of moving bodies.

The next letter that refers to this topic is his of 10 September 1899:

> "A good way of investigating how a body's relative motion with respect to the luminiferous ether affects the velocity of propagation of light in transparent bodies occurred *to me* in Aarau [a Swiss town Einstein had recently visited]. *I* have also thought of a theory on this subject that seems to me to be very plausible. But enough of this! (Vol. 1, p. 230, translation from Stachel, *Physics Today*, May 1987, p. 45).

I have added, and will continue to add my own emphasis to each use of "I," "my," etc. in his letters. If we are going to attach great significance to one use of "our" in this context, I insist that we attach similar significance to his many uses of first person singular pronouns in the same context. It should also be borne in mind that his letters cited were written during periods when Einstein and Marić were separated for some time; so there is a strong presumption that any new work he reports to her during these periods really is exclusively his.

A couple of weeks later, during the same period of separation, he writes:

> *I* also wrote to Professor [Wilhelm] Wien in Aachen about the work on the relative motion of the luminiferous ether with respect to ponderable matter, which "the boss" [Heinrich Friedrich Weber, Einstein's physics professor at the Polytechnical School] treated in such a stepmotherly fashion. *I* read a very interesting paper from the year 1893 by this man [Wien] on the same topic" (Vol. 1, pp. 233–234, translation from Stachel, *Physics Today*, May 1987, pp. 45–46).

The paper by Wien, entitled "On Questions Relating to the Translatory Motion of the Luminiferous Ether," discusses a number of experiments, including the renowned experiment by Michelson and Morley on this topic. I take this as evidence that Einstein knew something about this experiment by the time he wrote this letter. Even if Marić were the person who first informed Einstein about this experiment, as Dr. Troemel-Ploetz suggested in New Orleans on the basis of the flimsiest evidence, it would not have been a very significant input into the development of the special theory of relativity. But factually, this letter shows that Einstein would have found out about it on his own, in any case, by September 1899.

There is now a gap of over a year in references to this topic. The next reference is in Einstein's letter of 27 March 1901, the one about "our work" cited first above. The next letter that *may* refer to this topic is that of Marić to Einstein in early November 1901. Einstein had been befriended by Alfred Kleiner, Professor of Physics at the University of Zurich, and begun to explore the possibility of a doctoral thesis with Kleiner. In response to a missing letter of his, Marić writes Einstein:

> How pleased I am that Kleiner was nice to you! And during what holiday could you perhaps carry out the investigation? (Vol. 1, p. 316).

This *may* be a reference to an experiment that Einstein had just recently described in a letter to his friend and fellow student at the Poly, Marcel Grossmann:

> On the investigation of the relative motion of matter with respect to the luminiferous ether, a considerably simpler method has occured to *me*, which is based on customary interference experiments. If only relentless fate would give *me* the necessary time and peace! When we see each other, *I* will tell you more about it (Einstein to Marcel Grossmann, 6 September 1901, Vol. 1, p. 316, translation from Stachel, *Physics Today*, May 1987, p. 46).

On 17 December 1901, Einstein writes Marić:

> *I* am now working very eagerly on an electrodynamics of moving bodies, which promises to become a capital paper. *I* wrote you that *I* doubted the correctness of the ideas about relative motion [that letter has not been found]. But *my* doubts were based solely on simple mathematical error. Now *I* believe in it more than ever! (Vol. 1, pp. 325–326, translation from Stachel, *Physics Today*, May 1987, pp. 46–47).

Two days later Einstein wrote Marić that he had

> spent the whole afternoon with Kleiner in Zurich and explained *my* ideas on the electrodynamics of moving bodies to him. ... He advised *me* to publish *my* ideas about the electromagnetic theory of light for moving bodies together with the experimental method. He found the experimental method proposed by *me* to be the simplest and most appropriate one conceivable. ... *I* shall most certainly write the paper in the coming weeks (Vol. 1, p. 328, translation from Stachel, *Physics Today*, May 1987, p. 47).

Whatever Einstein may have written then, he did not publish anything on the topic until 1905. Perhaps part of the explanation may be found in his final reference to the topic, on 28 December 1901:

> *I* now want to buckle down to work and study what Lorentz and Drude have written on the electrodynamics of moving bodies. [Jakob] Ehrat [a friend and former fellow Polytechnical School student] must get the literature for me (Vol. 1, p. 330, translation from Stachel, *Physics Today*, May 1987, p. 47).

In summary, the letters to Marić show Einstein referring to *his* studies, *his* ideas, *his* work on the electrodynamics of moving bodies over a dozen times (and we may add a couple more if we include his letter to Grossmann), as compared to *one* reference to *our* work on the problem of relative motion. In the one case where we have a letter of Marić in direct response to one of Einstein's, where it would have been most natural for her to respond to his ideas on the electrodynamics of moving bodies, we find the same response to ideas in physics that we find in all her letters: silence. This *proves* nothing, as I emphasized in my paper, but it certainly must influence our estimate of the probability that Marić made a significant contribution.

Albert Einstein corresponded with his friend Michele Besso for about fifty years. Einstein's letters to Besso are filled with scientific references, many more and in much greater detail than in his letters to Marić. (For whatever reason, scientific comments are almost entirely lacking in Einstein's letters to Marić after their marriage.) Besso's letters to Einstein are similarly filled with scientific comments. (The Einstein–Besso correspondence has been published in German with a French translation, so these claims are easily checked.) Besso is also the only person Einstein thanks for help in his 1905 paper on special relativity. Yet Besso never wrote an important paper in physics, and his efforts at collaborative research in general relativity with Einstein came to naught. Late in his life, Einstein characterized Besso as an "eternal student." What does this mean? To me, it means that Besso was capable of understanding things that Einstein explained to him, and of asking intelligent questions that could help Einstein develop his own ideas (Einstein's ideas, that is)—but that Besso was not capable of any creative effort of his own. This is what I mean when I say that Besso acted as a sounding board for Einstein.

Now I challenge Walker and Troemel-Ploetz: On the strength of the Einstein–Besso letters, and the reference to Besso in Einstein's 1905 relativity paper, do you want to claim that Besso was the creative force behind Einstein, or even an equal scientific partner in any of his creative work? If so, please explain why you feel that Besso was, and where this leaves Marić. If not, please explain why you feel that there is a stronger case for Mileva Marić than for Besso. In her case, we have *no* published papers; *no* letters with a serious scientific content, either to Einstein nor to anyone else; nor any other *objective* evidence of her supposed creative talents. We do not even have hearsay accounts of conversations she had with anyone else that have a specific, scientific content, let alone a content claiming to report her ideas. (If you believe any of these assertions to be wrong, please cite the evidence for your belief.)

If objective evidence of Marić's talent is uncovered in the future, I shall gladly acknowledge her role as indicated by that evidence. Until then, I must continue to assert that the probabilities are against her having played a creative role, but rather favor her having played the role of a sounding board in the early years of her relationship with Einstein just as Besso did. The evidence also indicates that she ceased to play this role within a few years. Why this happened is a really interesting question, and I don't doubt it has something to do with Einstein's attitude to her. But exaggerated claims for her role on the basis of present evidence can only do a disservice to her memory.

Addendum to "The Young Einstein: Poetry and Truth"

In another paper on Einstein and Marić, which I decided not to include in this volume because of many overlaps with the two published here, I gave an example from Shakespeare on how love dissolves ego boundaries. In *The Merchant of Venice*, Portia is in love with Bassano, but cannot tell him in so many words before his choice of a casket that will decide her fate. She says (Act III, Scene 1):

> One half of me is yours, the other half yours,
> — Mine own I would say, but if mine then yours,
> And so all yours (Act III, Scene 1).

In his *Psychopathology of Everyday Life*, Freud cites this as a beautiful example of a parapraxis in the service of the ego. Later, after Bassano has chosen the right casket, she says (Act III, Scene 2):

> Myself, and that which is mine, to you and yours is now converted.

The bard has here, as in so many places, given poetical form to an all too-common human experience. Perhaps it is not too far-fetched to suggest that this was something that the young Einstein experienced.

Notes

[1] The only biography of Marić incorrectly suggests that she did not take the final diploma examination. It states: "Like Albert Einstein, Mileva Marić also obtained a leaving certificate [Abgangszeugnis] from the Polytechnic with the date of 2 August 1900. The reason why, in spite of her equivalent achievements in the last two years of study, she did not take the diploma examination like him is unknown. That she wanted to is indubitable" [Desanka Trbuhović-Gjurić, *Im Schatten Albert Einsteins/ Das tragische Leben der Mileva Einstein-Marić* (Bern/Stuttgart: Verlag Paul Haupt), 1983, p. 38].

[2] The first woman entered the Poly in 1871, the first woman was awarded a degree in 1877 (Gottfried Guggenbühl, *Geschichte der Eigenössischen Technischen Hochschule in Zürich* (Zurich: Buchverhandlung der Neuen Zürcher Zeitung, 1955), pp. 97–98). Both were Russians. The first Swiss woman graduated in 1895 (Klaus Urner, "Vom Polytechnikum zur Eidgenössichen Technischen Hochschule: Die ersten hundert Jahren 1855–1955 im Ueberblick." In Jean-François Bergier and Hans Werner Tobler, eds., *Eidgenössiche Technische Hochschule Zürich 1955–1980/ Festschrift zum 125 jährigen Bestehen* [Zurich: Verlag Neue Zürcher Zeitung, 1980], pp. 17–59; see pp. 52–53).

[3] The only work known to me on the subject is a typed and duplicated text entitled "Die ersten Diplomatinnen der ETH Zürich/ Provisorische Verzeichnis" (Zürich: Wissenschaftshistorische Sammlungen der ETH Bibliothek, 1987). I thank Dr. Beat Glaus of the ETH Bibliothek for sending me a copy of this document, and for informing me that it was prepared by Frau Verena Larcher, also of that library.

Albert Einstein and Mileva Marić: A Collaboration that Failed to Develop

John Stachel

On December 28, 1901, twenty-two-year-old Albert Einstein assured his fiancée Mileva Marić:[1] "When you're my dear little wife, we'll diligently work on science together so we don't become old philistines, right? My sister seemed so crass to me. You'd better not get that way—it would be terrible."[2] Yet, in almost two decades together,[3] during which he became a leading theoretical physicist and published dozens of papers,[4] he never acknowledged her help in any of them, nor did she publish anything of her own. What went wrong?

It has been suggested that Marić actually made major contributions, perhaps even doing the preponderance of the work in some cases, to important papers published in Einstein's name, contributions that he simply failed to acknowledge.[5]

The available evidence does not support such claims, as I have argued elsewhere[6] and will argue here. A sketch of Marić's life up to her separation from Einstein,[7] with emphasis on a discussion of her work in physics and its relation to his,[8] leads to the conclusion that she played a small but significant supporting role in his early work, a role that later diminished to the point that she felt excluded from his career. Finally, I shall consider some possible reasons why a full collaboration between them never developed.

Marić's Student Years

Mileva Marić was born in 1875 to a mother of Montenegrin extraction, in Titel, a town in the Vojvodina, then part of the Austro-Hungarian Empire. Her Serb father was a middle-level official in the Hungarian bureaucracy, who saw to it that she received an education quite unusual for a young woman of that time and place, including two years as a private pupil at the Royal Gymnasium in Zagreb, where

Helena M. Pycior, Nancy G. Slack, and Pnina G. Abir-Am, eds.
Creative Couples in the Sciences
pp. 207–219, 330–335 ©1996 Rutgers University Press

her father was then working. After receiving special permission to attend the otherwise all-male physics class, she got the highest grades in both physics and mathematics. She finished her secondary education in Zurich, graduating from a girls' school in 1896. After a semester of medical studies at the University of Zurich, she transferred to the Swiss Federal Polytechnical School, Poly for short, enrolling in Section VIA, which trained teachers of mathematics and physics.

Marić's move to Switzerland is not hard to understand. French universities were the first in Europe to admit women.[9] Switzerland was second, with Zurich in 1865, and other Swiss universities soon followed. The Poly did so in 1876, and the first woman graduated from Section VIA in 1894.[10] Young women in search of a higher education, many of them Slavs, went to Paris if comfortable with French, like Marie Sklodowska, or Switzerland if more at home in German, like Rosa Luxemburg.[11] Russians and South Slavs from the Austro-Hungarian Empire flocked to Switzerland.[12]

Einstein and Marić were the only two physics students to enter Section VIA in 1896. Both took basically the same required courses, but rather different electives.[13] During her second year, she went to Heidelberg to attend mathematics and physics lectures, returning after one term.[14] As a result, she passed the Poly's intermediate examinations a year later than he did, using his physics lecture notes to help prepare.[15]

After her return, the two became very closely attached, spending most of their time together. In spite of the firm opposition of his parents to the liaison[16]—an opposition that led to dramatic clashes between Einstein and his parents–the two lovers resolved to live together after graduation, marrying as soon as economic circumstances permitted. Their relationship included more than romance; to supplement the meager offerings of the Poly in theoretical physics, they jointly studied many classic works.[17] They also spent a great deal of time working in the well-equipped laboratories of Heinrich Friedrich Weber, senior of the two professors of physics.

In 1900 both took the final examinations. Her physics grades were comparable to his, but she got a decidedly lower grade in mathematics; he passed with an average of 4.91 out of a possible 6, while she failed with an average of 4.0[18] Still hopeful, she reregistered the next year to retake the final examinations.

Both saw Weber as their potential mentor in the process of gaining entry to the physics community. She continued to work on her diploma thesis in Weber's laboratory (see below), which she hoped to use as the basis for an eventual doctorate.[19] Einstein also expected to remain at the Poly as Weber's *Assistent* (the lowest rung on the European academic ladder) while working on a doctorate. But his failure to obtain this position, which he felt had been more or less promised to him by Weber, led to increasing friction between Weber and the young couple.[20] Einstein's efforts to get an assistantship in mathematics at the Poly, and in physics at numerous other universities also failed, and he and Marić tried to find other jobs, again without success.[21] For the next couple of years, he lived from hand to mouth, working at a series of temporary academic and tutoring jobs outside Zurich.

In the midst of this trying period, while studying to retake the final examinations, Marić became pregnant. After again failing,[22] she left for home vowing never again to work with Weber. It is quite likely that the friction over Einstein played a role in her estrangement from Weber. Cut off from the physics community, she was now entirely dependent on her relationship with Einstein for intellectual as well as emotional support. He, on the other hand, had found another mentor, Alfred Kleiner, professor of physics at Zurich University, and begun work on a doctoral thesis.[23]

During the remainder of her pregnancy, the couple were reunited only once, after his parents sent a letter to hers making their hostility painfully clear. Marić fled the ensuing crisis to be near Einstein, who was working as a substitute teacher. To preserve the proprieties, she stayed in a nearby town for a few weeks, meeting him only on weekends, and then returned home. Her letters to Albert during this period sound notes of real despair, while his reassure her of his devotion and depict a rosy future (see the opening quotation of this paper) once this difficult period in their lives has passed.

Their daughter, referred to as "Lieserl" in Einstein's letters, was born early in 1902.[24] The same year, Einstein moved to Bern to start work at the Swiss Patent Office, where he remained for seven years. Marić soon followed, but without Lieserl, and the couple married early in 1903.[25] It was not uncommon at the time to legitimize a birth by a subsequent marriage,[26] and Einstein had earlier resolved that the child would join them after theirs.[27] But Lieserl was never reunited with her parents, and, in spite of recent efforts to find more information, her ultimate fate remains unknown (see below).

The episode undoubtedly placed a great strain on their relationship, as their elder son, Hans Albert, seems to have later surmised. A biographer with unique access to information from him[28] reports:

> Friends had noticed a change in Mileva's attitude and thought the romance might be doomed. Something had happened between the two, but Mileva would only say that it was "intensely personal." Whatever it was, she brooded about it and Albert seemed to be in some way responsible. Friends encouraged Mileva to talk about her problem and get it out in the open. She insisted that it was too personal and kept it as a secret all her life. . . . Mileva married Albert despite the incident. . . . She did not think of the shadow her "experience" would cast over their life together.[29]

Married Life

"All happy families resemble one another, but each unhappy family is unhappy in its own way," wrote Tolstoy,[30] and so it came to be for the Einsteins. At first things went well for the newly married pair—at least in the picture both painted for their friends. Einstein wrote to his friend Michele Besso: "Well, now I'm an honorably married man, and lead a very nice, comfortable life with my wife. She takes care of everything exceptionally well, cooks well, and is always cheerful."[31] Shortly after, Marić wrote to her friend Helene Savić in a similar vein: "I am, if possible, even more attached to my dear treasure than I already was in the Zurich days.

He is my only companion and society and I am happiest when he is beside me." She also inquired about the possibility of teaching jobs for herself and Einstein in Belgrade, her last known reference to the possibility of a career for herself.[32]

In September 1903, while she was visiting her parents, Einstein wrote: "Now come back to me soon. 3 1/2 weeks have already passed and a good little wife shouldn't leave her husband alone any longer. Things don't look nearly as bad at home as you think. You'll be able to clean up in short order."[33] Marić presumably went to her parents to see to Leiserl's future, which Einstein discussed (the last known reference to her) in a way that suggests they had already decided not to keep her ("As what is the child registered? We must take precautions that problems don't arise for her later"). He mentioned a serious illness ("I'm very sorry about what has befallen Leiserl. It's so easy to suffer lasting effects from scarlet fever. If only this will pass"), and she may have died subsequently; but the reference to her future suggests that she survived. If so, she may have suffered permanent mental or physical damage and been placed in some institution. If she survived unharmed, she may have been adopted by a relative, or given up for a "normal" adoption.[34]

Einstein had apparently just learned about Marić's second pregnancy: "I'm not the least bit angry that poor Dollie [his nickname for Marić] is hatching a new chick. In fact, I'm happy about it and had already given some thought to whether I shouldn't see to it that you get a new Lieserl."[35] But there never was a "new Lieserl." The second child was a boy, Hans Albert, born in 1904; another son, Eduard, was born in 1910.

During his seven years as patent clerk, especially from 1905 on, Einstein produced a steady stream of scientific papers, which, by the end of the decade, gained him a reputation as one of the most promising young theoretical physicists. He left the Patent Office in 1909 to take his first full-time academic post, as an assistant professor at the University of Zurich. By this time the marriage was in trouble. Marić confided to Savić: "In mid-October on the 14th we leave Bern, where I have now spent 7 years, so many beautiful and, I must say, also bitter and difficult days."[36]

We have seen one source of her bitterness: the final decision about Lieserl, taken early in the Bern years. Another source was quite recent: a marital crisis, involving Einstein's friend, Anna Meyer-Schmid.[37] Suspecting her of designs on Albert, Marić wrote Meyer-Schmid's husband. Enraged, Einstein wrote Herr Schmid attributing Marić's conduct to unmotivated jealousy.[38]

Marić's letter to Savić goes on to boast of Einstein's success: "He is now counted among the leading German-speaking physicists and is being frightfully courted. I am very happy about his success, which he has really earned: I only hope and wish that fame does not exert a detrimental influence on his human side."[39] A letter written to Savić soon after Einstein and Marić settled in Zurich further explains her fears:

> You see, with such fame, not much time remains for his wife. I read a certain maliciousness between the lines when you wrote that I must be jealous of science, but what can one do, the pearls are given to one, to the other the case. ... I often ask myself

> . . . whether I am not rather a person who feels a great deal and passionately, fights a great deal and also suffers because of that; and out of pride or perhaps shyness puts on a haughty and superior air until he himself believes it to be genuine. And I must ask you, even if the latter were the case, and my innermost soul stood less proudly, even then could you love me? You see I am very starved for love and would be so overjoyed to hear a yes, that I almost believe wicked science is guilty, and I gladly accept the laughter over it.[40]

Einstein's academic star rose with dramatic speed: In 1911 he accepted a full professorship at the German University in Prague, and the next year was called back to a similar post at his alma mater in Zurich.[41] In 1914 he was named a member of the Prussian Academy of Sciences and head of the prestigious Kaiser Wilhelm Institute for Physics, moving to Berlin to take this full-time research job. A letter to Savić in 1911 gives further insight into Marić's feelings during this period:

> I . . . believe we women cling much longer to the memory of that remarkable period called youth, and involuntarily would like things always to remain that way. Don't you find that to be so; men always accommodate themselves better to the present moment. Things are going well for mine; he works very hard, gives courses that are very well liked and attended, as well as many lectures, which I never fail to attend. Since there are rather many musical occasions in our house, we really have very little time that we can pass together in privacy and tranquility.[42]

These touching and remarkably frank letters depict a woman who feels she is losing her husband, not least because of his successful career in science. They convey a growing sense of exclusion from that career, but no sense of deprivation of credit for his scientific work. Her own earlier ambitions seem completely subsumed by ambitions for him, ambitions that go hand in hand with forebodings of what his success augurs for their relationship.

The toll on Marić became apparent to those around her. Referring to the period around 1912, Peter Michelmore gives us an insight into how things appeared to their son, Hans Albert:

> Close friends . . . worried because [Marić's] dark moods were becoming more frequent. She was far too introverted. She never talked about herself. Even alone with the family, she had little to say and her long periods of silence irritated Albert. If they ever discussed the root of the trouble, that mysterious pre-marital incident, nobody knew about it.
>
> Hans Albert, then an eight-year-old with a distinct mind of his own, sensed the tension between his parents. But his father's personality assured him all would be well.[43]

In retrospect, Hans Albert evidently thought that the loss of Lieserl was at the root of the estrangement of his parents. At the time, he served as his mother's surrogate for the waning love of his father. In 1909 she wrote Savić: "[Hans Albert] should start school early next year, but unfortunately he entered the world a week too late and probably will not be accepted. Then he will stay with his mama for another year; we are actually inseparable and cling terribly to each other."[44]

By 1912, whether she knew it then or not, Marić was competing with more than science for Einstein's affections. During a visit to Berlin, he had started a romantic liaison with his cousin Elsa Löwenthal, a divorcée with two young daughters and literary aspirations, then living there with her parents.[45] His letters to her refer to Marić, often alluded to as "my cross," in increasingly bitter terms: "Miza [nickname for Marić] is the sourest sourpuss that has ever been. . . . I cannot be at ease at home . . . she herself is the most tormented one, and cannot understand that she herself creates the graveyard atmosphere. Miza is by nature unlovable and mistrustful. When one responds accordingly, she feels persecuted."[46]

By the end of 1913 Marić was aware to some extent of the situation, as he informed Löwenthal: "She [Marić] doesn't ask about you, but I believe she does not therefore underrate the significance that you have for me."[47]

Shortly after their move to Berlin in April 1914, Marić realized that one of its chief attractions for Einstein was cousin Elsa, and she returned to Zurich with the two boys, never again to live with Einstein as husband and wife.[48]

The Nature of Their Intellectual Relationship

From the letters Einstein and Marić exchanged as students, a picture emerges of two young people very much enchanted with each other, not least because of their common love for physics.[49] However, the contrast between their comments on this subject is striking. Einstein's show a young man passionately engaged with his subject, constantly telling Marić about his readings of both the classics and new papers. Rather than giving bare reports, he critically evaluates his readings, often adding ideas of his own bearing on their subject matter. Without the benefit of hindsight, one cannot point to anything in these early letters giving proof positive of budding genius, but they do convey the distinct impression of an original and imaginative mind at work.

Marić's comments depict an eager, hardworking student, but without a spark of originality, or more precisely, of scientific originality, for she does display flashes of literary talent, catching fire in some descriptive passages rather than in comments on physics.[50]

What was the nature of their intellectual relationship during the student years? They studied physics together, which was very important for Einstein, who at first was quite dependent on her. During the summer break of 1899 he wrote: "When I read Helmholtz for the first time I could not—and still cannot—believe that I was doing so without you sitting next to me. I enjoy working together very much, and find it soothing and less boring."[51] Later that year, Marić requested his help in preparing for her intermediate examinations, which she took a year after he did (see the previous section).[52] Einstein's physics notes contain a correction in her hand, confirming that she read them carefully.[53]

In discussing his ideas, Einstein occasionally called upon her for help, such as finding data to corroborate them (see next section); but the letters suggest that the most important role she played in their intellectual relationship during these years was "that of a sounding board for Einstein's ideas," as the editors of the

Collected Papers (myself included) put it. He had a strong need to clarify and develop his ideas in dialogue with others, a "role also played on occasion by his friends Michele Besso and Conrad Habicht" after his move to Bern.[54]

It is difficult to gauge the nature of her reponse to his ideas, since many of his letters and even more of her replies are lost. But fortunately we have her reply to the letter containing his most important original ideas. Almost half of Einstein's letter is devoted to his earliest discussion of the electrodynamics of moving bodies.[55] Her reply comments on every topic Einstein discussed in this and his previous letter (which she had received at the same time): family matters, vacation, examination preparations, and so on, with the sole exception of the electrodynamics of moving bodies. None of her ten other extant letters comments on his scientific ideas, so this exchange may be typical.

While we can never know their private conversations, later reminiscences suggest her taciturnity in discussions. Philipp Frank, who knew Einstein and questioned him extensively for a biography, discussed their student years: "For Einstein it had always been pleasant to think in society, or better perhaps, to become aware of his thoughts by putting them into words. Even though Mileva Maritsch [Marić] was extremely taciturn and rather unresponsive, Einstein in his zeal for his studies hardly noticed it."[56] Just after their marriage, Einstein and two friends set up the mock "Olympia Academy" to discuss topics in the foundations of science, usually holding their sessions in the Einsteins' home. Maurice Solovine recalled that "Mileva, intelligent and reserved, listened attentively to us, but never intervened in our discussion."[57]

Marić's Work in Physics

Marić's only comments on her own work concern her studies at the Poly. The most interesting concerns her diploma thesis, prepared as part of the final examinations.[58] "Prof. Weber has accepted my proposal for the diploma thesis, and was even very satisfied with it. I am very happy about the investigations I'll have to do for it. E[instein] has also chosen a very interesting topic."[59] Both of them carried out experimental studies of heat conduction, one of Weber's pet research topics, under his supervision. Einstein also commented on her work in a letter to Marić: "For the investigation of the Thomson effect I have again resorted to a different technique, which is similar to your method for determining the dependence of K [the coefficient of thermal conductivity] on T and which also presupposes such an investigation."[60] Weber graded her work 4 (out of 6) and his 4.5.[61] In retrospect, Einstein characterized the topic of their work harshly, as "totally uninteresting to me";[62] neither thesis led to a publication and the Poly routinely discarded such student theses, so an independent judgment is impossible. At any rate, evidence that Marić devised an experimental technique has no bearing on the question of her talent in theoretical physics, the area in which Einstein made his name. Marić's other references to her work in physics are limited to discussion of preparations for examinations (see previous section), as are Einstein's other comments on it.

More relevant are two letters to Savić discussing Einstein's work in theoretical physics. The first states:

> Albert has written a paper on physics that will probably soon be published in the physical Annals.[63] You can imagine how proud I am of my dear treasure. It is really no ordinary work, but very significant, on the theory of fluids. We have sent a private copy to [Ludwig] Boltzmann, and would really like to know what he thinks of it, hopefully he will write to us.[64]

The work in question is a theory of molecular forces. Discussing this work, Einstein wrote Marić:

> The results on capillarity I recently obtained in Zurich seem to be entirely new despite their simplicity. When we're back in Zurich we'll try to get some empirical data on this subject from [Professor] Kleiner [of the University of Zurich]. If this yields a law of nature, we'll send the results to Wiedemann's *Annalen* [*der Physik*].[65]

Marić's second letter discussed the doctoral thesis, based on the same theory, that Einstein submitted to the University of Zurich in 1902 and then withdrew:[66]

> Albert has written a splendid work that he has submitted as a dissertation. In a couple of months he will probably receive the doctorate. I have read it with great pleasure and true admiration for my dear little treasure, who has such a clever head. When it is printed, I will send you a copy. It deals with the investigation of molecular forces in gases on the basis of various known phenomena. He is really a splendid fellow.[67]

In both letters, Marić states that the works were written by Einstein, claiming no role in the formulation of the theory; he also speaks of his results.[68] Nevertheless, in discussing this work both slip easily into the "we" mode, which should be kept in mind when evaluating similar uses of the first-person plural in his letters.

The most notable of these is a reference to "our work" on a problem of much greater significance than his theory of molecular forces (see below), one of the complex of problems that led to the special theory of relativity; and the passage has been cited to support claims that Marić was coauthor of that theory.[69] Leaving aside the fact that his letter was written in 1901, whereas the theory was not finished until 1905, it is important to put the passage into context.

Physics aroused emotions in Einstein that, during the early stage of their courtship, he felt impelled to share with Marić, come what may. For example, soon after she told him she was pregnant—surely a difficult time for both—he opened a letter as follows: "I have just read a wonderful paper by Lenard. ... Under the influence of this beautiful piece I am filled with such happiness and such joy that I absolutely must share some of it with you. Be happy and don't fret, darling. I won't leave you and will bring everything to a happy conclusion."[70] It is striking how many of his few references to joint work were penned at difficult moments in their relationship, amid reassurances of his love and devotion.

For example, Einstein referred to "our work on relative motion" after he left Zurich to stay with his parents, whom she knew to be violently opposed to their engagement. Here is the context:

> You are and will remain a shrine for me to which no one has access; I also know that of all people, you love me the most, and understand me the best. I assure you that no one here would dare, or even want, to say anything bad about you. I'll be so happy and proud when we are together and can bring our work on relative motion to a successful conclusion! When I see other people I can really appreciate how special you are.[71]

His words here are moving in their emotional intensity, but provide no clue about her contribution to "our work." Elsewhere in his letters, he does mention specific ideas about "relative motion" and many other topics in physics, but he always refers to his own work. Here is an example: "I'm busily at work on an electrodynamics of moving bodies, which promises to be quite a capital piece of work. I wrote to you that I doubted the correctness of the ideas about relative motion. But my reservations were based on a simple calculational error. Now I believe in them more than ever."[72]

Naturally enough, their correspondence practically stops after Marić joins him in Bern. Their few letters from the crucial years 1903–1905 that led up to the final formulation of the theory of relativity contain nothing relevant, nor is there any other contemporary documentation. Later reminiscences suggest that she continued to play a modest role in his work. The one I find most significant comes indirectly from Hans Albert Einstein, presumably based on information he got from his parents (see note 28). Discussing Einstein's work on special relativity, Michelmore writes, "Mileva helped him solve certain mathematical problems, but nobody could assist with the creative work, the flow of fresh ideas. ... [After he wrote up his work] Mileva checked the article again and again, then mailed it. 'It's a very beautiful piece of work,' she told her husband."[73] The mathematics involved does not go beyond elementary calculus, and it seems unlikely that Marić contributed unique mathematical expertise to the paper; one may speculate that she might have suggested methods of proving certain results and/or checked calculations.

Einstein indeed does thank someone "who stood faithfully at my side and to whom I owe many valuable suggestions" at the end of his paper,[74] but it is his "friend and colleague M[ichele] Besso."[75] Taken together with his silence about Marić, this is interesting—if negative—evidence of his attitude toward her role in his work.

Other documents suggest that Marić played the role of amanuensis on occasion after 1905. Einstein's notebook for his lecture course on mechanics, given during the winter semester of 1909–1910 at the University of Zurich, includes "seven pages of notes in Mileva Einstein-Marić's handwriting, containing material very closely corresponding to the introductory sections of the first notebook, followed by an eighth page with a drawing of three intersecting circles, also in Einstein-Marić's hand."[76] And a document entitled "Reply to Planck's Manuscript," dated to 1909 or 1910 and included in a letter of Einstein to Planck, is also in Marić's hand.[77]

Another Einstein lecture notebook from 1910–1911 testifies not only to her familiarity with the notes but to her continued affection. She inserted the words: "Here give a dear little kiss to his [word not deciphered]."[78]

Her letters to Savić from 1909 on, cited in the last section, indicate that she attended his public lectures, but bear witness to her growing sense of isolation from his career, as does her only letter to Einstein from this period, written after the 1911 annual meeting of the Society of German Scientists and Physicians in Karlsruhe, which he attended: "It must surely have been very interesting in Karlsruhe; I would have all too gladly also listened a little, and seen all these grand people [*diese feinen Leute*]."[79]

To sum up, Marić seems to have encouraged and helped Einstein in a number of ways during their years together, notably as the alter ego to whom he could express his ideas freely while developing them in isolation from the physics community. She also appears to have helped by looking up data, suggesting proofs, checking calculations, and copying some of his notes and manuscripts. He never publicly acknowledged this help, nor did a true collaboration ever develop. As he took an increasingly prominent place in the physics community after 1909, she felt increasingly isolated from his work and threatened by his success.

Why Did a Real Collaboration Never Develop?

Was it possible for a married couple to successfully and publicly collaborate in physics at the beginning of the century? Two well-known contemporary couples managed to do so: Marie Sklodowska and Pierre Curie,[80] and Paul Ehrenfest and Tatiana Afanasieva.[81] There are interesting similarities between them and the Einsteins. All three wives were Slavs with a higher education living in milieus not free of prejudice against educated women.[82] All three husbands came from secular backgrounds; Einstein and Ehrenfest were Jews, raised in south German urban environments (Munich and Vienna respectively), who had yet to establish their careers when they married.[83]

There is also a striking contrast. In the case of the Curies and Ehrenfests, there is abundant contemporary evidence of the importance of the woman's role in their joint work, and each wife pursued a scientific career after her husband's death. Marić, of course, did not pursue a scientific career before or after her separation from Einstein, but we see it cannot have been because of the impossibility of such a career.

It also cannot have been because she lacked initial motivation and support. In the face of the many prejudices against and obstacles to women's higher education, particularly in the physical sciences, Marić possessed sufficient talent and drive, and got sufficient familial and institutional support, to successfully pursue an academic career that brought her to the brink of graduation from the Poly and pursuit of scientific research, alone or in collaboration with Einstein, or at the least a career as a teacher of science. What went wrong?

I suggest three interrelated factors may help to explain why Marić never pursued a scientific career, and in particular why a truly creative collaboration between Marić and Einstein never developed:

1. Her talents in physics were modest, so that she could not take advantage of the "exceptional" status sometimes granted even then to women of the stature

of Marie Curie and Lise Meitner in physics, of Sofia Kovalevskaia and Emmy Noether in mathematics.

2. She lost the inner self-confidence and drive necessary to pursue a career in science in the face of the many obstacles that women face.

3. Despite his earlier-expressed intentions, after their marriage Einstein failed to encourage her to pursue an independent career or to involve her in serious collaboration.

I shall elaborate a little on these factors. As we have seen, in spite of her early successes as a student, there is no evidence that Marić ever made the crucial transition from student of physics to independent research worker. There is no record of her original ideas, and her comments on the ideas of others are uncritical. For example, she vastly overrated Einstein's theory of molecular forces (see above) when a critical judgment might have helped him to discard it sooner than he did.[84]

There is nothing really surprising about this;[85] most physicists, male or female, would have had to play a subordinate role in collaborating with Einstein. The evidence suggests that she did so, even without public acknowledgment, which might have been painful to her had she not been so willing to acknowledge his superiority and subordinate her own career goals to his. But she accepted this role without complaint, and even accepted—but not without complaint—her growing exclusion from this modest role in his work. Part of her resignation may be ascribed to her great love and admiration for him. But I think there is more to the story.

Her early letters evidence the good cheer, drive, and talent necessary to enable a young woman of her generation to get from a remote region of the Balkans to Section VIA of the Poly. But by the end of her student days—at a crucial point in her intellectual development—she lost the inner self-confidence so vital to overcoming the considerable obstacles on her path to a career in physics, a self-confidence that Einstein possessed in abundance and never lost, even at the most desperate moments in his life. Two failures to pass the final examinations and the loss of Weber as a mentor were undoubtedly contributing factors, but her relationship with Einstein also played a role. The pressures on a woman to subordinate her intellectual to her emotional life were even stronger then than they are today. As her letters attest, she was painfully shy and fearful of criticism, and his parents' opposition, of which she was well aware even though she had never met them (rather than shield her, Einstein reported their comments),[86] must have afflicted her. Above all, her pregnancy out of wedlock and the fate of Lieserl seem to have contributed to an underlying depression that grew as the years passed. Perhaps partly in reaction to the loss, she became exceptionally devoted to her first son, born early in the second year of their marriage; and she was unable to find a way to combine her conception of the duties of motherhood with those of a career outside the home.

Given all these factors, she still might have played a more satisfying if subordinate collaborative role in his work, as did several male physicists during this period.[87] A married woman at that time was hardly likely to find another mentor than her husband, so her fate as a physicist was entirely dependent on Einstein.[88] But after their marriage, he failed to foster such a full collaboration. However modest her talents, he could have publicly acknowledged her contributions to his work, and helped her to enter the world of physics after he gained recognition. He is reported to have helped around the house,[89] but obviously he was not engulfed by household duties, and he could have done more to ensure that she was not. Instead, he seems to have been content to let her play the "philistine" role of Hausfrau, involving her in his work as little more than occasional amanuensis, and never publicly acknowledging her contributions. Again, there is an obvious contrast with Pierre Curie and Paul Ehrenfest, who took pains to assure that their wives' contributions to joint work were publicly acknowledged,[90] so that success was shared. Far from bringing Einstein and Marić closer, the widespread recognition of Einstein's scientific activities became an important factor in their ultimate estrangement.

NOTES

[1] She sometimes used Marity, the Hungarian form of her last name; she followed Swiss custom after her marriage, using Einstein-Marić or Einstein-Marity.

[2] *Albert Einstein and Mileva Marić, The Love Letters*, trans. Shawn Smith, ed. Jürgen Renn and Robert Schulmann (Princeton, 1992), 72–73, cited hereafter as *The Love Letters*. Einstein's correspondence, including letters to and from Marić, will also be cited from *The Collected Papers of Albert Einstein*, vol. 1, *The Early Years, 1879–1902*, ed. John Stachel et al. (Princeton, 1987), and vol. 5, *The Swiss Years: Correspondence, 1902–1914*, ed. Martin Klein et al. (Princeton, 1993); cited hereafter as *Collected Papers*, vols. 1 and 5.

[3] They met in 1896, married in 1903, separated in 1914, and divorced in 1919.

[4] For his publications during this period, see *The Collected Papers of Albert Einstein*, vol. 2, *The Swiss Years: Writings, 1900–1909*, ed. John Stachel et al. (Princeton, 1989); vol. 3, *The Swiss Years: Writings, 1909–1911*, ed. Martin Klein et al. (Princeton, 1993); and vol. 4, *The Swiss Years: Writings, 1912–1914*, ed. Martin Klein et al. (Princeton, 1995); cited hereafter as *Collected Papers*, vols. 2, 3, and 4.

[5] See Desanka Trbuhović-Gjurić, *Im Schatten Albert Einsteins/Das tragische Leben der Mileva Einstein-Marić* (Bern/Stuttgart, 1983), cited hereafter as *Im Schatten Albert Einsteins*; Senta Troemel-Ploetz, "Mileva Einstein-Marić: The Woman Who Did Einstein's Mathematics," *Women's Studies International Forum* 13 (1990): 415–432; Evan Harris Walker, "Did Einstein Espouse His Spouse's Ideas?" *Physics Today* 42, no. 2 (February 1989), 9–11 (for my comments, see ibid., 11–13); idem, "Ms. Einstein" (paper presented at the AAAS meeting, New Orleans, February 1990); and idem, "Mileva Marić's Relativistic Role" (presented at the AAAS Meeting, Washington, D.C., February 1991).

[6] "The Young Einstein: Poetry and Truth," this volume, pp. 21–38. See also Roger Highfield and Paul Carter, *The Private Lives of Albert Einstein* (London/Boston, 1993), cited hereafter as *Private Lives*, and Abraham Pais, *Einstein Lived Here* (Oxford/New York, 1994).

[7] Sources for information on her life include *Im Schatten Albert Einsteins*; Dorde [George] Krstić, "Mileva Einstein-Marić," Appendix A in Elizabeth Roboz Einstein, *Hans Albert Einstein: Reminiscences of His Life and Our Life Together* (Iowa City, 1991); her correspondence with Einstein in *Collected Papers*, vols. 1 and 5; and her letters to her

friend and confidante, Helene Savić, née Kaufler. Some excerpts from the Savić letters are cited from *Collected Papers*, vol. 1, and unpublished excerpts are cited (in my translations) from photocopies of originals, presented by Savić's grandson, Professor Milan Popović (Belgrade), to the editors of *The Collected Papers*. These copies will be cited as in the Einstein Papers Project Archives, Boston University. A useful synthesis of this material is found in *Private Lives*.

[8] Einstein is discussed here only insofar as is relevant to their intellectual relationship. For a differing account of their relationship, more skeptical of Einstein's early devotion to Marić, see *Private Lives*.

[9] See Phyllis Stock, *Better Than Rubies: A History of Women's Education* (New York, 1978), 166; cited hereafter as *Better Than Rubies*. There also may have been medical reasons for Marić's move, since she had been very ill with a lung disorder.

[10] See Schweizer Verband der Akademikerinnen, *Die Frauenstudium an der Schweizer Hochschulen* (Zurich, 1928), cited hereafter as *Die Frauenstudium*.

[11] For a discussion of the first generation of Russian women to study in Zurich, see Christine Johanson, *Women's Struggle for Higher Education in Russia, 1850–1900* (Kingston/Montreal, 1987), 51–58. According to Johanson, while many male students were hostile, "most professors allowed no sexual discrimination in the classroom" (53).

[12] Indeed, presure from Russian women prompted Zurich to open its doors (see *Better Than Rubies*, 145). In the first decades after the Swiss universities admitted women, the large majority were non-Swiss, mainly Slavs (see *Die Frauenstudium*).

[13] For his *Matrikel* (official record), see *Collected Papers*, vol. 1, doc. 28, pp. 45–40. Her *Matrikel* is in file no. 85, *Rektoratsarchiv*, Eidengenössische Technische Hochschule (ETH).

[14] Trbuhović-Gjurić suggsts, without any evidence, that Marić left the Poly in flight from her intense romantic relationship with Einstein (see *Im Schatten Albert Einsteins*). Their letters suggest that the relationship was not yet very intense (see *Collected Papers*, vol. 1, esp. docs. 36 and 39). The brevity of Marić's stay in Heidelberg may be explained by Kaplan's observation that "the first women students at Heidelberg ... suffered from extraordinary gender discrimination" (Marion Kaplan, *The Making of the Jewish Middle Class: Women, Families, and Identity in Imperial Germany* [New York, 1991], 149).

[15] For this information, see *Collected Papers*, vol. 1, esp. docs. 50, 52, and 53.

[16] His parents' opposition was based on Marić's age (she was four years older than Einstein), her intellectuality, and probably her Slavic origins. His mother made the first two objections explicit: "By the time you're 30 she'll be an old witch." "Like you, she is a book—but you ought to have a wife" (*The Love Letters*, 20). Anti-Slav prejudices are still common in Germany, and Einstein's parents had not objected to his earlier romance with a young teacher of Swiss-German background, who was also slightly older than he (see *Collected Papers*, vol. 1, docs. 15, 18, and 32).

[17] Einstein's letters to Marić mention treatises by Boltzmann, Drude, Helmholtz, Kirchhoff, and Mach (see *Collected Papers*, vol. 1).

[18] See *Collected Papers*, vol. 1, doc. 67, p. 247). The three mathematics students in VIA took different exams. Trbuhović-Gjurić (*Im Schatten Albert Einstein*) does not mention her failure to graduate; Troemel-Ploetz ("The Woman Who Did Einstein's Mathematics") ascribes it to discrimination against women at the Poly without mentioning her grades; while Walker ("Ms. Einstein") states, without citing evidence, that "Marks be-

low 5.00 were probably customarily below the passing grade." Einstein, with a total of 54 points out of a possible 66, was one point short of that average, while Marić, with a total of 44 points, was 11 points short.

[19] In mid-1900, she mentions "a large work ... that I have chosen for myself as a Diploma Thesis and probably also a Doctoral Thesis" (*Collected Papers*, vol. 1, 260, n. 5). In May 1901, Einstein asks about her doctoral thesis, advising her to use some of Weber's work in it, "even if you only seem to" (ibid., 305).

[20] In May 1901, Marić wrote Savić, "I have already quarreled a couple of times with Weber, but we're already used to that" (*Collected Papers*, vol. 1, doc. 109, p. 303, my translation).

[21] See *Collected Papers*, vol. 1, doc. 87, p. 275.

[22] See Protocol of Section VIA, July 26, 1901, ETH Library (Zurich). Her average was again 4.

[23] Einstein first mentions Kleiner in October 1900 (*Collected Papers*, vol. 1, 267); a year later, he discussed the completed dissertation (ibid., 321). He withdrew it in February 1902 (see ibid., doc. 132, p. 331), probably because of objections by Kleiner, but they stayed in contact. Einstein's successful 1905 doctoral dissertation was approved by Kleiner, who helped him obtain his first full-time academic post in 1909 (see below).

[24] Presumably, Lieserl was born at Marić's home. However, recent efforts to find civil or church records of the birth in her hometown or nearby failed.

[25] The delay was connected with the opposition of his family (see *Collected Papers*, vol. 1, doc. 138, p. 336). On his deathbed, Einstein's father gave his consent in October 1902, according to Abraham Pais, *"Subtle is the Lord ...": The Science and the Life of Albert Einstein* (Oxford, 1982), 47.

[26] see *Private Lives*, 90.

[27] Late in 1901, after he was assured of a Patent Office job, he wrote Marić: "The only problem that still needs to be resolved is how to keep our Lieserl with us; I wouldn't want to have to give her up. Ask your Papa, he's an exceptional man and knows the world better than your overworked, impractical Johnny" (*Collected Papers*, vol. 1, doc. 127, p. 324, translation from *The Love Letters*, 68).

[28] Peter Michelmore, *Einstein: Profile of the Man* (New York, 1962), states: "Hans Albert Einstein ... had never discussed his father before with any writer, at least not in depth. But he answered all my questions, and waited while I wrote down all the answers" (vii). Hans Albert inherited his mother's papers, and his first wife, Frieda Einstein-Knecht, transcribed excerpts from Einstein's letters discussing Lieserl. So, if not told earlier by either parent, Hans Albert knew about his sister by the time he spoke to Michelmore.

[29] Michelmore, *Einstein*, 42.

[30] Leo Tolstoy, *Anna Karenina*, trans. Louise and Aylmer Maude (London, 1965), 1.

[31] *Collected Papers*, vol. 5, doc. 5, letter of January 22, 1903, p. 10 (my translation).

[32] Marić to Savić, March 20, 1903, copy in Einstein Papers Project Archives, Boston University.

[33] *Collected Papers*, vol. 5, doc. 13, p. 22, translation modified from *The Love Letters*, 53.

[34] For further speculation, see *Private Lives*, 88–91.

[35] *Collected Papers*, vol. 5, doc. 13, p. 22, translation from *The Love Letters*, 53.

[36] Marić to Savić, September 3, 1909, copy in Einstein Papers Project Archives, Boston University.

[37] The flirtatious nature of their earlier relationship is apparent from a poem Albert wrote for her (*Collected Papers*, vol. 1, doc. 49, p. 220).

[38] See *Collected Papers*, vol. 5, 181, 198–199; Einstein-Marić to Georg Meyer, May 23, 1909, copy in the Archive of the Einstein-Gesellschaft, Swiss National Library (Bern). For a fuller account, see *Private Lives*, 124–126. Einstein's anger flared up again over forty years later, when he blamed Marić's pathological jealousy on "uncommon ugliness" (Einstein to Erika Scharer-Meyer [Meyer-Schmid's daughter], cited in *Collected Papers*, vol. 5, 199, n. 4).

[39] Marić to Savić, September 3, 1909, copy in Einstein Papers Project Archives, Boston University.

[40] Marić to Savić, n.d. [c. October 1909], copy in Einstein Papers Project Archives, Boston University.

[41] By this point, the Poly had been renamed the Eidgenössische Technische Hochschule, or ETH for short.

[42] Marić to Savić, n.d. [c. January 1911], copy in Einstein Papers Project Archives, Boston University.

[43] Michelmore, *Einstein*, 57.

[44] Marić to Savić, n.d. [c. October 1909], copy in Einstein Papers Project Archives, Boston University.

[45] As children, they were well acquainted, and her father (nicknamed "Rudolf the rich" by Einstein) was the chief creditor of his father's debts (see *Collected Papers*, vol. 1, doc. 93, p. 281); for their relationship, see his letters to her in *Collected Papers*, vol. 5; for her poetry reading, see Pais, *Einstein Lived Here*, 145.

[46] *Collected Papers*, vol. 5, 585, 587.

[47] Ibid., 558.

[48] After their divorce, he regularly stayed at Marić's house when visiting Zurich.

[49] See *Collected Papers*, vol. 1. For a more detailed discussion of their relationship up to 1905, see "The Young Einstein."

[50] For her most extensive comment on physics, see *Collected Papers*, vol. 1, doc. 36, last paragraph, p. 59. For an example of her descriptive powers, see ibid., doc. 109, pp. 301–302.

[51] *The Love Letters*, 9.

[52] Ibid., 12–13.

[53] See *Collected Papers*, vol. 1, doc. 37, p. 139.

[54] Ibid., vol. 1, xxxix–xi.

[55] "On the Electrodynamics of Moving Bodies" is the title of his famous 1905 paper on special relativity (*Collected Papers*, vol. 2, doc. 28). See the next section for further discussion of this topic.

[56] Philipp Frank, *Einstein: His Life and Times* (New York, 1953), 21.

[57] Albert Einstein, *Lettres à Maurice Solovine*, ed. Maurice Solovine (Paris, 1956), introduction, xii.

[58] This has sometimes been confused with a doctoral thesis. Marić hoped to use her diploma thesis work as the basis for a doctorate, but she was never a candidate for that degree.

[59] *Collected Papers*, vol. 1, doc. 63, pp. 243–244; translation from the supplementary *English Translation*, trans. Anna Beck (Princeton, 1987), 138.

[60] *The Love Letters*, 30.

[61] See *Collected Papers*, vol. 1, doc. 67.

[62] See ibid., note 33, 244.

[63] I.e., the *Annalen der Physik*; it became his first publication (see *Collected Papers*, vol. 2, doc. 1).

[64] *Collected Papers*, vol. 1, doc. 85, p. 273, my translation.

[65] Ibid., doc. 79, p. 267, my translation.

[66] See *Collected Papers*, vol. 1, doc. 132, p. 331.

[67] Ibid., doc. 125, p. 320, my translation.

[68] It has been suggested that she attributed *her* work to him. But it is hard to see why she would do so in private letters to a close personal friend. If the expressions of admiration in these letters were meant to characterize her own work, they would give a most unpleasant impression of her character. If we accept her word that she picked her final diploma thesis topic, I see no reason to doubt it when she says that he wrote the articles in question.

[69] See the articles by Walker and Troemel-Ploetz cited in note 5.

[70] *The Love Letters*, 54.

[71] Ibid., 39.

[72] Ibid., 69.

[73] Michelmore, *Einstein*, 45–46. Such comments, and similar (but less reliable) anecdotal accounts by Marić's relatives in the Vojvodina (see *Im Schatten Albert Einsteins*), led to Senta Troemel-Ploetz's appellation: "Mileva Marić: The Woman Who Did Einstein's Mathematics."

[74] See *Collected Papers*, vol. 3, doc. 1, p. 125, descriptive note.

[75] Ibid., 306. Besso's role is explained more precisely in later reminiscences by Einstein, notably his 1922 Kyoto lecture (see ibid., 264), and Michelmore also mentions it (*Einstein*, 45).

[76] *Collected Papers*, vol. 2, doc. 23, pp. 276–306.

[77] Ibid., doc. 3, pp. 177–178.

[78] Ibid., doc. 11, p. 321.

[79] Mileva Marić to Albert Einstein, October 4, 1911, in Einstein, *Collected Papers*, vol. 5, doc. 290, p. 331.

[80] Einstein and Marić met Marie Curie only after Pierre's death. For her life, see Eve Curie, *Madame Curie*, trans. Vincent Sheean (New York, 1937); Rosalind Pflaum, *Grand Obsession: Madame Curie and Her World* (New York, 1989); and Helena M. Pycior, "Marie Curie's 'Anti-natural Path': Time Only for Science and Family," in *Uneasy Careers and Intimate Lives: Women in Science, 1798–1979*, ed. Pnina G. Abir-Am and Dorinda Outram (New Brunswick, NJ, 1989), 191–214.

81 Both Einstein and Marić knew Ehrenfest and Afanasieva. For his life and their relationship, see Martin Klein, *Paul Ehrenfest*, vol. 1, *The Making of a Theoretical Physicist* (Amsterdam, 1970). Klein cites an obituary in Dutch, but there is no biography of Afanasieva.

82 Speaking of the German milieu, Kaplan notes "the popular stereotype of the Russian female student, who was portrayed as a radical, both politically and personally" (*The Making of the Jewish Middle Class*, 147); and she writes that "bourgeois parents displayed extraordinary ambivalence regarding their daughters' aspirations. . . . the fear lingered that educated daughters would educate themselves right out of the marriage market" (142).

83 Pierre had a well-established career in physics when he met Marie.

84 A few years later he referred to his first two papers as "worthless beginner's works" (see *Collected Papers*, vol. 5, doc. 66, p. 79).

85 "[O]ut of about one thousand [male] students there is hardly a single one who has the abilities for independent scientific accomplishment in the higher sense, so the demands on women at the least should not be set any higher" (Ella Wild, "Einleitung" to *Die Frauenstadium*, 15–16.

86 It seems plausible that he used Marić to help him break free of his family, especially his mother.

87 See, e.g., Lewis Pyenson, "Einstein's Early Scientific Collaboration," *Historical Studies in the Physical Sciences*, 7 (1976): 84–123.

88 I am indebted to Pnina Abir-Am for this insight.

89 See, e.g., the account by his son Hans Albert, cited in *Private Lives*, 129.

90 For the Curies, see Helena M. Pycior, "Reaping the Benefits of Collaboration While Avoiding Its Pitfalls: Marie Curie's Rise to Scientific Prominence," *Social Studies of Science* 3 (1993): 301–323. There is no study of the collaboration between the Ehrenfests, but I can cite a few indications of his efforts. Of the two articles they wrote jointly in 1906, the first is signed Tatiana and Paul Ehrenfest, the second is signed Paul and Tatiana Ehrenfest (see Paul Ehrenfest, *Collected Scientific Papers*, ed. Martin Klein [Amsterdam/New York, 1959], 107, 127). Their joint article on the foundations of statistical mechanics in the prestigious *Encyklopaedie der Mathematischen Wissenschaften* states: "The critical review and systematization of the results of all fundamental investigations was carried out by the authors in common work. P. Ehrenfest bears the ultimate responsibility for the final editing" (213).

Einstein's Jewish Identity

John Stachel

> *I am sure that Einstein's strongest source of identity, after science, was to be a Jew, increasingly as the years went by.*
>
> ABRAHAM PAIS (PAIS 1982, PP. 314–315)

Dedicated to the memory of Gerald Tauber[1]

In the first part of this paper I recount what is known about those events in Einstein's life that helped to shape his attitude to religion in general, and to Judaism in particular, and his approach to other issues involving the Jewish people — in short, the events that helped to shape his conscious self-identification as a Jew. This account stops with his 1923 trip to Palestine since I believe that his Jewish identity was essentially established by that time.

In the second part, I give an account of his mature views on Jews, Judaism, and Zionism. No attempt is made to recount his many activities in behalf of the Jewish people, individually and collectively, nor his reaction to the holocaust. Rather, the discussion centers on a critical analysis of his views in the twenties and thirties, their origins in the German cultural Zionist milieu, and the romantic elements in Einstein's outlook. His attempt to reconcile Zionism and internationalism, particularly as it found expression in his views on Jewish–Arab relations, is considered in some detail.

Part I. From Munich to Jerusalem

Both Einstein's paternal and maternal families came from small towns in southern Germany, where his ancestors had lived for many generations.[2] His paternal uncle Jakob was the first member of the family to get a higher education. After studying engineering at the Stuttgart Polytechnic, in 1876 he settled in Munich (then and still the largest metropolis in southern Germany),[3] where he soon went into the electrotechnical business, at that time a venture into the "high-tech" frontier. Einstein's father Hermann became a partner in Jakob's business,[4] and he and his family moved to Munich in 1880, when Einstein was a little over one year old.

Originally prepared for the
Symposium on "Einstein in Context," Jerusalem and Tel Aviv, 1990

Thereafter, the two brothers and their families lived in close proximity until 1894, when they moved their business to Italy.

The move from villages and small towns to larger cities was widespread among enterprising young Jews in south Germany during the latter half of the nineteenth century, with a markedly quickening pace after 1870.[5] The religious fervor of their forefathers had already disappeared or diminished among many of these Jews, who had begun the attempt to assimilate to life in a Christian environment.[6] In a biography read and approved by Einstein, his son-in-law Rudolf Kayser described the Munich milieu in which Einstein grew up:

> In the Munich of the [eighteen-]eighties ... there were to be observed few social and religious contrasts. Descent and religious denomination were not much questioned. But the Jews, though united in common interests with the Christians, lived, as everywhere and at all times, in a certain seclusion. The childhood of Albert Einstein was thus passed in a characteristically Jewish environment (Kayser 1930, p. 24).

Although they unhesitatingly identified themselves as Jews and continued certain Jewish customs, the parents of Einstein did not practice the Jewish religion or observe kosher dietary laws.[7] Their post-emancipation generation was even more assimilated to German culture, particularly in its love for the German literary classics, than had been previous generations of south German Jews.[8]

1. Einstein's Childhood

Einstein later recalled that "though a child of entirely irreligious (Jewish) parents," he "came ... to a deep religiousness..." in his early childhood (Einstein 1979; a fuller citation from this passage is given below). According to his sister, Maja Winteler–Einstein, the impulse to Einstein's religiosity came from his religious instruction, which began at age seven:[9]

> With his entry into public school, religious instruction, then compulsory in Bavaria, also had to begin. In the family itself there reigned a free-thinking spirit, free of dogma with respect to religion, that both parents had brought with them from their own homes. Religious matters and prescriptions were not discussed. Since Albert now had to have religious instruction on account of the law, it was imparted to him at home by a distant relative, thereby awakening a deep religious feeling in him. He heard about divine will and works pleasing to God, a way of life in accord with the divine will, without these teachings being couched in a specific dogma. Nevertheless he was so full of religious fervor that on his own he observed religious prescriptions in every detail. For example, he ate no pork for reasons of conscience, not because his family set the example. He remained true to this self-chosen way of life for years (Winteler–Einstein 1987, pp. liv-lx)

Kayser gives a similar account of this period, adding some interesting details:

> With religious awe he felt [nature's] presence and perceived [in] it the whole of God's majesty, which compelled adoration. Thus the boy revealed a religious frame of mind which, like so much that happened within him, was not properly understood by his parents.

> Albert longed for a religious life and for religious instruction. But he heard only ironic and unfriendly talk about dogmatic ritual. His father prided himself on the fact that he was a freethinker. ... Albert's father was proud that Jewish rites were not practiced in his house. The boy, however, grieved over the fact that the Jewish dietary laws were neglected. For him they were holy commandments which must not be despised or ridiculed.
>
> In his own way, he sought to give his religious temper expression, He wrote and set to music brief songs in praise of God and sang them in his home and on the street. He identified God with nature (Kayser 1930, pp. 28–29).

All Bavarian primary schools were denominational at the time, and the one Jewish school in Munich had been closed since 1872.[10] Einstein's parents sent him to a nearby Catholic elementary school; he was the only male Jewish pupil during the three years he attended the school.[11] As noted, religious education was compulsory, and in addition to his private Jewish instruction he attended the Catholic instruction at school. On the basis of conversations with Einstein, Moszkowski reports:

> [I]t had so happened that he received Jewish as well as Catholic religious instruction at the same time, and that he took from them what they had in common that was conducive to the strengthening of faith, and not what separated them (Moszkowski 1921, p. 219).

Kayser confirms that:

> He did not feel religious differences. On the contrary, he perceived the sameness of all religions. The stories of the Old Testament and Jesus' Way of Sorrow impressed him with equal power.[12]
>
> The Catholic teacher of religion liked him. But one day the same teacher brought a large nail to class and told the pupils that it was the nail with which the Jews had nailed Jesus to the cross. The incident stimulated in the pupils anti-Semitic feeling which was turned against their Jewish fellow student Einstein (Kayser 1930, p. 30).

This was a time of recrudescence of anti-Semitism in the new German Empire.[13] Indeed, Einstein's childhood in Munich was not free of directly anti-Semitic attacks. He recalled:

> Among the children, particularly in elementary school, anti-Semitism was rife. It was based on racial characterisics that, remarkably enough, were known to the children,[14] and on impressions from religious education. Actual attacks and insults on the way to school were frequent, but generally not too bad. At any rate, they sufficed to instill in the child a vivid feeling of alienation (draft letter of 3 April 1920).

At the age of eight, Einstein transferred to the prestigious Luitpold Gymnasium,[15] at which he received advanced primary and secondary schooling until he left Germany seven years later. His curriculum included instruction in the Jewish religion and the Hebrew language from Heinrich Friedmann, who also prepared him to be bar mitzvah. Later, his religious instruction came from Dr. Joseph Perles, the Munich rabbi, and Dr. Cossmann Werner, who succeeded Perles as rabbi.[16] Replying to fiftieth birthday congratulations from Friedmann in 1929, Einstein reminisced:

> How often have I regretted not to have been more diligent in the language and literature of our forefathers. I often read the Bible, but the original text has remained inaccessible to me. It is really not your fault; you fought valiantly against laziness and frivolity (Einstein to Heinrich Friedmann, 12 March 1929).

Kayser states that:

> [Perles's] elucidations of the Proverbs of Solomon and the ethics of the Fathers were of such profound religious significance that Einstein could never forget them (Kayser 1930, p. 38).

According to the curriculum, during his last term at the Luitpold he studied the Psalms under Werner's tutelage.[17] Years later Einstein paid tribute in words that suggest the profound influence their study had on him:

> The Jewish tradition ... contains something ... which finds splendid expression in many of the Psalms, namely, a sort of intoxicated joy and amazement at the beauty and grandeur of this world, of which man can form just a faint notion. This joy is the feeling from which true scientific research draws its spiritual sustenance ... To tack this feeling to the idea of God seems mere childish absurdity ("Is there a Jewish Point of View?", cited from Einstein 1954, p. 186)

Moszkowski suggests that, for some time, the Bible constituted Einstein's chief reading matter.[18] Friendship with Max Talmey, a medical student ten years his senior, which started when Einstein was about ten-and-a-half, played a major role in changing his reading habits. Talmey introduced Einstein to books such as Büchner's *Kraft und Stoff* [*Force and Matter*], which made a great impression on the young boy.[19] As Einstein recalled in his *Autobiographical Notes*, written in 1949 when he was almost seventy, his religious phase:

> reached an abrupt end at the age of twelve. Through the reading of popular scientific books I soon reached the conviction that much in the stories of the Bible could not be true. The consequence was a positively fanatic freethinking, coupled with the impression that youth is intentionally being deceived by the state through lies; it was a crushing impression. Mistrust of every kind of authority grew out of this experience, a skeptical attitude toward the convictions that were alive in any specific social environment—an attitude that has never again left me, even though, later on, it has been tempered by a better insight into the causal connections (cited from Einstein 1979, pp. 3, 5).

2. A "Konfessionslos" Jew

Einstein first registered officially as *konfessionslos* [without religious affiliation] when he was 16,[20] and generally continued to do so thereafter. However, he did not hestitate to register as Jewish when necessity demanded it. In 1911, he had to switch his religious affiliation to "*Mosaisch*" [Jewish] in order to obtain a professorship at the German University in Prague, at that time located in the Austro-Hungarian Empire, since *konfessionslos* was not an acceptable option.[21] A year later, when Einstein left Prague for a better position at the Swiss Federal Polytechnic (ETH) in Zurich, he vainly urged his friend Paul Ehrenfest to make a similar change in religious registration in order to be eligible for the now-vacant Prague job.[22]

While Einstein remained "*konfessionslos*" throughout his later years, he often described his own belief system by the phrase "cosmic religion."[23] What did he mean by this concept? Here is one answer he gave:

> The most beautiful experience we can have is the mysterious. It is the fundamental emotion which stands at the cradle of true art and true science. Whoever does not know it and can no longer wonder, no longer marvel, is as good as dead, and his eyes are dimmed. It was the experience of mystery–even if mixed with fear—that engendered religion. A knowledge of the existence of something we cannot penetrate, our perceptions of the profoundest reason and the most radiant beauty, which only in their most primitive forms are accessible to our minds—it is this knowledge and this emotion that constitute true religiosity; in this sense, and in this alone, I am a deeply religious man. I cannot conceive of a God who rewards and punishes his creatures, or has a will of the kind that we experience in ourselves. Neither can I nor would I want to conceive of an individual that survives his physical death; let feeble souls, from fear or absurd egoism, cherish such thoughts. I am satisfied with the mystery of the eternity of life and with the awareness and a glimpse of the marvelous structure of the existing world, together with the devoted striving to comprehend a portion, be it ever so tiny, of the Reason that manifests itself in nature ("The World As I See It," written in 1931, cited from Einstein 1954, p. 9).

3. German Anti-Semitism

Although he left Germany as a boy of 15, Einstein remained vividly aware of the role of anti-Semitism in German life. In a letter to his financée (who was not Jewish) in 1901, describing his vain efforts to get a job after graduating from the ETH the previous year, Einstein referred to "the anti-Semitism . . . in the German-speaking countries, which is just as unpleasant to me as it is detrimental" as a "principal difficulty" in his search for an academic position (Einstein 1987, p. 282).[24] Unable to find such a position, he finally secured work as a patent clerk in the Swiss Patent Office with the help of the father of a Swiss fellow-student, Marcel Grossmann.

I do not mean to suggest that anti-Semitism was the *only* reason for Einstein's difficulty in finding an academic job. "Protection" was at least as important then as it still is today in many places, and aside from Grossmann's father he had precious little support from anyone of influence. In addition, he was not considered an easy person to get along with. His fiancée, Mileva Marić, writing to a close friend in 1901 about his difficulties in getting a job, remarked: "You know that my treasure has a bad mouth and on top of that is a Jew" (Einstein 1987, p. 320). While it makes clear that his personality did not help, her comment also confirms that anti-Semitism played a role in his difficulties.

It also played a role later when, in 1909, after a number of epoch-making papers had made his name well known among physicists, Einstein was finally appointed to an academic position. In recommending his appointment, the Dean of the University of Zurich's Philosophical Faculty felt obliged to comment:

> Dr. Einstein is an Israelite, and ... precisely to the Israelites among scholars are ascribed all sorts of unpleasant peculiarities of character, such as pushiness, impudence, petty-mindedness in the perception of their academic position, and the like, and indeed in numerous cases not entirely unjustly.
>
> But it must be said that there are also among the Israelites people who do not have a trace of these unpleasant qualities, and therefore that it is not proper to disqualify a man just because he happens to be an Israelite.[25]

The recommendation goes on to absolve Dr. Einstein from possession of these unpleasant Jewish qualities—but perhaps enough has been quoted already.

4. Return to Germany, Turn to Zionism

Einstein presumably never saw this document since he later wrote that, during the years he lived in Switzerland, "There was nothing that called forth any Jewish sentiments in me" ("Wie ich Zionist wurde," *Jüdische Rundschau*, 1921, translation cited from 1931b, p. 39). If we are to believe Michele Besso, his close friend and confidant, Einstein even considered renouncing his Jewish identity entirely at about this time.[26]

After his comment on Switzerland, Einstein immediately added: "When I moved to Berlin [in 1914] all that changed" (ibid.). Elsewhere he explained the nature of this change:

> I discovered for the first time that I was a Jew, and I owe this discovery more to Gentiles than to Jews. ... I saw worthy Jews basely caricatured, and the sight made my heart bleed. I saw how schools, comic papers, and innumerable other forces of the Gentile majority undermined the confidence even of the best of my fellow-Jews, and felt that this could not be allowed to continue (letter to Prof. Hellpach, 8 October 1929, cited from Einstein 1954, p. 171).

Shortly after the move to Berlin, Einstein took his first recorded public political action, an act of protest as a Jew. He was invited by the Russian Academy of Sciences to visit their country in connection with an attempt by German astronomers to check his prediction of the gravitational deflection of light during the solar eclipse of August 1914, which could best be observed from the Ukraine. Einstein refused the invitation, because: "It is repugnant to me to travel needlessly to a country in which my people [*Stammesgenossen*] are so brutally persecuted" (letter of Einstein to P. P. Lazarev, 16 May 1914, printed in Kirsten and Treder 1983, p. 354). Since the First World War broke out just before the eclipse, his refusal probably saved him from brief internment along with several German astronomers who had hoped to observe the eclipse.[27]

After settling in Berlin, Einstein became especially upset by the treatment of the many young Jews from Eastern Europe who came to Germany for higher education.[28]

> When I moved to Berlin... I realized the difficulties with which many young Jews were confronted. I saw how, amid anti-Semitic surroundings, systematic study, and with it the road to a safe existence, was made impossible for them. This refers especially to the East-European-born Jews in Germany, who were continually exposed

> to provocation.... These Eastern-born Jews are made the scapegoat of all the ills of present-day German political life and all the after-effects of the war. When the Government contemplated the expulsion of these Jews, I stood up for them and pointed out... the inhumanity and folly of such a measure.[29] ... These and similar happenings have awakened in me the Jewish national sentiment ("Wie ich Zionist wurde," *Jüdische Rundschau*, 1921, translation cited form 1931b, pp. 39–41)

During the First World War Einstein was involved in a number of pacifist activities, and played a small part in the partially-aborted democratic revolution of 1918–1919 that followed the German surrender.[30] The upsurge of German anti-Semitism in the wake of these events led Einstein to comment:

> Here there is strong anti-Semitism and savage reaction, at least among the "cultivated." (Einstein to Paul Ehrenfest, 4 Dec. 1919).

In February 1919, Kurt Blumenfeld, one of the leading German Zionists, approached Einstein, among other Jewish scholars, to try to win him for the Zionist cause. Blumenfeld recounted their conversations in letters and later reminiscences;[31] he described his method as helping people like Einstein to uncover their Jewishness rather than trying to indoctrinate them, "attempting to bring out of people what is hidden in them, and never attempting to introduce something that does not correspond to a person's nature."[32] He was notably successful with Einstein who, by the end of the year, was actively involved in efforts to gain support from his fellow Jewish scientists for the projected Hebrew University in Jerusalem.[33]

After he gained public fame in Germany in December 1919,[34] Einstein became the subject of attacks on the theory of relativity as well as his own person, attacks that often had a pronounced anti-Semitic undertone. A public meeting against the theory was held at the Berlin Philharmonic in August 1920, addressed by Paul Weyland, an engineer whose speech was particularly demagogic, and Otto Gehrke, a physicist who had long been an opponent of relativity.[35] Einstein attended, and wrote a reply to his critics for a leading Berlin newspaper. While confining himself in the main to scientific issues, he prefaced his reply with the comment:

> I have good grounds to believe that other motives than the search for truth form the basis of this enterprise. (If I were a German nationalist with or without the swastika instead of a Jew of liberal, internationalist convictions, then ...)[36]

A discussion of the theory of relativity that had been scheduled for the following month at the Bad Nauheim meeting of the German Association of Scientists and Physicians (the German equivalent of our AAAS) threatened to become the occasion for another anti-Semitic demonstration. The leaders of the German physics community were alarmed by this possibility, and careful management of the discussion by Max Planck, the chair of the session, kept its surface tone within bounds.[37] But Einstein, who took part in the discussion, was still upset by the anti-Semitically tinged campaign against him. The following year he wrote to Fritz Haber, the famed chemist who was a strongly assimilated Jew, that:

> Other events of the last year... must impel a self-esteeming Jew to take Jewish solidarity more seriously than would have appeared appropriate and natural in earlier times.

> Think of Rothe,[38] Willamovitz-Möllendorf,[39] and the famous Nauheim brigade, which only got rid of the fool Wieland [Weyland] for opportunistic reasons (Einstein to Fritz Haber, 9 March 1921).

Einstein wrote Haber in response to the latter's attempt to dissuade him from joining Chaim Weizmann on a trip to the United States to raise money for the projected University in Jerusalem, and—perhaps as importantly for Weizmann if not for Einstein—to assert the former's authority as leader of the world Zionist movement in the face of a strong challenge from one wing of American Zionists.[40] In spite of Einstein's initial reluctance to go, he was persuaded by Blumenfeld to accompany and support Weizmann.[41] The trip had a profound effect on him:

> It was in America that I first discovered the Jewish people. I have seen any number of Jews, but the Jewish people I had never met either in Berlin or elsewhere in Germany. This Jewish people, which I found in America, came from Russia, Poland and Eastern Europe generally. These men and women still retain a healthy national feeling; it has not yet been destroyed by the process of atomisation and dispersion. I found these people extraordinarily ready for self-sacrifice and practically creative. They have, for example, managed in a short time to secure the future of the projected University in Jersualem.. . . I also found that it was mostly the middle classes and the ordinary folk, and not those enjoying a high social position or any natural advantages, who had most conspicuously preserved the healthy feeling of belonging together and the willingness to make sacrifices. (1921, translated in Einstein 1931b, pp. 48–49).

Thus, it was concern for the establishment of the Hebrew University that motivated his first public actions in support of the Zionist cause.

On his return from another trip, this time to the Far East in 1922–1923, Einstein spent the first two weeks of February touring Palestine, then under British rule under a mandate from the League of Nations. On February 7 he formally inaugurated the Hebrew University in a ceremony held in a temporary building on Mount Scopus. He was introduced with the words: "Mount the platform which has been waiting for you for 2000 years." Einstein's words were not recorded, but it is known that he opened with a sentence in Hebrew that he had memorized, in order that the first official words spoken at the University be in that language.[42] In spite of the intermittent efforts to pursuade him to join the faculty of the University, he never seriously considered settling in Palestine, (by contrast, Blumenfeld's Zionism demanded a personal commitment to *aliyah*),[43] nor did Einstein ever expect the majority of Jews to settle in the Holy Land.

Part II. Einstein's Judaism, Einstein's Zionism

Starting about 1921, Einstein began to publish a number of articles about his views on the Jewish question, and after his visit to Palestine he started to express himself about the future of the *Yishuv*. Before turning to his views, however, it is necessary to say something about the then-prevalent German version of Zionism in order to put these views in their proper context.

5. German "Volkishness" and German Zionism

The historian George Mosse has chronicled in detail how, during the latter half of the nineteenth century, one strain of German romanticism developed into what he calls the Volkish outlook, which according to Mosse was characterized by the following traits. It:

> looked back nostalgically to the historical period in which the soul of the Volk still had free play. ... Such a life was not merely that of the peasant tilling the soil, however, but also encompassed a true human creativity. National unity was conceived in cultural terms and the binding force of a common language continued to play a major role in Volkish thought. ... The Volk must grow like a tree from its roots in historical soil, striving toward a genuine creativity within the collective whole. ... Not all Volkish thought was aggressive. At the end of the nineteenth century, for example, middle-class German youth organized its own Youth Movement, which envisaged a world wherein all separate peoples would exist peacefully side by side—each with its own unique concept of the Volk. Small wonder that Zionist youth were attracted to this kind of nationalism. ... [E]mphasis on the soil and landscape as the unchanging sources of Volkish inspiration ... A concern on the part of the intellectuals for the uprootedness of the growing middle classes ... The Volkish soul was embattled against those who sought to destroy it, but they were more often internal than foreign adversaries. (Mosse 1970, pp. 9–15 passim).
>
> Racist ideas ... were often but by no means inevitably an ingredient of such enthusiasm [for the Volk] ... [T]he urge to break with the bourgeois world, to revitalize a culture that seems to have lost its vitality (ibid., pp. 78–79). They rediscovered the emotional basis of human nature... (ibid., p. 82).

Mosse emphasized that—in spite of its habitual (but not universal) association with pronounced anti-Semitic sentiments—the Volkish outlook strongly influenced German Jewry;[44] in particular, it had such an influence on German Zionism.[45] In his history of the latter, Stephen Poppel characterized as "remarkable ... the extent to which German Zionism shared the rhetoric and assumptions of the distinctively German *völkisch* ideology" (Poppel 1976, p. 127). He went on to give numerous examples, as had Mosse.

Neither Mosse nor Poppel mention Einstein, but their characterizations of German-Zionist Volkishness well fit Einstein's comments on Judaism and Zionism during the twenties:

> Zionism springs from an even deeper motive than Jewish suffering. It is rooted in a Jewish spiritual tradition, whose maintenance and development are for Jews the *raison d'être* of their continued existence as a community. In the re-establishment of the Jewish nation in the ancient home of the race, where Jewish spiritual values could again be developed in a Jewish atmosphere, the most enlightened representatives of Jewish individuality see the essential preliminary to the regeneration of the race and the setting free of its spiritual creativeness (letter to the *Manchester Guardian*, 12 October 1929, cited from Einstein 1931b, pp. 78–79).
>
> In Palestine it is not our aim to create another people of city-dwellers, leading the same life as in the European cities, and governed by the standards and conceptions of the European bourgeoisie. We aim at creating a people of workers, at creating the Jewish village in the first place, and we desire that the treasures of culture should

> be accessible to our laboring class, especially since, as we know, Jews in all circumstances place education above all else ("Ein Wort auf den Weg," *Jüdische Rundschau*, 3 April 1925, translation cited from Einstein 1931b, pp. 65–66).

His words are not the casual products of an occasion, but summarize a consistent outlook, if one that is not unique in the context of German Zionism.[46] How should we characterize Einstein's outlook?

6. A Secular, Universalist Ideology

Sixty years ago, Salo Baron pointed out that:

> [T]he *idealistic* type of historiography [which] has dominated Jewish historical writing for centuries, in fact, for milennia ... essentially accepted the primacy of ... "inner" factors. [Its adherents] set up a sort of autonomous national will which was the driving force in shaping the destinies of the people and which, in the supreme interest of national self-preservation, made all the necessary adjustments required in the different periods and regions. ... Some will perhaps concede the shortcomings of such an [idealist] approach to the history of any other nation, but will none the less insist that in the case of the Jews, whose general psychological and social make-up throughout history was so largely determined by their own religion and by the religious attitudes of their neighbors, no other explanation can do equal justice to the specific nature of the subject (Baron 1939, cited from Baron 1964, pp. 75–77).

Maxime Rodinson,[47] after citing Baron, proposes a four-fold classification of such idealist ideologies, based on whether they are religious or secular, nationalist or universalist in nature. There is a "religious ... nationalist ideology, in which the universal God is specially concerned with the survival of the chosen people," and a "secular nationalist ideology that recognizes the Jewish nation as the sole supreme value" (Rodinson 1983, p. 74).

For the religious universalists, "the election of Israel can be strictly subordinated to the divine plan directed for the good of humanity." For secular universalists, "although the idolatry of the ethnic group is eschewed, the idea that the strictly Jewish entity could be dissolved in any form is repudiated. One is then drawn to seek and define a substratum of permanent values bound up with the existence of the various Jewish entities of the past, and to proclaim, for the past and future alike, the necessity of the bond between a given collection of values and a minimum Jewish grouping. It then supposedly follows that humanity as a whole has an interest in the perpetuation of this Jewish group, so that the worship of these values may be maintained" (ibid., pp. 74–75).

It is clear from the citations at the end of the last section that Einstein espoused an idealist ideology. As one might expect, his views on the Jewish question place him in the secular universalist category. We have already noted his opposition to conventional religious concepts. He was also strongly inclined to internationalism.

> If we did not have to live among intolerant, narrow-minded, and violent people, I should be the first to throw over all nationalism in favor of universal humanity (letter to Prof. Hellpach, 8 October 1929, cited from Einstein 1954, p. 172)

In 1925, he explained how he reconciled his belief in internationalism with his support of Zionism:

> The existence of different nationalities and consequently of mutually antagonistic nationalisms, both within and without Europe, must be considered a misfortune in my opinion. . . .
>
> On the other hand, there is a fact that cannot be ignored: The Jews are almost everywhere treated as members of a group that is clearly characterized nationally. This may appear deplorable to Jews, like myself, who consider membership in a single human nationality as a possible ideal, even though difficult to attain.
>
> Nietzsche said that one of the peculiarities of the Jewish people consists in the ability to recognize and realize "the subtle utilization of misfortune."
>
> Jews must also make their nationality useful. May they do so for the sake of universal welfare.
>
> It is their responsibility to develop from within themselves those virtues and those beliefs that are indispensible for those who would serve humanity. Since the disappearance of the Jewish nationality, for the moment at least, appears impossible, the Jews must justify its existence. And for this reason they must, without ridiculous arrogance, regain consciousness of the human values that they represent. By study of their past, by a better understanding of the spirit [Geist] that accords with their race, they must learn to know anew the mission that they are capable of fulfilling (Einstein 1925).

Einstein saw Zionism as the chief instrument in helping Jews to forge the solidarity necessary to fulfill this mission.

> Inspired by the mystique of Zionism, perhaps they will finally be able to fulfill the tasks that are incumbent upon them, and which demand the high-principled exertions and single-minded labor of Israel. . . . A Jew who strives to impregnate his spirit with humanitarian ideals can declare himself a Zionist without contradiction.
>
> What one must be thankful to Zionism for is the fact that it is the only movement that has given many Jews a justified pride, that it has once again given a despairing race the necessary faith, if I may so express myself, given new flesh to an exhausted people [Volk] (ibid.).

7. Einstein's Cultural Zionism

Let us now look more closely at some of the elements of Einstein's Zionism. Einstein believed that a major source of the problems of German Jewry was their loss of a sense of community. In 1921, he gave the following characterization of Jewish life in Germany before emancipation:

> A century ago our forefathers, with few exceptions, lived in the ghetto. They were poor, without political rights, separated from the Gentiles by a barrier of religious traditions, habits of life, and legal restrictions; their intellectual development was restricted to their own literature, and they had remained almost unaffected by the mighty advance of the European intellect which dates from the Renaissance. And yet these obscure, humble people had one great advantage over us: each of them belonged in every fiber of his being to a community in which he was completely absorbed, in which he felt himself a fully privileged member, and which demanded nothing of him

that was contrary to his natural habits of thought. Our forefathers in those days were pretty poor specimens intellectually and physically, but socially speaking they enjoyed an enviable spiritual equilibrium (cited from Einstein 1954, p. 181)

Einstein looked not just to the past for his Jewish communal ideal, he also looked to the East. In a passage already cited, he wrote of his encounter with Jews who:

came from Russia, Poland and Eastern Europe generally. These men and women still retain a healthy national feeling; it has not yet been destroyed by the process of atomisation and dispersion. I found these people extraordinarily ready for self-sacrifice and practically creative (written in 1921, cited from Einstein 1931b, pp. 48–49).

The point is not, of course, the accuracy of these descriptions, which certainly involve a good deal of idealization (for example, they neglect the role of women and of social tensions within the Jewish community), but how well this account fits Volkish patterns of thought described in the previous section. Einstein's views place him squarely in the tradition of German cultural Zionism.

Cultural Zionism, first espoused by Ahad Ha'am and Martin Buber (who played an important role in introducing the Volkish element into German Zionism),[48] emphasized the cultural and spiritual renewal of the Jewish people. It saw itself in opposition to political Zionism, as espoused by Herzl, which focused on the establishment of a Jewish state.[49] Einstein's statement, "The object which the leaders of Zionism have before their eyes is not a political but a social and cultural one" (1921, translation cited from Einstein 1954, p. 179), makes clear that he identified with the cultural Zionists. German cultural Zionists like Adolf Friedmann distinguished between the "*materielle Judennot*, a material distress" of the East European Jews, and the "*geistige Judennot*, the spiritual, intellectual, and emotional distress" of the German Jews (Poppel 1976, p. 28; see also the quotations from Franz Oppenheimer in ibid., p. 58). Like them, Einstein saw Zionism as the primary remedy for this *geistige Judennot* through its efforts to re-create the spirit of the Jewish Volk. But what makes a Jew a Jew?

8. What is a Jew?

As Mosse noted (see above), Volkishness does not necessarily imply racism. While he lived in Germany, however, Einstein seems to have accepted the then-prevalent racist mode of thought, often invoking such concepts as "race" and "instinct," and the idea that the Jews form a race. In 1920 he wrote:

Anti-Semitism as a psychological phenomenon will always be with us so long as Jews and non-Jews are thrown together. But where is the harm? It may be thanks to anti-Semitism that we are able to preserve our existence as a race; that at any rate is my belief (translation cited from Einstein 1931b, p. 35).

He elucidated in 1921:

While in America ... anti-Semitism knows only social forms, in Germany communal anti-Semitism is much stronger even than social. As I view the matter, the fact

> of the racial peculiarities of Jews is bound to have an influence on their social intercourse. I believe that German Judaism is thus being influenced to a great extent by anti-Semitism. With increasing wealth and increasing education, the religious customs which formerly prevented the mixing of Jews with Gentiles have tended to disappear. There was thus nothing but the antithesis which Jews represent, and which is called anti-Semitism, to preserve Jewish separateness. Without this antithesis, assimilation in Germany would have been completed long ago (translation cited from Einstein 1931b, pp. 38–39).

In the same year he wrote:

> Nations with racial differences appear to have instincts which work against their fusion. The assimilation of the Jews to the European nations among whom they lived, in language, in customs, and to some extent even in the forms of religious organisation, could not eradicate the feeling of lack of kinship between them and those among whom they lived. In the last resort, this instinctive feeling of lack of kinship is referable to the law of conservation of energy. For this reason it cannot be eradicated by any amount of well-meant pressure. Nationalities do not want to be fused; they want to go each its own way. A state of peace can be brought about only if they mutually tolerate and respect each other (translation cited from Einstein 1931b, pp. 46–47).

Here, Einstein slipped without comment from "nations with racial differences" to "nationalities" without qualification as groups opposing fusion.

After living in the United States for several years, and presumably after he became aware of the campaign by Franz Boas and his fellow anthropologists against the concepts of racial purity and racial instinct, exploited so devastatingly by the Nazis, Einstein rejected any racial or other biological sanction for Judaism. Even before that, Einstein had defined Judaism as primarily a certain attitude to life:

> Judaism seems to me to be concerned almost exclusively with the moral attitude in life and to life. I look upon it as the essence of an attitude to life which is incarnate in the Jewish people rather than the essence of the laws laid down in the Torah and interpreted in the Talmud ... The essence of that conception seems to me to lie in an affirmative attitude to the life of all creation. The life of the individual only has meaning in so far as it aids in making the life of every living thing nobler and more beautiful. Life is sacred, that is to say, it is the supreme value, to which all other values are subordinate ("Is There a Jewish View of Life?", *Opinion*, 26 September 1932, translation cited from Einstein 1954, p. 185).

By 1938, he decisively rejected any taint of racism in his concept of Judaism:

> I have conceived of Judaism as a community of tradition. Both friend and foe, on the other hand, have often asserted that the Jews represent a race; that their characteristic behavior is the result of innate qualities transmitted by *heredity* from one generation to the next. This opinion gains weight from the fact that the Jews for thousands of years have predominantly married within their own group. Such a custom may indeed *preserve* a homogeneous race—if it existed originally; it cannot *produce* uniformity of the race—if there was originally a racial intermixture. The Jews, however, are beyond doubt a mixed race, just as are all other groups of our civilization. Sincere anthropologists are agreed on this point; assertions to the contrary all belong to the field of political propaganda and must be rated accordingly ("Why do They Hate the Jews?", written in 1938, cited from Einstein 1954, p. 196)

However, aside from this anti-racist emendation, his views remained basically unchanged:

> The members of any group existing in a nation are more closely bound to one another than they are to the remaining population. Hence a nation will never be free of friction while such groups continue to be distinguishable. ... There will always be friction between such groups—the same sort of aversion and rivalry that exists between individuals. ... The Jews, too, form such a group with definite characteristics of its own, and anti-Semitism is nothing but the antagonistic attitude produced in non-Jews by the Jewish group. This is a normal social reaction (ibid., pp. 193–194).

Having now accepted the need for a purely "cultural" definition of Judaism, Einstein spelled out the "definite characteristics" of this "community of tradition," singling out "two traditional traits, which seem to me the most basic:"

> The bond that has united the Jews for thousands of years and that unites them today is, above all, the democratic ideal of social justice, coupled with the ideal of mutual aid and tolerance among all men.
>
> The high regard in which [the Jewish community] holds every form of intellectual aspiration and spiritual effort. I am convinced that this great respect for intellectual striving is solely responsible for the contributions that the Jews have made toward the progress of knowledge... I am convinced that this is not due to any special wealth of endowment ... At the same time a strong critical spirit prevents blind obeisance to any mortal authority (ibid., p. 195).

Elsewhere he put this point as follows:

> The pursuit of knowledge for its own sake, an almost fanatical love of justice and striving for personal independence— these are the features of the tradition of the Jewish people, which make me experience my membership in it as a gift of fate (*Mein Weltbild*, 1934).

9. Einstein, the Yishuv, and the "Arab Question"

The need to strengthen the sense of community was his main motive for supporting the Jewish colonization of Palestine.

> It is for me beyond any shadow of doubt that in present circumstances the rebuilding of Palestine is the only object which has a sufficiently strong appeal to stimulate the Jews to effective corporate action (ibid.).

After his return from Palestine in 1923, Einstein never wavered in his support for the Jewish community in Palestine. He never expected that all Jews would—or should—emigrate there, but, as the last quotation indicates, saw in their collective efforts to support this community, the most potent cohesive force between Jews everywhere.

As is well known, within the Zionist movement, Einstein was a strong supporter of both labor Zionism and of efforts to foster Jewish-Arab cooperation. But the emphasis of labor Zionists on the need for Jews to do all manual labor needed in factory and on field threatened the future of Arab labor in Palestine. As Walter Lacqueur put it in his history of Zionism:

> It is one of the tragic ironies of the history of Zionism that those who wanted close relationships with the Arabs contributed, albeit unwittingly, to the sharpening of the conflict ... [T]heir fanatical insistence on manual labor ("redemption through toil") seemed to confirm Arab suspicions about Jewish separatism and the displacement of Arab peasants and workers (Lacqueur 1972, pp. 220–221).

Initially, according to Blumenfeld, Einstein had doubts about the Zionist emphasis on manual labor. During their 1919 conversations:

> With extreme naiveté he asked questions, and his comments on the answers were simple and unconventional. "Is it a good idea to eliminate the Jews from the spiritual calling to which they were born? Is it not a retrograde step to put manual capabilities, and above all agriculture, at the center of everything Zionism does?" (Blumenfeld 1962, translation cited from Clark 1971, pp. 377–378).

Whether due to Blumenfeld's persuasive powers or some other cause, Einstein soon wholeheartedly supported the efforts of the Labor Zionists. Blumenfeld reported to Weizmann in 1921 that Einstein was "strongly inclined to socialism" and the "issue of Jewish labor and of Jewish workers influenced him strongly" (Blumenfeld 1976, p. 67). In 1925 Einstein wrote:

> We aim at creating a people of workers, at creating the Jewish village in the first place" ("Ein Wort auf den Weg," *Jüdische Rundschau*, 3 April 1925).

In "Working Palestine" he wrote of

> the most valuable class of people living there, namely, those who are transforming deserts into flourishing settlements by the labor of their hands ... an elite composed of strong, confident and unselfish people" (cited from Einstein 1954, p. 183).

But he never seemed to fully face the tragic contradiction between his support for Jewish labor and his support for Arab-Jewish reconciliation. Indeed, before 1929 he never seemed to face the fact that Zionism, in fostering Jewish emigration to Palestine, was also fostering a nascent Palestinian Arab nationalism. He was among those who, as Lacqueur put it:

> thought that the growth of Arab nationalism and anti-Zionist attacks were the result of the activities of individual villains, the effendis (who were annoyed because the Jews had spoiled the fellahin by paying them higher wages) (Lacqueur 1972, p. 232).

Lacqueur adds that

> It was more than a little naive to put the blame for Arab anti-Zionism on professional inciters, frustrated Arab notables, and the notorious urban riff-raff, for there were basic clashes between the two national movements (ibid., p. 234).

It was such a naive approach to the problem of Palestinian nationalism that Einstein initially took.[50] Before his visit to Palestine in 1923 he does not seem to have written anything about the "Arab question," as Zionists referred to the issue. Reporting on his visit, he listed debts and malaria as the main difficulties encountered by "our workers on the land," adding:

> In comparison with these two evils the Arab question becomes as nothing ... I have myself seen more than once instances of friendly relations between Jewish and Arab

> workers. I believe that most of the difficulty comes from the intellectuals—and, at that, not from the Arab intellectuals alone (Einstein 1923).

In 1927, he wrote:

> At no time did I get the impression that the Arab problem might threaten the development of the Palestine project. I believe rather that, among the working classes especially, Jew and Arab on the whole get on excellently together.[51] The difficulties which are as it were inherent in the situation do not rise above the threshold of consciousness when one is on the spot. The problem of the rehabilitation and sanitation of the country seems incomparably more difficult (cited from Einstein 1931b, p. 58).

Although there had been a number of Arab-Jewish clashes over the years, including quite serious ones in 1920 and 1921,[52] it took the major anti-Jewish disturbances of 1929 to make Einstein aware of the gravity of the problem. Then he warned:

> that [the Zionist] movement [must] avoid the danger of degenerating into a blind nationalism. In my opinion, we must endeavor above all that psychological understanding and an honorable will towards cooperation take the place of resentment towards the Arabs. The overcoming of this difficulty will, in my opinion, be the touchstone that our community has a right to existence in the higher sense. I must unfortunately openly acknowledge that the attitude of our [Zionist] officialdom, as well as the majority of public expressions in this connection, appear to me to leave much to be desired (Einstein to Heinrich York-Steiner, 19 November 1929).

Thereafter, he made repeated efforts to combat the more extreme forms of Jewish nationalism and to foster Jewish-Arab dialogue. During the thirties, he supported the idea of a bi-national state in Palestine as the way to reconcile the rights of both peoples.

In spite of these efforts, and the sincere good will behind them, in my opinion Einstein's approach to the issue of Jewish-Arab relationships still left much to be desired. Since at least 1920, the Palestinian Arab leadership had been working to define Jewish-Arab relations in Palestine as a zero-sum game.[53] Cultural Zionists had tried to restate the issues in such a way that their disputes could be resolved to the mutual advantage of both parties. Yet there remained a major obstacle: the rapid growth of the *Yishuv* in the 1930s, due primarily to the rising tide of European fascism and anti-Semitism, was lessening the margins, within which such a redefinition was possible—to say nothing of the lack of interest in such a redefinition on the part of many political Zionists. It was becoming clear to most Jewish leaders that, barring the unlikely eventuality that the Arabs altered their intransigent opposition, the real choice was between persistence in imposing a Zionist solution on them, which meant accepting the need to use force; or renunciation of any solution based on the use of force, which meant renunciation of a Zionist solution. Resolute thinkers on either side made their choice.

Einstein could not accept either horn of this dilemma. In 1944, he and his Princeton neighbor, the writer Erich Kahler, undertook to reply to the Congressional testimony of Philip Hitti, the Arabist scholar also of Princeton, on the future of Palestine. At this late date, they replied to his avowals of Arab nationalism in a disconcertingly familiar vein:

> Jewish youth . . . took over from the period of Arabian predominance deserts and rocks and barren soil and turned them into flowering farms and plantations, into forests and modern cities. They created new forms of cooperative settlements and raised the living standard of the Arabian and Jewish population alike. . . . They offer their assistance and their experience for the economic and scientific advancement of the Arab countries, for the lifting of their population to a modern standard of living.
>
> But this, unfortunately, is just what the Arabs leaders do not want. For the true source of Arab resistance and hostility toward a Jewish Palestine is neither religious nor political, but social and economic. . . . The rich Arabian landowners did nothing to improve the nature, the civilization, or the living standards of their countries.. . . [T]he big Effendis fear the example and the impulse which the Jewish colonization of Palestine presents to the peoples of the Near East, they resent the social and economic uplift of the Arabian workers in Palestine (cited from Kahler 1967, pp. 136–137, which reprints Hitti's testimony and reply, as well as Einstein and Kahler's response and concluding reply).

When it came to the crunch, in spite of his earlier fears about the "inner damage Judaism will sustain [from the creation] of a Jewish state with borders, an army, and a measure of temporal power no matter how modest," he wholeheartedly supported the creation of the state of Israel. No doubt, this was basically because he saw no other viable solution for the remnants of European Jewry after the Holocaust. Yet he never ceased to worry about the effects of state power on the attitude of the Jewish community. Three months before his death, Einstein wrote:

> The most important aspect of our policy must be our ever-present, manifest desire to institute complete equality for the Arab citizens living in our midst, and to appreciate the difficulties of their present situation... The attitude we adopt toward the Arab minority will provide the real test of our moral standards as a people (Einstein to Zvi Lurie, 4 January 1955).

10. The Romantic Strain in Einstein

Why did Einstein so readily accept the Volkish-tinged German variant of cultural Zionism? I suspect that it appealed to a little-noted element in his nature. Einstein is generally regarded as a rationalist, a child of the Enlightenment, and I do not want to minimize this "classical" element in his outlook on life. Rather, I want to stress the existence of another, "romantic" element in his nature that coexisted with the "classical" element, sometimes harmoniously and sometimes with considerable tension.[54] I shall discuss Einstein's romanticism in more detail elsewhere; here I shall just recall the central significance of his emotional longing for unity with nature and his fellow man.

We have seen that the young Einstein had developed an emotionally-tinged religious pantheism long before he read Spinoza. Einstein stressed the abruptness with which the conventionally religious elements in this pantheism were sloughed off when he came into contact with popular-scientific descriptions of the world;

but he also stressed repeatedly that such pantheistic feelings continued to play a central role in motivating his devotion to scientific research. Even the words with which he discussed this devotion reveal the romantic side of his temperament. At the age of eighteen he wrote:

> Strenuous intellectual work and the contemplation of God's Nature are the angels that will lead me consolingly, strengtheningly and yet unsparingly through all the troubles of this life. (Einstein to Pauline Winteler, May 1987, Einstein 1987, pp. 55–56).

Two decades later he said:

> I believe with Schopenhauer that one of the strongest motives that leads men to art and science is escape from everyday life with its painful crudity and hopeless dreariness, from the fetters of one's own ever shifting desires. A finely tempered nature longs to escape from personal life into the world of objective perception and thought; this desire may be compared with the townsman's irresistible longing to escape from his noisy, cramped surroundings into the silence of high mountains, where the eye ranges freely through the still, pure air and fondly traces out the restful contours apparently built for eternity ("Motive des Forschens," 1919, cited from Einstein 1954, p. 225).

In one of his credos he affirmed:

> The most beautiful experience we can have is the mysterious. This is the true emotion that stands at the cradle of true art and true science ("The World as I See It," 1931, cited from Einstein 1954, p. 11).

Together with his feeling for the unity of nature, and his longing for unity with nature, went an intense, romantic longing for some ideal form of human interdependence, of human community—a longing that managed to coexist with his equally intense (and more realistic) need for independence and solitude.[55]

> [W]ithout deeper reflection one knows from daily life that one exists for other people—first of all for those upon whose smiles and well-being our own happiness is wholly dependent, and then for the many, unknown to us, to whose destinies we are bound by ties of sympathy. A hundred times every day I remind myself that my inner and outer life are based on the labors of other men, living and dead, and that I must exert myself in order to give in the same measure as I have received and am still receiving ("The World as I See It," 1931, translation cited from Einstein 1954, p. 8).

In the same essay, Einstein wrote:

> My passionate sense of social justice and social responsibility has always contrasted oddly with my pronounced lack of need for direct contact with other human beings and human communities (ibid., p. 9).

I suggest that Einstein's renewed experience of German anti-Semitism after his move to Berlin may also have reawakened his childhood memories of devotion to Judaism and experiences of anti-Semitism. He was thus well prepared when Blumenfeld began to stimulate his interest in Zionism. In particular, the romantic elements of his world view, especially the longing for an idealized human community, prepared him to accept the variant of cultural Zionism then prevalent in Germany. As long as his need for solitude—and the lack of any sense of urgency—kept him out of contact with the realities of day-to-day Zionist politics, he could maintain this stance.[56] But with the mounting wave of German anti-Semitism after Hitler's accession to power, particularly after the outbreak of the Second World War and the Nazi attempt to exterminate European Jewry, he felt compelled by events to concede more and more to the program of the political Zionists, which demanded a state at any cost, even as he remained aware of the cost to his ideals.

11. Conclusion

One who takes a non-idealist approach to history, as I do,[57] is left uneasy by any idealist account of Jewish history. To quote Rodinson:

> All these [idealist] interpretations of Jewish history ... are ideological. By that I mean they are inspired by the desire to demonstrate (or at least to suggest) what they postulate, and that what they postulate corresponds to exigencies that are not scientific but instead pragmatic and vital for the consciousness of an individual or a group. We then have people who *need* to found their existence on the notion of the necessary permanence of Jewry as a community, whether religious or temporal. In either case, it seems to me, socio-historical vision is distorted (Rodinson 1983, p. 74).

Such ideologies seek to explain the history of the many really-existing Jewish communities as the evolution of some "essence of Judaism," detached from any concrete basis in the widely varying life circumstances of the groups that have described themselves as Jewish over the two millenia since the diaspora. To cite Rodinson again:

> The Jews have indeed constituted specific groups and categories, perhaps even exceptional ones in the sense that a set of laws and conjunctures gave rise to types of formations and evolution not encountered elsewhere. But they are not exceptional in the sense that the general laws that govern the history of human groups do not apply to them. ... [N]o substratum of empirical forces can be discovered to account for the action of any 'spirit,' or immutable 'essence' characteristic of a people or civilization regardless of the situations in which they find themselves (Rodinson 1983, pp. 75–76).

So I find the basic premises of Einstein's views on the Jewish people quite unsatisfactory. As a Jew, I share his secularism and universalism, but I find the basically idealist and romanticist elements in his outlook quite unsympathetic. Nevertheless, to cite Rodinson one final time

> No universalist 'materialist' can view with an identical regard religious or secular nationalist ideologies on the one hand and those that ascribe primacy to the service to humanity on the other (ibid., p. 75).

Einstein's lifelong devotion to humanist ideals and his attempts to apply them to the complex social problems of his time demand the highest respect. In spite of any weaknesses one may find in his outlook, Einstein's call for Jewish self-respect in the lands of the diaspora, his support of a secular, humanistic Judaism, his conciliatory views on Jewish-Arab relations, and his suggestions on finding the path to peace in the Middle East are still of more than purely historical interest—particularly to his fellow Jews. Rather than unthinking adulation of his every word, or cynical manipulation of the Einstein myth, we can honor him best by reading and pondering his words, modifying or rejecting what we find to be obsolete, and using in our current struggles what we find to be of lasting value.

NOTES

[1] When he died, Professor Tauber was in the midst of assembling a complete collection of Einstein's writings on Judaism, Zionism, and Jewish–Arab relations.

[2] For information on Einstein's ancestry, see Tänzer 1931. For a charming description of German–Jewish rural life in the nineteenth century, see Schwab 1956.

[3] The population of Munich in 1880 was about 230,000 and rapidly growing—it had increased by about 50% by 1890. The Jewish population, which in 1880 was about 4,100 (almost 2%), also increased by about 50% in the next decade (see the article "München," in Kayser and Stoob 1974, pp. 394–455). Baerwald observes: "In spite of their relatively small numbers, the Jews soon played an important role in the city [of Munich], especially in its economic life. It is remarkable what a significant part they had, not only in petty and wholesale trade, but also in business and industry (Baerwald 1982, p. 22).

For information on Jakob Einstein, see Winteler–Einstein 1987, note [9], pp. l–li.

[4] For information on the Einstein family and its business affairs, see Pyenson 1985, and Winteler–Einstein 1987, including the editorial annotations.

[5] Cahnman emphasizes that this move was in some sense a return. The village Jews of south Germany "were not peasants of a different ethnicity. They were urbanites transmuted into rural folk [by their earlier expulsion]. ... In the nineteenth century, the village Jews moved back to the country towns and from there to the larger urban centers or to America" (Cahnman 1989, p. 60). For the more rapid pace of urbanisation after 1870, see Lowenstein 1976, p. 52.

The timing of the Einstein brothers' move to Munich was also not exceptional. Before 1861, the system of Jewish registration (*Judenmatrikel*) in Bavaria limited freedom of movement there for Jews. After it was abolished, "Industrious men came to Munich in large numbers from the villages and little towns of the Swabian and Franconian countryside, further from the Upper and Rhine Palatinates, from Württemberg, and many other parts of Germany" (Cahnman 1989, p. 97). For a history of Bavarian Jewry up to the time of emancipation, see Schwarz 1963.

[6] Cahnman, describing "a small town Jewish community in Hesse," observed that "what had been a divine commandment for the grandfathers tended to become a hallowed usage for the fathers and finally was performed merely perfunctorily because public opinion still seemed to insist on it" (Cahnman 1989, p. 54).

Bach observed that "the passionate concern of the first half of the century for the Jewish faith and its compatibility with modern life belonged to the past. In the age of positivism, religion as a whole was suspected of clashing with the hard-and-dominant faith of both Gentiles and Jews. Thus it was as well to hold Judaism to be a historical and rational religion that implied optimistic belief in moral progress and social improvement ... and leave it at that" (Bach 1984, p. 112).

[7] See Winteler–Einstein 1987, quoted below. She states that this "free-thinking spirit" had come from the homes of their parents. The given names of Einstein's parents, Hermann and Pauline, indicate the degree of assimilation of Einstein's grandparents. A generation earlier, few German Jews would have borne such "Aryan" names. Of Hermann's five brothers and sisters, only two (Jakob and Jette) had traditional Jewish given names.

[8] Cahnman described the Jews of south and west Germany: "Thoroughly integrated into the rural communities where their ancestors had appeared as 'wandering Jews,' they were an inseparable, though segregated part of the environment, and did not constitute a distinctive society ... Both culturally and commercially, the Jews of south and west Germany were closer to Paris than to Berlin or the despised 'colonial' regions of the East" (Cahnman 1989, pp. 8–9).

Philipp Frank noted that: "In the Jewish families [of small-town south Germany] the German classics took their place as teachers of ethics and moral behavior alongside the prophets. Friedrich Schiller, Lessing and Heine were honored as were the Psalms of Solomon and the Book of Job" (Frank 1979, p. 16). Einstein's father "loved literature and in the lamplight, evenings, read aloud Schiller and Heine" (Kayser 1930, p. 26). See also Katz 1985.

[9] Kayser 1930 maintains that this period started before Einstein began school, but his sister's account seems more probable.

[10] For the Jewish school system in Munich, see Dingfelder 1927. In 1850, there had been 150 Jewish schools in Bavaria. By 1870 there were still 124. Presumably the closing of many Jewish schools after this date was connected with the attainment of full legal equality by Bavarian Jews after Bavaria's incorporation into the German Empire in 1871 (for a translation of the decree, see Mendes-Flor and Reinharz 1980, p. 139). However, in 1910 there were still 84 such schools (figures cited from Lowenstein 1976, p. 43).

[11] See Winteler–Einstein 1987, especially notes [38], p. lvii, [40], p. lviii, and [44], p. lx.

[12] In later life, Einstein still saw no conflict. In "Christianity and Judaism" he wrote: "If one purges the Judaism of the Prophets and Christianity as Jesus Christ taught it of all subsequent additions, especially those of the priests, one is left with a teaching which is capable of curing all the social ills of humanity" (Einstein 1954, pp. 184–185).

[13] See, for example, Bach 1984, pp. 122–135, Bieber 1979.

[14] Moses Hess had already commented in 1862 that: "The German hates the Jewish religion less than the race; he objects less to the Jews' peculiar beliefs than to their peculiar noses" (*Rom und Jerusalem*, translation cited from Hess 1918, p. 58).

[15] Gymnasium education still centered around study of the Greek and Latin classics in their original languages. At the time, it was more common for well-to-do Jewish families to send their children to *Realschulen*, which offered a more practical education centered around modern culture and the sciences (see Mosse 1985, p. 3).

[16] For Einstein's activities at the Luitpold Gymnasium, see Winteler–Einstein 1987, pp. lx–lxiv; for his curriculum, see Einstein 1987, Appendix B, pp. 346–355.

For an account of the Jewish community in Munich, with discussions of Perles and Werner, see Baerwald 1982.

[17] See Einstein 1987, Appendix B, p. 351.

[18] See Moszkowski 1921, p. 220.

[19] In later years Einstein was less impressed with Büchner's work. Other popular-scientific books that he read at this time are Aaron Bernstein's *Naturwissenschafliche Volksbücher*, and Alexander von Humboldt's *Kosmos*. See Einstein 1979, Winteler–Einstein 1987, p. lxii, and Talmey 1932, pp. 161–163.

[20] See Einstein 1987, p. 20, note [1] to Doc. 16.

[21] See Frank 1979, p. 137.

[22] See Klein 1970, pp. 180–181.

[23] See, for example, the collection of his writings under that title: Einstein 1931a.

[24] Switzerland was evidently not exempt from his strictures. It was the last country in Western Europe to grant civil and political rights to Jews, who for a long time were only permitted to reside in the Canton of Aargau. A referendum in 1862 to remove legal restrictions on the Aargau Jews was defeated by a large majority; all such restrictions were finally abolished by the Federal Constitution of 1874 (see Dubnov 1973, pp. 378–381).

[25] C. Stoll to H. Ernst, 4 March, 1909. The full letter appears in vol. 5 of *The Collected Papers of Albert Einstein*. A longer excerpt is translated in Pais 1982, pp. 185–186.

[26] Besso wrote to Einstein: "I believe it was ... around about 1908 or '09 ... that I explained to you that your idea of breaking free from Israel wouldn't work. It was not a matter of this or that outlook or viewpoint or of a conscious adherence—I believe I said, but of the atmosphere of things taken for granted that surrounds the child from birth on, even before birth, in the mother's body" (Einstein and Besso 1972, letter of 10 October 1939, p. 346). This was about the time that Einstein is reported to have told a Jewish friend that his wife had had their two boys baptized while visiting her family, adding "Well, it's all the same to me!" ["Na ja, mir kann's egal sein!"] (Seelig 1952, p. 133). Besso had also written Einstein: "perhaps it is due in part to me, with my defense of Judaism and the Jewish family, that your family life took the turn that it did, and that I had to bring Mileva back from Berlin to Zurich" (Einstein and Besso 1972, letter of 17 January 1928, p. 238). This is a reference to Einstein's separation from his first wife in 1914.

[27] For the fate of the German solar eclipse team, see Clark 1971, pp. 174–176.

[28] Einstein's close association with Jakob Grommer, a highly gifted young Jewish mathematician from Brest-Litovsk, which lasted from about 1917 to 1927, may have played a role in alerting him to the problems of East European Jewish students (for information about Grommer, see Pais 1982, pp. 487–488).

[29] In an article in the *Berliner Tageblatt* "Die Zuwanderung aus dem Osten," Dec. 30, 1919, Einstein 2001, Doc. 30.

[30] For a documented account of some of his political activities from 1914–1919, see Nathan and Norden 1960, pp. 1–35.

[31] Blumenfeld 1956, 1962, 1976.

[32] Blumenfeld 1956, p. 75. These words are particularly significant in view of Blumenfeld's assertion that Einstein called upon him to draft many statements on Zionism, particularly while both were still in Germany. It is thus quite possible that many of the words cited in this article as Einstein's were actually written by Blumenfeld. Nevertheless, as Einstein wrote to Blumenfeld, "you know how to copy my style so effortlessly, and indeed so well, that after some time I myself cannot distinguish which of us was actually the writer" (letter of 1 June 1944).

[33] See e.g., Hugo Bergmann to Einstein, 22 October 1919, asking him, as "the one whom the world correctly describes as the greatest Jewish researcher," for his help; and Einstein's reply of 5 November 1919: "I take a warm interest in the affairs of the new colony in Palestine and especially in the university to be founded there."

[34] The results of the British eclipse expeditions to test Einstein's revised light-deflection prediction were officially announced on November 6 (see, e.g., Clark 1971; p. 230). On December 14, his picture appeared on the cover of the *Berliner Illustrierte Zeitung*.

[35] For an account of this meeting and Einstein's reaction, see Clark 1971, pp. 256–258.

36 "Meine Antwort. Ueber die anti-relativitätstheoretische G.M.b.H," *Berliner Tageblatt*, 27 August 1920, Einstein 2001, Doc. 45. This article is translated in Tauber 1979, pp. 97–99.

37 For a discussion of the Bad Nauheim meeting, see Einstein 2001, Editorial Note, "Einstein's Encounters with German Anti-Relativists," and Doc. 46.

38 Probably a reference to Gustav Roethe, Secretary of the German Academy of Sciences. Roethe resisted suggestions that the Academy, of which Einstein was a leading member, come to his defense. See Roethe to Max Planck, 10 September 1920 in Kirsten and Treder 1979, p. 205.

39 A reference to Ulrich von Wilamowitz-Moellendorff, the noted classical philologist, whose article against the pacifist G. F. Nicolai Einstein cites in a letter (Einstein to Hans Delbrück, 26 January 1920, Kirsten and Treder 1979, pp. 200–201), and to whom Einstein wrote: "I understand that you are reluctant to be one of the sponsors of the Anglo-American Literary Aid campaign for Central Europe because you do not wish to appear on the same list with me" (April 19, 1920, cited from Nathan and Norden 1960, p. 39).

40 For discussions of the fight between the pro-Weizmann faction and the Brandeis-Mack faction of American Zionists, see e.g., Halpern 1987, Chapter 4, "Encounter and Clash," pp. 171–232; Urofsky 1975, Chapter 7, "Schism," pp. 246–298.

41 According to Shapiro, "The attempts of the American [Zionist] leaders to dissuade Einstein from coming to the country as a member of the delegation failed. He had come to raise money for the future Hebrew University and was advised by them that he would be more successful if he did not come as an official member of the Zionist delegation. The argument did not impress him. . ." (Shapiro 1971, p. 172).

Kurt Blumenfeld wrote to Chaim Weizmann, 15 Mar. 1921: "Einstein is a man of the highest honor and solidity and is interested in our cause most strongly because of his revulsion from assimilatory Jewry. I consider it impossible that our American opponents can make an impression on him with their views." He added "Berger tells me that you are expecting speeches from Einstein. Please be very careful in this respect. Einstein is a poor speaker and often says things out of naiveté that are unwelcome to us" (Blumenfeld 1976, pp. 65–66).

For an account of Einstein's trip with Weizmann, see Clark 1971, pp. 382–391. Weizmann apparently took Blumenfeld's advice, and Einstein was called upon to provide effective if laconic support. "A few days after his arrival . . . the great physicist gave one of the shortest speeches on record, and certainly the shortest one that evening [at a Zionist rally in New York]. After a grandiose introduction, he stood up and declared: "Unser and euer führer, Dr. Weizmann, hat geshproken; folgt ihm!" ("Your leader and ours, Dr. Weizmann has spoken; heed him!") and sat down" (Urofsky 1975, p. 287). Einstein spoke in German not Yiddish, but otherwise the report seems basically accurate (Shapiro 1971, p. 172 cites another variant of the speech).

42 For an account of this trip, see Clark 1971, pp. 393–396.

43 The Blumenfeld radical faction had moved the so-called Posen Resolution adopted at the 1912 German Zionist Congress, which declared it "the obligation of every Zionist—above all the financially independent ones—to incorporate emigration to Palestine in his life program" (quoted from Poppel 1976). Einstein's unwillingness to emigrate is presumably one of the main reasons that Blumenfeld states that "there were limits to [Einstein's] Zionism that were conditioned by his nature." He reports that they often discussed the

question of *aliyah*, which Einstein "rejected on personal grounds that were decisive for him" (Blumenfeld 1956, p. 85).

[44] Eva Reichmann has pointed out that: "It was one of the characteristics of German Jewry that it utilised the intellectual trends which it absorbed from the Gentile world for effecting a spiritual regeneration of Judaism, which found expression in the Science of Judaism, Jewish religious liberalism, Samson Raphael Hirsch's system of Europeanised orthodoxy, Gabriel Riesser's political fight for Jewish rights, and the Zionist renaissance (Reichmann 1951, p. 170).

[45] See Mosse 1964, Part I, pp. 13–125; and Mosse 1970, Chapter 4, "The Influence of the Volkish Idea on German Jewry," pp. 77–115. Mosse signals the importance of the Bar Kochba group of Prague Jews (notably Buber) for the development of German-Jewish Volkishness. Since Einstein was in contact with this group during his year in Prague, it is possible that he adopted some of their attitudes at the time, or—more likely, if he was influenced at all—he recognized their merits in retrospect after his move to Berlin (for some speculations on the influence of this group on Einstein, see Feuer 1982, pp. xii-xix). The only preserved almost-contemporary comment on the group by Einstein, in 1916, suggests that he was not yet aware of any such influence. He spoke of "a small circle ... of philosophical and Zionist enthusiasts, which was loosely grouped around the university philosophers, a medieval-like band of unworldly people..." (Born 1971, p. 4).

[46] For the outlook of the German Zionist movement in the twenties, see Poppel 1976, Chaps. 6–10, pp. 85–157.

[47] Rodinson is an independent Marxist of French-Jewish origins. He has been bitterly attacked for his views on Zionism, particularly in France. I wonder if he has been as carefully read?

[48] Robert Weltsch, one of the leaders of German Zionism, wrote: "We few Western Zionists had no spiritual orientation, no inner confidence in Zionism ... [Buber] formulated the question that had tormented us, and gave an answer. The affirmation of Jewish nationality ... of 'blood' as the formative power of our experience, gave us a new orientation in time and space. It indicated to us our historical place." Cited from Poppel 1976, p. 130.

[49] For the two forms of Zionism and the clash between them, see, for example, Lacqueur 1972, pp. 162–171.

[50] I say "naive," because it was not impossible to foresee the turn events would take if current Zionist policies were pursued, as the example of Judah Magnes proves. See his 1920 letter to a "Dear Friend," Magnes 1982, Doc. 33, pp. 183–190, especially pp. 187–189.

[51] This belief had already been refuted by events years before. As Lacqueur notes, "it was precisely the influx of Jewish workers into Palestine with the second Aliya which aggravated the conflict. ... [In 1908] Levontin, the director of the local Anglo-Palestine bank ... wrote ... that the Zionist labour leaders were sowing hatred against Zionism in the heart of the local population by speaking and writing against giving jobs to the Arab workers, reported ... in 1911 that he too was continually trying to impress on them the need to refrain from acts of hostility in their relation with the Arabs" (Lacqueur 1972, pp. 218–219).

[52] There was a serious clash in May 1921, for example, during the Einstein–Weizmann visit to the United States. It is hard to believe that Einstein could have remained unaware of it.

[53] I have taken this characterization from Schlaim 1987, p. 7: "The view of the conflict between the Palestinian and Jewish national movements that [Hajj Amin, the mufti of Jerusalem,] held and vigorously propagated might in today's jargon be called a zero-sum game, that is to say a game in which every gain by one party is necessarily at the expense of the other."

[54] Einstein himself acknowledged "The contrasts and contradictions that can permanently live peacefully side by side in a skull. . ." ("Aphorisms for Leo Baeck," 1953, cited from Einstein 1954, p. 28).

[55] In his essay "Why Socialism?" (1949), Einstein wrote: "Man is, at one and the same time, a solitary being and a social being" (cited from Einstein 1954, p. 153).

[56] What happened when he was drawn into day-to-day Zionist activities, as for example in the struggles over the development of the Hebrew University, is another story.

[57] I would gladly call my approach a Marxist view, if that term had a clear and unambiguous meaning today. However, I believe that many "non-Marxists" would agree on the weaknesses of any "essentialist" account of historical phenomena—and that some "Marxists" would not.

References

Baerwald, Leo (1982). "Juden und jüdische Gemeinden in München vom 12. bis 20. Jahrhundert." In *Vergangene Tagen/Jüdische Kultur in München*, Hans Lamm, ed. München/Wien: Langen/Müller Verlag, pp. 19–30.

Bach, Hans Israel (1984). *The German Jew/A Synthesis of Judaism and Western Civilization 1730–1930*. Oxford: The Littman Library, Oxford University Press.

Baron, Salo Wittmayer (1939). "Emphases in Jewish History." *Jewish Social Studies*, vol. 1. Reprinted in Baron 1964, pp. 65–89.

—— (1964). *History and Jewish Historians/Essays and Addresses*. Philadelphia: Jewish Publication Society of America.

Bieber, Hans-Joachim (1979). "Anti-Semitism as a Reflection of Social, Economic and Political Tension in Germany: 1880–1930." In *Jews and Germans from 1860 to 1933/The Problematic Synthesis*. David Bronsen, ed. Heidelberg: Carl Winter Universitätsverlag, pp. 33–77.

Blumenfeld, Kurt (1956). "Einstein's Beziehungen zum Zionismus und zu Israel." In *Helle Zeit– Dunkle Zeit/ in Memoriam Albert Einstein*. Carl Seelig, ed. Zürich/-Stuttgart/Wien: Europa Verlag, pp. 74–86.

—— (1962). *Eriebte Judenfrage/Ein Vierteljahrhundert deutscher Zionismus*. Stuttgart: Deutsche Verlagsanstalt.

—— (1976). *Kamp um den Zionismus/Briefe aus fünf Jahrzehnten*. Miriam Sambursky and Jochanan Ginat, eds. Stuttgart: Deutsche Verlagsanstalt.

Cahnman, Werner J. (1989). *German Jewry/Its History and Sociology*. New Brunswick/Oxford: Transaction Publications.

Clark, Ronald W. (1971). *Einstein/The Life and Times*. New York/Cleveland: World Publishing Co.

Dingfelder, Simon (1927). "Aus der Geschichte des jüdischen Schulwesens in München 1800–1872." *Bayrische Israelitische Gemeindezeitung*, vol. 3: 354–357.

Dubnov, Simon (1973). *History of the Jews*, Vol. 5, rev. ed. *From the Congress of Vienna to the Emergence of Hitler*. Moshe Spiegel, ed. South Brunswick/New York/London: Thomas Yoseloff.

Einstein, Albert (1923). "My Impressions of Palestine." *New Palestine*, vol. 4: 341.

—— (1925). "Botschaft." *Jüdische Rundschau*, vol. 30: 129. French translation in *La revue juive*, vol. 1: 14–16.

—— (1931a). *Cosmic Religion, with Other Opinions and Aphorisms*. New York: Covici-Friede.

—— (1931b). *About Zionism/Speeches and Letters*. New York: Macmillan.

—— (1954). *Ideas and Opinions*. New York: Crown.

—— (1979) *Autobiographical Notes/A Centennial Edition*. Paul Arthur Schilpp, ed. LaSalle/Chicago: Open Court.

—— (1987). *The Collected Papers of Albert Einstein*, vol. 1. John Stachel et al., eds. Princeton: Princeton University Press, 1987.

—— (2001). *Ibid.*, vol. 7, Michael Jansen et al., eds. Princeton: Princeton University Press, 2001.

Ettinger, Shmuel (1976). "The Modern Period." In *A History of the Jewish People*, Haim Hillel Ben-Sasson, ed. Cambridge: Harvard University Press, pp. 727–1096.

Frank, Philipp (1979). *Albert Einstein/Sein Leben und seine Zeit*. Braunschweig/Wiesbaden: Vieweg.

Halpern, Ben (1987). *A Clash of Heroes/Brandeis, Weizmann, and American Zionism*. New York/Oxford: Oxford University Press.

Hess (1862). *Rom und Jerusalem*. Leipzig: Eduard Wengler.

—— (1918). *Rome and Jerusalem*. Meyer Waxman, tr. New York: Bloch Publ. Company.

Kahler, Erich (1967). *The Jews Among the Nations*. New York: Frederick Unger.

Katz, Jakob (1985). "German Culture and the Jews." In Reinharz and Schatzenberg 1985, pp. 85–99.

Kayser, Rudolf [pseud. Anton Reiser] (1930). *Albert Einstein—A Biographical Portrait*. New York: Albert & Charles Boni.

Kayser, Rudolf, Stoob, Heinz, eds. (1974). *Bayerisches Städtebuch*, Part 2. Stuttgart/Berlin/Köln/Mainz: W. Kohlhammer, 1974.

Kirsten, Christa, Treder, Hans-Jürgen, eds. (1979). *Albert Einstein in Berlin 1913–1933*. Part 1. *Darstellung und Dokumente*. Berlin: Akademie-Verlag.

Kirsten, Christa, Treder, Hans-Jürgen (1983). "Albert Einstein 1879–1955." In *Wegbereiter der deutsch-slawischen Wechselseitigkeit*. Eduard Winter and Günther Jarosch, eds. Berlin: Akademie-Verlag, pp. 349–363.

Klein, Martin J. (1970). *Paul Ehrenfest*, vol. 1. *The Making of a Theoretical Physicist*. Amsterdam/London/New York: North Holland and American Elsevier.

Lacqueur, Walter (1972). *A History of Zionism*. New York: Schocken Books.

Lowenstein, Steven M. (1976). "The Pace of Modernisation of German Jewry in the Nineteenth Century." In Leo Baeck Institute, *Year Book XXI*. London: Secker & Warburg, pp. 41–56.

Magnes, Judah Leon (1982). *Dissenter in Zion/From the Writings of Judah L. Magnes*. Arthur A. Goren, ed. Cambridge/London: Harvard University Press.

Mosse, George L. (1964). *The Crisis of German Ideology/Intellectual Origins of the Third Reich*. New York: Grosset and Dunlap.

—— (1970). *Germans and Jews*. New York: Howard Fertig.

—— (1985). "Jewish Emancipation/Between *Bildung* and Respectability." In Reinharz and Schatzberg, pp. 1–16.

Moszkowski, Alexander (1921). *Einstein/Einblicke in seine Gedankenwelt*. Hamburg/Berlin: Hoffmann and Campe, F. Fontane & Co.

Nathan, Otto, Norden, Heinz, eds. (1960). *Einstein on Peace*. New York: Schocken Books.

Pais, Abraham (1982). *'Subtle Is the Lord' . . . The Science and the Life of Albert Einstein*. Oxford: Clarendon Press/New York: Oxford University Press.

Poppel, Stephen M. (1976). *Zionism in Germany 1897–1933*. Philadelphia: Jewish Publication Society of America.

Pyenson, Lewis (1985). "Audacious Enterprise: The Einsteins and electrotechnology in late nineteenth-century Munich." In *The Young Einstein*. Boston: Adam Hilger, pp. 35–57.

Reichmann, Eva G. (1951). *Hostages of Civilization/The Social Sources of National Socialist Anti-Semitism*. Boston: Beacon Press.

Reinharz, Jehuda, Schatzberg, Walter, eds. (1985). *The Jewish Response to German Culture*. Hanover, NH/London: University Press of New England.

Rodinson, Maxime (1983). *Cult, Ghetto, and State/The Persistence of the Jewish Question*. London: Al Saqi Books.

Schlaim, Avi (1987). *Collusion Across the Jordan/King Abdullah, the Zionist Movement, and the Partition of Palestine*. New York: Columbia University Press.

Schwab, Hermann (1956). *Jewish Rural Communities in Germany*. London: Cooper Book Co.

Schwarz, Stefan (1963). *Die Juden in Bayern im Wandel der Zeiten*. München/Wien: Günter Olzog Verlag.

Seelig, Carl (1952). *Albert Einstein/Eine Dokumentarische Biographie*. Zürich/Stuttgart/Vienna: Europa Verlag.

Shapiro, Yonathan (1971). *Leadership of the American Zionist Organization 1897–1930*. Urbana/Chicago/London: University of Illinois Press.

Talmey, Max (1932). *Relativity Theory Simplified and the Formative Period of its Inventor*. New York: Falcon Press.

Tänzer, Aron (1931). "Der Stammbaum Prof. Albert Einstein." *Jüdische Familienforschung*, vol. 28: 419–421.

Tauber, Gerald, ed. (1979). *Albert Einstein's Theory of General Relativity*. New York: Crown.

Urofsky, Melvin I. (1975). *American Zionism from Hertzl to the Holocaust*. Garden City: Doubleday.

Winteler–Einstein, Maja (1987). "Albert Einstein — Beitrag für sein Lebensbild" (Excerpt). In *Einstein* 1987, pp. xlviii–lxvi.

Einstein on Civil Liberty

John Stachel

Even if later work as a physicist had not brought me into contact with the towering scientific achievements of Albert Einstein, I would always have revered him for his role as a symbol of steadfast resistance to the modern inquisition which threatened to destroy civil liberties in this country during the cold war years. At a time when so many institutions and individuals seemed to vacillate or give way before the onslaught of the witch-hunters in and out of government, Einstein stood like a beacon, seeming to say to those of my generation: stand fast—there is hope!

It is probably hard for young people of this generation to realize what the psychological atmosphere was at that time. Today almost everyone is skeptical of official pronouncements. The idea that our government might act unjustly towards some—or all—of its citizens is commonplace. That this country faced a real threat of loss of basic liberties during the Nixon era, under cover of protection from subversive dangers from abroad and at home; that opposition to the Vietnam War could be a form of patriotism—today these are widely-accepted views. But during the Truman–Eisenhower Era, during the Korean War, similar thoughts were in themselves proof of a subversive bent—best not thought and certainly not given voice! Better to go along with what government, press and other media parroted incessantly.

Einstein quickly saw the dangers of the kind of thought control with which his adopted country was threatened and started to speak out on issue after issue, advocating "heretical" ideas and above all defending the right to hold such ideas. He challenged the intellectuals of his country: If you do not set an example, if you do not resist this inquistion, then you merit the intellectual slavery being prepared for you. We shall never know exactly how many hearts were gladdened, how many backbones were stiffened, by this challenge; but surely they were many. Einstein's correspondence preserves hundreds of letters attesting to his inspirational role.

Rights
25:6 (1979)

The list of civil liberties issues on which he spoke out during this period is long indeed—so long that I cannot attempt to cite it here. Those interested will find a partial account in *Einstein on Peace*, edited by Otto Nathan and Heinz Norden. Detractors sometimes imply that, in his naiveté, Einstein would sign almost any appeal placed before him. His correspondence shows this was far from being the case. He weighed such appeals carefully. Sometimes he declined to sign because he had such contempt for the authorities being petitioned that he regarded it as undignified to appeal to such persons. He looked at the sponsorship of each appeal, not wanting to appear to be an adherent of the Communist Party on the one hand, nor a mouthpiece for the current anti-communist hysteria on the other. He was also careful in choosing his issues because he was afraid that too frequent an appearance of his name would only serve to diminish its impact, without materially aiding the cause in question.

Of course, Einstein did not escape the fury of the would-be executioners of the Bill of Rights. He received not a few letters filled with hate, attacking him as a radical, as a Jew, as a pacifist (of course, less polite language was usually used!). Public attacks were also frequent. He was not always defended by those quick to praise him today. It is amusing to compare what leading press organs, such as *The New York Times*, had to say about Einstein when he advised non-cooperation with Congressional witch-hunters with what they had to say about him this Centennial year. It is always safer to praise dead defenders of freedom!

No doubt, part of the reason for Einstein's quick and firm reaction to threats to civil liberties in the United States, his adopted country, was that he had seen similar attacks succeed before. When the democratic forces in Weimar Germany were going down to defeat before the Nazi onslaught, Einstein had played a prominent role in trying to rally German intellectuals to the defense of their freedoms. But throughout his life, his fundamental credo, based on his deepest feelings and instinctive reactions, had always made him sensitive to violations of the rights of the individual anywhere in the world. Einstein, who referred to himself as "an incorrigible nonconformist," nevertheless saw the individual as having profound obligations towards the social matrix within which he or she moved. He saw every society as characterized by tension between the need of the individual for solitude and inner independence of spirit; and a need for nurturing by his or her surroundings, as well as to contribute something to others in recompense for that nurturing. If society, instead of nurturing the individual and fostering inner independence, demanded blind obedience and conformity of thought, social tyranny was the result. If society encouraged the individual to forget his or her social ties and develop an egoistic individualism, social atomism would result. It is in the narrow space between these two dangers that Einstein saw the need for each individual and each democratic society to maneuver. I think we must say, unfortunately, it is not a form of maneuvering at which any contemporary society has shown itself particularly adept. The example of Einstein—his deeds as well as his words—stands as a challenge to each of us, and to our societies; to try to find the way to preserve and extend our social achievements without encroaching upon vital individual liberties.

Einstein and the "Research Passion"

John Stachel

I shall present Einstein's views on the nature of the creative process, the kinship and differences between artistic and scientific creativity, his own creative method and the criteria for judging artistic works and scientific theories. I shall primarily quote his own words, without attempting to evaluate them. It seems to me that his views have an independent value, certainly a much higher one than mine. I shall end with Einstein's own views on the attempts to find direct links between his theories and certain trends in modern art.

In his often-quoted address to Max Planck on his 60th birthday in 1918, Einstein expressed his vision of the deepest wellspring of human creativity:

> Man tries to make for himself in the fashion that suits him best a simplified and intelligible picture of the world; he then tries to some extent to substitute this cosmos of his for the world of experience, and thus to overcome it. This is what the painter, the poet, the speculative philosopher, and the natural scientist do, each in his own fashion. Each makes this cosmos and its construction the pivot of his emotional life, in order to find in this way the peace and security which he cannot find in the narrow whirlpool of personal experience.[1]

Just before this paragraph he said:

> I believe with Schopenhauer that one of the strongest motives that leads men to art and science is escape from everyday life with its painful crudity and hopeless dreariness, from the fetters of one's own ever shifting desires. A finely tempered nature longs to escape from personal life into the world of objective perception and thought; this desire may be compared with the townsman's irresistible longing to escape from his noisy, cramped surroundings into the silence of high mountains, where the eye ranges freely through the still, pure air and fondly traces out the restful contours apparently built for eternity.

Talk given at the Annual Meeting of the
American Association for the Advancement of Science
in Detroit, May 1983

It is clear from other evidence, if it is not clear enough from the vehemence of these statements, that he included himself in this description. He wrote to Hermann Broch about the latter's *Vergil*:[2]

> The book shows me clearly what I fled from when I sold myself body and soul to Science—the flight from the I and We to the It.

Einstein even attaches an ethical character to this flight:

> The true value of a human being is determined primarily by the measure and the sense in which he has attained liberation from the self.

Another feature art and science shared was emphasized in an article on tolerance that he wrote in 1934. Due to the intolerance of the editors of the magazine for which it was written (they wanted changes and cuts) it was not published during his lifetime:[3]

> Whether it be a work of art or a significant scientific achievement, that which is great and noble comes from the solitary personality. European culture made its most important break away from stifling stagnation when the Renaissance offered the individual the possibility of unfettered development.

The emphasis on truly creative accomplishment as an individual effort does not mean that Einstein was unaware of the social matrix that makes possible the emergence of such creative individuals. In a 1932 essay on "Society and Personality" devoted to this interrelationship, he wrote:[4]

> The individual is what he is and has the significance that he has not so much in virtue of his individuality but rather as a member of a great human community.

In a 1949 essay on "Why Socialism?", he put it as follows:[5]

> Man is at one and the same time a solitary being and a social being.

What did Einstein see as the main difference between artistic and scientific creativity? He touched that question briefly in a contribution he reluctantly made in 1921 to a German magazine on modern art:[6]

> Although I am aware that I have nothing original, let alone worthy of publication, to say on the theme you mention, I send you the enclosed aphoristic utterance to demonstrate my good will. If my ink had been less viscid, I would have done justice to the wish expressed in your friendly letter by sending a more sumptuous opus.

The "aphoristic utterance" is entitled "What Artistic and Scientific Experience Have in Common":

> Where the world ceases to be the scene of our personal hopes and wishes, where we face it as free beings admiring, asking, and observing, there we enter the realm of Art and Science. If what is seen and experienced is portrayed in the language of logic, we are engaged in science. If it is communicated through forms whose connections are not accessible to the conscious mind but are recognized intuitively as meaningful, then we are engaged in art. Common to both is the loving devotion to that which transcends personal concerns and volition.

Other remarks by Einstein suggest that the gap between logic and intuition is not as wide as this aphorism might suggest. In a letter of 1949, for example, he writes:[7]

> A new idea comes suddenly and in a rather intuitive way. That means it is not reached by conscious logical conclusions. But thinking it through afterwards you can always discover the reasons which have led you unconsciously to your guess and you will find a logical way to justify it. Intuition is nothing but the outcome of accumulated earlier intellectual experience.

He is even willing to speak of "artistic desire" in characterizing outstanding scientific achievement. Speaking of Planck in 1913, he wrote:[8]

> The pleasure with which one ever and again picks up these books [of Planck] is not least due to the direct, truly artistic style which is characteristic of all of Planck's works; above all one has the impression in studying Planck's writings that the artistic desire forms one of the most powerful driving forces of his creativity. Not without good reason one is told that, even after graduating from the gymnasium, Planck was in doubt whether to devote himself to the study of mathematics and physics or to the study of music.

While ready to admit that someone like Planck could have chosen, at a certain point, between an artistic and a scientific career, he did not feel in his own case that there was any direct connection between his musicality and his scientific work. Replying to a question on such a possible connection he wrote:[9]

> Music does not *influence* research work, but both are nourished by the same source of longing and they complement one another in the release they offer.

Now let me turn to Einstein's views on the sources and nature of his own creative work. In a little known address to the students of UCLA in 1932, Einstein expressed himself at some length on this question.[10]

> Science as something already in existence, already completed, is the most objective, impersonal thing that we humans know. Science as something coming into being, as a goal, however, is just as subjectively, psychologically conditioned as are all other human endeavors. This is so much the case that the question of the goal and meaning of science will receive quite different answers at different times and from different personalities.
>
> To be sure, all are agreed that science must establish a connection between facts of experience such that on the basis of experienced facts we are enabled to predict other such facts. According to the view of many positivists the most complete possible solution of this task is the only goal of science.
>
> I do not believe, however, that so primitive an ideal would really permit the kindling to a high degree of that researcher's passion from which really great accomplishments arise. A stronger, but also more obscure, drive lies behind the tireless exertions tied to such achievements: one wants to comprehend being, reality. One shrinks from using such words, indeed, because he is perplexed if called upon to explain what is really to be understood by 'reality' and by 'comprehend' in this general statement.
>
> If we divest the statement of its mystical elements, then we mean: we are looking for a system of concepts of the greatest possible simplicity which links together the observable facts. Here we do not mean by simplicity, however, that the assimilation

of this system shall give the learner the least possible difficulty; but rather that the system shall contain the least possible number of independent assumptions or axioms, for the totality of logically mutually independent axioms signifies the remainder of the uncomprehended.

When one speaks of scientific subjects, then the little word 'I' should play no role in his exposition. If one speaks, however, about aims and strivings in science, then one should be allowed to treat of himself. For one experiences no other strivings and purpose so immediately as his own. My own strivings were concerned quite predominantly with logical unification in physics. First of all, I was disturbed that electrodynamics favored *one* state of motion above all others, without any sort of occasion for this preference being present in empirical experiences. Thus the special theory of relativity arose, which moreover fused together the electric and magnetic fields, as well as mass and energy and momentum and energy into conceptual unities. Then, from the endeavor to comprehend inertia and gravitation as essentially identical, the general theory of relativity arose, which moreover avoided those implicit axioms which are concealed by the use of special coordinate systems in the formulation of fundamental laws.

It is especially disturbing currently that the gravitational and the electric fields enter into theory as logically independent fundamental concepts. After many years of effort a satisfactory logical unification has been reached—so I believe—through the use of a new mathematical method that I have devised together with my distinguished collaborator Dr. W. Mayer.

An important task of this type still remains outstanding, which has been proposed by many, but so far has found no satisfactory solution: the reduction of atomistic structure to the field concept. At the basis of all such attempts lies the belief that being is completely harmonious in its structure. Today we have less ground than ever before to allow ourselves to stray from this wondrous belief.

His descriptions of his own creative process are loaded with emotional terms which testify to the appositeness of his reference to a "research (or "researcher's" to be more accurate) passion." Speaking of the origin of the special theory of relativity, for example he wrote:[11]

I will give you as an example the situation which led to the setting up (*Aufstellung*) of the special theory of relativity. Mechanically all inertial systems are equivalent. Experience shows that this equivalence also extends to optics and electrodynamics. But this equivalence appeared unattainable in the theory of the latter. Early on I came to the conviction that this had its basis in a deep incompleteness of the theoretical system. The desire to discover and remove this generated a state of physical tension in me that was resolved after seven years of vain searching by the relativising of the concepts of time and length... It was always the striving for a logically simple interpretation of empirically known connections, supported by the conviction of the existence of a logically simple interpretation.

Einstein once attempted to describe in some detail the exact nature of his creative mental process:[12]

The words or the language, as they are written or spoken, do not seem to play any role in my mechanism of thought. The psychical entities which seem to serve as elements in thought are certain signs and more or less clear images which can be "voluntarily" reproduced and combined.

> There is, of course, a certain connection between those elements and relevant logical concepts. It is also clear that the desire to arrive finally at logically connected concepts is the emotional basis of this rather vague play with the above-mentioned elements. But taken from a psychological viewpoint, this combinatory play seems to be the essential feature in productive thought—before there is any connection with logical construction in words or other kinds of signs which can be communicated to others.
>
> The above-mentioned elements are, in my case, of visual and some of muscular type. Conventional words or other signs have to be sought for laboriously only in a secondary stage, when the mentioned associative play is sufficiently established and can be reproduced at will.
>
> According to what has been said, the play with the mentioned elements is aimed to be analogous to certain logical connections one is searching for. . . . In a stage when words intervene at all, they are, in my case, purely auditive, but they interfere only in a secondary stage, as already mentioned.

It would be interesting to compare this account with similar ones by outstanding artistic figures, if any such exist. I suspect many similarities—as well as differences—would emerge. At any rate, it seems a decisive refutation of the oft-repeated idea that thinking requires words.

His later views on quantum theory were also accompanied by an emotional cathexis that occasionally breaks through in his private letters:[13]

> I find the idea that there should not be laws for being but only laws for probabilities simply disgusting (a nauseatingly indirect description).

So far I have spoken about the process of theory-creation, but Einstein felt that the judgment of already-completed theories, these "most objective impersonal things that we humans know," is subject to criteria which, to me, seem related to aesthetic ones by which a work of art might be judged. In speaking of "the points of view from which physical theories may be analyzed critically at all, "Einstein distinguishes two:[14]

> The first point of view is obvious: the theory must not contradict empirical facts. However evident this demand may in the first place appear, its application turns out to be quite delicate. For it is often, perhaps even always, possible to retain a general theoretical foundation by adapting it to the facts by means of artificial additional assumptions. In any case, however, this first point of view is concerned with the confirmation of the theoretical foundation by the available empirical facts.
>
> The second point of view is not concerned with the relationship to the observations but with the premises of the theory itself, with what may briefly but vaguely be characterized as the "naturalness" or "logical simplicity" of the premises (the basic concepts and the relations between these). This point of view, whose exact formulation meets with great difficulties, has played an important role in the selection and evaluation of theories from time immemorial. The problem here is not simply one of a kind of enumeration of the logically independent premises (if anything like this were at all possible without ambiguity), but one of a kind of reciprocal weighing of incommensurable qualities. Furthermore, among theories with equally "simple" foundations; that one is to be taken as superior which most sharply delimits the otherwise feasible qualities of systems (i.e., contains the most specific claims). Of the "scope"

of theories I need not speak here, inasmuch as we are confining ourselves to such theories as have for their object the *totality* of all physical phenomena. The second point of view may briefly be characterized as concerned with the "inner perfection" of the theory, whereas the first point of view refers to the "external confirmation." The following I reckon as also belonging to the "inner perfection" of a theory: We prize a theory more highly if, from the logical standpoint, it does not involve an arbitrary choice among theories that are equivalent and possess analogous structures.

I shall not attempt to excuse the lack of precision of the assertions contained in the last two paragraphs on the grounds of insufficient space at my disposal; I must confess herewith that I cannot at this point, and perhaps not at all, replace these hints by more precise definitions. I believe, however, that a sharper formulation would be possible. In any case it turns out that among the "oracles" there usually is agreement in judging the "inner perfection" of the theories and even more so concerning the degree of "external confirmation."

Indeed, the trend of development of modern theoretical physics suggested to Einstein that such aesthetic criteria would more and more guide the development of physical theories:[15]

The theory of relativity is a fine example of the fundamental character of the modern development of theoretical science. The initial hypotheses become steadily more abstract and remote from experience. On the other hand, it gets nearer to the grand aim of all science, which is to cover the greatest possible number of empirical facts by logical deduction from the smallest possible number of hypotheses or axioms. Meanwhile, the train of thought leading from the axioms to the empirical facts or verifiable consequences gets steadily longer and more subtle. The theoretical scientist is compelled in an increasing degree to be guided by purely mathematical, formal considerations in his search for a theory, because the physical experience of the experimenter cannot lead him up to the regions of highest abstraction. The predominantly inductive methods appropriate to the youth of science are giving place to tentative deduction. Such a theoretical structure needs to be very thoroughly elaborated before it can lead to conclusions which can be compared with experience. Here, too, the observed fact is undoubtedly the supreme arbiter; but it cannot pronounce sentence until the wide chasm separating the axioms from their verifiable consequences has been bridged by much intense, hard thinking. The theorist has to set about this Herculean task fully aware that his efforts may only be destined to prepare the death blow to his theory. The theorist who undertakes such a labor should not be carped at as "fanciful"; on the contrary, he should be granted the right to give free reign to his fancy, for there is no other way to the goal. His is no idle daydreaming, but a search for the logically simplest possibilities and their consequences.

Is there any justification for speaking of a similarity in criteria in the evaluation of a work of art and a scientific theory? In a letter, Einstein indicates what he saw as the similarities and the differences:[16]

If I disregard the practical value of a science, then indeed I do see something similar in scientific and artistic endeavors. Both attempt to put together a whole, which is in itself obscure, out of parts in such a way that the underlying order creates lucidity and clarity. The lucidity and clarity thus attained produce great satisfaction in us. This occurs just as well in art as in science. In science this unity-creating ordering principle is logical connection, while in art the ordering principle remains anchored in the unconscious. In the latter case it is always a matter of modes of connection given by tradition, which are experienced as equally compelling by those living within this tradition as logical connection is by scientifically oriented persons.

The essence of traditional modes of connection in art is seen clearly in simple art forms, e.g., musical melody and in ornamentation based on an intuitively graspable understandable regularity. In both cases the means of achieving are experienced as necessary in a way similar to that in which logical deduction is in mathematics. In more complex art forms these traditionally anchored means of creating clarity or "unity" are less easily grasped.

A work of art can thus only be experienced and evaluated by one in whom the relevant traditional modes of connection are active. There is no other legitimation of the latter than their living existence. If they are given, then it is with respect to them that the work is good or bad, according to the perfection with which, on the basis of the traditional modes of connection, the impression of lucid unity is attained.

If what has been said is correct, then it is absurd to attempt to evaluate traditional modes of connection (the languages of an artistic epoch, so to speak) in any sort of comparative relation to each other.

Notes

[1] "Principles of Research," in Albert Einstein, *Ideas and Opinions*, New York: Crown 1954, p. 225.

[2] Einstein to Hermann Broch, 1945, quoted in Banesh Hoffmann and Helen Dukas, *Albert Einstein: Creator and Rebel*, New York: Viking 1972, p. 254.

[3] Helen Dukas and Banesh Hoffmann, eds., *Albert Einstein: The Human Side*, Princeton: Princeton University Press 1979, p. 89.

[4] *Ideas and Opinions*, p. 13.

[5] *Ideas and Opinions*, p. 153.

[6] *The Human Side*, pp. 37–38.

[7] Albert Einstein to Dr. H. L. Gordon.

[8] "Max Planck als Forscher," *Die Naturwissenschaften 1* (1913), 1077–1079, reproduced in Martin J. Klein et al., eds., *The Collected Papers of Albert Einstein*, vol. 4, *The Swiss Years/Writing 1912–1914*, pp. 561–563. The quotation is from p. 1079.

[9] *The Human Side*, p. 78. The letter is from 1928.

[10] Albert Einstein, "Address to the Students of the University of California at Los Angeles" in *Builders of the Universe*, Los Angeles: U.S. Library Assocation 1932.

[11] Albert Einstein to Erika Oppenheimer, 13 Sept. 1932. Ms. Oppenheimer was a doctoral student in Wetheimer's Psychology Institute in Frankfurt and his letter was written in response to an enquiry into the motivations for his research activities.

[12] Albert Einstein, "A Mathematician's Mind," in *Ideas and Opinions*, pp. 25–56. His words were in reply to questions from Jacques Hadamard who was preparing *An Essay on the Psychology of Invention in the Mathematical Field*, Princeton: Princeton University Press 1945.

[13] Albert Einstein to Tatyana Ehrenfest-Afanasieva, 1938. The letter also comments on the Munich sell-out of Czechoslovakia, which may help to account for Einstein's foul mood.

[14] Albert Einstein, *Autobiographical Notes*, LaSalle/Chicago: Open Court 1979, pp. 21, 23.

[15] Albert Einstein, "The Problem of Space, Ether and the Field in Physics," *Ideas and Opinions*, pp. 276–285. The quotation is from p. 282.

[16] Albert Einstein to Paul Laporte, May 4, 1946. It is translated in Laporte's article "Cubism and Relativity," *Art Journal*, 1966.

Part II

Editing the Einstein Papers

“A Man of My Type” — Editing the Einstein Papers[1]

John Stachel

Introduction: Confessions of an Ex-Editor

Towards the end of the career of many a distinguished scientist, or shortly afterwards, an edition of his or her works is published under the title: “The Collected Papers of *X*.” In addition to officially sanctifying *X*’s importance, the edition is intended primarily as a research aid, making *X*’s published papers easily accessible to scholars and students working in *X*’s field. Little help is given in providing a wider context for these writings: the circumstances in the field leading to their creation, how these works fit into *X*’s life, how they were assimilated into the field, and their role in later developments. While not denying the importance of such rites of passage or minimizing the real utility of their products, their aims must be clearly distinguished from those of a project like the *Collected Papers of Albert Einstein* (*The Einstein Papers* for short), which is based on the historical editing tradition.

Soon after being appointed editor,[2] it was my good fortune to attend the National Historical Publication and Records Commission’s 1977 summer school on documentary editing, hosted by the Laurens Papers project[3] at the University of South Carolina in Columbia. Under the patient tutelage of Mary Jo Kline and David Chestnutt, a historically illiterate physicist learned the techniques of the documentary editing tradition usually called “historical editing” in the United States. As the name suggests, the method was developed by historians working in the fields of political and social history.[4] Julian Boyd’s edition of *The Papers of Thomas Jefferson*[5] founded the modern phase of this tradition, setting standards

The revised version was first given at the
Symposium on Critical Editing in Philosophy and Science,
Boston Colloquium for the Philosophy of Science,
March 18, 1993.

that have served as benchmarks, accepted or modified in a variety of ways by later editors. Historians of science have hardly begun to apply the historical editing method. When I started work, Nathan Reingold's edition of *The Papers of Joseph Henry*[6] was the only such example. It was intended from the outset to be selective, while my mandate was to produce a substantially complete edition of *The Einstein Papers*, to date the only such edition of a scientist's papers.[7]

Early on, I had the privilege of meeting Julian Boyd, Nate Reingold, and Arthur Link,[8] and benefiting from their advice and examples in many ways.[9] One negative lesson was later to stand me in good stead: the danger of over-identifying with the subject of one's labors. From the outset I realized the importance of clearly distinguishing between my life and that of Albert Einstein. To help maintain this distinction, I resolved to continue independent scholarly work.[10] Such work was incomparably less important to the world than Einstein's, but very important for my own psychic equilibrium while serving as editor. In addition I felt that, if circumstances should ever so dictate, I could then always walk away from the project and resume my professional career.[11] I was soon able to test these ideas when the project almost walked away from me,[12] and for about two years I could not proceed with any editorial work.

After these problems had been resolved, I found that I was able to throw myself single-mindedly and whole-heartedly into the work, yet without blurring the line between where Stachel ended and Einstein began, which I did for the next eight years. After turning sixty, with the project well-established at Boston University and the first two volumes of *The Einstein Papers* published, I decided to resign and devote my remaining active years to a full-time career as a teacher and scholar, a decision I have not regretted.[13]

Returning to the early days of the project, I knew and admired the Trustees of the Einstein Estate, Dr. Otto Nathan and Miss Helen Dukas, before becoming editor. Without losing my friendly feelings, I soon became aware of another peril involving loss of boundaries: the danger of becoming a "keeper of the flame" rather than a seeker of the truth. Those who have been close to The Loved One often maintain an image or, perhaps more accurately, want the world to maintain an image that may not be compatible with all of the surviving documentary evidence. The conclusion seems obvious: the refractory documents must be withheld from publication, and scholars who do not subscribe to the "party line" on The Loved One denied access to his/her papers.[14] When it became clear to the Trustees that I took literally my mandate to prepare a complete edition of the Papers of Albert Einstein, and in the meantime wanted to grant access to the Einstein Archives to all qualified scholars, conflict was inevitable. Without entering into tedious details, let me observe that while not by nature bloodthirsty, my experiences (and stories I have heard from other editors) have led me to look upon the ancient custom of hecatomb, the ritual slaughter of a defunct chieftain's surviving retinue, with a lot more sympathy than I would once have imagined possible.

More seriously, I think that death cancels all injunctions to silence by the living. Max Brod certainly did the right thing by publishing rather than destroying Kafka's manuscripts. The only reason for temporarily withholding material from publication that I can accept is to spare unnecessary pain to the living. If close relatives would be hurt by the publication of some document, it is perfectly reasonable to postpone its publication. As long as the document is preserved, time has its way of solving such problems.

Although I did not know it when I started work, the Trustees were aware of the existence of letters by Einstein, some of them since published,[15] others to be published later, that conflict with the "plaster saint" image they wished to preserve, and—the more serious problem, in my opinion—that might have hurt Margot Einstein, Albert's surviving step-daughter. They took every measure to assure no such letters saw the light of day,[16] but since all were not in their possession, were always afraid one might make an unwelcome appearance. When they finally started to be published, those who might really have been hurt (as opposed to those whose image of the plaster saint would be shattered) were dead, and surviving members of the Einstein family couldn't have cared less.[17] Other letters in the possession of Margot Einstein were placed in the Hebrew University Library under seal for twenty years after her death, but again time will solve that problem. Since I made copies of some of the earlier letters while they were still in the Einstein Archive, including six of which the originals seem to have vanished, this will not even impede publication.

Which brings me to a final lesson that stood me in good stead in the search for additional documents, a lesson that can be learned from a contrary-minded reading of Henry James' story, *The Aspern Papers*.[18] In order to get hold of these papers, love letters of a long-dead American poet to Juliana Bordereau, the narrator of the story introduces himself under false pretenses into the household of the now-aged Miss Bordereau, feigns affection for her spinster niece; and having finally located their hiding place, he tries to steal the papers in the middle of the night. In spite of these reprehensible deeds, we are supposed to feel some sympathy for a man whose better instincts thwart his quest: Caught in the act by the beldam, he confesses the look in her eyes "made me horribly ashamed." After her aunt's death, the niece offers him the letters in exchange for marriage, but he flees; by the time he plucks up courage to return, she has burned them.

The story is meant to be read as (in Leon Edel's words) "a moral fable for historians and biographers," a warning of the dangers of what James calls "literary concupiscence," that is, the obsessive pursuit of needed documents. I read it as a warning of the dangers of allowing the nobler human sentiments to get in the way of such "literary concupiscence." The narrator's failing was to flinch at deeds any good confidence man must be prepared to commit in the normal course of his profession, and what is an editor in search of documents but a confidence man? Miss Bordereau's words when she catches the narrator red-handed provide the historical editing profession with its enduring motto: "Ah you publishing scoundrel!"

Historical Editing

I shall briefly recall four features of the historical editing approach that helped to shape the Einstein edition: comprehensiveness, chronological organization, literality of transcription and annotation as an aid to interpretation.

**Comprehensiveness* The ideal is to include all significant words from the hand (or vocal chords, if recorded) of the subject: writings, published and unpublished, including variations from manuscript to manuscript and/or edition to edition; notebooks and drawings; written accounts of oral presentations; correspondence, incoming as well as outgoing; and such third-party documents (i.e., neither written by nor addressed to the subject) as are felt to be necessary for a fuller understanding of the other documents. Naturally, some selectivity is unavoidable in practice; but the reasons for departures from the ideal should be clearly stated and should conform to well-defined principles.

**Chronological organization* As opposed to an attempt to organize this large body of material by subject matter, for example, it is presented chronologically. This sequence provides a record of the development of the subject's idea and activities that reflects the many complexities, contradictions, gaps and multiplicities inherent in such a genetic record, without attempting to impose the neat patterns that often characterize synthetic accounts of a subject's life and work.

**Literality of transcription* The publication of such a comprehensive edition implies that most scholars will consult it, rather than go to the original manuscripts or editions. These scholarly users—and such an edition is intended primarily for them—will be served best by a literal transcription of manuscript documents. While departures from strict literality are inherent in any printed version of a manuscript, they should be kept to a minimum and the principles on which departure are based should be carefully noted. Such a text will also aid those scholars who need to consult the original manuscripts.

**Annotation as an aid to interpretation* Such an edition is not meant to provide an "authoritative" interpretation of the subject's words and deeds. Rather, it is intended to prove the reader with texts which are indispensible for his or her own interpretation. This requires a certain necessary minimum of annotation of documents in the many instances in which they do not "speak for themselves" to a contemporary reader. Of course, many delicate questions of the borderline between annotation and interpretation arise in the attempt to implement this principle: any comment on a document adds an element of interpretation; however, in cases such as the explanation of a name or geographical reference, the provision of a date, etc., these elements are rather small.

Is the historical editing model appropriate for the works of a scientist like Einstein? A first and obvious answer is that Einstein was more than a scientist. After the 1919 announcement of observation of gravitational displacement of stellar images during a solar eclipse by two English expeditions, Einstein became a world celebrity. More and more a cultural artifact, his image became the subject of a mythical process. He was caught up in a number of political, social and cultural activities. He rubbed shoulders—and corresponded—with many of the leading non-scientific figures of his time. To a large segment of the public and the press he became an oracle, to be consulted on the most varied topics—and rallied to the most varied causes. His replies to such queries are often just as illuminating—and often more so—as his letters to more celebrated correspondents; he took his duties as a public figure quite seriously, even if he occasionally larded his replies with humour.

Even his scientific career in the narrower sense of the word has a public aspect that is of great interest. Rejected by the scientific community as a young man in Switzerland, later taken to its bosom in late Wilhelmine and Weimar Germany, during the last stage of his career in Princeton he came to be revered more as a totem than taken seriously by his fellow physicists. After 1919 he became a living myth and a symbol of science to the wider world. How did the German-speaking and later the international physics community come to terms with this most unusual personality in their midst? How did he come to terms with the dramatic changes in his scientific and social role? One cannot hope to deal with such questions adequately until Einstein's correspondence and related third-party documents become available.

However, even if we leave these so-called "external" aspects of his scientific activities out of account (to use the sometime fashionable but always misleading distinction between "internal" and "external" aspects of the development of science), there are good "internal" reasons why a critical edition of Einstein's papers is needed. Einstein himself stated some of the reasons why we need to investigate its subjective aspects if we are to understand the development of science:

> Science as something already in existence, already completed, is the most objective, impersonal thing that we humans know. Science as something coming into being, as a goal, is just as subjectively, psychologically conditioned as are all other human endeavors... When one speaks of scientific subjects, then the little word "I" should play no role in his exposition. If one speaks, however, about aims and strivings in science, then one should be allowed to treat of himself. For one experiences no other strivings and purposes so immediately as his own.[19]

It was quite characteristic of Einstein's attitude to science, and to his own role in science, that he consented to write a scientific autobiography near the end of his career.[20] He was an exceedingly self-absorbed and self-conscious person. Early in his life he made the search for "a simplified and intelligible picture of the world... the pivot of his emotional life," as he once put it,[21] a tendency that grew with the years. His writings and correspondence are filled with reflections on his own work, as well as the work of other scientists, reflections which contribute a great deal to understanding the development of twentieth-century physics from as near to an insider's view of one of its chief architects as we can get.

There are other ways in which a critical edition of his papers will contribute to an understanding of his science, ways that Einstein did not see so clearly. Einstein included very little information about his personal life in his scientific autobiography, offering the following justification:

> The essential in the being of a man of my type lies precisely in *what* he thinks and *how* he thinks, not in what he does or suffers. Consequently, the obituary can limit

itself in the main to the communicating of thoughts which have played a considerable role in my endeavors.[22]

Yet this begs a crucial question: the problem of trying to understand *how* "a man of my type" came into being. Einstein remarks early in the *Autobiographical Notes* that "today's person of sixty-seven is by no means the same as was the one of fifty, of thirty or of twenty," nor of ten, we may add. Unless one subscribes to a "homunculus" theory of mind, according to which pre-formed traits writ small in the brain of baby Albert slowly but steadily found expression in the course of his life, it is clear that one must look to the events of that life in order to begin to understand how a man of this type was shaped—on the basis of biological dispositions, no doubt—by his experiences. Just *because* his type is so unusual, every scrap of evidence collected about his development is potentially precious. We are far from understanding the normal creative process well enough, let alone that in a scientific genius, to know what evidence will prove significant for future insights into its nature. Therefore, even the personal aspects of Einstein's life, apart from their obvious biographical interest, are of potential scientific importance.

After he became a celebrity, Einstein made great efforts to preserve some privacy, indeed one can say that he sometimes passionately longed for it. However, as discussed earlier, one must recognize that, legitimate as such a desire was during his lifetime, death effectively wipes it out. In contrast to other recent editions,[23] so-called "personal" letters are an integral part of the Einstein edition. It must be noted that the editors cannot make the final decision on whether any material should be excluded: under Einstein's will, the Hebrew University of Jerusalem, as successor to the now-dissolved Estate of Albert Einstein, is so empowered. This power has not been exercised in the case of material that the Estate certainly would have withheld from publication, so there is good reason to expect a very restricted exercise, if any, of this power in the future. The passage of time, again, will make it easier to resolve such questions.

There is another way in which the edition can contribute to an understanding of science that differs from Einstein's. In a passage cited earlier, Einstein spoke of science as "the most objective, impersonal thing that we humans know." Indeed, Einstein's concept of "cosmic religion" was based on his belief in the existence of objective, totally impersonal cosmic laws (or rather one law, since his goal was always ultimate unification of these laws). The aim of the theorist was to uncover these laws, guided by "experiment... an inexorable and not very friendly judge of his work."[24] This view of the scientific enterprise undoubtedly guided and continues to guide the work of most scientists, and has also dominated much past work in history and philosophy of science. Although perhaps not an inevitable consequence, a Whiggish outlook on the development of science, i.e., a tendency to judge all past scientific activities by standards set by currently prevailing theories, is certainly suggested by this view of science.

New currents in the history and philosophy of science are flowing today, strongly influenced by recent sociological studies of the field. It is certainly much too early (assuming that it will ever be possible) to try to make a definitive synthesis, and it would certainly be unjustified to suggest that it will exclude some form of critical-realistic interpretation of the scientific enterprise. Yet it does not seem premature to suggest that it will assign a much more active role to both theorist and experimentalist in the shaping and elaboration of the conceptual structure of scientific theories than do traditional empiricist accounts. Study of the correspondence, unpublished manuscripts, discussion comments, and laboratory notebooks, in addition to the entire corpus of the published writings of key scientists can contribute a great deal to a deeper understanding of the interplay between theorist and experimentalist and how it shapes current as well as past scientific development. While a few scholars will go to the sources, which often include hard-to-decipher manuscript documents and hard-to-get journals, a critical edition of these papers enables many more people to take part in this exciting work.

It has recently been argued that modern technological developments such as electronic archives will make editorial projects like *The Einstein Papers* obsolete. While I strongly favor such projects for the Einstein Archive as well as others, I find it hard to believe that they can replace the historical editing process in the near future. Even with access to a copy of the Einstein Archive, I find myself constantly using the already-published volumes in the course of my own scholarly work for ease of reference, assured transcriptions, helpful annotations, and bibliographical information, to name just a few reasons. Whether an electronic version of the editorial output could replace printed books is another matter. For my generation, I doubt it, but perhaps younger scholars might feel sufficiently at home to dispense entirely with hard copy.

Recent editions of letters as well as published and unpublished writings of key figures in the origins of quantum mechanics have given a glimpse of the rich material available to historians and philosophers of science attempting to reconstruct and understand the development of that theory.[25] *The Einstein Papers* is already contributing and will continue to contribute to an understanding of the development of the "old" and "new" quantum theories, as well as the theory of relativity, modern differential geometry, statistical mechanics, and the search for a unified field theory.

Einstein on His Works

How would Einstein have reacted to such an edition? Of course, it is impossible to really answer such a question, but we can look at the record. Comparatively early in his career, while visiting Japan he sanctioned a Japanese edition of his published scientific papers. In the *Foreword* to that edition, written in 1923, he wrote:

> Our science advances so rapidly that original papers generally lose their current significance very quickly and appear outdated. On the other hand, however, there is always a peculiar fascination in following the development of a theory with the help of the original papers; and it is not rare for such a study to offer a deeper insight into the material than a systematic account of the finished product, polished by the work of many contemporaries.[26]

Thirty years later, however, when a collected Spanish edition of his papers was suggested, Einstein replied:

> It is an intolerable idea for me that all the things that I have written should be republished. You know very well that even those publications which had value at the time of their appearance have lost their current interest with the later progress of science. For this reason the idea is quite repulsive to me.[27]

One can only speculate on what took place in the intervening years to cause such a drastic change in Einstein's attitude towards the republication of his scientific writings; but regardless of Einstein's opinion, interest in these writings has remained strong among physicists, historians of science, philosophers of science and students. Many editions of individual works or collections remain in print in a number of languages, including English—not always in the best possible translations, it must be noted. Collections of his non-scientific and popular-scientific writings are also perennnially reprinted to meet the equally perennial demand.

Einstein also discussed the fate of his correspondence. In 1919, he was asked by Ludwig Darmstadter, founder of the scientific manuscript collection in the Prussian State Library, to give his correspondence to what became the Darmstadter Collection. He replied:

> I cannot bring myself to make publicly available in my lifetime, or that of the writers, letters sent to me personally. I am, however, entirely agreed that after my death, those letters in my possession worthy of preservation be entrusted to your Collection.[28]

The course of German history combined with Einstein's growing attachment to the Zionist cause to change the ultimate repository of his papers from Berlin to the Hebrew University Library in Jerusalem. The provisions of his will, which established an Estate to look after the preservation and use of his manuscripts and correspondence, among other duties, and bequeathed his papers to the Hebrew University after final dissolution of the Estate, testify to his continued recognition of the importance of the fate of his papers.

Two people published editions of their correspondence with Einstein after his death, and several other individual correspondences were published after the death of both parties. Many other letters have been published in whole or in part, in a great many scattered books and periodicals.[29] Yet, taken altogether, the published portion of Einstein's correspondence represents no more than the tip of the iceberg.

The Collected Papers

The edition will include all of Einstein's writings, published and unpublished, that can be located. Variants between manuscripts and published texts, and between various editions printed during Einstein's lifetime are noted. Existing research and lecture notebooks will also be published. Journalistic accounts of speeches will be included, as will transcriptions of the few recordings of his voice. All available letters by Einstein are being printed, as well as significant letters to him, when permission to print can be obtained. Other known letters to him will be summarized, published in excerpt, or calendared.

Third-party documents deemed important for understanding his education, public career or personal life during his early years, from which so little contemporary documentation survives—and about which so many myths abound—were published in the first volume (*The Early Years, 1879–1902*), which includes all non-trivial contemporary references to Einstein in the letters of his family and his future wife, Mileva Marić, that we have been unable to locate. In addition, it contains the relevant portions of an unpublished biographical memoir, covering his youth, by his sister Maja. Extensive annotations draw upon new documentary sources and numerous biographical memoirs by his friends. At the risk of appearing immodest, I will cite the judgement of Highfield and Carter that "publication [of vol. 1] revolutionized our understanding of Einstein's early years."[30]

Later volumes will provide important third-party documents much more selectively, especially for the period after Einstein became world-famous. However interesting and important such a work would be, the edition cannot provide a documentary record of Einstein's influence on the twentieth century.

After the first volume which contains no published writings, the edition divides the *Papers* into two series: *Writings* and *Correspondence*. In addition to letters and third-party documents, the correspondence volumes include chronologies of Einstein's life, and brief biographies of otherwise unknown individuals who started to play a significant role in it, during the period covered. After *The Early Years*, his life has been divided into three periods: *The Swiss Years* (1902–1914), *The Berlin Years* (1914–1933), and *The Princeton Years* (1933–1955). The already-published *Writings* and *Correspondence* from *The Swiss Years* have been edited in parallel, with numerous cross-references between the two series.[31]

Published *Writings* are reproduced in fascimile from the original editions, with variants in manuscripts and later editions footnoted. Literal transcriptions of manuscript documents are given; errors and other peculiarities of spelling, punctuation, grammer, etc., are reproduced without comment, unless the text would be confusing to someone who reads German (but is not necessarily a native-speaker).

Annotation of texts takes several forms: *The General Introduction* and *Introduction to Volume One* provide, respectively, an overall survey of the nature of the edition and of Einstein's early scientific work, relationship with Marić, and the search for his first permanent position. The *Introduction* to vols. 2, 3, and 4 survey the writings included in those volumes, and the *Introduction* to vol. 5 provides a similar survey of the correspondence. Similar introductions are planned for the *Berlin* and *Princeton Years*. Editorial headnotes serve to introduce particular documents, or groups of documents linked by some common theme. For example, in vol. 1 there are headnotes on: *Einstein's First Scientific Essay*, *The Swiss Federal Polytechnical School (ETH)*, *Einstein on Electrodynamics of Moving Bodies*, *Swiss Citizenship*, among other topics. Footnotes offer more detailed information on individual points raised by the texts, and appendices provide additional relevant material. For example, appendices to the first volume provide information about the curriculum of, and texts used in, the schools that Einstein attended, from his Munich Volksschule (public primary school) to the ETH (Swiss Federal Polytechnical School). Wherever possible, information in the annotations concerning Einstein is based on primary sources.

It goes without saying that a critical edition must present all texts in their original languages. Every translation represents an interpretation, if not worse. *Traduttore traditore* goes the Italian saying, and many English and French translations of Einstein's writings could be cited in support of this maxim. However, the facts of life must be faced: no one has the gift of all tongues. *The Einstein Papers* will include translations of texts not in German, French or English.

Ideally, it might have been nice to include canonical English translations of all non-English texts. However, practical considerations such as the delay in publication and increase in cost that this would involve suggested another solution. A separately-edited paperback supplement to each volume is available, providing serviceable translations of non-English texts. Placing the translations in a separate and clearly subsidiary position has two advantages: literal but not literarily polished translations can be used, and it reminds readers that the German text is primary. German was Einstein's language, which he habitually used when he wanted to express himself precisely, so a struggle with the German text, using the translation as a pony, is highly preferable to the illusion that any translation can fully convey the import and all the nuances of Einstein's deceptively simple German.

What sort of new results can be expected from the approach adopted? I shall give a few examples. Over half of the documents included in the first volume were not originally in the Einstein Archive, but were found in the course of a search for new material. These include: a previously unknown exchange of love letters with Marie Winteler, the daughter of his host while he lived in Aarau attending the cantonal secondary school; letters exchanged with Mileva Marić, soon to become his first wife, and her letters to a mutual friend. The majority of the other items in the volume are also previously unpublished. Volume Three includes unpublished lecture notes that Einstein prepared for his first lecture courses at the University of Zürich, and Volume Four includes important unpublished materials such as: the most comprehensive review of the special theory of relativity that Einstein ever wrote, dating from 1912; a research notebook depicting the development of the general theory of relativity during the crucial period when Einstein adopted a non-flat metric tensor field to describe gravitation and began his search for the field equations of general relativity; as well as further notes prepared for his lecture courses on mechanics, statistical mechanics, electricity and magnetism, and relativity. As noted earlier, starting with Volumes One and Five, the vast bulk of his correspondence has started to appear in print for the first time, including important and extensive scientific exchanges with Wien, Lorentz, Ehrenfest and Mach, as well as extremely revealing personal letters to his first and second wives.

Writers of overall biographies of Einstein have already begun to make use of the new material;[32] but my feeling is that it is premature to attempt such an overall re-evaluation of Einstein. On the other hand, with the completion of the series of volumes on *The Swiss Years*, the time is ripe for a comprehensive study of the first half of Einstein's life, from his birth in 1879 to his move to Berlin in 1914. There is always a tendency to read the young Einstein in the light of the ancient sage of legend,[33] and it would be extremely interesting to re-evaluate the first thirty-five years of his life that brought him to the center of the physics community, without making any *a priori* attempt to integrate it with the as-yet insufficiently documented later years of fame, influence and then isolation.

Finally, let me indicate how comparison of Einstein's manuscripts with the standard editions of his writings is beginning to turn up errors in printed texts. This might seem a trivial matter, but such errors are not always insignificant, as shown by the following example: Einstein based his presentation of the special theory of relativity on two principles: the principle of relativity and the principle of the constancy of the velocity of light. The correct interpretation of the latter principle has been the subject of some controversy. One formulation by Einstein, which seemed to be quite different from almost every other one he ever gave, occurs in his article: "What is the theory of relativity," originally written for the *Times* of London in 1919. It has been reprinted innumerable times in various editions of *Mein Weltbild*, and in English translation in *The World As I See It* and *Ideas and Opinions*. In the German editions the statement of the principle reads:[34]

> Das Liche hat in Vakuum stets eine bestimmte Ausbreitungsgeschwindigkeit, unabhängig vom Bewegungszustand und von der Lichtquelle.

The English text provides an example of a translator adding something not in the German text (the addition is placed in parentheses):

> . . .light in vacuo always has a definite velocity of propagation independent of the state of motion (of the observer) or of the source of the light.

None of Einstein's other statements of the principle make reference to the state of motion of the observer, or anything but the state of motion of the light source. Fortunately, a manuscript exists, which reads:[35]

> Das Licht hat in Vakuum stets eine bestimmte Ausbreitungs Geschwindigkeit (unabhängig vom Bewegungszustand der Lichtquelle).

This statement of the light principle does not differ at all from the others, and hence cannot be cited—as has been the mistranslated version of the incorrect text—to support differing interpretations of the meaning of the principle.

Here is a more amusing example of what such comparisons between manuscript and published text can reveal. When the Russian mathematician, Friedmann, first published his solutions to the Einstein gravitational field equations representing expanding cosmological models, Einstein (who then favoured a static model) thought he had found an error in Friedmann's calculations. Einstein published a note to that effect,[36] but was soon convinced that the error was his. He published a retraction of his first note, admitting the complete correctness of Friedmann's result.[37] However, in manuscript he added the following final sentence:[38]

> It follows that the field equations, besides the static solution, permit dynamic (that is, varying with the time coordinate) spherically symmetric solutions for the spatial structure [to which a physical significance can hardly be ascribed].

The final phrase (in brackets) is neatly crossed out in the manuscript, and does not appear in the printed text. Einstein was thus saved from having rejected as unphysical the cosmological model that, within a few years, was to become the basis of modern cosmology!

Examination of the entire corpus of Einstein's work also helps to establish connections between documents that otherwise might easily escape notice. For example, even an attentive reader of a paper on the nature of intermolecular forces that Einstein published in 1911[39] might not realize that in it he silently corrects a fundamental error he made in his first paper, published a decade earlier.[40] Since Einstein generally acknowledged his errors quite freely (as the previous example demonstrates), this oversight may be taken as evidence of how much he wanted—and wanted others—to forget what he referred to in 1907 as 'my two worthless beginner's works.'[41]

Notes

[1] This talk is a revised and updated version of a paper with the same title published in the *British Journal for the History of Science*, Vol. 20, pp. 57–66 (1987).

[2] The project was sponsored jointly by the Estate of Albert Einstein and the Princeton University Press. As provided in Einstein's will, after dissolution of the Estate, the Hebrew University of Jerusalem took over all its rights and responsibilities, including cosponsorship of *The Einstein Papers*.

[3] David R. Chestnutt, ed., *The Papers of Henry Laurens*, vols. 1–10, University of South Carolina Press, 1968–1985.

[4] For a comprehensive discussion of documentary editing, including the differences between the literary and historical editing traditions and methods, see Mary-Jo Kline, *A Guide to Documentary Editing*, Johns Hopkins Press, 1987.

[5] Julian P. Boyd, ed., *The Papers of Thomas Jefferson*, vols. 1–21, Princeton University Press, 1950–1982. The series had reached 1791 when Boyd died. It was continued by Charles Cullen (vols. 22–23) and John Catanzariti (vols. 24–28), Barbara Oberg (vol. 29). The series has now reached 1797 (2001).

[6] Nathan Reingold, ed., *The Papers of Joseph Henry*, vols. 1–8, Smithsonian Institution Press, 1972–1998. Reingold resigned after completing volume five. It was continued by Marc Rothenberg (vols. 6–8).

[7] The other major edition to have been started recently, *The Collected Works of Niels Bohr*, is being edited on rather different principles. In particular, the published papers and selected manuscripts are being issued in a series of volumes organized on a combination of chronological and topical principles, with a selection of related correspondence included. Neither a calendar of the complete correspondence nor an indication of the principles guiding the selection of manuscripts and letters is included.

[8] His edition of the Wilson Papers is one of the most successful examples of the method (Arthur S. Link, ed., *The Papers of Woodrow Wilson*, vols. 1–64, Princeton University Press, 1966–1991). He is the only editor to start and complete an edition of Presidential Papers. (Julian Boyd died before he even reached Jefferson's Presidency!)

[9] Their working methods allowed one to observe the evolution of historical editing from handicraft (Boyd wrote things out), via manufacture (Link used typewriters) to modern

industry (Reingold was introducing computers). *The Einstein Papers* has adopted word processing techniques as they developed, and is now producing camera-ready copy.

[10] I am grateful to Herb Bailey, then Director of the Princeton University Press, for understanding this need and making provision for it in my letter of appointment.

[11] Like *The Einstein Papers*, many historical editing projects are sponsored by groups whose aims generally do not coincide completely with those of the editors, and it is almost inevitable that conflicts of interest, large or small, will arise. When they do, it is always valuable if the editor can make a credible threat to resign based upon an independent scholarly position. Boston University very generously granted me leave for seven years while I worked on the project in Princeton, and allowed me to continue full-time work on it after the move to Boston.

[12] Differences soon arose between Herb Bailey, the Director of the Princeton University Press, and the Trustees of the Einstein Estate (see below) over whether I should continue as sole editor of *The Einstein Papers*, or become part of a troika. The project was at a standstill for a couple of years until this and other issues were settled through arbitration and litigation. For an account of these matters, see Roger Highfield and Paul Carter, *The Private Lives of Albert Einstein.* (London/Boston: Faber and Faber, 1993), Chapter 12.

[13] Here again, as in so many other respects, I benefitted from the example of Nathan Reingold, who earlier made a similar decision.

[14] In the aftermath of the Nixon years, Miss Dukas used to say "I have my enemies list."

[15] See *The Collected Papers of Albert Einstein.* (Princeton: Princeton University Press), vol. 1, John Stachel et al., eds., *The Early Years, 1879–1902.* (1987), vol. 5, Martin Klein et al., eds., *The Swiss Years: Correspondence, 1902–1914.* (1993), vol. 8. Robert Schulmann et al., eds., *The Berlin Years: Correspondence 1914–1918* (1998). The Hebrew University actually forbade publication of some letters in vol. 8.

[16] When Albert's elder son, Hans Albert, and his first wife, Frida Knecht, tried to publish a book of letters exchanged between Albert and Mileva Marić, his first wife and Hans Albert's mother, the Trustees successfully blocked publication of the book (see Highfield and Carter, op. cit., pp. 267–269). Excisions in the published version of *Albert Einstein/Michele Besso Correspondance 1903–1955*, Pierre Speziali, ed., Hermann, 1972, were requested by the Estate.

[17] Generational distance seems to produce emotional distance, at least in Eurocentric cultures. A person who would be upset to learn that his father was a horsethief, will be a bit troubled to learn the same about his grandfather, and when it comes to his great-grandfather . . .

[18] There are actually two slightly different texts of the story. I consulted the 1878 original from *The Turn of the Screw and Other Short Novels*, New American Library, 1962, pp. 153–251; and the 1907–1909 revision from *The Aspen Papers and the Turn of the Screw*, Penguin Books, 1984, pp. 43–142.

[19] From an address to students of UCLA in 1932. The German text and an English translation appear in *Builders of the Universe*, U.S. Library Association, 1932. I have retranslated the German text from pp. 94–96.

[20] The *Autobiographical Notes* were first published in 1949. I shall quote the revised English translation in Albert Einstein, *Autobiographical Notes/A Centennial Edition*, Paul A. Schilpp, ed., Open Court, 1979.

[21] He was referring specifically to Max Planck, but more generally to a certain type of scientist, among whom he clearly included himself. See "Principles of Research," in Albert Einstein, *Ideas and Opinions*, Crown, 1954, pp. 224–227.

[22] *Autobiographical Notes*, op. cit.

[23] The admirable edition of the Pauli correspondence is explicitly denoted "Wissenschaftliche Briefwechsel," and the selected correspondence in the Bohr edition is of a similar nature (for references see note 25).

[24] From an inscription in memory of Kammerlingh-Onnes. See Helen Dukas and Banesh Hoffman (ed.), *Albert Einstein/The Human Side*, Princeton, 1979, p. 18.

[25] See Karl von Meyenn (ed.), *Wolfgang Pauli/Wissenschaftliche Briefwechsel mit Bohr, Einstein, Heisenberg, u.a.*, Springer, 1979; Leon Rosenfeld and Erik Rudinger (eds.), *Niels Bohr/Collected Works*, North-Holland, 1972; W. Blum, H. P. Durr and H. Rechenberg (eds.), *Werner Heisenberg/Gesammelte Werke, Abteilungen A.B.*, Springer, 1984; *Abteilung C.*, Piper, 1984.

[26] The German text of the *Foreword* is published in the Japanese edition. I have translated from a photocopy in the Einstein Archive.

[27] Albert Einstein to David Bohm, 28 October 1954 (Item 8-050 in the Control Index to the Einstein Archive).

[28] Albert Einstein to Ludwig Darmstadter, 29 December 1919 (item 43-509 in the Control Index to the Einstein Archive).

[29] Solovine and Born published their own correspondence: Maurice Solovine (ed.), *Albert Einstein/Lettres à Maurice Solovine*, Gauthier-Villars, 1956; Max Born (ed.), *Albert Einstein/Max Born Briefwechsel 1916–1955*, Nymphenburger, 1969. Other published correspondences include Pierre Speziali (ed.), *Albert Einstein/Michele Besso/Correspondance 1903–1955*, Hermann, 1972; Armin Hermann (ed.), *Albert Einstein/Arnold Sommerfeld/ Briefwechsel*, Schabe & Co., 1968; Robert Debever (ed.), *Elie Cartan–Albert Einstein/Letters on Absolute Parallelism 1929–1932*, Princeton, 1979. For a list of some other published Einstein letters see Bruce R. Wheaton and J. L. Heilbron, *An Inventory of Published Letters To and From Physicists 1900–1950*, Office for History of Science and Technology, Univ. of California, 1982.

[30] Highfield and Carter, op. cit., p. 181.

[31] In addition to vol. 2 cited earlier, see vol. 3, *Writings, 1909–1911*, Martin Klein et al., eds., 1993, vol. 4, *Writings, 1912–1914*, ibid., 1994, and vol. 5, *Correspondence, 1902–1914*, ibid., 1993.

[32] In addition to Highfield and Carter, cited above, another biography that incorporates the newly-discovered material is Albert Fölsing, *Albert Einstein Eine Biographie*, Suhrkamp, 1993. See also Abraham Pais, "In the Shadow of Albert Einstein," in *Einstein Lived Here*, Oxford University Press, 1992.

[33] Or perhaps now, with all the rumors about his love life, I should say the aged satyr of new but equally legendary fashion.

[34] See, for example, Albert Einstein, *Mein Weltbild*, Ullstein, 1981; and Albert Einstein, *Ideas and Opinions*, Crown, 1954; for editions currently in print. The essay in ques-

tion is printed on pp. 127–131 of the German edition, and pp. 227–232 of the English edition.

[35] Item 1-002 in the Control Index to the Einstein Archive.

[36] A. Einstein, "Bemerkung zu der Arbeit von A. Friedmann: *Über die Krummung des Raumes," Zeitschrift fur Physik*, **11** (1922), p. 326.

[37] A. Einstein, "Notiz zu der Bemerkung der Arbeit von A. Friedmann: *Über die Krummung des Raumes," Zeitschrift fur Physik*, **16** (1922), p. 228.

[38] Item 1-026 in the Control Index to the Einstein Archive. The translation is my own.

[39] See *The Einstein Papers*, vol 3, Doc. 12, "Comment on Eötvos's Law," pp. 401–406, especially note 1, p. 406.

[40] See *The Einstein Papers*, vol. 2, Doc. 1, "Conclusions Drawn from the Phenomenon of Capillarity," pp. 9–21, especially note 10, p. 21. For a discussion, see *ibid.*, [Editorial Note], "Einstein on the Nature of Molecular Forces," pp. 3–8, especially p. 8.

[41] See Albert Einstein to Johannes Stark, 7 December 1907, in *The Einstein Papers*, vol. 5, Doc. 66, p. 79.

Introduction to the Guide to the Duplicate Einstein Archive and Control Index

John Stachel

Introduction

The Einstein Archive is housed at the Institute for Advanced Study. At the center of work on the Archives has always been—and hopefully will remain for a long time—Miss Helen Dukas. Starting in 1928 as Prof. Einstein's secretary, she remained with him until his death, and was personally involved in the preparation and handling of most of the documents in the Archive after that date. Her excellent memory constitutes the single most important resource of the Archive; her amiable and helpful manner provide one of the great pleasures of working with it.

After Prof. Einstein's death, as one of the two Trustees of his Estate (the other is Dr. Otto Nathan), she set about the enormous task of annotating, cataloguing, copying and making available to the scholarly community the vast riches of the Einstein *Nachlass*.

Among those who contributed to these efforts over the intervening years one name stands out: that of Prof. Gerald Holton of Harvard University. Over the years since Einstein's death, but especially during the decade of the 1960s, he played a crucial role in the task of ordering, cataloguing, abstracting, and microfilming a major portion of the documents in the Archive, as well as searching for additional documents. I hope that some day he will tell this story himself, but I must here record at least his successful efforts to secure funding, twice from the Rockefeller Foundation and later from the Sources for the History of Quantum Mechanics Project of the American Institute of Physics, to support various aspects of the work of the Archive. Prof. John Wheeler, then at the Princeton Physics

Department, encouraged and supported these efforts from the start. His interest in making the Einstein papers available to the scholarly community has continued down to the present. He not only helped find funding, but also candidates from Princeton graduate students for work on the cataloguing and description of a large part of the collection under the supervision of Miss Dukas and Prof. Holton. Among the products of these efforts were a catalogue of a large part of the scientific correspondence and manuscripts, with a brief description of each item; and an eight-reel microfilm of a portion of the collection. Prof. Martin Klein of Yale later prepared a chronological copy of this catalogue.

In 1971, the Estate of Albert Einstein and the Princeton University Press reached an agreement to publish the Complete Writings of Albert Einstein. Mr. Herbert S. Bailey, Jr., the Director of the Press, had long wanted such an edition as a fitting continuation of Einstein's association with Princeton during the last two decades of his life. His enthusiasm for and devotion to this project have served to move it forward over the decade since, and kept it alive through many difficult moments.

A complete microfilm of the Archive was made under the auspices of the Trustees of the Estate, with the cooperation of the Institute for Advanced Study and the Princeton University Library. The work was carried out by the Princeton Microfilm Corporation at the Institute for Advanced Study under the supervision of Miss Dukas, and resulted in 61 reels of microfilm.

Three copies of the microfilm were made: one was deposited in the Firestone Library of Princeton University; one in the Library of the Institute for Advanced Study; and one was reserved for the Editor of the Einstein Papers.

The Estate and the Press secured a planning grant from the Division of History and Philosophy of Science of the National Science Foundation to enable work to begin on the planning of a complete edition of the writings of Albert Einstein. Work since then has been financed by this grant and its various renewals, by contributions provided by the Estate and the Press, and by a grant from the Sloan Foundation.

I started work on the Project in January 1977. Office space and all its other facilities were made available by the Institute for Advanced Study. I should like to acknowledge warmly the hospitality and help provided by the Institute and its Director, Dr. Harry Woolf, during the three years I worked there. It became clear that a duplicate of the Archive would have to be made, in order to respect the legitimate desire of the Estate that the original documents be handled as little as necessary in the course of the editorial work. It was decided to make a copyflo from the microfilm negative, and then to conform each page of the copyflo to the corresponding page (or pages) of the original document, noting any peculiarities of the original on the copy, and directly copying illegible pages, missing pages, or documents added to the Archive since the microfilming was done. The aim was to construct as exact a duplicate of the Archive as possible, with each document filed in a folder corresponding to the folder in the Archive, to facilitate access to the original should the particular copy prove inadequate for any later editorial purpose.

The copyflo was made, and I started to work on it in mid-1977, with the help of my assistant at that time, Ms. Franny Blasky, and—during the summer of that year—of Mr. Peter Galison, then just graduating from the Harvard University History of Science program. At first, all three of us worked on conforming the copies to the originals. I had direct access to most of the Archive (there was one file kept under combination lock, to which I had access only through Miss Dukas). Ms. Blasky and Mr. Galison received original documents from and returned them to Miss Dukas. However, Dr. Nathan raised objection to this procedure, and insisted that only the Editor have access to the originals. From August 1977, I alone did all conforming of copies to the originals.

While I was doing this, Ms. Blasky, and after her departure in April 1978, my new and most capable assistant, Mrs. Alice Calaprice, gave each page control numbers, and prepared preliminary descriptions for our computerized Control Index. A description of the control numbering system and indexing procedure will be found in the sections: *Control Numbers* and *Control Index* on pp. 12–14.

It was originally estimated, on the basis of the number of documents in the early reels, as well as a "guesstimate" of the total number of frames in the microfilm, that there were about 10,000 documents in the Archive. Much of the early planning was based on this estimate, which proved to be off by a factor of four. We originally had ambitious plans to develop a content as well as a control index, and early decided that the index should be computerized. A visit to Dr. Nathan Rheingold at the Joseph Henry papers in the Smithsonian Institution, who very kindly showed me his system and provided material describing it, helped shape this decision. With the help of Mr. Robert Stachel, my son and then a student at Brandeis University, we worked out a preliminary plan for coding the information for the computer in a very flexible form, which would enable us to employ a relatively large number of untrained people to do the indexing work. We were never able to realize these plans; but the decisions made then influenced the course of the later work.

As it turned out, under the constraints that will be explained in the following paragraphs, it was only possible to prepare a control index, not an index of contents. Moreover, it developed that there were some 42,000 rather than 10,000 documents. This put an unexpected strain on our efforts to complete the work in the designated time.

By September 1977 it had become clear that there was a sharp difference of opinion between the Press and the Estate as to whether I should continue as Editor of the Einstein papers. In order to allow preliminary work to proceed under grants from the National Science Foundation, it was finally agreed that I should be appointed Editor *pro tem* for a term to end in July 1979.

In the autumn of 1978, as the much larger number of documents in the Archive than was first estimated became clear, it was decided that additional help was needed. Mrs. Edith Laznovsky started work on the indexing, assisting Mrs. Calaprice. By then it had also become clear that, with the expiration of my term as Editor *pro tem* in mid-July 1979, access to the original documents might be denied (as indeed proved to be the case). Thus, my first priority became completion of the conforming. Originally, I had hoped to be able to check preliminary descriptions of all documents in the control index made by Ms. Blasky and later by Mrs. Calaprice and Mrs. Laznovsky, while continuing to conform the copies. And indeed, this was done for the first few reels. Now I had to drop that aspect of the work, to assure completion of the conforming by July 14, 1979; and the checking of the control index was put off until afterwards. The duplicate Archive was completely conformed by this deadline, and I was able to start work on checking the index entries. By this time Mrs. Calaprice and Mrs. Laznovsky had completed the preliminary indexing, and we decided to share the rechecking of the entries; I did the first thirty reels, and they divided up the other thirty-one, each checking the other's work. We finished early in January 1980.

The National Science Foundation, in accordance with an agreement between the Press and the Estate, provided funding for a complete copy of the duplicate Archive to be made for the use of scholars. This work was carried out under the able supervision of Mr. John Leypoldt of the Princeton University Library. The work on the copy of the Archive now in the Mudd Library was done by Liliana Margarita Costanza, Chin Hyo Kahng, Sung Ja Kim, Paul Miller, and Jong Oak Kyung. A copy of the Control Index is also available for use with this copy of the Archive.

As will be evident from this history, the control index was originally conceived for use as part of the editorial work of the Einstein Project. Although we have tried to make it as useful as we could to outside scholars, it still bears traces of its original purpose. I am very happy that a copy of the Archive is now available to all interested scholars, together with the Index and this Guide.

In any undertaking involving the indexing of some 42,000 items, there are bound to be errors. In particular, because of the constraints of time and access to the original Archive, we have not been able to do all the checking we would like to have done, or to prepare a content index. We shall be grateful to any users who discover errors if they will (after cussing us out for wasting more of their time) fill out one of the "ERROR REPORT FORMS" available from the Librarian. You have our thanks in advance, and the thanks of the users who follow you.

★ ★ ★★

It remains only to thank again the many people who have worked on the Archive, its duplicates and the index: the help provided to me by Herbert S. Bailey, Jr. and Alice Calaprice stands out especially in my mind; and gratefully to acknowledge the cooperation and support of the Division of History and Philosophy of Science of the National Science Foundation, and later the Sloan Foundation, which have provided funding for this project since 1977. Dr. Ronald Overman, Assistant Program Director for History and Philosophy of Science, and Dr. Betsy Clark, both of the National Science Foundation, and Mr. Steven White of the Sloan Foundation played especially helpful roles.

Grateful acknowledgment is also due to the Princeton University Cyclotron staff, particularly to Prof. Frank Calaprice for offering virtually free use of the computer facility; to Dr. Richard Kouzes for invaluable technical assistance and guidance and for providing the final computer programs for the index; and to the physics graduate students who were so willing to share the computer with Mrs. Calaprice and patiently gave help when needed.

Addendum

As the reader will have seen, this Guide was prepared at a time when the fate of the Einstein Papers Project hung in the balance. The rest of the story is told in *'A Man of My Type'—Editing the Einstein Papers*, but let me note here that the Einstein Archive is now housed in the Jewish National and Hebrew Library, Hebrew University, Jerusalem, where a good deal of work has been done on perfecting and adding to the Control Index. Another duplicate of the Archive was made for the Boston University Library, where it is housed in Special Collections and is open to interested scholars.

Last year (2000), the Einstein Project itself, together with the copy of the Einstein Archive used by the editors, moved to the California Institute of Technology. Interested readers should consult their website:

`http://www.einstein.caltech.edu`

Part III

Surveys of Einstein's Work

Introduction to Einstein: The Formative Years

John Stachel

I.

In an attempt to characterize his scientific work as a whole, Einstein wrote: "The real goal of my research has always been the simplification and unification of the system of theoretical physics" (from the response to a questionnaire submitted to Einstein in 1932; Dukas and Hoffmann 1979, p. 11). Of course, such a sweeping retrospective assertion should be examined carefully before it is accepted; but, as we shall see, Einstein's earliest endeavors do provide considerable evidence of his effort to simplify and unify "the theoretical physicist's picture of the world" (Einstein 1918, p. 108), as he later called it,[1] an effort that came to dominate his outlook on physics. Fred Gregory (this volume, Chapter 2) gives us an insight into one of the popular-scientific sources of this effort, while Mara Beller (this volume, Chapter 4) gives us insights into one of its philosophical sources as well as into the ongoing regulative role it played in Einstein's work.

On occasion, Einstein's sometimes dazzling technical virtuosity was deployed to solve problems unconnected with his fundamental concerns. But by and large, his skills served his passion for a more profound understanding of the physical world. "From youth onwards my entire scientific effort was directed to deepening the foundations of physics" (from a recording made on 6 February 1924, in Herneck 1966). Sahotra Sarkar's description of Einstein's brilliantly simple technique for analyzing Brownian motion in order to provide direct evidence for atomism well illustrates this point (this volume, Chapter 7).

Einstein succinctly stated his criteria for judging a physical theory: "A theory is the more impressive the greater the simplicity of its premises, the more different kinds of things it relates, and the more extended its area of applicability" (Einstein 1979, p. 31). Such an increase in theoretical simplicity and comprehensiveness

Einstein Studies, Vol. 8: *Einstein: The Formative Years*, pp. 1–22

may be the result of any of several factors. First of all, it may result from a reduction in the number of fundamental concepts, or independent relations between these concepts.[2] The young Einstein was impressed, for example, by Hertz's attempt to eliminate the concept of force from the foundations of mechanics, and tried to emulate it in his own work.[3]

It also may result from developing a common theoretical explanation for a number of phenomena that previously appeared to be unrelated. As a student, Einstein was impressed by the quantitative relation between the coefficients of viscosity, heat conduction and diffusion established by the kinetic theory of gases.[4] As early as 1901, in commenting on his first paper, Einstein wrote: "It is a glorious feeling to recognize the unification of a complex of phenomena that appear to direct sense experience as completely separate things" (Einstein to Marcel Grossmann, 14 April 1901, Einstein 1987, Doc. 100).[5]

Finally, the increase may result from the development of a single, coherent theoretical structure, which embraces a number of less general theories that originally were independent of—or even in conflict with—each other. An outstanding example is Maxwell's theory (see Einstein 1979, pp. 30, 32), which subsumed the previously independent theories of electricity, magnetism, and optics; and which Einstein found "the most fascinating subject at the time I was a student" (Einstein 1979, p. 30).

Of course, the search for a unified theoretical basis for all of physics did not originate with Einstein. Three such foundational programs were current during his formative years, each with universal and exclusive pretensions: the traditional mechanical world-view based upon classical mechanics, the energeticist program based upon thermodynamics, and the electromagnetic world-view, based upon Maxwell's theory.[6]

From the time of Newton until the close of the nineteenth century, the prevalent view among physicists was that mechanical concepts ultimately would suffice to explain all physical phenomena, an outlook often referred to as the mechanical world-view.[7] While still quite young, Einstein became acquainted with this through the reading of several popular-scientific works.[8] His first scientific essay, written at the age of sixteen, takes for granted the possibility of a mechanical explanation of electromagnetic phenomena (see Einstein 1987, Doc. 5). He later recalled that, as an ETH student, he was profoundly impressed by the ability of mechanics to provide explanations "in areas that apparently had nothing to do with mechanics." As examples, he cites the mechanical theory of light, the kinetic theory of gases, and the deduction of the laws of thermodynamics from "the statistical theory of classical mechanics" (Einstein 1979, p. 18). However, he adds that Mach's critical discussion of the foundations of mechanics, which he first read around 1897, helped protect him from dogmatic adherence to the mechanical world-view (see Section II).[9]

During the last decades of the nineteenth century, a few physicists and physical chemists claimed that it was possible to derive the laws of mechanics themselves from the principle of conservation of energy. By means of thermodynamical techniques, they hoped to base all of physics on the concept of energy, and hence referred to their program as energetics.[10] Skeptical of all forms of atomism, they were particularly critical of the kinetic theory of gases.[11] Einstein, who had adopted the atomistic outlook and accepted the kinetic theory while still a student (see Einstein 1987, the editorial notes "Einstein on Thermal, Electrical and Radiation Phenomena" and "Einstein on Molecular Forces"), does not appear to have been seriously influenced by this approach,[12] except negatively: The challenge to atomism spurred his search for more direct arguments for the existence of atoms and molecules.[13] The energeticist viewpoint, never widespread, did not long survive the theoretical attacks on its foundations around the turn of the century, and the new experimental evidence for atomism a few years later.[14]

By the time Einstein became acquainted with Maxwell's theory, physicists were gradually becoming accustomed to treating electric and magnetic fields as independent entities, to be employed—at least provisionally—on a par with mechanical ones.[15] When Maxwell formulated his theory, he still accepted the ultimate goal of a dynamical explanation of the electromagnetic field. The lack of any clear-cut success in the numerous attempts to provide such an explanation led by the turn of the century to skepticism about the entire enterprise. H.A. Lorentz's influential interpretation of Maxwell's theory greatly contributed to the acceptance of the field as an independent concept (see, for example, Lorentz 1895). Lorentz distinguished the physics of ponderable matter, composed of particles, some of which carry electric charges, from the physics of the ether, the exclusive seat of electromagnetic fields. The ether, present everywhere—even within matter—is totally immobile, does not interact mechanically with matter, but exerts electric and magnetic forces on charged particles. After the discovery of the electron, a charge carrier much less massive than any atom or ion, there were attempts to explain its inertial mass as only "apparent," that is, entirely due to the behavior of its electromagnetic field. The partial success of such attempts suggested the so-called electromagnetic world-view: the program of explaining all mechanical concepts and laws by means of Maxwell's theory.[16]

Yet however much he aspired to a unified approach, Lorentz was forced to employ a dualistic mixture of mechanical and electromagnetic concepts in order to formulate his theory of electrons. Einstein was presumably recalling his own acceptance at the turn of the century of such a dualistic outlook when he stated in 1914:

> About fifteen years ago, no one yet doubted that a correct account of the electrical, optical and thermal properties of matter was possible on the basis of Galilei-Newtonian mechanics applied to molecular motions and of Maxwell's theory of the electromagnetic field. (Einstein 1914, p. 740)

About the turn of the century, Einstein embarked on several lines of investigation that by 1905 led him to challenge the unrestricted validity of *both* Galilei-Newtonian mechanics and Maxwell's theory. But beyond his detailed criticisms of each of these theories, he was troubled by the dualism inherent in the contrast "between the theoretical concepts that physicists have formed regarding gases and other ponderable bodies and the Maxwell theory of electromagnetic processes in so-called empty space" (Einstein 1905a, p. 132). Einstein did not accept either the mechanical or the electromagnetic world-view as currently formulated; but his striving for simplicity and unification led him to anticipate the ultimate elimination of this dualism.[17] There is a discussion in Section III of his first, tentative proposals for its elimination; here we shall indicate some of his reasons for rejecting the unlimited validity of both Newtonian mechanics and Maxwell's electrodynamics.

Two problem complexes, each connected with the properties of electromagnetic radiation, led to these rejections. Einstein's early attempts to set up a consistent electrodynamics of moving bodies (see the editorial note in Einstein 1987, "Einstein on the Electrodynamics of Moving Bodies," pp. 223–225), were stymied by "the apparent incompatibility of the law of light propagation or rather of Lorentz's theory with the experimentally valid equivalence of all inertial systems" (from a recording made on 6 February 1924, in Herneck 1966, p. 134). In 1905 he succeeded in removing this apparent incompatibility by means of the theory of relativity. While Maxwell's equations are compatible with the principles of the theory, Einstein found that the Newtonian equations of motion for a particle had to be modified.

An even more "revolutionary" (Einstein to Conrad Habicht, 18 or 25 May 1905, Einstein 1993a, Doc. 27, pp. 31–32) challenge to the classical outlook emerged from Einstein's study of blackbody radiation (see the editorial note in Einstein 1987, "Einstein on Thermal, Electrical and Radiation Phenomena," pp. 235–237). Originally "stimulated by Wien's and Planck's investigations" at the turn of the century, by the middle of the decade Einstein had concluded "that mechanics and electrodynamics stand in insoluble contradiction to the facts of experience" (Herneck 1966, p. 134).[18]

Einstein was the first physicist to argue that the fundamental concepts of both mechanics and electrodynamics require drastic modification (see Einstein 1905a, Einstein 1906). But even though he regarded the dualistic conceptual basis he was forced to use as inadequate and provisional, and sought ways to modify the fundamental concepts to bring them into a harmonious unity, he was not disposed to prematurely jettison either the concepts of mechanics or of electrodynamics.[19] He was a master at finding his way among contradictory concepts of limited validity. He recognized that there are broad domains of phenomena, to the understanding of which classic mechanics and/or Maxwell's electrodynamics continue to provide reliable guidance.[20]

So he continued to investigate novel consequences of each theory within its domain of validity. In addition, he discovered ways to take advantage of contradictions between concepts by using one theory within its domain of validity to probe the limits of the other.

II.

Einstein's comments on foundational problems manifest an informed concern, which he (and others) later attributed in good part to his extensive readings in this area, as well as discussions with friends and fellow physicists. While a student, he read two of Ernst Mach's historical-critical studies, the *Mechanik* and *Wärmelehre* (see Einstein to Marić, 10 September 1899, Einstein 1987, Doc. 54, pp. 229–231, especially note 8), at the suggestion of his friend Michele Besso, with whom he discussed foundational questions.[21] In 1909, Einstein wrote to Mach:

> Naturally I know your major works quite well, among which I particularly admire the *Mechanics*. You have had such an influence on the epistemological outlook of the younger generation of physicists, that even your present-day opponents, such as Planck, without doubt would be regarded by one of those physicists, such as they were by-and-large a couple of decades ago, as "Machians." (Einstein to Ernst Mach, 9 August 1909, Einstein 1993a, Doc. 174, p. 204)

Commenting later on the prevailing mechanistic outlook in physics at this time, Einstein stated: "It was Ernst Mach who, in his *History of Mechanics*, upset this dogmatic faith; this book exercised a profound influence upon me in this regard while I was a student" (Einstein 1979, p. 18)[22] The works of the "masters of theoretical physics" (Einstein 1955, p. 146) that he studied during his student years,[23] in particular those of Boltzmann, Helmholtz, Kirchhoff, and Hertz,[24] also contain sophisticated discussions of many foundational and methodological questions. His letters to her suggest that he discussed them with his fellow-student and fiancee, Mileva Marić. After graduation, Einstein appears to have discussed some of these questions with Alfred Kleiner, Professor of Physics at the University of Zurich, who became his *Doktorvater* and whose views may have influenced him.[25]

The wide-ranging "regular reading and discussion evenings" (Einstein to Michele Besso, 6 March 1952, Speziali 1972, p. 464) of Einstein and his friends Maurice Solovine and Conrad Habicht, who banded together in 1902 to form the playfully-named "Olympia Academy,"[26] undoubtedly played a significant role in the development of his ideas on foundational issues.[27] According to Solovine, "The end of the 19th and beginning of the 20th century was the heroic age of research on the foundations and principles of the sciences, and this is what was also our constant preoccupation" (Solovine 1956, p. viii). Solovine gave an extensive list of the readings that formed the basis of their discussions (Solovine 1956, pp. vii–viii).[28] He and Einstein started to read Karl Pearson's *Grammar of Science* (Pearson 1900) before Konrad Habicht joined them. After that, the three of them read and discussed: Mach's *Analyse der Empfindungen*[29] and *Mechanik*,[30] Mill's *Logic*,[31] Hume's *A Treatise of Human Nature*,[32] Spinoza's *Ethica* (Spinoza 1677),[33] unspecified memoirs and lectures by Helmholtz,[34] several chapters of Ampère's *l'Essai sur la philosophie des sciences* (Ampère 1834), Riemann's *Über die Hypothesen, welche der Geometrie zu Grunde liegen* (Riemann 1854), several chapters of Avenarius's *Kritik der reinen Erfahrung* (Avenarius 1888, 1890), Clifford's *On the Nature of Things-in-Themselves*,[35] Dedekind's *Was sind und was sollen die Zahlen?* (Dedekind 1893), and Poincaré's *La science et l'hypothèse* (Poincaré 1902).[36] Solovine describes Poincaré's book as one "that profoundly impressed us and kept us breathless for many weeks" (Solovine 1956, p. viii), and a number of references to Poincaré in Einstein's later writings suggest the influence of this careful reading (see, for example, Einstein 1917, §31, p. 72; Einstein 1921, pp. 122–123; Einstein to Michele Besso, 6 March 1952, Speziali 1972, p. 464).

Einstein's work on the theory of relativity constitutes a striking example of the fruitfulness of his epistemological concerns. The introduction to Einstein 1905b, his first paper on the subject, closes with a methodological comment that reflects these concerns:

> The theory to be developed here—like every other electrodynamics—is based upon rigid-body kinematics, since the assertions of any such theory concern relations between rigid bodies (coordinate systems), clocks and electromagnetic processes. Insufficient consideration of this circumstance is the root of the difficulties, with which the electrodynamics of moving bodies currently has to contend (Einstein 1905b, p. 892).

By directing Einstein's attention to the problem of the formation of scientific concepts (including the concepts of space and time), the role of conventions in scientific thought, and the place of formal principles in the structure of scientific theories, to name only a few relevant topics, his readings and discussions helped prepare him for the task of revising the kinematical foundations of physics. Discussing "the axiom of the absolute character of time, or rather of simultaneity," Einstein later stated:

> To recognize clearly this axiom and its arbitrary character already implies the essentials of the solution of the problem. The type of critical reasoning required for the discovery of this central point was decisively furthered, in my case, especially by the reading of David Hume's and Ernst Mach's philosophical writings (Einstein 1979, p. 50).[37]

Elsewhere, in citing readings of the Olympia Academy that were "rather influential in my development," Einstein listed Poincaré together with Hume and Mach (Einstein to Besso, 6 March 1952, Speziali 1972, p. 464).

III.

Einstein's work during the period discussed in this volume may be classified under three broad headings:

(1) Efforts to extend and perfect the classical (Galilei-Newtonian) mechanical approach, including its statistical extensions (his work on molecular forces, the foundations of statistical physics, molecular dimensions, and Brownian motion). The papers of Sahotra Sarkar and Jürgen Renn deal in detail with this aspect of Einstein's work.

(2) Efforts to extend and perfect Maxwell's electrodynamics, as interpreted in Lorentz's electron theory, while modifying classical mechanics to make it cohere with electrodynamics (his work on the theory of relativity and the electrodynamics of moving media). Robert Rynasiewicz's paper deals in detail with this aspect of Einstein's work.

(3) Demonstrations of the limited validity of both classical mechanics and Maxwell's electromagnetic theory and efforts to comprehend phenomena that cannot be explained on the basis of these theories (his work on the quantum hypothesis). John Stachel's other paper discusses some aspects of this work.

Here, I shall outline Einstein's work in each of these categories. The works published through 1909 are all reprinted in Einstein 1989, and I include references to the very useful headnotes in that volume.

(1) *Efforts to Extend and Perfect the Classical-mechanical Approach*

By the turn of the century, Einstein was already occupied with the problems that were to take him beyond classical physics (see Section I). Yet his first five papers, published between 1901 and 1904, fall within the framework of Galilei-Newtonian mechanics and kinetic-molecular extensions of it. In the first two, Einstein attempted to explain several apparently disparate phenomena in liquids and solutions with the help of a simple hypothesis on the nature of intermolecular forces and their variation with the chemical composition of molecules (see Einstein 1989, editorial note, "Einstein on the Nature of Molecular Forces," pp. 3–8). Einstein hoped that his work might help decide a long-standing conjecture about a common basis for molecular and gravitational forces—an indication of his strong ambition, at the outset of his career, to contribute to the unification of physics.

Einstein made extensive use of thermodynamical arguments in the two papers, and such arguments played an important role in all his early work (see Klein 1967). The second paper raises a question about the relation between the thermodynamic and kinetic-molecular approaches to thermal phenomena, which is answered in the following paper, the first of three, published between 1902 and 1904, devoted to the atomistic foundations of thermodynamics (see Einstein 1989, editorial note, "Einstein on the Foundations of Statistical Physics," pp. 41–55). Einstein attempted to formulate the minimal assumptions about a mechanical system needed to derive the basic concepts and principles of thermodynamics. Presumably because he derived it from such general assumptions, he regarded the second law as a "necessary consequence of the mechanical world-view" (Einstein 1902, p. 432). He derived an equation for the mean square energy fluctuations of a system in thermal equilibrium that, in spite of its mechanical origins, involves only thermodynamical quantities. In his first mention in print of the topic, Einstein proceeded to apply this equation to black body radiation, the only system for which it was clear to him on what length scale such fluctuations should be significant. The application suggests that Einstein may have been attempting to treat radiation as a mechanical system.

In his doctoral dissertation, submitted in 1905, Einstein used classical-hydrodynamical methods to obtain an estimate of the size of large molecules in solution (see Einstein 1989, editorial note, "Einstein's Dissertation on the Determination of Molecular Dimensions," pp. 170–182). His papers on Brownian motion, the first of which appeared in the same year, also extend the scope of applicability of classical mechanical concepts (see Einstein 1989, editorial note, "Einstein on

Brownian Motion," pp. 206–222). Einstein noted that, if the kinetic-molecular theory of heat is correct, the laws of thermodynamics cannot be universally valid, since fluctuations must give rise to microscopically visible violations of the second law. Indeed, such fluctuations provide an explanation of the well-known Brownian motion of microscopic particles suspended in a liquid. Einstein regarded his work as establishing the limits of validity, within which thermodynamics could be applied with complete confidence (see Einstein 1907c, p. 373).

(2) *Efforts to Extend and Perfect Maxwell's Electrodynamics and Modify Classical Mechanics to Cohere with It*

Well before 1905, Einstein apparently was aware of a number of experiments suggesting that the mechanical principle of relativity (the equivalence of all inertial frames for the description of mechanical phenomena) should be extended to optical and electromagnetic phenomena. However, such an extension was in conflict with what he regarded as the best current electrodynamical theory, Lorentz's electron theory, which grants a privileged status to one inertial frame: the ether rest frame. In 1905 Einstein succeeded in resolving this conflict through a critical analysis of the kinematical foundations of mechanics, which also underlie any theory of electrodynamics (see above). This analysis led to the recognition that the Lorentz transformations are kinematical in nature, and hence, any acceptable physical theory must be invariant under these transformations. Maxwell's equations, suitably reinterpreted after eliminating the concept of the ether, meet this requirement; but Newton's equations of motion need to be revised. Thus, Einstein's work on the theory of relativity provides an example of his use of one theory—Maxwell's electrodynamics—to find the limits of validity of another—Galileo-Newtonian mechanics, even though he was already aware of the limited validity of the former (see 3 below).

One of the major accomplishments of Einstein's approach, which his contemporaries found difficult to appreciate, is that relativistic kinematics is independent of the theories that impelled its formulation. Einstein had not only developed a coherent kinematical basis for both mechanics and electrodynamics, but (leaving aside the problem of gravitation) for any new physical theories that might be introduced. To use terms that he employed later, Einstein had created a theory of principle, rather than a constructive theory.[38] At the time he expressed the distinction in these words:

> One is in no way dealing here ... with a "system," in which the individual laws would implicitly be contained and could be found merely by deduction therefrom, but only with a principle that (in a way similar to the second law of thermodynamics) permits the reduction of certain laws to others. (Einstein 1907b, p. 207)

The principles of such a theory, of which thermodynamics is his prime example, are universal assertions based upon a large amount of empirical data; they do not purport to constitute explanations of the phenomena on which they are based. In contrast, constructive theories, such as the kinetic theory of gases, do attempt to explain some class of phenomena, such as the gas laws on the basis of hypothetical entities employed to construct the explanation, such as atoms in motion. It is well known that elements of the distinction between principle and constructive theories occur in Poincaré's writings (see, for example, Poincaré 1904b).[39] Two other sources that may have inspired Einstein's emphasis on the role of principles in physics are Violle[40] and Kleiner.[41]

In spite of the merits of the theory of relativity, however, Einstein was of the opinion that

> a physical theory can only be satisfactory, if its structures are composed of *elementary* foundations. The theory of relativity is just as little ultimately satisfactory as, for example, classical thermodynamics was before Boltzmann had interpreted the entropy as probability. (Einstein to Arnold Sommerfeld, 14 January 190, Einstein 1993a, Doc. 73, pp. 86–89)[42]

Einstein's 1905 paper considered Maxwell's equations only with convection currents in empty space. The problem of formulating macroscopic, relativistically invariant equations for electromagnetic fields in moving media, including conduction currents, was first discussed by Minkowski in 1907 (see Minkowski 1908). In 1908, Einstein worked on this problem in collaboration with Jakob Laub (see Einstein 1989, editorial note, "Einstein and Laub on the Electrodynamcs of Moving Media," pp. 503–507).

(3) *Demonstrations of the Limited Validity of Both Classical Mechanics and Maxwell's Electromagnetic Theory and Attempts to Comprehend Phenomena that Cannot Be Explained by these Theories*

Einstein's efforts to perfect classical mechanics and Maxwell's electrodynamics, and to make both theories compatible, may still be regarded, in the broadest sense, as extensions of the classical approach to physics. However original his contributions in these areas, however revolutionary his conclusions about space and time appeared to his contemporaries, however fruitful his work proved to be for the exploration of new areas of physics, he was still engaged in drawing the ultimate consequences from conceptual structures that were well established by the end of the nineteenth century. What is unique about his stance during the first decade of this century is his unwavering conviction that the concepts of classical mechanics and those of Maxwell's electrodynamics, even after his modifications of them, are incapable of explaining a growing list of phenomena involving the behavior and interactions of matter and radiation. Einstein constantly reminded his colleagues of the need to introduce radically new concepts to explain the structure of both matter and radiation. He introduced some of these new concepts himself, notably the quantum hypothesis, even though he was unable to integrate them into a coherent theory (see Einstein 1989, editorial note, "Einstein's Early Work on the Quantum Hypothesis," pp. 134–148).

His first paper on the quantum hypothesis provides a striking example of his style, mingling critique of the old concepts with the search for new ones. It opens by demonstrating that the equipartition theorem together with Maxwell's equations leads to a definite formula for the black body radiation spectrum, now known as the Rayleigh-Jeans distribution. This distribution, which at low frequencies matches the empirically-validated Planck distribution, cannot hold at all frequencies, since it implies a divergent total energy. (He soon gave a similar argument based on the equipartition theorem, demonstrating that classical mechanics cannot explain the thermal or optical properties of a solid, represented as a lattice of atomic or ionic oscillators; see Einstein 1907a.)

Einstein next investigated the high frequency (Wien) region, where the classically-derived distribution breaks down most dramatically. In this region, he showed, the entropy of a sample of monochromatic radiation with a fixed temperature depends on its volume in exactly the same way as does the entropy of a sample of a gas composed of statistically independent particles, no matter what their law of motion. In short, monochromatic radiation in the Wien limit behaves thermodynamically as if it were composed of mutually independent quanta of energy, each with an energy proportional to its frequency. Emboldened by this parallel, he took the final step, proposing the "very revolutionary" hypothesis that matter and radiation can only interact through the exchange of such energy quanta. He demonstrated that this hypothesis can be used to explain a number of apparently disparate phenomena involving the interaction of matter and radiation, notably the photoelectric effect.

In 1905, Einstein did not use Planck's distribution law to explore the nature of radiation. The following year, he showed that Planck's derivation of this law implicity depends on the assumption that charged oscillators can have only energies that are integral multiples of the quantum of energy, and hence only exchange energy with the radiation field in such quanta. In 1907, Einstein argued that uncharged oscillators should be similarly quantized, thereby explaining both the success of the DuLong-Petit law for most solids at ordinary temperatures, and the anomalously low values of the specific heats of certain substances. He related the temperature at which departures from the DuLong-Petit law become significant to the fundamental frequency of the atomic oscillators, and hence to the optical absorption spectrum of the solid (see Einstein 1907a).

In spite of his conviction of its fundamental inadequacy, Einstein continued to manipulate still-reliable aspects of classical mechanics with remarkable skill to explore further the structure of electromagnetic radiation. In 1909 he applied his theory of Brownian motion to a two-sided mirror immersed in thermal radiation. He showed that the mirror would be unable to carry out a Brownian motion indefinitely, if the fluctuations in the radiation pressure on its surfaces were solely due to the effects of random waves, as predicted by Maxwell's theory. Only the presence of an additional term, corresponding to pressure fluctuations due to the impact of random particles, guarantees the mirror's continued Brownian motion. Einstein showed that similar fluctuation terms in the energy are consequences of Planck's law. He regarded such fluctuation phenomena as his strongest argument for ascribing physical significance to the hypothetical light quanta.

Einstein was far from regarding his work on the quantum hypothesis as constituting a satisfactory theory of radiation or matter. As noted above, he emphasized that a physical theory is satisfactory only "if its structures are composed of *elementary* foundations," adding "that we are still far from having satisfactory elementary foundations for electrical and mechanical processes" (Einstein to Arnold

Sommerfeld, 14 January 1908, Einstein 1993a, Doc. 73, pp. 86–89). Einstein felt that he had not reached a real understanding of quantum phenomena because (in contrast to his satisfactory interpretation of Boltzmann's constant as setting the scale of statistical fluctuations) he had been unable to interpret Planck's constant "in an intuitive way" (ibid.). The quantum of electric charge also remained "a stranger" to theory (see Einstein 1909a, p. 192). He was convinced that a satisfactory theory of matter and radiation should *construct* the quanta of electricity and of radiation, not simply *postulate* them.[43]

As a theory of principle (see above), the theory of relativity provides important guidelines in the search for such a theory. Einstein anticipated the ultimate construction of "a complete world-view that is in accord with the principle of relativity" (Einstein 1907c, p. 372). In the meantime, the theory offered clues to the constructions of such a world-view. One clue concerns the structure of electromagnetic radiation. Not only is the theory compatible with an emission theory of radiation, it requires that radiation transfer mass between an emitter and an absorber, reinforcing Einstein's hypothesis that radiation manifests a particulate structure under certain circumstances. He maintained that "the next phase in the development of theoretical physics will bring us a theory of light, which may be regarded as a sort of fusion of the undulatory and emission theories of light" (Einstein 1909b, pp. 482–483). Other principles that Einstein regarded as reliable guides in the search for an understanding of quantum phenomena are the conservation of energy and Boltzmann's principle.[44]

Einstein anticipated that: "The same theoretical modification that leads to the elementary unit [of charge] will also lead to the quantum structure of radiation as a consequence" (Einstein 1909a, pp. 192–193). In 1909 he made his first known attempt to find a field theory that would explain both the structure of matter (the electron) and of radiation (the light quantum). After investigating relativistically invariant, non-linear generalizations of Maxwell's equations, he wrote:

> I have not succeeded . . . in finding a system of equations, which I could see was suited to the construction of the elementary quantum of electricity and the light quantum. The manifold of possibilities does not seem to be so large, however, that one need draw back in fright from the task. (Einstein 1909b, p. 193)[45]

This attempt may be regarded as the forerunner of his later search for a unified field theory of gravitation and electromagnetism, which would also explain quantum phenomena.

Einstein's attempt to incorporate gravitation into the theory of relativity led him to recognize a new formal principle, the principle of equivalence, which he interpreted as demonstrating the need to extend the relativity principle if gravitation is to be included in its scope (see Einstein 1989, editorial note, "Einstein on the Theory of Relativity," pp. 253–274).[46] He found that, when gravitational effects are taken into account, it is impossible to maintain the privileged role that the inertial frames of reference still hold in the original relativity theory (see Einstein 1907d, p. 454). He started to search for a group of transformations wider than the Lorentz group, under which the laws of physics would remain invariant when gravitation is included, a search that lasted until the end of 1915 and led to what he considered his greatest achievement, the general theory of relativity.

By 1909 then, Einstein at the age of 30 had already embarked on the search for a relativistic theory of gravitation, and begun the search for a theoretical foundation for quantum phenomena and a unified theory of matter and radiation, searches that were to occupy him for the rest of his life.

Notes

[1] For a discussion of the significance of the "Weltbild" concept for Einstein, see Holton 1981.

[2] "A theory obviously only gains scientific value through having simpler fundamental assumptions, i.e., of lesser multiplicity, than its consequences that are comparable with experience" (Einstein 1915, p. 254). Later, he spoke of "use of a minimum of primary concepts and relations."

[3] See Einstein 1979, pp. 28, 30. Einstein regarded a similar "elimination of the concept of force" as a major accomplishment of one of his first papers (Einstein to Michele Besso, 17 March 1903, Einstein 1993a, Doc. 7, p. 18).

[4] See, for example, Einstein 1987, Doc. 18. Einstein commented on the relation that the kinetic theory established between these quantities in 1901 (see Einstein to Mileva Mirać, 15 April 1901, Einstein 1987, Doc. 101). In Einstein 1915, pp. 256–258, he discussed the three phenomena in some detail.

[5] This quotation shows a striking resemblance to the statement by Kant quoted by Beller (this volume, p. 93).

[6] For a survey of these three programs, see Jungnickel and McCormmach 1986, Chap. 24, pp. 212–245.

[7] The term is used in Einstein 1902, p. 432. For his later account of the mechanical world-view, see Einstein 1979, pp. 16–30. For a discussion of the status of the mechanical outlook at the end of the nineteenth century, see Klein 1972.

[8] There is evidence that he read Alexander von Humboldt's *Kosmos*, Ludwig Büchner's *Kraft und Stoff*, and Aaron Bernstein's *Naturwissenschaftliche Volksbücher*. See Einstein 1987, editorial note on "Einstein's First Scientific Essay," pp. 5–6. For Aaron Bernstein, see Fred Gregory's article in this volume.

[9] Einstein presumably read the third edition of Mach's *Mechanik*, Mach 1897 (see Einstein 1987, p. 230, note 8). For his rereading of the *Mechanik*, see Section II, p. 125 and note 30 below.

[10] For a detailed survey of the energetics movement and its critics, see Deltete 1983.

[11] Not all critics of atomism were energeticists. Mach's critical stance, for example, was rooted in his general distrust of concepts not directly accessible to experience.

[12] In 1913, he wrote in praise of Planck's 1896 essay against the energeticists (Planck, 1896), "in which it is shown that energetics is worthless as a heuristic method, indeed, that it even operates with untenable concepts" (Einstein 1913, p. 1077).

[13] He does not mention energetics in his reminiscences, although he does refer to the existence of skepticism about the reality of atoms (see Einstein 1979, pp. 44, 46), and to his goal, "to find facts that would guarantee as much as possible the existence of atoms of definite finite size" (Einstein 1979, p. 44).

[14] For a discussion of the theoretical attacks, see Deltete 1983. For a discussion of the experimental studies, see Nye 1972.

[15] For Einstein's later account of the challenge to the mechanical world-view presented by Maxwell's theory, see Einstein 1979, pp. 30–34. The German quotation in the text is from p. 30, the translation from p. 31.

[16] For a discussion of this program, see McCormmach 1970.

[17] Einstein 1907c discusses the defects of "our current electromechanical world-view," and looks forward to "a comprehensive world-view, corresponding to the relativity principle." See Holton 1981 for a discussion of Einstein's "Weltbild."

[18] For a more detailed later account by Einstein of the crisis in the foundations of physics initiated by Planck's studies of thermal radiation, see Einstein 1979, pp. 34–42. For his reaction to this crisis, see pp. 42–44. For Einstein's early study of blackbody theory, see Einstein 1989, editorial note, "Einstein's Early Work on the Quantum Hypothesis," pp. 134–148.

[19] In 1909 he noted "that it will only be a matter of a modification of our present theories, and not a complete abandonment of them" (Einstein 1909a, p. 192).

[20] See Einstein 1907c, pp. 372–373, for Einstein's comments on the range of applicability of Maxwell's equations. In the *Autobiographical Notes*, he commented: "The success of the theory of Brownian motion showed again conclusively that classical mechanics always led to trustworthy results whenever it was applied to motions in which the higher time derivatives of the velocity are negligible" (Einstein 1979, p. 46).

[21] For Besso's recommendation of Mach's books, see Einstein to Mileva Mirać, 10 September 1899, Einstein 1987, Doc. 54, pp. 229–231, note 8. For evidence of a discussion with Besso, see Einstein to Marić, 4 April 1901, Einstein 1987, Doc. 96, pp. 284–286.

[22] Einstein probably read the third edition, Mach 1897, while a student. A copy was in his library until Helen Dukas gave it to Nathan Rosen. A more specific acknowledgment of Mach's role in the development of the theory of relativity is cited later in this section. For discussions of the relationship between Einstein and Mach, see Herneck 1966 and Holton 1968.

[23] Einstein speaks of studying the masters "with sacred enthusiam."

[24] Einstein mentions his study of Kirchhoff, Helmholtz, and Hertz in his *Autobiographical Notes* (Einstein 1979, p. 14). For contemporary evidence of his reading of these authors, as well as of Boltzmann's *Gastheorie*, see Einstein 1987: Einstein to Marić, 10 August 1899 (Doc. 52), 10 September 1899 (Doc. 54), 1 August 1900 (Doc. 69), 13 September 1900 (Doc. 75), and 15 April 1901 (Doc. 101).

[25] For Einstein's discussions with Kleiner, see Einstein to Mileva Marić, 19 December 1901, Einstein 1987, Doc. 130, pp. 328–329. For a possible influence of Kleiner's views on Einstein see note 42 below.

[26] The earliest mention of the Academy is in a postcard of 30 November 1903 from Einstein to "Herr Dr. Conrad Habicht missing member of the A[cademy] of S[cience] [+] O[lympia]" (Einstein 1993a, Doc. 15, p. 24).

[27] For reminiscences of Einstein, which include a discussion of the Olympia Academy, see Solovine 1956, pp. v–xiii.

[28] I mention the works in the order in which Solovine lists them. He indicates that they read many other works on similar themes, in addition to a number of literary works. In a

letter of 14 April 1952 to Carl Seelig (SzZE Bibliothek, Hs 304: 1006), Solovine includes Poincaré 1905 in an otherwise similar, but less complete list of the readings of the Olympia Academy.

[29] Three editions came out within a few years: Mach 1900, Mach 1902, and Mach 1903.

[30] Solovine adds: "which Einstein already had gone through previously" (Solovine 1956, p. viii). Einstein probably read the third edition, Mach 1897, while he was a student. The Olympia Academy members may have used the fourth or fifth editions, Mach 1901, Mach 1904.

[31] Solovine notes: "Book III of Mill's Logic concerning induction also received our attention for a considerable period" (Solovine 1956, p. viii). The 8th ed., Mill 1872, was translated into German twice: Mill 1877, Mill 1884–1887.

[32] Solovine states: "For weeks we discussed the singularly sagacious critique of the notions of substance and of causality made by David Hume" (Solovine 1956, p. viii). For Einstein's comments on Hume see the text and note 38 below.

[33] Einstein presumably used one of the German translations: Spinoza 1887, Spinoza 1893.

[34] Helmholtz's scientific papers are published in three volumes: Helmholtz 1882, 1883, 1895 (the epistemological papers are included in Helmholtz 1883). His more popular writings are published in Helmholtz 1884. The fact that the Academy members read Riemann 1854 (see below) suggests that they also read Helmholtz's writings on the foundations of geometry, such as Helmholtz 1868.

[35] Clifford's essay was issued as a book in German, Clifford 1903, which Einstein presumably used.

[36] Einstein may have used the German edition, Poincaré 1904a, which contains extensive additional notes by Ferdinand Lindemann.

[37] In 1915, Einstein wrote to Schlick: "You have also correctly seen that this current of thought [positivism] had a great influence on my efforts, and indeed E. Mach and even more Hume, whose Treatise on Understanding I studied with fervor and admiration shortly before the discovery of the theory of relativity. It is very well possible that without these philosophical studies I would not have arrived at the solution" (Einstein to Moritz Schlick, 14 December 1915 (Einstein 1998, Doc. 165, p. 220). The reference to "Treatise on Understanding" is a conflation of the titles of two of Hume's works. Solovine's memoir suggests that it was the *Treatise on Human Nature* that Einstein read (see p. 7). See also Einstein to Michele Besso, 6 January 1948, and 6 March 1952, p. 7 (Speziali 1972, p. 464). In the latter, Einstein stated that he read Hume "in a quite good German edition," presumably Hume 1895, a translation of the first part of Hume 1739.

[38] For the distinction between theories of principle and constructive theories, see Einstein 1919. He later reminisced about the origins of the special theory: "Gradually I despaired of the possibility of discovering the true laws by means of constructive efforts based on known facts. The longer and the more desperately I tried, the more I came to the conviction that only the discovery of a universal formal principle could lead us to assured results. The example I saw before me was thermodynamics" (Einstein 1979, p. 48). For several years after 1905, Einstein referred to the relativity principle, rather than to the theory of relativity.

[39] For a study of the physics of principles in Poincaré, and its antecedents, see Giedymin 1982, chap. 2, "The Physics of Principles and its Philosophy: Hamilton, Poincaré and Ramsey," pp. 42–89.

[40] The German edition of Violle's textbook, Violle 1892, which Einstein studied (see Einstein 1987, p. lxiv), characterizes physical principles as "originally hav[ing] developed from experience, but now occupy[ing] a much higher position; they are a type of postulate, the correctness of which is proved only by the continually perceived reliability of the conclusions to which they lead" (p. 90).

[41] Alfred Kleiner, with whom Einstein had been in contact since 1901, when he began to discuss physics with him (see Einstein to Mileva Marić, 19 December 1901, Einstein 1987, Vol. 1, Doc. 130), argued for the importance of such general principles as guides in the search for new theories, especially where commonly accepted constructive hypotheses encounter seemingly insurmountable obstacles (Kleiner had in mind primarily difficulties with mechanical models of the ether). In 1901 he wrote: "It must be emphasized that, in the extension of the sciences, we are directed to the application of general principles, if we want to ... go beyond the sphere of the momentary state science has reached." He cited as examples of such principles: "The law of the constancy of matter, the principle of conservation of energy, and second law of thermodynamics, the ether hypothesis, etc." (Kleiner 1901, pp. 131–132).

[42] A decade later, Einstein elaborated this idea, introducing the concept of "constructive theories": "These attempts to construct a picture of more complex phenomena from a relatively simple underlying formalism. ... When we say that we have succeeded in understanding a group of natural processes, we always mean by this that a constructive theory has been found, which embraces the processes in question" (Einstein 1919).

[43] "A satisfactory theory should in my opinion be so constructed that the electron appears as a solution, so that external fictions are not needed, in order not to have to accept that its electrical charges fly apart" (Einstein to Sommerfeld, 14 January 1908, Einstein 1993a, Doc. 73, pp. 86–89). This is presumably a reference to the Poincaré stresses.

[44] See Einstein 1911 for a discussion of these two principles.

[45] This attempt seems to represent Einstein's first step toward a field ontology.

[46] His analogy between relativity theory and thermodynamics as theories of principle may have prepared him to accept that, like the second law of thermodynamics, the original principle of relativity has its limits. [See pp. 191–214, this volume].

References

Ampère, André-Marie (1834). *Essai sur la philosophie des sciences ou exposition analytique d'une classification naturelle de toutes les connaissances humaines*. Paris: Bachelier.

Avenarius, Richard (1888). *Kritik der reinen Erfahrung*. Vol. 1. Leipzig: O.R. Reisland.

Avenarius, Richard (1890). *Kritik der reinen Erfahrung*. Vol. 2. Leipzig: O.R. Reisland.

Clifford, William Kingdon (1903). *Von der Natur der Dinge an sich*. Hans Kleinpeter, trans. and ed. Leipzig: Johann Ambrosius Barth.

Dedekind, Richard (1893). *Was sind und was sollen die Zahlen*? 2nd ed. Brauhnschweig: Friedrich Vieweg und Sohn.

Deltete, Robert John (1983). "The Energetics Controversy in Late Nineteenth-Century Germany: Helm, Ostwald and Their Critics." 2 vols. Ph.D. dissertation. Yale University.

Dukas, Helen, and Hoffman, Banesh, eds. (1979). *Albert Einstein, the Human Side: New Glimpses from his Archives*. Princeton: Princeton University Press.

Einstein, Albert (1902). "Kinetische Theorie des Wärmegleichgewichtes and des zweiten Hauptsatzes der Thermodynamik." *Annalen der Physik* 9: 417–433. Reprinted in Einstein 1989, Doc. 3, pp. 57–73.

—— (1905a). "Über einen die Erzeugung und Verwandlung des Lichtes betreffende heuristischen Gesichtspunkt." *Annalen der Physik* 17: 132–148. Reprinted in Einstein 1989, Doc. 14, pp. 150–166.

—— (1905b). "Zur Elektrodynamik Bewegter Körper." *Annalen der Physik* 17: 891–921. Reprinted in Einstein 1989, Doc. 23, pp. 276–306.

—— (1906). "Zur Theorie der Lichterzeugung und Lichtabsorption." *Annalen der Physik* 20: 199–206. Reprinted in Einstein 1989, Doc. 34, pp. 350–357.

—— (1907a). "Die Plancksche Theorie der Strahlung und die Theorie der Spezifischen Wärme." *Annalen der Physik* 228: 180–190. Reprinted in Einstein 1989, Doc. 38, pp. 379–389.

—— (1907b). "Bemerkungen zu der Notiz von. Hrn. Paul Ehrenfest: 'Die Translation deformierbarer Elektronen und der Flächensatz." *Annalen der Physik* 23: 206–208. Reprinted in Einstein 1989, Doc. 44, pp. 410–412.

—— (1907c). "Über die vom Relativitätsprinzip geforderte Trägheit der Energie." *Annalen der Physik* 23: 371–384. Reprinted in Einstein 1989, Doc. 45, pp. 414–427.

—— (1907d). "Über das Relativitätsprinzip and die aus demselben gezogenen Folgerungen." *Jahrbuch der Radioaktivität and Elektronik* 4: 411–462. Reprinted in Einstein 1989, Doc. 47, pp. 433–484.

—— (1909a). "Zum gegenwärtigen Stand des Strahlungsproblems." *Physikalische Zeitschrift* 10: 185–193. Reprinted in Einstein 1989, Doc. 56, pp. 542–550.

—— (1909b). "Über die Entwickelung unserer Anschauungen über das Wesen und die Konstitution der Strahlung." *Deutsche Physikalische Gesellschaft. Verhandlungen* 11: 482–500. Reprinted in *Physikalische Zeitschrift* 10 (1909): 817–825 and in Einstein 1989, Doc. 60, pp. 564–582.

—— (1911). L'état actuel du problème des chaleurs spécifiques." In *La théorie du rayonnement et les quanta. Rapports et discussions de la réunion tenue à Bruxelles, du 30 octobre au 3 novembre 1911. Sous les auspices de M. E. Solvay*. P. Langevin and M. de Broglie, eds. Paris: Gauthier-Villars, pp. 407–435; German text is published in Einstein 1993b, Doc. 26, pp. 521–543.

—— (1913). "Max Planck als Forscher." *Die Naturwissenschaften* 1: 1077–1079. Reprinted in Einstein 1995, Doc. 23, pp. 561–563.

—— (1914). "Antrittsrede des Hrn. Einstein." *Königlich Preussische Akademie der Wissenschaften* (Berlin). *Sitzungsberichte*: 739–742. Reprinted in Einstein 1996, Doc. 3, pp. 20–23.

—— (1915). "Theoretische Atomistik." In *Die Kultur der Gegenwart. Ihre Entwicklung und ihre Ziele*. Paul Hinneberg, ed. Part 3, sec. 3, vol. 1, Physik. Emil Warburg, ed. Leipzig: B.G. Teubner, pp. 251–263. Reprinted in Einstein 1995, Doc. 20, pp. 521–533.

—— (1917). *Über die spezielle und die allgemeine Relativitätstheorie. (Gemeinverständlich).* Braunschweig: Friedrich Vieweg und Sohn. Reprinted in Einstein 1996, Doc. 42, pp. 421–507.

—— (1918). "Motiv des Forschens." Cited from the version published under the title: "Prinzipien der Forschung." In *Mein Weltbild*, Carl Seelig, ed. Zürich: Europa Verlag, 1981, pp. 107–110.

—— (1919). "Time, Space and Gravitation." *Times* (London). 28 November, p. 13.

—— (1921). *Geometrie und Erfahrung. Erweiterte Fassung des Festvortrages gehalten an der Preussischen Akademie der Wissenschaften zu Belin am 27. Januar 1921.* Berlin: Julius Springer. Cited from the reprint in: *Mein Weltbild.* Carl Seelig, ed. Zürich: Europa Verlag, 1981, pp. 119–127.

—— (1955). "Erinnerungen-Souvenirs." *Schweizerische Hochschulzeitung* 28 (*Sonderheft*): 143–153.

—— (1979). *Autobiographical Notes. A Centennial Edition.* P.A. Schilpp, ed. La Salle, Ill.: Open Court, 1979.

—— (1987). *The Collected Papers of Albert Einstein.* Vol. 1, *The Early Years, 1879–1902.* John Stachel et al., eds. Princeton: Princeton University Press.

—— (1989). *The Collected Papers of Albert Einstein.* Vol. 2, *The Swiss Years: Writings, 1900–1909.* John Stachel et al., eds. Princeton: Princeton University Press.

—— (1993a). *The Collected Papers of Albert Einstein.* Vol. 5, *The Swiss Years: Correspondence, 1902–1914.* Martin J. Klein et al., eds. Princeton: Princeton University Press.

—— (1993b). *The Collected Papers of Albert Einstein.* Vol. 3, *The Swiss Years: Writings, 1909–1911.* Martin J. Klein et al., eds. Princeton: Princeton University Press.

—— (1995). *The Collected Papers of Albert Einstein.* Vol. 4, *The Swiss Years: Writings, 1912–1914.* Martin J. Klein et al., eds. Princeton: Princeton University Press.

—— (1996). *The Collected Papers of Albert Einstein.* Vol. 6. *The Berlin Years: Writings, 1914–1917.* A. J. Kox et al., eds. Princeton: Princeton University Press.

—— (1998). *The Collected Papers of Albert Einstein.* Vol. 8. *The Berlin Years: Correspondence, 1914–1918.* Robert Schulmann et al., eds. Princeton: Princeton University Press.

Giedymin, Jerzy (1982). *Science and Convention: Essays on Henri Poincaré's Philosophy of Science and the Conventionalist Tradition.* Oxford and New York: Pergamon Press.

Helmholtz, Hermann von (1868). "Ueber die Thatsachen, die der Geometrie zum Grunde liegen." *Königliche Gesellschaft der Wissenschaften zu Göttingen. Nachrichten.* No. 9 (3 June): 193–221. Reprinted in Helmholtz 1883, pp. 618–639.

—— (1882). *Wissenschaftliche Abhandlungen.* Vol. 1. Leipzig: Johann Ambrosius Barth.

—— (1883). *Wissenschaftliche Abhandlungen.* Vol. 2. Leipzig: Johann Ambrosius Barth.

—— (1884). *Vorträge und Reden.* 2 vols. Braunschweig: Friedrich Vieweg und Sohn.

—— (1895). *Wissenschaftliche Abhandlungen.* Vol. 3. Leipzig: Johann Ambrosius Barth.

Herneck, Friedrich (1966). "Zwei Tondokumente Einsteins zur Relativitätstheorie." *Forschungen und Fortschritte* 40: 133–135.

Holton, Gerald (1968). "Mach, Einstein, and the Search for Reality," *Daedelus* 97: 636–673.

—— (1981). "Einstein's Search for the *Weltbild*," *Proceedings of the American Philosophical Society* 125: 1–15.

Hume, David (1739). *A Treatise of Human Nature: Being an Attempt to Introduce the Experimental Method of Reasoning into Moral Subjects*. Book 1, *Of the Understanding*. London: John Noon.

—— (1895). *Ein Traktat über die menschliche Natur*. Part 1, *Über den Verstand*. E. Köttgen, trans., Theodor Lipps, ed. Hamburg: Leopold Voss.

Jungnickel, Christa, and McCormmach, Russell (1986). *Intellectual Mastery of Nature: Theoretical Physics from Ohm to Einstein*. Vol. 1, *The Torch of Mathematics 1800–1870*. Vol. 2, *The Now Mighty Theoretical Physics* 1870–1925. Chicago: University of Chicago Press.

Klein, Martin (1967). "Thermodynamics in Einstein's Thought." *Science* 157: 509–516.

—— (1972). "Mechanical Explanation at the End of the Nineteenth Century." *Centaurus* 17: 58–82.

Kleiner, Alfred (1901). "Über die Wandlungen in den physikalischen Grundanschauungen." In *Verhandlungen der Schweizerischen Naturforschenden Gesellschaft bei ihrer Versammlung zu Zofingen den 4., 5. und 6. August 1901 (84. Jahresversammlung)*. Zofingen: P. Ringier, pp. 3–31.

Lorentz, Hendrik Antoon (1895). *Versuch einer Theorie der electrischen und optischen Erscheinungen in bewegter Körpern*. Leiden: E.J. Brill.

Mach, Ernst (1897). *Die Mechanik in ihrer Entwicklung. Historisch-kritisch dargestellt*. 3rd ed. Leipzig: F.A. Brockhaus.

—— (1900). *Die Analyse der Empfindungen und das Verhältnis des Physischen zum Psychischen*. 2nd enl. ed. Jena: Gustav Fischer.

—— (1901). *Die Mechanik in ihrer Entwicklung. Historisch-kritisch dargestellt*. 4th ed. Leipzig: F.A. Brockhaus.

—— (1902). *Die Analyse der Empfindungen und das Verhältnis des Physischen zum Psychischen*. 3rd enl. ed. Jena: Gustav Fischer.

—— (1903). *Die Analyse der Empfindungen und das Verhältnis des Physischen zum Psychischen*. 4th enl. ed. Jena: Gustav Fischer.

—— (1904). *Die Mechanik in ihrer Entwicklung. Historisch-kritisch dargestellt*. 5th ed. Leipzig: F.A. Brockhaus.

McCormmach, Russell (1970). "H.A. Lorentz and the Electromagnetic View of Nature." *Isis* 61: 459–497.

Mill, John Stuart (1872). *A System of Logic Ratiocinative and Inductive: Being a Connected View of the Principles of Evidence and the Methods of Scientific Investigation*. 2 vols. 8th ed. London: Longmans, Green, Reader, and Dyer.

—— (1877). *System der deductiven und inductiven Logik. Eine Darlegung der Principien wissenschaftlicher Forschung, insbesondere der Naturforschung*. 4th ed. J. Schiel, trans. Braunschweig: Friedrich Vieweg und Sohn.

—— (1884–1887). *System der deductiven und inductiven Logik. Eine Darlegung der Principien wissenschaftlicher Forschung, insbesondere der Naturforschung*. 3 vols. 2nd ed. Theodor Gomperz, trans. Leipzig: Fues.

Minkowski, Hermann (1908). "Die Grundgleichungen für die elektromagnetischen Vorgänge in bewegter Körpern." *Königliche Gesellschaft der Wissenschaften zu Göttingen. Mathematische Klasse. Nachrichten*: 53–111.

Nye, Mary Jo (1972). *Molecular Reality: A Perspective on the Scientific Work of Jean Perrin*. London: Macdonald.

Pearson, Karl (1900). *The Grammar of Science*. 2nd ed. London: Adam & Charles Black.

Planck, Max (1896). "Gegen die neuere Energetik." *Annalen der Physik und Chemie* 57: 72–78.

Poincaré, Henri (1902). *La science et l'hypothèse*. Paris: E. Flammarion.

—— (1904a). *Wissenschaft und Hypothese*. Ferdinand and Lisbeth Lindemann, trans. Annotations by Ferdinand Lindemann. Leipzig: B.G. Teubner.

—— (1904b). "L'état actuel et l'avenir de la physique mathématique." *Bulletin des sciences mathématiques* 28: 302–324.

—— (1905). *La valeur de la science*. Paris: E. Flammarion.

Riemann, Bernhard (1854). "Ueber die Hypothesen, welche der Geometrie zu Grunde liegen." *Königliche Gesellschaft der Wissenschaften und der Georg-Augusts-Universität* (Göttingen). *Mathematische Classe*. Abhandlungen 13 (1867): 133–152. Lecture, Göttingen, 10 June 1854.

Solovine, Maurice, ed. and trans. (1956). *Albert Einstein: Lettres à Maurice Solovine*. Paris: Gauthier-Villars.

Speziali, Pierre, ed. (1972). *Albert Einstein-Michele Besso. Correspondance, 1903–1955*. Paris: Hermann.

Spinoza, Baruch (1677). "Ethica ordine geometrico demonstrata." In *Opera postuma*. Amsterdam: J. Rieuwertsz., pp. 1–264.

—— (1887). *Die Ethik*. J. Stern, trans. Leipzig: Reclam.

—— (1893). *Benedict von Spinoza's Ethik*. 5th ed. J.H. von Kirchmann, trans. Berlin: Philosophisch-Historischer Verlag.

Violle, Jules (1892). *Lehrbuch der Physik*. German edition by E. Gumlich et al. Part 1, *Mechanik*. Vol. 1, *Allgemeine Mechanik und Mechanik der festen Körper*. Berlin: Julius Springer.

The Other Einstein: Einstein Contra Field Theory

John Stachel

Besides the well-known advocate of unified field theories, there was "another Einstein," who was skeptical of the continuum as a foundational element in physics. This paper presents evidence for the existence of this "other Einstein," and of the debate between the two Einsteins that lasted most of Einstein's life.

I

Albert Einstein is often taken by physicists and philosophers of science as the exemplar par excellence of the field-theoretical outlook. On the level of theoretical physics, this point of view implies that the ultimate goal of physics is some sort of unified field theory; particles and their properties are to be derived in one way or another from fields and their properties.[1] On the level of philosophy, the field viewpoint implies that the ultimate ontology is a field ontology: particles and their properties can be reduced to fields and their properties.

At first glance, the major themes of Einstein's lifework seem to fall within such a program, in both its physical and its philosophical aspects. On the basis of the special theory of relativity (erroneously, as it happens), Einstein concluded that direct-interaction theories of particles without the mediation of fields are no longer tenable once one recognizes the existence of a maximum signal velocity (see, e.g., Einstein 1949, 61). Einstein's first attempt at a unified field theory, around 1909, was a nonlinear theory of the electromagnetic field, from which he hoped to derive the properties of both matter (electrons) and radiation (light quanta, later photons) (see Einstein 1989, Doc. 60, 581–82). The general theory of relativity provides a field-theoretical account of gravitation; moreover, through its incorporation of the principle of general covariance, it seems to demand that any future physical theory that includes gravitation be built on the foundation of the space-time continuum. Finally, one remembers Einstein's almost forty-year-long search for a unified field theory of gravitation and electromagnetism, which

Reprinted from *Science in Context*
Vol. 6, 1993, pp. 275–290

he hoped would also provide an explanation of quantum phenomena that would undercut the foundational pretensions of quantum mechanics.

One might object that atomism was also a major theme in Einstein's lifework—both classical atomism as embodied in his work on Brownian motion, for example, and what we may call "quantum atomism" as embodied in his proposal of the light quantum hypothesis, his contribution to the development of Bose–Einstein statistics, and so on. The themes of field and atom do form an opposing pair in the work of Einstein, as Gerald Holton has justly emphasized ([1972] 1988). Nevertheless, in another respect they form a unity: both continuous field and discontinuous atom are grounded in the space-time continuum as their home. Opposed as they appear to be in one sense, in this deeper sense neither member of the pair really escapes from the continuum. One member—the field—takes the entire space-time continuum as its *actual* home; the other member—the atom—though confined in each particular realization to a timelike world tube, takes the entire space-time continuum as its *potential* home. While each member of the pair has universal pretensions, they have sustained a dualistic coexistence—and competition for hegemony—in the space-time continuum for a century and a half, from Faraday's time until our own. Einstein was well aware of this unifying role of the continuum: "Physics up to now is naturally in its essence a continuum physics, in spite of the use of the material point, which looks like a discontinuous conceptual element" (Einstein to H. S. Joachim, 24 August 1954, as translated in Stachel 1986, 380–81 [a fuller extract from this important letter is given below]).

Beyond the opposition between field-continuum and atom-discontinuum, with both unified in the space-time continuum, there lies the possibility of an even more profound, truly exclusive opposition: the opposition between a physics based in one way or another on the space-time continuum and a physics that is not based at all on a continuum. To give a more positive name to such a noncontinuum physics, I shall follow Einstein's lead and refer to it as a purely algebraic physics (see below).

Before going into the opposition between continuum physics and purely algebraic physics, let me first suggest that the reason why Einstein more or less identified continuum physics with the field alternative of the field-atom opposition was that his work revolutionized the concept of continuum physics.

Before Einstein's work on the general theory of relativity, the space-time continuum could be taken as the stage upon which the drama of physics was enacted, to use a by now hackneyed metaphor. Be the play comedy or tragedy, in verse or prose, be the scenery elaborate or minimalist, the stage was unchanged. As the stage, space-time had a profound influence on the drama of physics enacted upon it; but the drama itself left the stage unaltered. In general relativity, the stage itself becomes a part of the action of the drama, changing with the play. This interaction between space-time and the rest of physics comes about because the fundamental field in general relativity, the space-time metric, plays a dual role in the theory: in addition to determining the geometrical and chronometrical properties of the world, the metric tensor field plays the role of the gravitational field potentials. As such, they are coupled to all other aspects of physics through the gravitational field equations. As John Wheeler likes to put it: matter and energy here curve space and time there. So much is recognized by most people familiar with general relativity, and it is a remarkable enough departure from the traditional role of space and time in prerelativistic physics.

Yet Einstein drew an even more remarkable conclusion from the general covariance of his theory: in "empty" regions of space-time (i.e., regions where there is nothing but gravitational field), the spatio-temporal aspects of the world, including even the physical identity of its points (events), are no more than properties of the metrical-gravitational field. That is, if there were no such field, there would be no events—and *a fortiori* no manifold of events—no anything (for further discussion, see Stachel 1987b, 1989, 1991a, 1991b). To continue with my hackneyed metaphor, the play of the gravitational field not only *modifies* the stage—without the play, there would *be* no stage!

Suppose one adopts this point of view about the general theory of relativity, as Einstein certainly did; and suppose one believes that future physics will continue to employ the space-time continuum as a fundamental element, as Einstein often did. Then, if further progress in physics is to incorporate the dynamical view of space-time inherent in the general theory, it is hard to see how it can escape from the field viewpoint. How else could one express general covariance? Of course, one could just abandon the dynamical view of the space-time structure, and return to the pre-general-relativistic concept of this structure as a given, nondynamical one. Indeed, this is the route that (special-)relativistic quantum field theory has taken. I need hardly add that to Einstein such an abandonment represented not progress but a singularly dangerous regression.

> You consider the transition to special relativity as the most essential thought of relativity, not the transition to general relativity. I consider the reverse to be correct. I see the most essential thing in the overcoming of the inertial system, a thing that acts upon all processes but undergoes no reaction. This concept is in principle no better than that of the center of the universe in Aristotelian physics. (Einstein to Georg Jaffe, 19 January 1954, as translated in Stachel 1986, 377)

Were I addressing a community of physicists, I would discuss reasons why a physicist today might be inclined to accept or reject Einstein's viewpoint (see Stachel 1991a). But in this context and since our subject is Einstein, the important point to emphasize is that this *was* undoubtedly Einstein's final viewpoint about the progress of physics. *If* the space-time continuum is to remain a fundamental element of future physics, some sort of generally-covariant field-theoretical generalization of the general theory of relativity is required. As mentioned above, he devoted almost forty years of his life to the search for such a generalized theory.

I am sure you caught the caveat in my last remark: "*If* the space-time continuum is to remain a fundamental part of future physics." This caveat brings me, finally, to the subject of this paper: "The Other Einstein." If one looks at Einstein's work carefully, including his published writings but particularly his correspondence and reminiscences of conversations with him, one finds strong evidence for the existence of *another* Einstein, one who questioned the fundamental significance of the space-time continuum. His skepticism seems to have deepened over the years, resulting late in his life in a profound pessimism about the field-theoretical program, even as he continued to pursue it.

But now let us return to the beginning of his career and, on the basis of admittedly fragmentary evidence, try to discern how his skepticism about the continuum may have arisen. Although he used both field and atom in his work, Einstein was always uncomfortable with the current dualism in physics and believed that one would ultimately have to choose between the two. Although atomism grew up within the context of the continuum, Einstein came to believe that, if one were to give up the field concept in favor of atomism, it should be a radical atomism that renounced the continuum as well. So I shall begin with his early doubts about the field concept, doubts first expressed as skepticism about the ether, and then in his light quantum hypothesis.

II

Our first glimpse of Einstein's views on physics occurs in an essay sent to his favorite uncle in 1895 (see Einstein 1987, Doc. 5, 6–9). The essay proves that at sixteen he believed unquestioningly in the mechanistic concept of the ether, alterations in the elastic state of which constitute electric and magnetic fields. Our next glimpse of his views comes in a letter written four years later to his fellow physics student and fiancée Mileva Marić; it shows a dramatically-changed outlook. He had clearly studied Maxwell's theory in the interval and, in particular, read Heinrich Hertz' classic accounts of electrodynamics (really more reorganizations than expositions of Maxwell's own version of the theory). He wrote:

> I am more and more convinced that the electrodynamics of moving bodies, as currently presented, is not correct, and that it should be possible to present it in a simpler way. The introduction of the term "ether" into theories of electricity and magnetism leads to the notion of a medium of whose motion one can speak without, I believe, being able to associate any physical meaning with such a statement. I believe that electrical forces are directly definable only for empty space, as Hertz also emphasizes. Furthermore, electrical currents are not to be regarded as "vanishing of electric polarization in time," but rather as motions of real electric charges. . . . Electrodynamics would then become the theory of the movements of moving electric and magnetic charges in empty space. Which of the two conceptions would have to be adopted, would have to result from experiments on radiation. (*Ibid.*, Doc. 52, 225–27)

I shall leave aside discussion of the obvious implications of this letter for Einstein's work on the electrodynamics of moving bodies, work that culminated six years later in the special theory of relativity (for some comments, see Stachel 1987a). However, it is worth noting that in this first extant letter in which Einstein discusses his ideas about physics he already uses a mode of thought that recurs over and over again in his work: the simultaneous consideration of two different conceptions of some physical phenomenon, emphasizing the contrast between—or even mutual exclusivity of—these conceptions.[2] In this case, it is two mutually exclusive conceptions of the nature of electric currents that he emphasizes. One conceives of them as displacement currents in some medium,[3] while the other conceives of them as convective currents—that is, "motions of real electric charges... in empty space." The first point of view requires a material medium for such currents, whereas the second needs only "empty space." Einstein's inclination toward the second point of view—known today as the Lorentz interpretation of Maxwell's theory—led him to suggest at this early date that the ether may be a superfluous element in electrodynamics, which might better be formulated exclusively in terms of charged particles moving in empty space. In view of the then-prevailing identification of the electromagnetic field with the state of the ether, regarded as a "medium of whose motion one can speak," is it going too far to suggest that in 1899 there were already hints of Einstein's skepticism about the field concept?

I know of no further evidence bearing on this question until his two 1905 papers on quanta and relativity (see Einstein 1989, Doc. 14, 149–69, and Doc. 23, 275–310, respectively). In the relativity paper, we find a by-now definite rejection of the ether concept. Perhaps even more relevant, the quantum paper contains his demonstration that the Maxwell field concept, together with the equipartition theorem, leads to an untenable conclusion: the infinite energy content of a unit volume of radiation in thermal equilibrium at any finite temperature. In the very opening lines of the paper, Einstein suggests that this paradoxical conclusion may arise from the field concept itself, which ascribes an infinite number of degrees of freedom to the electromagnetic field within any finite region of space, however small—an ascription he contrasts with the finite number of degrees of freedom associated with a system of particles in a finite volume, however large. Even though he is well aware of its limitations, he is bold enough to suggest a "heuristic viewpoint": under certain circumstances, radiation can be treated as if it consists of a sort of gas of independent particles, which he calls light quanta.

Over the next decade his views on the nature of these light quanta fluctuated: are they fundamental entities, or must they be constructed as solutions to nonlinear field equations? Indeed, whether or not they exist at all is a question on which he expressed contrary opinions at different times. Yet he does seem to have held firmly to one idea that he expressed in 1910 in the following words: "Without the ether, energy continuously distributed in space seems to me an absurdity" (Einstein 1993, Doc. 3). I shall not go into the details of his changing views, nor into the reasons that finally led him after 1915 to accord full citizenship rights to light quanta (see Pais 1982, sec. VI; Stachel 1986). A constant element in all these endeavors is the continuum associated with the space-time structure of the special theory of relativity, so they provide no evidence that Einstein had transcended the field-atom opposition.

So far, the first known evidence of such an advance is in a letter of 1916 to Walter Dällenbach, a former student of Einstein's at the ETH:

> You have correctly grasped the drawback that the continuum brings. If the molecular view of matter is the correct (appropriate) one; i.e., if a part of the universe is to be represented by a finite number of moving points, then the continuum of the present theory contains too great a manifold of possibilities. I also believe that this "too great" is responsible for the fact that our present means of description miscarry with the

> quantum theory. The problem seems to me [to be] how one can formulate statements about a discontinuum without calling upon a continuum (space-time) as an aid; the latter should be banned from the theory as a supplementary construction not justified by the essence of the problem, [a construction] which corresponds to nothing "real." But we still lack the mathematical structure unfortunately. How much have I already plagued myself in this way!

Further passages from this truly remarkable letter will be cited and discussed in section IV. But let us first pause to consider the following question. His own words, as well as the sharpness and clarity of Einstein's formulations, give evidence of considerable prior thought about the problem of the continuum. What could have stimulated Einstein to so "plague" himself with this question? By the end of 1912, Einstein had begun that phase of his work on the general theory of relativity that led him to the novel conclusion, discussed above, that the space-time structure is not independent of the gravitational field. Perhaps his work on the general theory, focusing his attention on the space-time manifold and its properties, combined with his long-standing concern with a fundamental understanding of the quantum, first stimulated Einstein to consider the possibility of eliminating the continuum completely. On the other hand, Einstein's references to the atomic structure of matter in general and to quantum phenomena in particular in support of his skeptical views on the continuum suggest that he might have started to think about these questions well before his work on general relativity, in connection with his work on the quantum problem. However plausible these speculations may be, it seems likely that the origins of Einstein's preoccupation with the problem of the continuum are to be found even earlier, as is suggested in the following section.

III

Around 1902, Einstein undertook extensive readings and discussions on the foundations of science with his fellow members of the mock "Olympia Academy," Maurice Solovine and Paul Habicht. Thanks to Solovine's reminiscences, we have a rather extensive list of their readings.[4] From this list it is clear that, long before his work on general relativity, the three friends discussed a number of works that could have drawn Einstein's attention to the contrast between mathematical continua and discreta; works that also make clear the significance of this distinction for the problem of physical space. Not only Solovine but Einstein himself testifies to their lengthy discussions of Hume's *Treatise of Human Nature*.[5] Part 2 of the *Treatise*, "Of the Ideas of Space and Time," opens with two sections that argue against the infinite divisibility of both our notions of space and time and of space and time themselves.[6] Hume summarizes these arguments in Section 4:

> Our system concerning space and time consists of two parts, which are intimately connected together. The first depends on this chain of reasoning. The capacity of the mind is not infinite; consequently no idea of extension or duration consists of an infinite number of parts or inferior ideas, but of a finite number, and these simple and indivisible: 'Tis therefore possible for space and time to exist conformable to this idea: And if it be possible, 'tis certain they actually are conformable to it; since their infinite divisibility is utterly impossible and contradictory.

> The other part of our system is a consequence of this. The parts, into which the idea of space and time resolve themselves, become at last indivisible; and these indivisible parts, being nothing in themselves, are inconceivable when not filled with something real and existent. The ideas of space and time are therefore no separate or distinct ideas, but merely those of the manner or order, in which objects exist. (Hume 1969, 88)

The point is not how convincing Einstein would have found the first part of Hume's argument (although I believe he must have found the second part, concerning the relational nature of space and time, quite convincing), but whether it might have stimulated him to question the necessity of treating space and time as continua.

Two other books on Solovine's Olympia Academy reading list also discuss the question of the infinite divisibility of space. Karl Pearson's *The Grammar of Science* uses the distinction between the limited divisibility of perceptible objects and unlimited possibilities for division of conceptual objects such as geometrical space to criticize Hume's treatment of the question. Poincaré's *La science et l'hypothèse* contains a lengthy discussion of what he calls the physical continuum and the mathematical continuum, and the relation between the two.

More directly relevant to our theme, however, are two other Olympia readings, Riemann's "Ueber die Hypothesen, welche der Geometrie zu Grunde liegen" (1854) and Dedekind's *Was sind und was sollen die Zahlen*?, that explicitly distinguish between geometries based respectively on continua and on non-continua. Einstein's statement that in 1912 he was not yet acquainted with "Riemann's and Ricci's or Levi-Civita's work" (Einstein 1921) is in apparent conflict with Solovine's testimony that the Academy members read Riemann's "Ueber die Hypothesen, welche der Geometrie zu Grund liegen." It is possible to reconcile the two assertions by assuming—as the association of Riemann with Ricci and Levi-Civita suggests—that Einstein was referring to the mathematically more technical aspects of Riemann's paper, whereas the Olympia Academy had confined its discussions to its more philosophical aspects. At any rate, the indisputable fact is that by 1916 Einstein had read Riemann 1854, so I shall cite the relevant passages. After noting that: "Notions of quantity are possible only where there exists already a general concept which allows various modes of determination," Riemann distinguishes between "a continuous or a discrete manifold" depending on whether "there is or is not found among these modes of determination a continuous transition from one to another" (Riemann 1854, 273, as translated in Smith [1929] 1959, 412). He notes that space provides an example of a continuous, threefold extended manifold, and goes on to discuss continuous manifolds with applications to space. The following passage occurs near the end of the essay:

> The question of the validity of the postulates of geometry in the indefinitely small is involved in the question concerning the ultimate basis of relations of size in space. In connection with this question, ... the above remark is applicable, namely that while in a discrete manifold the principle of metric relations is implicit in the notion of manifold, it must come from somewhere else in the case of a continuous manifold. Either then the actual things forming the groundwork of space must constitute a discrete manifold, or else the basis of metrical relations must be sought for outside that actu-

> ality, in colligating forces that operate upon it. (Riemann 1854, 285–86, as translated in Smith [1929] 1959, 424–25)

Leaving aside the well-known anticipation of some aspects of general relativity in this passage, it clearly raises the alternatives continuum-discretum, later explored by "the two Einsteins."

In addition to Solovine's testimony, there is good evidence that Einstein must have read Dedekind 1893 at some point. In 1921, Einstein started to use the phrase "freie Schöpfungen des menschlichen Geistes" (free creations of the human mind) to characterize axioms and other concepts, a usage that recurs often thereafter in his writings.[7] This is just the phrase that Dedekind uses to describe numbers in the foreword to the first edition of *Was sind und was sollen die Zahlen*? However, there is a passage in the foreword that is much more significant for present concerns.

> For a great part of the science of space the continuity of its configuration is not even a necessary condition. ... To explain this matter more clearly I note the following example: If we select three non-collinear points A, B, C at pleasure, with the single limitation that the ratios of the distances AB, AC, BC are algebraic numbers, and regard as existing only those points M, for which the ratios of AM, BM, CM to AB are likewise algebraic numbers, then the space made up of points M, as is easy to see, is everywhere discontinuous: but in spite of this discontinuity, and despite the existence of gaps in this space, all constructions that occur in Euclid's *Elements* can, so far as I can see, be just as accurately effected as in perfectly continuous space; the discontinuity of this space would not be noticed in Euclid's science, would not be felt at all. If anyone should say that we cannot conceive of space as anything else than continuous, I should venture to doubt it and to call attention to the fact that a far advanced, refined scientific training is demanded in order to perceive clearly the essence of continuity and to comprehend that besides rational quantitative relations, also irrational, and besides algebraic, also transcendental quantitative relations are conceivable. (Dedekind 1893, xii–xiii, as translated in Dedekind [1901] 1963, 37–38)

It is even possible that Einstein's use of the phrase "algebraic physics" harks back to Dedekind's example of a non-continuous Euclidean geometry based on the algebraic numbers,[8] although it must be admitted that there are many other possible sources for this term.

IV

Just when Einstein started to consider the problem of "how one can formulate statements about a discontinuum without calling upon a continuum (space-time) as an aid" is an intriguing question. Even more intriguing is the problem itself, which still confronts anyone trying to set up a purely algebraic physics. It is remarkable that Einstein formulated the issue so sharply in 1916. In the second paragraph of the letter to Dällenbach cited above, we find the two Einsteins debating:

> Yet I see difficulties of principle here too. The electrons (as points) would be the ultimate entities in such a system (building blocks). *Are there* indeed such building

blocks? Why are they all of equal magnitude? Is it satisfactory to say: God in his wisdom made them all equally big, each like every other, because he wanted it that way; if it had pleased him, he could also have created them different. With the continuum viewpoint one is better off in this respect, because one doesn't have to prescribe elementary building blocks from the beginning. Further, the old question of the vacuum! But these considerations must pale beside the overwhelming fact: The continuum is more ample than the things to be described. (Einstein to Walter Dällenbach, November 1916, as translated in Stachel 1986, 379)

As noted earlier, in 1909 the "continuum Einstein" had already discussed the possibility of constructing particles (electrons and light quanta) as solutions to nonlinear equations for the electromagnetic field (see his Salzburg lecture, reproduced in Einstein 1989, Doc. 60, 563–83). He seems to have had this possibility in mind in 1916 when he alluded to the advantages of not having to "prescribe elementary building blocks from the beginning" if one adopts the continuum viewpoint. Nevertheless, in the final sentence quoted, the other Einstein seems to have got the better of the argument.

The internal debate must have continued, and evidence of it over the next few years may well exist in correspondence that I have not examined. The next relevant item that I found dates from 1923, when Einstein first confronted the question in public, in a paper entitled: "Does Field Theory Offer Possibilities for the Solution of the Quantum Problem?" Einstein points out that the great successes of quantum theory over the last quarter of a century should not be allowed to conceal the lack of any logical foundation for the theory—a point he and others had made numerous times over the years (see, e.g., the evidence cited in Stachel 1986). The facts encompassed by the quantum rules, he adds, must lead one to doubt whether the difficulties could be overcome by any consistent development of earlier theories.

The essential element of the previous theoretical development, which is characterized by the headings mechanics, Maxwell–Lorentz electrodynamics, theory of relativity, lies in the circumstance that they work with differential equations that uniquely determine events [*das Geschehen*] in a four-dimensional spatio-temporal continuum if they are known for a spatial cross-section. In the unique determination of the temporal continuation of events by means of partial differential equations lies the method, through which we take into account the law of causality. In view of the existing difficulties, one has despaired of the possibility of describing the actual processes by means of differential equations. Even beyond that, the possibility of a complete [*lückenlos*] extension of the law of causality on the basis of the four-dimensional space-time continuum has been doubted. All of these doubts are epistemologically permissible and quite understandable in view of the existing deep difficulties. Before we seriously consider such extreme possibilities, however, we have to test whether one must conclude from the previous facts and arguments that it is really impossible to succeed with partial differential equations (Einstein 1923, 359; my translation, J.S.).

Einstein goes on to describe a scheme (overdetermination) for setting up systems of field equations that only have solutions for discrete sets of initial conditions.

But the important point for present purposes is that the debate between the two Einsteins has now been brought into the open.

From this point on my survey can make no claim to being systematic. I shall merely present some examples indicating that "the other Einstein" was alive and well for the remainder of both Einstein's lives, and indeed seems to be getting the better of the battle with his continuum opponent as the years go by.

In 1935, he explained to his old friend Paul Langevin some difficulties in accepting quantum mechanics as the ultimate foundational theory. He went on:

> In any case one does not have the right today to maintain that the foundation must consist in a *field theory* in the sense of Maxwell. The other possibility, however, leads in my opinion to a renunciation of the time-space continuum and to a purely algebraic physics. Logically this is quite possible (the system is described by a number of integers; "time" is only a possible viewpoint [*Gesichtspunkt*], from which the other "observables" can be considered—an observable logically coordinated to all the others. Such a theory doesn't have to be based upon the probability concept. For the present, however, instinct rebels against such a theory (Einstein to Paul Langevin, 3 October 1935, as translated in Stachel 1986, 379–80).

This is the earliest use of the phrase "a purely algebraic physics" that I have found so far. I also cannot forbear adding that the viewpoint about time expressed in this letter is now being actively investigated by an important group of workers in quantum gravity (for a survey of this work, see in Ashtekar and Stachel 1991, "The Issue of Time in Quantum Gravity," 126–296), who had no idea that they were working on an idea anticipated by Einstein until it was pointed out to them. (In the main, they are still working from the continuum viewpoint, I hasten to note.)

Shortly after this, in March 1936, Einstein expressed similar thoughts in print:

> It has been suggested that, in view of the molecular structure of all events in the small, the introduction of a space-time continuum may be considered as contrary to nature. Perhaps the success of Heisenberg's method points to a purely algebraical method of description of nature, to the elimination of continuous funtions from physics. Then, however, we must also give up, on principle, the utilization of the space-time continuum. It is not inconceivable that human ingenuity will some day find methods that will make it possible to proceed along this path. Meanwhile, however, this project resembles the attempt to breathe in an airless space ("Physics and Reality," cited from Einstein 1954, 319, translation modified).

By the 1950s a definite note of pessimism has crept into Einstein's comments on the prospects of the field-theoretical approach:

> In present-day physics there is manifested a kind of battle between the particle-concept and the field-concept for leadership, which will probably not be decided for a long time. It is even doubtful if one of the two rivals finally will be able to maintain itself as a fundamental concept (Einstein to Herbert Kondo, 11 August 1952, as translated in Stachel 1986, 380).

He was fully aware that the collapse of the continuum concept would take more with it than his hopes. He wrote to his oldest living friend, Michele Besso, in 1954:

> I consider it entirely possible that physics cannot be based upon the field concept, that is on continuous structures. Then *nothing* will remain of my whole castle in the air including the theory of gravitation, but also nothing of the rest of contemporary physics (Einstein to Besso, 10 August 1954, as translated in ibid., 380).

In the same year, he wrote to David Bohm:

> I must confess that I was not able to find a way to explain the atomistic character of nature. My opinion is that if the objective description through the field as an elementary concept is not possible, then one has to find a possibility to avoid the continuum (together with space and time) altogether. But I have not the slightest idea what kind of elementary concepts could be used in such a theory (Einstein to David Bohm, 28 October 1954, as translated in ibid., 380).

The last point explains why Einstein hesitated to publicize his skeptical views. He thought it pointless to do so unless one had some idea as to how to proceed.

V

However, at least once in his correspondence Einstein went into a little more detail about the nature of such a purely algebraic physics:

> The alternative continuum-discontinuum seems to me to be a real alternative; i.e., there is here no compromise. By discontinuum theory I understand one in which there are no differential quotients. In such a theory space and time cannot occur, but only numbers and number-fields and rules for the formation of such on the basis of algebraic rules with exclusion of limiting processes. Which way will prove itself, only success can teach us.
>
> Physics up to now is naturally in its essence a continuum physics, in spite of the use of the material point, which looks like a discontinuous conceptual element, and has no more right of existence in field description. Its strength lies in the fact that it posits parts which exist quasi-independently, *beside* one another. Upon this rests the fact that there are reasonable laws, that is rules which can be formulated and tested for the individual parts. Its weakness lies in the fact that it has not been possible up to now to see how that atomistic aspect including quantum relations can result as a consequence. On the other hand dimensionality (as four-dimensionality) lies at the foundation of the theory.
>
> An algebraic theory of physics is affected with just the inverted advantages and weaknesses, aside from the fact that no one has been able to propose a possible logical schema for such a theory. It would be especially difficult to derive something like a spatio-temporal quasi-order from such a schema. I cannot imagine how the axiomatic framework of such a physics would appear, and I don't like it when one talks about it in dark apostrophes [*Anredungen*]. But I hold it entirely possible that the development will lead there; for it seems that the state of any finite spatially limited system may be fully characterized by a finite number of numbers. This speaks against the continuum with its infinitely many degrees of freedom. The objection is not decisive only because one doesn't know, in the contemporary state of mathematics, in what way the demand for freedom from singularity (in the continuum theory) limits the manifold of solutions (Einstein to H. S. Joachim, 24 August 1954, as translated in ibid., 581).

This letter also bears comparison with the 1916 letter to Dällenbach for its remarkable insights into the continuing debate between the two Einsteins.

To the end of his life, Einstein was still on the lookout for new mathematical tools that might help turn such speculations, which he thought it best to keep private, into the basis of a real theory. The noted mathematician Abraham Fraenkel reports a conversation that he had with Einstein in 1951. You will see at once why Einstein responded with such interest to what Fraenkel told him:

> In December 1951 I had the privilege of talking to Professor Einstein and describing the recent controversies between the (neo-) intuitionists and their "formalistic" and "logistic" antagonists; I pointed out that the first attitude would mean a kind of atomistic theory of functions, comparable to the atomistic structure of matter and energy. Einstein showed a lively interest in the subject and pointed out that to the physicist such a theory would seem by far preferable to the classical theory of continuity. I objected by stressing the main difficulty, namely, the fact that the procedures of mathematical analysis, e.g., of differential equations, are based on the assumption of mathematical continuity, while a modification sufficient to cover an intuitionistic-discrete medium cannot easily be imagined. Einstein did not share this pessimism and urged mathematicians to try to develop suitable new methods not based on continuity (Fraenkel 1954).

It is now time to bring my story to a close. Perhaps I may best do so by reminding the reader of Einstein's last published words—the final words of the posthumously published "Appendix Two" to the fifth edition of *The Meaning of Relativity*:

> One can give good reasons why reality cannot at all be represented by a continuous field. From the quantum phenomena it appears to follow with certainty that a finite system of finite energy can be completely described by a finite set of numbers (quantum numbers). This does not seem to be in accordance with a continuum theory, and must lead to an attempt to find a purely algebraic theory for the description of reality. But nobody knows how to obtain the basis of such a theory. (Einstein 1955, 166).

Einstein had decided to voice his skepticism about the continuum at the very moment when he was presenting the last version of his final unified field theory. The Other Einstein had the last word.

Acknowledgment: I thank Professor Jürgen Renn for a critical reading of the original paper and for several helpful suggestions for revision.

Notes

[1] The way will differ markedly depending on whether, like Einstein, one stays at the level of classical field theory, or accepts quantum field theory.

[2] Many other examples could be given: the contrast between the finite number of degrees of freedom of mechanical systems and the infinite number of degrees of freedom of the electromagnetic field (see Einstein 1989, Doc. 14, 149–69), which is discussed below; the contrast between the wave and particle aspects of the behavior of thermal radiation (see *ibid.*, Doc. 60, 563–83), etc.

[3] As Lorentz noted, "Helmholtz and following him Hertz and Cohn use the word polarization in another sense, namely in the sense of our displacement **D**" (Lorentz 1903, 94). Einstein seems to have followed Hertz' usage in this letter.

[4] For discussions of the Olympia Academy and Solovine's account of its readings, see Einstein 1989, xxiv–xxv.

[5] For Einstein's acknowledgement of the influence of Hume, see Einstein 1989, xxiii–xxiv.

[6] It is worth noting that Einstein actually recorded an encounter with the problem of infinite divisibility of matter at a much earlier date. In undertaking a self-study of the calculus some time between the ages of twelve and sixteen, Einstein used a textbook that extensively cited Leibniz and Herbart's views on this question. He made marginal notes that indicate his interest in the issue (see Einstein 1987, Doc. 4, 4).

[7] See, for example, the following items: "On the Method of Theoretical Physics" (Einstein [1934] 1954, 272), "Physics and Reality" (Einstein [1936] 1954, 291).

[8] An algebraic number is one that satisfies some algebraic equation with integral coefficients.

References

Ashtekar, Abhay and John Stachel, eds. 1991. *Conceptual Problems of Quantum Gravity, Einstein Studies*, vol. 2. Boston: Birkhäuser.

Dedekind, Richard, 1893. *Was sind und was sollen die Zahlen*? 2nd ed. Braunschweig: Vieweg und Sohn.

—— [1901] 1963. "The Nature and Meaning of Numbers." In *Essays on the Theory of Numbers*, translated by Wooster Woodruff Beman, 29–115. Chicago: Open Court. Reprinted edition. New York: Dover.

Einstein, Albert, 1921. "Geometrie und Erfahrung," *Preussische Akademie der Wissenschaften, physikalisch-mathematische Klasse. Sitzungsberichte*, [n.v.], 123–30.

—— 1923. "Bietet die Feldtheorie Möglichkeiten für die Lösung des Quantenproblems?," *Preussische Akademie der Wissenschaften, physikalisch-mathematische Klasse. Sitzungsberichte*, [n.v.], 359–64.

—— 1949. "Autobiographical Notes." In *Albert Einstein: Philosopher-Scientist*, edited by Paul Arthur Schilpp. LaSalle, Ill.: Open Court.

—— 1954. *Ideas and Opinions*. New York: Crown.

—— 1955. "Appendix II: Relativistic Theory of the Non-Symmetric Field." In *The Meaning of Relativity*, 133–66, 5th ed. Princeton, NJ: Princeton University Press.

—— 1987. *The Collected Papers of Albert Einstein*, Vol. 1: *The Early Years, 1879–1902*, edited by John Stachel et al. Princeton, NJ: Princeton University Press.

—— 1989. *The Collected Papers of Albert Einstein*, Vol. 2: *Writings, 1900–1909*, edited by John Stachel et al. Princeton, NJ: Princeton University Press.

—— 1993. *The Collected Papers of Albert Einstein*, Vol. 3: *Writings, 1909–1911*, edited by Martin Klein et al. Princeton, NJ: Princeton University Press.

Fraenkel, Abraham H. 1954. "The Intuitionistic Revolution in Mathematics and Logic," *Bulletin of the Research Council of Israel* 3:283–89.

Holton, Gerald, [1972] 1988. "On Trying to Understand Scientific Genius." In Holton, *Thematic Origins of Scientific Thought*, 2nd ed., 371–98. Cambridge, MA: Harvard University Press. (Originally published in *The American Scholar* 41:95–110.)

Hume, David, 1969. *A Treatise of Human Nature*. Edited by Ernest C. Mosser. Harmondsworth: Penguin Books.

Lorentz, Hendrik Antoon, 1903. "Maxwells elektromagnetische Theorie." In *Encyklopädie der Mathematischen Wissenschaften*, vol. 5, *Physik: Zweiter Teil*, edited by Arnold Sommerfeld, 67–144. Leipzig: B. G. Teubner.

Pais, Abraham, 1982. *"Subtle Is the Lord . . ." The Science and the Life of Albert Einstein.* Oxford: Oxford University Press.

Riemann, Bernhard, [1854] 1953. "Ueber die Hypothesen, welche der Geometrie zu Grunde liegen," *Königliche Gesellschaft der Wissenschaften und der Georg-Augusts-Universität* (Göttingen), *Mathematische Klasse, Abhandlungen* 13:133–52. Cited from B. Riemann, *Gesammelte mathematische Werke*, 2nd ed., edited by Heinrich Weber, Leipzig: B. G. Teubner, 1902. Reprinted edition, 272–87. New York: Dover.

Smith, David Eugene, [1929] 1959. *A Source Book in Mathematics*, New York: Dover.

Stachel, John, 1986. "Einstein and the Quantum: Fifty Years of Struggle." In *From Quarks to Quasars: Philosophical Problems of Modern Physics*, edited by Robert G. Colodny, 349–85. Pittsburgh: University of Pittsburgh Press. [See pp. 367–402 in this volume].

—— 1987a. "Einstein and Ether Drift Experiments." *Physics Today* 40:45–47. [See pp. 171–176].

—— 1987b. "How Einstein Discovered General Relativity: A Historical Tale with Some Contemporary Morals." In *General Relativity and Gravitation: Proceedings of the Eleventh International Conference on General Relativity and Gravitation, Stockholm, 6–12 July 1986*, edited by Malcolm A. H. MacCallum, 200–208. Cambridge: Cambridge University Press. [See pp. 293–300].

—— 1989. "Einstein's Search for General Covariance, 1912–1915." In *Einstein and the History of General Relativity, Einstein Studies*, Vol. 1, edited by Don Howard and John Stachel, 63–100. Boston: Birkhäuser. [See pp. 301–338].

—— 1991a. "Einstein and Quantum Mechanics." In *Conceptual Problems of Quantum Gravity*, edited by A. Ashtekar and J. Stachel, 13–42. Boston: Birkhäuser. [See pp. 403–426].

—— 1991b. "The Meaning of General Covariance: The Hole Story." *Philosophical Problems of the Internal and External World/Essays on the Philosophy of Adolf Grünbaum*, edited by John Earman et al., 129–160. Pittsburgh: University of Pittsburgh Press/Konstanz: Universitätsverlag Konstanz.

Part IV

Special Relativity

"What Song the Syrens Sang":* How Did Einstein Discover Special Relativity?

John Stachel

If you have read Edgar Allen Poe's "The Murders in the Rue Morgue," perhaps you remember the epigraph that Poe chose for this pioneer detective story:

> What song the Syrens sang, or what name Achilles assumed when he hid himself among women, though puzzling questions, are not beyond all conjecture.[1]

I believe that the problem of how Einstein discovered the special theory of relativity (SRT) falls into this category of "puzzling questions," that "are not beyond *all* conjecture."[2] Let me begin by explaining why.

When I started work on the Einstein Papers, there was already a large literature on the origins of SRT compared, say, to the rather scanty amount published on the origins of the general theory of relativity (GRT). So I assumed that the development of SRT must be fairly clear. However, I soon learned that the amount of work published on the origin of SRT and GRT are just about inversely proportional to the available primary source material. For GRT, we have a series of Einstein's papers from 1907 to 1915, capturing the successive steps of his search for the final version of the theory. In addition, there is extensive contemporary correspondence on the subject, several research notebooks, records of lectures given by Einstein during this period, not to mention a number of later reminiscences and historical remarks by Einstein.[3]

For SRT we have the paper *On the Electrodynamics of Moving Bodies*, in which the theory was first set forth in 1905 in its finished form, indeed a rather

L'Opera di Einstein, Umberto Curi, ed., pp. 21–37

English text previously unpublished

*English text of "Quale canzone cantarono le sirene: come scopro Einstein la teoria speciale della relatività?"

polished form (which is not to say that it bears no traces of its gestation process). The only earlier documentary evidence consists of literally a couple of sentences to be found in the handful of preserved early Einstein letters (I will quote both sentences later). We do have a number of later historical remarks by Einstein himself, sometimes transmitted by others (Wertheimer, Reiser-Kayser, Shankland, Ishiwara, for example), which raise many problems of authenticity and accuracy; and some very late Einstein letters, answering questions such as whether he had prior knowledge of the Michelson–Morley experiment, what works by Lorentz he had read, the influence of Poincaré, Mach, Hume, etc., on his ideas; Einstein's replies are not always self-consistent, it must be noted.[4]

Yet the urge to provide an answer to the question of the discovery of SRT has proven irresistable to many scholars. It is not hard to see why: A twenty-six year old patent expert (third class), largely self-taught in physics, who had never seen a theoretical physicist (as he later put it), let alone worked with one, author of several competent but not particularly distinguished papers, Einstein produced four extraordinary works in the year 1905, only one of which (not the relativity paper) seemed obviously related to his earlier papers. These works exerted the most profound influence on the development of physics in the 20th Century. How did Einstein do it? Small wonder that Tetu Hirosige, Gerald Holton, Authur I. Miller, Abraham Pais, John Earman, Clark Glymour, Stanley Goldberg, Robert Rynasiewicz, Roberto Torretti, *et al.*, have been moved to study this question. I shall not try to record my debts to and differences with each of these scholars, lest this survey become even longer and more tedious than it is already; but must at least acknowledge the influence of their work on my own.[5] I resisted the urge to conjecture for some years, but have finally succumbed, so I can well understand the temptation.

Contrary to my original, naive expectation, no general consensus has emerged from all this work. Given the nature of the available documentation and the difficulty of understanding any creative process—let alone that of a genius—this really is not surprising. I now believe that the most one can hope to do in discussing the discovery of SRT is to construct a plausible conjecture. Such a conjecture will be based upon a certain weighting of the scanty evidence we possess, based upon certain methodological hypotheses, as well as the imagination of the conjecturer.[6] There are bound to be differences of opinion in these matters. All one can demand is that it be made clear on what methodological hypotheses a conjecture is based, and a demonstration that the conjecture is in accord with the available evidence when the latter is weighted in accord with these hypotheses.

Let me emphasize that no such account can hope to encompass those elements of the creative process that Einstein referred as as "the irrational, the inconsistent, the droll, even the insane, which nature, inexhaustibly operative, implants into the individual, seemingly for her own amusement," for "These things are singled out only in the crucible of one's own mind." Yet one may draw courage for the type of conjecture I have in mind from another remark of Einstein: "A new idea comes suddenly and in a rather intuitive way. That means it is not reached by conscious logical conclusions. But, thinking it through afterwards, you can always discover the reasons which have led you unconsciously to your guess and you will find a logical way to justify it. Intuition is nothing but the outcome of earlier intellectual experience."

I shall discuss only this intellectual, logical side of Einstein's struggles. Before trying to reconstruct these struggles, it is well to note that his outward existence was far from tranquil during the period when he was developing SRT. While attending the Polytechnic at Zurich, thanks to the support of maternal relatives, he was plagued by the thought that he was unable to help his family, which was in dire financial straits due to constant business reverses. He was the only graduate in his section (VIA) not to get an academic post, and lived a hand-to-mouth existence for almost two years, until he got a job at the Swiss Patent Office thanks to help from a friend's father. During this period he was under severe family pressure to break with his fiancée, whom he only married in 1903 after his father's death. His first child was born in 1904, and he had to support wife and child on his modest income from the Patent Office, while his mother found work as a housekeeper. So one must not think of Einstein as a tranquil academic, brooding at leisure on weighty intellectual problems. Rather one must imagine him fitting his intellectual work into the interstices of a professional career and personal life that might have overwhelmed someone with a different nature.

The main methodological hypothesis guiding my conjecture was stated by Hans Reichenbach some time ago: "...the logical schema of the theory of relativity corresponds surprisingly with the program which controlled its discovery." To put it in more hifalutin' terms, also due to Reichenbach, I believe that "the context of justification" of SRT used by Einstein can shed light on "the context of its discovery."[7] This hypothesis suggests that we can learn a good deal about the development of the theory by paying close attention to the logical structure of its initial presentation in 1905, and to the many accounts of the theory that Einstein gave afterwards. Of course, I have tried not to neglect any scrap of evidence known to me, including the pitifully small amount of contemporary documentation and the later reminiscences. But I have given special weight to Einstein's early papers, letters, and lectures, in which he sought to justify the theory to his contemporaries. Intellectually, Einstein was an exceedingly self-absorbed person, willing to go over and over the grounds for the theory again and again. These accounts, given over a number of years, are remarkably self-consistent. They provide evidence for a number of conjectures about the course of development of his own ideas, and occasionally even include explicit statements about it. I assume that by and large memory tends to deteriorate with time, and (worse) that pseudo-"memories" tend to develop and even displace correct recollections. So, a second methodological hypothesis which I shall adopt is that, in case of discrepancies between such accounts, earlier ones are to be given greater weight than later ones. Explicit remarks that Einstein makes about the discovery of SRT in the course of his later expositions must always be given great weight, but the earlier he made them the greater the weight I give to them. Of course, if some feature of Einstein's accounts remains unchanged over many years, I take this as evidence for giving such a point the most weight.[8]

It follows from these methodological assumptions that I must preface my conjectures with a brief résumé of the "logical schema of the theory of relativity" as it was first published in the 1905 paper. In this paper, as in almost all subsequent accounts, Einstein bases SRT on two fundamental principles: the principle of relativity and the principle of the constancy of the velocity of light. The principle of relativity originated in Galileian-Newtonian mechanics: Any frame of reference in which Newton's law of inertia holds (for some period of time) is now called an inertial frame of reference. From the laws of mechanics it follows that, if one such inertial frame exists, then an infinity of them must: All frames of reference (and only such frames) moving with constant velocity with respect to a given inertial frame are also inertial frames. All mechanical experiments and observations proved to be in accord with the (mechanical) principle of relativity: the laws of mechanics take the same form in any of these inertial frames. The principle of relativity, as Einstein stated it in 1905, asserts that *all* the laws of physics take the same form in any inertial frame—in particular, the laws of electricity, magnetism, and optics in addition to those of mechanics.

The second of Einstein's principles is based on an important consequence of Maxwell's laws of electricity, magnetism, and optics, as interpreted by H. A. Lorentz near the end of the nineteenth century. Maxwell had unified optics with electricity and magnetism in a single theory, in which light is just one type of electromagnetic wave. It was then believed that any wave must propagate through some mechanical medium. Since light waves easily propagate through the vacuum of interstellar space, it was assumed that any vacuum, though empty of ordinary, ponderable matter, was actually filled by such a medium, to which our senses did not respond: the ether. The question then arose, how does this medium behave when ordinary matter is present? In particular, is it dragged along by the motion of matter? Various possible answers were considered in the course of the nineteenth century, but finally only one view seemed compatible with (almost) all the known experimental results, that of H. A. Lorentz: The ether is present everywhere. Ordinary matter is made up of electrically charged particles, which can move through the ether, which is basically immobile. These charged particles, then called "electrons" or "ions", produce all electric and magnetic fields (including the electromagnetic waves we perceive as light), which are nothing but certain excited states of the immovable ether. The important experimental problem then arose of detecting the motion of ponderable matter—the earth in particular—through the ether.

No other theory came remotely close to Lorentz's in accounting for so many electromagnetic and especially optical phenomena. This is not just my view of Lorentz's theory, it was Einstein's view. In particular, he again and again cites the abberation of starlight and the results of Fizeau's experiment on the velocity of light in flowing water as *decisive* evidence in favor of Lorentz's interpretation of Maxwell's equations.

A direct consequence of Lorentz's conception of the stationary ether is that the velocity of light with respect to the ether is a constant, independent of the motion of the source of light (or its frequency, amplitude, or direction of propagation in the ether, etc.).

Einstein adopted a slightly—but crucially—modified version of this conclusion as his second principle: There is an *inertial frame* in which the speed of light is a constant, independent of the velocity of its source. A Lorentzian ether theorist could agree at once to this statement, since it was always tacitly assumed that the ether rest frame is an inertial frame of reference and Einstein had "only" substituted "inertial frame" for "ether." But Einstein's omission of the ether was deliberate and crucial: by the time he formulated SRT he did not believe in its existence. For Einstein a principle was just that: a principle—a starting point for a process of deduction, not a deduction from any (ether) theory. (I am here getting ahead of my story and will return to this point later.) The Lorentzian ether theorist would add that there can only be *one* inertial frame in which the light principle holds. If the speed of light is a constant in the ether frame, it must be non-constant in every other inertial frame, as follows from the (Newtonian) law of addition of velocities. The light principle hence *seems* to be incompatible with the relativity principle. For, according to the relativity principle, *all* the laws of physics must be the same in any inertial frame. So, if the speed of light is constant in one inertial frame, and that frame is not physically singled out by being the rest frame of some medium (the ether), then the speed of light *must* be the same (universal) constant in every other inertial frame (otherwise the democracy of inertial frames is violated). As Einstein put it in 1905, his two principles are "apparently incompatible." Of course, if they really were incompatible logically or physically, that would be the end of SRT.[9]

Einstein showed that they are not only logically compatible, but compatible with the results of all optical and other experiments performed up to 1905 (and since, we may add). He was able to show their logical compatibility by an analysis of the concepts of time, simultaneity, and length, which demonstrated that the speed of light really could have the privileged status, implied by his two principles, of being a universal speed, the same in every inertial frame of reference.[10]

Now I shall begin my conjecture about Einstein's discovery of SRT. In a 1921 lecture, Einstein stated that SRT originated from his interest in the problem of the optics of moving bodies. He seems to have been fascinated from an early age by the nature of light, a fascination that persisted throughout his life. From an essay he wrote in 1895, (at age 16), we know that he then believed in the ether, and had heard of Hertz's experiments on the propagation of electromagnetic waves; but he does not show any knowledge of Maxwell's theory. In much later reminiscences, he reports that during the following year (1895–1896) he conceived of a thought-experiment: what would happen if an observer tried to chase a light wave? Could s/he catch up with it? If so, s/he ought to see a non-moving light wave form, which somehow seemed strange to him. In retrospect, he called this "the first childish thought-experiment that was related to the special theory of relativity." Reliable accounts inform us that during his second year (1897–98) at the Swiss Federal Technical Institute, or Poly as it was then called, he tried to design an experiment to measure the velocity of the earth through the ether, being then unacquainted with either the theoretical work on this problem by Lorentz or the experiment of Michelson and Morley (M–M). A precious bit of contemporary documentary evidence reinforces this later account. In a letter to his schoolmate and friend Marcel Grossmann, written in the summer of 1901 (by then both had graduated from the Poly), Einstein wrote:

> A considerably simpler method for the investigation of the relative motion of matter with respect to the light ether has again occured to me, which is based on ordinary interference experiments. If only inexorable destiny gives me the time and peace necessary to carry it out.

At first sight, it would seem remarkable for Einstein to have written these words (which also show that he had not yet abandoned the concept of the ether), if he knew about the M–M experiment at this time.

However, while still at the Poly (i.e., before 1901) he appears to have studied Maxwell's theory (not covered in his school lectures) on his own, perhaps from the new textbook of August Föppl (which, in various reincarnations, such as Föppl–Abraham, Abraham–Becker, Becker–Sauter, has stayed in print to this day). Föppl discusses a problem which evidently made a strong and lasting impression on Einstein, since he opens the 1905 paper with a discussion of it. This is the problem of the relative motion of a magnet and a conducting wire loop. If the loop is at rest in the ether and the magnet is moved with a given velocity, a certain electric current is induced in the loop. If the magnet is at rest, and the loop moves with the *same* relative velocity, a current of the same magnitude and direction is induced in the loop. However, the ether theory gives a different explanation for the origin of this current in the two cases. In the first case an electric field is supposed to be created in the ether by the motion of the magnet relative to it (Faraday's law of induction). In the second case, no such electric field is supposed to be present since the magnet is at rest in the ether, but the current results from the motion of the loop through the magnetic field (Lorentz force law). This asymmetry of explanation, not reflected in any difference in the phenomena observed, must already have been troubling to Einstein. Even more troubling was the knowledge, when he acquired it, that all attempts to detect the motion of ponderable matter through the ether had failed. This was an "intolerable" (his word, about 1920) situation. Observable electromagnetic phenomena depend only on the *relative* motions of ponderable matter; their explanations differ, however, depending on the presumed state of motion of that matter relative to the hypothetical ether; yet all attempts to detect this presumed motion of ordinary matter relative to the ether end in failure! He later (c. 1920) recalled that the phenomenon of electromagnetic induction compelled him to adopt the relativity principle.

In 1938 he wrote "The empirically suggested non-existence of such an [ether wind] is the main starting point [point of departure] for the special theory of relativity."[11] It is not clear when the significance of the failure of all attempts to detect the motion of ordinary matter through the ether first struck him. The letter quoted above suggests that it was after the summer of 1901. We know from a letter to another friend, Michele Besso, dating from early 1903, that he had decided to "carry out comprehensive studies in electron theory." No later than that, and quite possibly earlier, he read Lorentz's 1895 book, "Attempt at a Theory of Electrical and Optical Phenomena in Moving Bodies." Einstein surely learned about the many such failures by reading this book, since one of its main purposes was to show that such failures were compatible with Lorentz's stationary ether theory. His later comments suggest that study of this book (Einstein says this is the only work by Lorentz he read before 1905) convinced him of the essential superiority of Lorentz' approach to the optics of moving bodies; yet it also convinced him that the Lorentz theory was still not fully satisfactory. Lorentz could explain away the failure to detect motion of matter relative to the ether convincingly to Einstein in all cases but one: the M–M experiment. To explain this, Lorentz had to introduce a special hypothesis, which to Einstein seemed completely unconnected with the rest of the theory: the famous Lorentz contraction. To Einstein, such an approach was not a satisfactory way out of the "intolerable dilemma." It seemed preferable to him to accept at face value the failure of the M–M and all similar experiments to detect motion of matter relative to the ether. Taken by themselves, these negative results suggested to Einstein that the relativity principle applied to electromagnetism, while the ether should be dropped as superfluous. There has been some confusion on this important point, so I shall expand on it. Sometimes the case is presented in such a way as to suggest that it was the "philosophical concept" of the relativity of all motion, as Einstein once called it, which was the key step in his rejection of the ether. But the concept of a stationary ether, as well as of a moving ether, is quite compatible with this philosophical concept of the relativity of motion: one need only assume that motions relative to the ether in the first case, as well as relative motions of the parts of the ether in the second, have physical efficacy. The leading advocates of both the dragged-along and the immovable ether concepts, Hertz and Lorentz, respectively, both understood this and both were read by Einstein.[12]

By the time he gave up the ether concept, Einstein most likely took this philosophical conception of the relativity of all motion for granted, presumably under the influence of his early reading of Mach's *Mechanics* (around 1897). What bothered him *now* was that no phenomenon existed that could be interpreted as empirical evidence for the physical efficacy of the motion of ordinary matter relative to the ether, in spite of repeated efforts to find one. Yet the best available theory—Lorentz's theory—could only attempt to explain away such failures. These explanations were satisfactory, within the framework of Lorentz' theory, in almost all known cases (i.e., for all experiments sensitive only to order v/c), and Einstein even seems to have been tempted to give up what we may call his *physical* relativity principle (with no ether needed). But Lorentz's explanation of the M–M experiment seemed to Einstein so artificial that he resisted this temptation, opting for the physical relativity principle. After eliminating the ether from the story altogether, one can simply take the results of the M–M and similar experiments as empirical evidence for the equivalence of all inertial frames for the laws of electricity, magnetism and optics as well as those of mechanics. I believe Einstein gave up the ether concept and definitely opted for the physical relativity principle at least a couple of years before the final formulation of SRT, perhaps even earlier. At any rate, at some point well before the 1905 formulation of the theory, he made this choice and adhered to it thereafter.

There was a related motive for his skepticism with regard to the ether, which I shall now mention. Not only was Einstein working on problems of the optics of moving bodies, he was also working on problems related to the emission and absorption of light by matter and of the equilibrium behavior of electromagnetic radiation confined in a cavity—the so-called black body radiation problem. He was using Maxwell's and Boltzmann's statistical methods, which he had redeveloped and refined in several earlier papers, to analyze this problem. This was itself a daring step, since these methods had been developed to help understand the behavior of ordinary matter while Einstein was applying them to the apparently quite different field of electromagnetic radiation.[13] The "revolutionary" conclusion to which he came was that, in certain respects, electromagnetic radiation behaved more like a collection of particles than like a wave. He announced this result in a paper published in 1905, three months before his SRT paper. The idea that a light beam consisted of a stream of particles had been espoused by Newton and maintained its popularity into the middle of the 19th century. It was called the "emission theory" of light, a phrase I shall use. The need to explain the phenomena of interference, diffraction and polarization of light gradually led physicists to abandon the emission theory in favor of the competing wave theory, previously its less-favored rival. Maxwell's explanation of light as a type of electromagnetic wave seemed to end the controversy with a definitive victory of the wave theory. However, if Einstein was right (as events slowly proved he was) the story must be much more complicated. Einstein was aware of the difficulties with Maxwell's theory—and of the need for what we now call a quantum theory of electromagnetic radiation—well before publishing his SRT paper. He regarded Maxwell's equations as some sort of statistical average—of what he did not know, of course—which worked very well to explain many optical phenomena, but could not be used to explain all the interactions of light and matter. A notable feature of his first light quantum paper is that it almost completely avoids mention of the ether, even in discussing Maxwell's theory. Giving up the ether concept allowed Einstein to envisage the possibility that a beam of light was "an independent structure," as he put it a few years later, "which is radiated by the light source, just as in Newton's emission theory of light."

So abandonment of the concept of the ether was a most important act of liberation for Einstein's thought in two respects: It allowed Einstein to speculate more boldly on the nature of light and it opened the way for adoption of his relativity principle as a fundamental criterion for all physical laws. I must add a word about Einstein's use of such principles as a guide to further research. In 1919 he explicitly formulated a broad distinction between constructive theories and theories of principle. Constructive theories attempt to explain some limited group of phenomena by means of some model, some set of postulated theoretical entities. For example, many aspects of the behavior of a gas could be explained by assuming that it was composed of an immense number of constantly colliding molecules. Theories of principle formulate broad regularities, presumably obeyed by all physical phenomena, making these principles criteria ("rules of the game") that any constructive theory must satisfy. For example, the principles of thermodynamics are presumed to govern all macroscopic phenomena. They say nothing about the micro-structure or detailed behavior of any particular gas, but do constitute limitations on any acceptable constructive theory of such a gas. Any theory not conserving the energy of the gas, for example, would be immediately rejected. Since the turn of the century, Einstein had been searching for a constructive theory of light, capable of explaining all of its properties on the basis of some model, and was to continue the search to the end of his days. But, "Despair [ing] of the possibility of discovering the true answer by constructive efforts," as he later put it, he decided that the only possible way of making progress in the absence of such a constructive theory was to find some set of principles that could serve to limit and guide the search for a constructive theory.[14] There is no contemporary evidence showing when Einstein adopted this point of view (he first indicated it in print as early as 1907). I believe he had done so by 1905. The structure of the 1905 SRT paper is certainly compatible with his having done so. It is based on the statement of two such principles, deduction of various kinematic consequences from them, and their application to Maxwell's electrical and optical theory.

To return to the main thread of my conjecture, I believe that Einstein dropped the ether hypothesis and adopted his relativity principle by 1903 or 1904 at the latest. This is by no means the end of the story. It seemed that he must then drop Lorentz's version of Maxwell's theory, based as it was on the ether hypothesis. With what was he to replace it? There is good evidence suggesting he spent a great deal of effort trying to replace it with an emission theory of light—the sort of theory suggested by his concurrent researches into the quantum nature of light.[15] An emission theory is perfectly compatible with the relativity principle. Thus, the M–M experiment presented no problem; nor is stellar abberation difficult to explain on this basis.[16]

Einstein seems to have wrestled with the problems of an emission theory of light for some time, looking for a set of differential equations describing such a theory that could replace the Maxwell–Lorentz equations; and trying to explain a number of optical experiments, notably the Fizeau experiment, based on some version of the emission theory. He could not find any such equations, and his attempt to explain the Fizeau experiment led him to more and more bizarre assumptions to avoid an outright contradiction. So he more-or-less abandoned this approach (you will soon see why I say more-or-less), after perhaps a year or more of effort, and returned to a reconsideration of the Maxwell–Lorentz equations. Perhaps there was a way of making these equations compatible with the relativity principle once one abandoned Lorentz's interpretation via the ether concept.

But here he ran into the most blatant-seeming contradiction, which I mentioned earlier when first discussing the two principles. As noted then, the Maxwell–Lorentz equations imply that there exists (at least) one inertial frame in which the speed of light is a constant regardless of the motion of the light source. Einstein's version of the relativity principle (minus the ether) requires that, if this is true for one inertial frame, it must be true for all inertial frames. But this seems to be nonsense. How can it happen that the speed of light relative to an observer cannot be increased or decreased if that observer moves towards or away from a light beam? Einstein states that he wrestled with this problem over a lengthy period of time, to the point of dispair. We have no details of this struggle, unfortunately. Finally, after a day spent wrestling once more with the problem in the company of his friend and patent office colleague Michele Besso, the only person thanked in the 1905 SRT paper, there came a moment of crucial insight. In all of his struggles with the emission theory as well as with Lorentz's theory, he had been assuming that the ordinary Newtonian law of addition of velocities was unproblematic. It is this law of addition of velocities that allows one to "prove" that, if the velocity of light is constant with respect to one inertial frame, it cannot be constant with respect to any other inertial frame moving with respect to the first. It suddenly dawned on Einstein that this "obvious" law was based on certain assumptions about the nature of time always tacitly made. In particular, the concept of the velocity of an object with respect to an inertial frame depends on time readings made at two different places in that inertial frame. (He later referred to this moment of illumination as "the step.")[17] How do we know that time readings at two such distant places are properly correlated? Ultimately this boils down to the question: how do we decide when events at two different places in the same frame of reference occur at the same time, i.e., simultaneously? Isn't universal simultaneity an intuitively obvious property of time? Here, I believe, Einstein was really helped by his philosophical readings. He undoubedly got some help from his readings of Mach and Poincaré, but we know that he was engaged in a careful reading of Hume at about this time; and his later reminiscences attribute great significance to his reading of Hume's *Treatise on Human Nature*. What could he have gotten from Hume? I think it was a relational—as opposed to an absolute—concept of time and space. This is the view that time and space are not to be regarded as self-subsistent entities; rather one should speak of the temporal and spatial aspects of physical processes; "The doctrine," as Hume puts it, "that time is nothing but the manner, in which some real object exists." I believe the adoption of such a relational concept of time was a crucial step in freeing Einstein's outlook, enabling him to consider critically the tacit assumptions about time going into the usual arguments for the "obvious" velocity addition law. This was the second great moment of liberation of his thought. I shall not rehearse Einstein's arguments here, but it led to the radically novel idea that, once one physically defines simultaneity of two distant events relative to one inertial frame of reference, it by no means follows that these two events will be simultaneous when the *same* definition is used relative to another inertial frame moving with respect to the first. It is not logically excluded that they *are* simultaneous relative to all inertial frames. If we make that assumption, we are led back to Newtonian kinematics and the usual velocity addition law, which is logically quite consistent. However, if we adopt the two Einstein principles, then we are led to a new kinematics of time and space, in which the velocity of light is a universal constant, while simultaneity is different with respect to different inertial frames; this is also logically quite consistent. The usual velocity addition law is then replaced by a new one, in which the velocity of light "added" to any other velocity ("added" in a new sense—it would be better to say "compounded with") does not increase, but stays the same! The Maxwell–Lorentz equations, when examined with the aid of this new kinematics, prove to take the same form in every inertial frame. They are, therefore, quite compatible with the relativity principle, which demands that the laws of electricity, magnetism and optics have this property. The presence or absence of an electric or magnetic field, is then also found to be relative to an inertial frame, allowing a completely satisfactory relativistic analysis of the example of the conducting wire loop and magnet in relative motion. Within six weeks of taking "the step," Einstein later recalled, he had worked out all of these consequences and submitted the 1905 SRT paper to *Annalen der Physik.*

This does not imply that Lorentz's equations are adequate to explain all the features of light, of course. Einstein already knew they did not always correctly do so—in particular in the processes of its emission, absorption and its behavior in black body radiation. Indeed, his new velocity addition law is also compatible with an *emission* theory of light, just *because* the speed of light compounded with any lesser velocity still yields the same value. If we model a beam of light as a stream of particles, the two principles can still be obeyed. A few years later (1909), Einstein first publicly expressed the view that an adequate future theory of light would have to be some sort of fusion of the wave and emission theories. This is an example of how the special theory of relativity functioned as a theory of principle, limiting but not fixing the choice of a constructive theory of light.

Here I shall end my conjectures on how Einstein arrived at SRT. To briefly recapitulate, I believe that the first principle, the relativity principle, recapitulates his struggles with the mechanical ether concept which led finally to the first crucial liberation of his thought—the abandonment of the ether. The second principle, the principle of the constancy of the speed of light, recapitulates his struggle, once he had definitely opted for the relativity principle, first to evade the Maxwell–Lorentz theory by an emission theory; then to isolate what was still valid in the Maxwell–Lorentz theory after giving up the ether concept and abandoning absolute faith in the wave theory of light. The struggle to reconcile the two principles could only end successfully after the second great liberation of his thought: the relativisation of the concept of time. The resulting theory did not force him to choose between wave and emission theories of light, but rather led him to look forward to a synthesis of the two. This synthesis was finally achieved, over twenty years later, in the quantum theory of fields, to the satisfaction of most physicists, but ironically, never to that of Einstein.

I cannot ask you to accept my conjectures after all of my warnings at the outset of this paper, but will be content if you say "Si non è vero, è ben trovato," "If it isn't true, it's well contrived."

NOTES

[1] Poe is quoting Sir Thomas Browne's *Hydrotaphia.*

[2] A preliminary question is raised by my use of the word "discovery." Is it better to speak of the "discovery" or the "creation" of a theory like SRT? "Discovery" suggests the finding of some pre-existent, objective structure, as when we say "Columbus discovered America." "Creation" suggests an individual, subjective act, as when we say "Tolstoy created *Anna Karenina*." Neither word seems really appropriate to describe what goes on in the scientific endeavor. Einstein apparently preferred the word "Erfindung" (invention) to describe how scientific theories come into being. Speaking of Mach, Einstein says: "Er meinte gewissermassen, dass Theorien durch *Entdeckung* und nicht durch *Erfindung* entstehen." (*Einstein–Besso Correspondance* (Hermann, Paris 1972), p. 191, dated January 6, 1948.

[3] In the study of the discovery of GRT, therefore, one may hope to formulate conjectures which can be either confirmed or refuted. For example: A study of Einstein's published papers and private correspondence between 1912-1915 convinced me that the standard explanation for his failure to arrive at the correct gravitational field equations until the end of this period—namely, his presumed lack of understanding of the meaning of freedom of coordinate transformations in a generally covariant theory and the ability to impose coordinate conditions that this freedom implied—could not be correct (see "Einstein's Search for General Covariance, 1912–1915," presented at the Ninth International Conference on General Relativity and Gravitation, July 17, 1980, this volume, pp. 301–338). On the basis of his study of a research notebook of Einstein from the early part of this period, John Norton was able to prove that Einstein already was aware of the possibility of imposing coordinate conditions on a set of field equations, and indeed had used the harmonic coordinate conditions (see John Norton, "How Einstein found his field equations: 1912–1915," *Historical Studies in the Physical Sciences 14*, 253 (1984). For reasons discussed in the text, one cannot hope to confirm or disconfirm most conjectures about the origins of SRT.

[4] For a survey of this material for the period up to 1923, see J. Stachel, "Einstein and Michelson: The Context of Discovery and the Context of Justification," Astron. Nachricht. **303**, 47 (1982). Unless otherwise noted, quotations from Einstein are cited from this paper, which gives the full references. [See this volume, pp. 177–190].

[5] See Arthur I. Miller, *Albert Einstein's Special Theory of Relativity* (Addison-Wesley, Reading 1981), which contains references to his earlier papers as well as those of Holton, Hirosige and many others; Abraham Pais, *'Subtle is the Lord...' The Science and the Life of Albert Einstein* (Oxford U.P., New York 1982); Stanley Goldberg, *Understanding Relativity* (Birkhäuser, Boston 1984); Roberto Torretti, *Relativity and Geometry* (Pergamon, Oxford 1983). Earman, Glymour and Rynasiewicz have not yet published a full account of their views; I thank them for making available copies of several preprints on this subject.

[6] A popular epigram among historians runs: "God is omnipotent, but even He cannot change the past. That is why He created historians."

[7] See the reference in footnote 4 for the source of the citations from Reichenbach. If my thesis here is correct, this argues against the still widely held view that these two contexts should be rigorously separated. But in this paper I shall not elaborate on the wider issue.

[8] For example, Einstein's statements of the second principle of SRT, the light principle, remained remarkably consistent throughout his lifetime (see the discussion of this principle below). Indeed, an apparent exception in the printed text of his article "What is the Theory of Relativity?," published originally in English translation in the *Times* of London in 1919, proved to be based upon an incorrect transcription of his manuscript.

[9] Much of the anti-relativity literature, which still continues to grow in volume if not in weight, is based on attempts to show that the two principles are indeed logically incompatible.

[10] Sometimes (e.g., by Pais and Goldberg), this consequence of Einstein's two principles is asserted to be his second principle. This is incorrect factually (Einstein's account of the second principle is one of the most consistent features of his discussions of SRT over the years—see footnote 8), and disturbing for several reasons: a) it makes it impossible to explain why Einstein refers to the two principles as apparently contradictory. There is no contradiction apparent between the relativity principle and this deduction from it; b) it is logically defective, since the two principles would no longer be logically independent, as they are in Einstein's formulation; c) most important for present purposes, this formulation deprives us of important clues to Einstein's reasoning that led to the development of SRT.

[11] Einstein to Max Talmey, June 6, 1938. The German text reads: "Die empirisch suggerierte Nichtexistenz einer solchen bevorzugten 'Wind-Richtung' ist der Haupt-Ausgangspunkt der speziellen Relativitätstheorie."

[12] Hertz said: "... the absolute motion of a rigid system of bodies has no effect upon any internal electromagnetic processes whatever in it, provided that all the bodies under consideration, including the ether as well, actually share the motion." (*Electromagnetic Waves*, p. 246). Lorentz said:

> That one cannot speak of the absolute rest of the ether, is self-evident indeed; the expression wouldn't even have any meaning. If I say for short, the ether is at rest, this only means that one part of this medium is not displaced with respect to the others and that all perceptible movements of the heavenly bodies are relative movements with respect to the ether. *Versuch*, p. 4 (1895).

[13] He was not alone in transferring statistical methods from ordinary matter to radiation. Planck had already done so, but Einstein did not see the relation of his work to Planck's until after publishing his first paper on the subject.

[14] See Albert Einstein, *Autobiographical Notes* (Open Court, LaSalle 1979), pp. 48 (German text) and 49 (English translation).

[15] One such piece of evidence, not cited in my earlier paper (see footnote 4), has only recently come to light. It occurs in the most complete review of SRT that Einstein ever wrote. It was prepared in 1912 but never published, and is still in private hands. Luckily, a copy has come into the possession of the Einstein Archive. In it, Einstein explains at some length the difficulties that are encountered (and presumably these are the ones he had encountered), if one tries to explain the results of the Fizeau experiment on the basis of an emission theory of light combined with the relativity principle and Galilei–Newtonian kinematics. [See The *Collected Papers of Albert Einstein*, vol. 4. *The Swiss Years: Writings 1912–1914* (Princeton University Press, Princeton 1995), Doc. 1, "Manuscript on the Special Theory of Relativity," pp. 32–36].

[16] Indeed, the earliest explanation of stellar abberation had been based on the emission theory.

[17] Abraham Pais has mentioned this in describing his conversations with Einstein.

Einstein and Ether Drift Experiments

John Stachel

Recently discovered letters, written at the turn of the century to his fiancée, shed new light on the origin of the special theory of relativity.

Volume 1 of *The Collected Papers of Albert Einstein* contains a number of previously unpublished lecture notes, examination papers and letters by Einstein.[1] Among the most notable new items are 42 letters written between 1898 and 1902 to his fiancée Mileva Marić, whom he met while they were fellow students of physics at the Swiss Polytechnical School in Zurich, which both entered in 1896.

These letters confirm Einstein's later recollection that he had begun to work on the electrodynamics of moving bodies many years before submitting his epochal 1905 paper on special relativity to *Annalen der Physik*.[2] They also record Einstein's continued interest, in the years between 1899 and 1901, in designing an optical experiment to test the putative motion of the Earth through the ether—which should have been detectable according to the then-prevalent interpretation of Maxwell's theory.

While there is no mention of Albert A. Michelson in any of the letters in volume 1, which covers the period from Einstein's birth until he got an appointment as patent clerk at the Swiss Patent Office, there is strong indirect evidence that he must have known of the Michelson–Morley experiment by 1899.[3] Here I will review briefly the new evidence of Einstein's early theoretical and experimental work on the electrodynamics of moving bodies.

Einstein's first comments on the subject, which appear in a remarkable letter that has been dated to August 1899, were inspired by a reading of Heinrich Hertz's basic papers on Maxwell's electrodynamics.

> I am more and more convinced that the electrodynamics of moving bodies, as currently presented, is not correct, and that it should be possible to present it in a simpler way. The introduction of the term "ether" into theories of electricity leads to the notion of a medium of whose motion one can speak without, I believe, being able to associate any physical meaning with such a statement.

Einstein is clearly skeptical about the concept of a movable ether, a concept that was basic to Hertz's theory of the electrodynamics of moving bodies. Whether

Physics Today
Vol. 40, pp. 45–47
1987

this skepticism already extended to the concept of the ether itself, as was certainly the case by 1905, is more doubtful. On the whole, Einstein's views in this letter seem similar in many ways to those of Hendrik A. Lorentz, who postulated a universal but immobile ether. But there is no mention of Lorentz in any surviving letter of Einstein's until December 1901, when he states that he intends to study what Lorentz and Paul Drude have written on the subject. So it is entirely possible that Einstein arrived at his views in 1899 independently of Lorentz.

Einstein refers in this letter to the need for "radiation experiments" to decide between various views of electrodynamics. In September of 1899 he writes:

> A good way of investigating how a body's relative motion with respect to the luminiferous ether affects the velocity of propagation of light in transparent bodies occurred to me in Aarau [a Swiss town Einstein had recently visited]. I have also thought of a theory on this subject that seems to me to very plausible. But enough of this!

Einstein goes on to commiserate with Marić, who was preparing at the Polytechnical School for a set of examinations that Einstein had already passed. There is no further evidence in the letters about the nature of Einstein's experiment or of his theoretical ideas.

A couple of weeks later, he informs Marić that

> I also wrote to Professor [Wilhelm] Wien in Aachen about the work on the relative motion of the luminiferous ether with respect to ponderable matter, which "the boss" [Heinrich Friedrich Weber, Einstein's physics professor at the Polytechnical School] treated in such a stepmotherly fashion.

This remark partially confirms the narrative that Rudolf Kayser, Einstein's son-in-law, gives in his 1930 biography of Einstein (written with Einstein's cooperation and approval):[4]

> He encountered at once, in his second year of college [1897–98], the problem of light, ether and the Earth's movement. This problem never left him. He wanted to construct an apparatus which would accurately measure the Earth's movement against the ether. That his intention was that of other important theorists, Einstein did not yet know. He was at that time unacquainted with the positive contributions, of some years back, of the great Dutch physicist Hendrik Lorentz, and with the subsequently famous attempt of Michelson. [Michelson first performed his experiment in 1881, and repeated it in 1887, partly in response to a criticism by Lorentz, before Lorentz's first major work on the electrodynamics of moving bodies in 1892.] He wanted to proceed quite empirically, to suit his scientific feeling of the time, and believed that an apparatus such as he sought would lead him to the solution of a problem, whose far-reaching perspectives he already sensed.
>
> But there was no chance to build this apparatus. The skepticism of his teachers was too great, the spirit of enterprise too small.

Kayser's account still does not offer any clues to what Einstein's experimental design could have been. The only evidence known to me on this question is the record of Einstein's 1922 lecture at Kyoto University, "How I created the theory of relativity," kept by the physicist Jun Ishiwara.[5] He was the first Japanese to publish on the theory of relativity, had visited Einstein on a trip to the West, was instrumental in getting Einstein to come to Japan and acted as translator of his lectures. Ishiwara's record, in Japanese, of Einstein's lecture includes Einstein's description of an experiment that occurred to him while he was a student:

> So I wanted to demonstrate by some means this motion of the Earth relative to the ether. . . . At the time when I posed this problem to myself, I never doubted the existence of the ether and the motion of the Earth. Thus, I predicted that if light from a source is reflected by a mirror, it should have different energies depending on whether it is propagated parallel or antiparallel to the direction of motion of the Earth; and I proposed verifying this with two thermocouples, by measuring the difference in the heat produced in each.

This may well be a description of the idea that Einstein had in Aarau, although his reference to "the velocity of propagation of light in transparent bodies" suggests that he may have had in mind some variant of Armand Fizeau's well-known experiment on this subject. Indeed, it is curious to note that in 1854 Fizeau had proposed an experiment on the difference in energy between light rays moving in opposite directions, which was actually performed in 1902 by Nordmeyer.[6]

In the "Aarau" letter of September 1899, Einstein explains why he turned to Wien for support of his ideas: "I read a very interesting paper from the year 1898 by this man [Wien] on the same topic." The paper was the text of Wien's report to the Society of German Scientists and Physicians, "On questions relating to the translatory motion of the luminiferous ether."[7] Here Wien discussed both Hertz's concept of a moving ether and Lorentz's concept of an immobile ether, and he briefly considers 13 experiments bearing on the question. The last one he mentions is the Michelson–Morley experiment. It is reasonable to conjecture that Einstein read this account in 1899, and that it thus represents the minimum information he had about the experiment by then. Here is Wien's account:

> *The Michelson–Morley experiment.* If the ether is at rest, then the time a light ray needs to travel back and forth between two glass plates must change if the plates are moving. The change depends on the quantity v^2A^2 [v is the velocity of the plates; A is the reciprocal of the speed of light], but should be observable by the application of interferometry.
>
> The negative result is incompatible with the assumption of an ether at rest. This assumption can only be maintained by means of the hypothesis that the linear dimensions of rigid bodies are altered by motion through the resting ether in the same ratio, so as to compensate for the lengthening of the path of the light ray.

Wien does not make it clear that interference between two perpendicular rays is the basis of the Michelson–Morley experiment. It is possible that Einstein did not see a more detailed account of the experiment until he read Lorentz's 1895 monograph[8] or Drude's book on optics,[9] which contains a summary of Lorentz's theory; both books include detailed discussions of the Michelson–Morley experiment. Just when Einstein read Lorentz and Drude is not clear, although as noted above, in a letter from December 1901 he states his intention to study their work.

Einstein's next comment on relative motion occurs in a letter to Marić written in March 1901: "How happy and proud I will be when the two of us together will have brought our work on relative motion to a successful conclusion." This comment raises the intriguing question of the nature of Marić's role in their collaboration. Her letters to Einstein (only ten from the period of the first volume have been found) contain no substantial references at all to physics. His letters to her contain references to joint study of books, requests for her to look up data, and one or two other mentions of joint work; but these letters give no indication of any ideas she contributed to their work.

Writing to his friend and former fellow-student Marcel Grossmann in September 1901, Einstein returns to the subject of ether drift experiments:

> On the investigation of the relative motion of matter with respect to the luminiferous ether, a considerably simpler method has occurred to me, which is based on customary interference experiments. If only relentless fate would give me the necessary time and peace! When we see each other, I will tell you about it.

The reference to "customary" [*gewöhnlich*] interference experiments in this letter is intriguing but puzzling. Any suggestion that Einstein had in mind nothing more than a repeat of the Michelson–Morley experiment seems to be ruled out by Einstein's report, in a subsequent letter, that Alfred Kleiner of the University of Zurich was enthusiastic about his experimental proposal. Kleiner was a well-informed experimenter, who later wrote a number of surveys of the then-current state of physics. It is hard to believe that he would not have known enough about the Michelson–Morley experiment to recognize a description of it.

Einstein was also developing his theoretical ideas on electrodynamics during this period. During the same month, he wrote Marić:

> I am now working very eagerly on an electrodynamics of moving bodies, which promises to become a capital paper. I wrote you that I doubted the correctness of the ideas about relative motion [that letter has not been found]. But my doubts were based solely on a simple mathematical error. Now I believe in it more than ever!

This passage suggests that Einstein had already adopted some version of the relativity principle—which is not to say that he had yet disentangled his ideas on relative motion from their electrodynamical background, let alone given them the kinematical foundation that proved essential to the formulation of the special theory of relativity. But the passage does suggest that he may have fully expected the outcome of his experiment to be negative.

In December, as mentioned above, Einstein wrote that he had

> spent the whole afternoon with Kleiner in Zurich and explained my ideas on the electrodynamics of moving bodies to him... He advised me to publish my ideas about the electromagnetic theory of light for moving bodies together with the experimental method. He found the experimental method proposed by me to be the simplest and most appropriate one conceivable... I shall most certainly write the paper in the coming weeks.

A few days later in December, he wrote Marić:

> I now want to buckle down to work and study what Lorentz and Drude have written on the electrodynamics of moving bodies. [Jakob] Ehrat [a friend and former fellow Polytechnical School student, who was now an Assistant there] must get the literature for me.

Whatever reading and writing he may have done at this time, Einstein published nothing on the subject for $3\frac{1}{2}$ years. Surviving correspondence sheds very little light on what happened. Perhaps a reading of Lorentz's work temporarily shook his faith in the relativity principle; perhaps he saw that the problems involved in upholding it were greater than he had anticipated. I have speculated elsewhere on the question of what happened between 1902 and 1905,[10] but there are unfortunately no relevant new letters from this period.

In summary, the newly discovered correspondence with Marić proves that Einstein was concerned with the theoretical and experimental aspects of the electrodynamics of moving bodies from at least 1899 on. He was very much interested in ether drift experiments, and appears to have designed at least two, which he hoped to carry out himself. While he was almost certainly aware in a general way of the existence of the Michelson–Morley experiment from late 1899 on, it is not mentioned at all in his surviving letters from that period. The new evidence thus serves to confirm, at least for the period 1899–1902, Gerald Holton's conclusion that the experiment did not play a significant role in Einstein's work. But ideas about ether drift experiments did form an important strand in his thinking about the complex of problems that ultimately led him to develop the special theory of relativity.

Notes

[1] J. Stachel et al., eds., *The Collected Papers of Albert Einstein: The Early Years (1879–1902)*, Princeton U. P., Princeton, NJ (1987). All Einstein quotations are translated from this volume with the kind permission of the Hebrew University of Jerusalem.

[2] A. Einstein, *Ann. Phys.* (Leipzig) **17** (1905), 891.

[3] For studies of the relationship of the Michelson–Morley experiment to Einstein's work, see the fundamental article by G. Holton, reprinted in G. Holton, *Thematic Origins of Scientific Thought*, Harvard U. P., Cambridge, Mass. (1973), p. 261. See also J. Stachel, *Astron. Nachr.* **47** (1982), 303. [See this volume, p. 177].

[4] R. Kayser [under the pseudonym A. Reiser], *Albert Einstein: A Biographical Portrait*, Boni, New York (1930), p. 52.

[5] J. Ishiwara, *Einstein Kyözyu-Koen-roku* [The Record of Professor Einstein's Lectures], Kabushika Kaisha, Tokyo (1971), p. 79. Widely differing English translations of the relevant passages on the origins of special relativity have appeared. [See, for example, *Physics Today*, August 1982, p. 45, and the letter by Arthur Miller on page 9 of this issue.] Fortunately, they all agree more or less closely on the passage cited. (I have also consulted a German translation prepared by H. J. Haubold and E. Yasui, whom I thank for making it available to me.) For the translation used here, see J. Stachel, *Astron. Nachr.* **47** (1982), 303.

[6] See J. Stachel, *Astron. Nachr.* **47** (1982), 303 for references and details.

[7] W. Wien, *Ann. Phys.* (Leipzig **65**(3), Beilage (1898), p. i.

[8] H. A. Lorentz, *Versuch einer Theorie der elektrischen und optischen Erscheinungen in bewegten Körpern*, Brill, Leiden (1895). Einstein later recalled that this was the only work by Lorentz he read before writing his paper on special relativity.

[9] P. Drude, *Lehrbuch der Optik*, Hirzel, Leipzig (1900), Chapter VIII of section 2 of the part on physical optics is entitled "Bewegte Körper."

[10] See J. Stachel, *Astron. Nachr.* **47** (1982), 303; and "What Song the Syrens Sang: How Did Einstein Discover Special Relativity?" [See this volume, pp. 157–169].

Einstein and Michelson: The Context of Discovery and the Context of Justification

John Stachel

The philosopher of science is not much interested in the thought processes which lead to scientific discoveries; he looks for a logical analysis of the completed theory, including the relationships establishing its validity. That is, he is not interested in the context of discovery, but in the context of justification . . . it appears amazing to what extent the logical analysis of relativity coincides with the original interpretation by its author, as far as it can be constructed from the scanty remarks in Einstein's publications. In contradistinction to some developments in quantum theory, the logical schema of the theory of relativity corresponds surprisingly with the program which controlled its discovery.[1]

Although the distinction between context of discovery and context of justification has been much discussed—and disputed—by philosophers of science, my paper is not primarily a contribution to that discussion. Rather I seek to follow the hint in the last part of the quotation from Reichenbach by asking: What can we learn from an examination of Einstein's mention of—or silence about—the Michelson–Morley experiment in the "context of justification" after the discovery of the special theory of relativity about the role of that experiment in the "context of discovery"? One may take courage in searching for such a connection from the following remark by Einstein:[2]

> A new idea comes suddenly and in a rather intuitive way. That means it is not reached by conscious logical conclusions. But thinking it through afterwards you can always discover the reasons which have led you unconsciously to your guess and you will find a logical way to justify it. Intuition is nothing but the outcome of accumulated earlier intellectual experience.

All of Einstein's known[3] expositions of the special theory of relativity from his 1905 paper up to his 1922 Kyoto lecture have been examined. Dr. Jun Ishiwara (or

Astronomische Nachrichten
Vol. 303, pp. 47–53
1982

Ishihara in another transliteration) acted as Einstein's interpreter during his 1922 lecture tour in Japan and published *The Record of Professor Einstein's Lectures.*[4] Ishiwara was a Japanese physicist who published articles on the theory of relativity as early as 1909, and spent time in Western Europe in 1912–14 working with a number of the leading theoretical physicists, including Einstein. He was involved in arranging Einstein's 1922 trip to Japan, and the two men were very close during this period, as a testimonial by Einstein makes clear. The *Record of Professor Einstein's Lectures* is based on the notes Ishiwara took during the lectures for his oral translations of them;[5] they were first published within a year of the Einstein trip. One of the lectures, delivered in Kyoto on December 14, 1922 was an impromptu discussion of "How I Created the Theory of Relativity", delivered in response to a request by Professor Nishida to speak on this topic. Since it is the first recorded personal account by Einstein, it forms a convenient point at which to end this survey of Einstein's early discussions of the special theory of relativity. In this paper I am confining myself to these early accounts, on the hypothesis that they are at least as likely (if not perhaps more so) as later accounts to accurately reflect the reasoning that led to the discovery of the special theory.[6]

These papers include:[7]

"Zur Elektrodynamik bewegter Körper," *Ann. d. Physik* **17**, 891 (1905) (Readex 9)

"Über das Relativitätsprinzip und die aus demselben gezogenen Folgerungen", *Jahrbuch der Radioaktivität und Elektronik* **4**, 411 (1907) (Readex 20)

"Über die Entwicklung unserer Anschauungen über das Wesen und die Konstitution der Strahlung," *Physikalische Zeitschrift* **10**, 817 (1909) (Readex 28)

"Principe de relativité et ses conséquences dans la physique moderne," *Archives des sciences physiques et naturelles* **29**, 5–18, 125–144 (1910) (Readex 32)

"Die Relativitätstheorie," *Naturforschende Gesellschaft Vierteljahresschrift (Zürich)* **56**, 1 (1911) (Readex 40)

"Zum Relativitätsproblem," *Scientia* **15**, 337 (1914) (Readex 66)

"Vom Relativitätstheorie," *Vossische Zeitung* (26 April 1914), 33, (Readex 67)

"Die Relativitätstheorie," in E. Lechner (ed.), *Die Physik*, Leipzig: Teubner, Vol. 1, 703 (1915) (Readex 70)

Über die spezielle und die allgemeine Relativitätstheorie, Gemeinverständlich, Braunschweig: Vieweg (1917) (Readex 91)

"My Theory," *Times* (London) (28 November 1919) (Readex 113)

The Meaning of Relativity: Four Lectures delivered at Princeton University, Princeton, NJ: Princeton University Press (1921) (Readex 121)

"A Brief Outline of the Development of the Theory of Relativity," *Nature* **106**, 782 (1921) (Readex 124)

"Über Relativitätstheorie, eine Londoner Rede," delivered at King's College in 1921; first printed in *Mein Weltbild* (Querido, Amsterdam, 1934) (Readex 357)

"How I Created the Theory of Relativity," see footnote 4 for references; delivered in 1922 and referred to hereafter as the Kyoto lecture

I shall first discuss these accounts of special relativity, given in the "context of justification." Then I shall discuss some other relevant items that give clues to Einstein's knowledge or goals when he was developing the special theory. Finally, I shall discuss how the evidence from the "context of justification" may be fitted into an account of the "context of discovery."

The Context of Justification

The Michelson–Morley experiment is cited in Readex 20 (1907), Readex 28 (1909), Readex 32 (1910), Readex 70 (1915), Readex 91 (1917), Readex 121 (1921), Readex 124 (1921), "Ueber Relativitätstheorie" ("On the Theory of Relativity") in Readex 357 (1921), and the Kyoto Lecture (1922). It is not mentioned directly in Readex 9 (1905), Readex 40 (1911) or Readex 66 (1914); Readex 9 (1905) refers, as is well known, to "unsuccessful attempts to discover any motion of the earth relative to the 'light medium' " ("... die mißlungenen Versuche, eine Bewegung der Erde relativ zum 'Lichtmedium' zu konstatieren ..."); and similar general references to the unsuccessful search for effects of the earth's motion occur in Readex 40 (1911) and Readex 66 (1914). Readex 67 (1914) is Einstein's first, short, newspaper article on the theory of relativity and makes no mention of any experiments. Readex 113 (1919), another, longer newspaper article is primarily an exposition of the general theory; it also makes no mention of experiments in the brief account of special relativity it contains.

Thus we may say the large majority (nine) of Einstein's early expositions of special relativity mention the Michelson–Morley experiment. Of the papers that do not, three speak generally of experiments to detect the motion of the earth, while two brief newspaper discussions of special relativity refer to no experiments at all.

Are there any common features to Einstein's mentions of the Michelson–Morley experiment? Yes: *Without exception*, it is cited as evidence for the relativity principle, and is *never* cited as evidence for the principle of the constancy of the velocity of light.[8]

Here are a few typical quotations:

> Der Versuch von Michelson und Morley hatte eben gezeigt, daß Erscheinungen auch da dem Relativitätsprinzip entsprechen, wo dies nach der Lorentzschen Theorie nicht einzusehen war. Es hatte daher den Anschein, als ob die Lorentzsche Theorie wieder verlassen und durch eine Theorie ersetzt werden müsse, deren Grundlagen dem Relativitätsprinzip entsprechen, denn eine solche Theorie ließe das negative Ergebnis des Versuches von Michelson und Morley ohne weiteres voraussehen. (Readex 20, 1907)[9]
>
> Der Michelsonsche Versuch legte die Voraussetzung nahe, daß alle Erscheinungen relativ zu einem mit der Erde bewegten Koordinatensystem allgemeiner überhaupt relativ zu jedem beschleunigungsfrei bewegten System nach genau den gleichen Gesetzen verlaufen. Diese Voraussetzung wollen wir im folgenden kurz "Relativitätsprinzip" nenne. (Readex 28, 1909)[10]

In this connection, all experience also in the realm of electrodynamics (in particular Michelson's experiment) supported the idea of equivalence of all inertial systems, i.e., was in favor of the special principle of relativity. (Readex 124, 1921)

Das durch die Entwicklung der Elektrodynamik und Optik erhärtete Gesetz der Konstanz der Lichtgeschwindigkeit im leeren Raum in Verbindung mit der durch Michelson's berühmten Versuch besonders scharf dargetanen Gleichberechtigung aller Inertialsystems (spezielles Relativitätsprinzip) ... (1921 in Readex 347, p. 216)[11]

Perhaps these quotations, the earliest two and two of the latest from the period surveyed, suffice to show the constancy of Einstein's use of the Michelson–Morley experiment in the "context of justification."

There is a second important feature common to many of Einstein's references to the Michelson–Morley experiment in these expositions. Einstein stresses a number of times that it was only the negative result, from the point of view of the Maxwell–Lorentz theory, of the Michelson–Morley experiment that prevented full acceptance of that theory even at the expense of the relativity principle:

Aber das negative Resultat des Experiments von Michelson und Morley zeigte, daß in einem bestinumten Falle auch ein Effekt zweiter Ordnung (proportional v^2/c^2) nicht vorhanden war, trotzdem er nach den Grundlagen der Lorentzschen Theorie bei dem Versuche sich hätte bemerkbar machen müssen. (Readex 20, 1907)[12]

Similar comments are found in Readex 28 (1909), Readex 32 (1910), etc. The most striking statement comes in Readex 70 (1915): After remarking that:

Die Lorentzsche Theorie erweckt nun unser Mißtrauen dadurch daß sie dem Relativitätsprinzip zu widersprechen scheint,[13]

Einstein states:

Die Erfolge der Lorentzschen Theorie waren so bedeutende, daß die Physiker unbedenklich das Relativitätsprinzip fallen gelassen hätten, wenn nicht ein wichtiges experimentelles Resultat vorgelegen wäre, von dem wir nun sprechen müssen, nämlich das Experiment von Michelson.[14]

Other Evidence

Before trying to draw any conclusions from the evidence of the "context of justification," I shall discuss some other relevant items. We know that there was a time when Einstein believed in the existence of the ether,[15] and even thought about experiments to investigate the effects of the motion of the earth through the ether.

In the Kyoto lecture (1922), Einstein says:

It was about 17 years ago when the idea to try to develop the principle of relativity first came to me. Of course, I cannot definitely say from where this idea came. I am certain, however, that it originated from the problem of the optics of moving bodies. Light propagates through the sea of ether. The earth is also moving through this ether. If observed from earth, the ether is moving relative to it. However, I could not find any evidence in the physics literature which clearly demonstrated this motion of the ether. So I wanted to demonstrate by some means this motion of the ether relative to the earth, namely the motion of the earth [through the ether]. At the time when I posed

> this problem to myself, I never doubted the existence of the ether, and the motion of the earth [through it]. Thus, I predicted that, if light from a source is reflected by a mirror, it should have different energies depending upon whether it propagated parallel or anti-parallel to the direction of motion of the earth; and I proposed verifying this with two thermocouples, by measuring the difference in the heat produced in each. This idea is similar to that of Michelson's experiment, but I did not carry out this experiment.[16]

There is a piece of contemporary evidence confirming Einstein's interest in this problem, and showing that the thermocouple experiment was not the only one that occurred to him. In a letter to Marcel Grossmann (his former classmate at the ETH, soon to be instrumental, with his father's help, in getting Einstein a job at the Swiss Patent Office) Einstein says:[17]

> Zur Erforschung der Relativbewegung der Materie gegen den Lichtäther ist mir wieder eine erheblich einfachere Methode in den Sinn gekommen, welche auf gewöhnlichen Interferenzversuchen beruht. Wenn mir nur einmal das unerbittliche Schicksal die zur Ausführung nötige Zeit und Ruhe gibt! Wenn wir uns wieder einmal sehen, werd'ich Dir darüber berichten.

This letter can be reliably dated on internal evidence to the summer of 1901. This suggests quite strongly that at that time Einstein had not yet heard of the Michelson–Morley experiment. Otherwise, why would he have talked of "a considerably simpler method . . . based on ordinary interference experiments" as something still to be done?

Even more interesting is the question of Einstein's thermocouple experiment. Such an experiment to detect the motion of the earth through the ether had actually been proposed by Fizeau in the middle of the nineteenth century.[18] It is possible that Einstein, in the course of his search for "evidence in the physics literature which clearly demonstrated this motion" of the earth through the ether, came across some account of Fizeau's proposal. In 1902 Lorentz published a paper[19] demonstrating, on the basis of his theory, that the experiment should fail to first order in (v/c). The experiment was actually performed by Nordmeyer in 1902, with negative result.[20] Einstein never seems to have mentioned this experiment, although it was mentioned in the first review paper on experimental evidence for the relativity principle by his collaborator Jacob Laub.[21] It is possible that Einstein saw the articles in the *Annalen*, or otherwise heard of the experiment before his 1905 paper; if so, it might have been one of the "unsuccessful attempts to discover any motion of the earth relative to the 'light medium' " alluded to in that paper.

Returning to evidence of Einstein's interests before 1905, there is a line in a letter to Michele Besso, dating from early 1903 which states:[22]

> In der nächsten Zeit will ich mich mit den Molekularkräften in Gasen abgeben, und dann umfassende Studien in Elektronentheorie machen.

The scope of those "comprehensive studies in electron theory" is not certain. They surely included the reading of Lorentz's 1895 *Versuch*,[23] from which Einstein could hardly have avoided learning about the Michelson–Morley experiment, if he had not done so earlier. The first fruits of this study were presented in a lecture to the Naturforschende Gesellschaft of Bern on "Theorie der elektromagnetischen Wellen."[24]

At some time during this period Einstein adopted the relativity principle for fundamental reasons having to do with the analysis of electromagnetic induction. Gerald Holton has stressed the importance of these considerations to Einstein.[25] In an unpublished manuscript Holton brought to light,[26] Einstein says:

> Bei der Aufstellung der speziellen Relativitätstheorie hat für mich der folgende, hier noch nicht erwähnte Gedanke über die Faradaysche magnet-elektrische Induktion eine führende Rolle gespielt.

Then he contrasts the symmetry of the effects when a magnet is moved towards a conductor or the conductor is moved towards the magnet with the asymmetry in the explanation offered by the Maxwell–Lorentz theory: The current produced is equal if the relative motion is the same in both cases; but in the first case an electric field is supposed to produce the current, while in the second an electromotive force on the electrons is credited with the production of the current. Einstein had offered essentially the same argument in his original 1905 paper (Readex 9), and discussed this example at even greater length in his 1910 semi-technical review (Readex 32). Einstein adds:[27]

> Der Gedanke, daß es sich hier um zwei wesens-verschiedene Fälle handle, war mir aber unerträglich. Der Unterschied zwischen beiden konnte nach meiner Überzeugung nur ein Unterschied in der Wahl des Standpunktes sein, nicht aber ein realer Unterschied ... Die Erscheinung der magnetelektrischen Induktion zwang mich dazu, das (spezielle) Relativitätsprinzip zu postulieren. [Footnote:] Die zu überwindende Schwierigkeit lag dann in der Konstanz der Vakuum-Lichtgeschwindigkeit, die ich zunächst aufgeben zu müssen glaubte. Erst nach jahrelangen Tasten bemerkte ich, daß die Schwierigkeit auf der Willkür der kinematischen Grundbegriffe beruhte.

The next problem Einstein faced was finding a treatment of electrodynamics consistent with the relativity principle. He focussed on the problem of the velocity of light. In a letter to Paul Ehrenfest written in 1912 Einstein stated:[28]

> Über Ihre Notiz[29] habe ich mich nicht im mindesten geärgert. Im Gegenteil. Solche Überlegungen sind mir ganz geläufig aus der vorrelativistischen Zeit. Ich wußte wohl, daß das Prinzip von der Konstanz der Lichtgeschwindigkeit etwas von dem Relativitätspostulat ganz Unabhängiges ist, und ich erwog, was wahrscheinlicher sei, das Prinzip von der Konstanz von c, wie es von der Maxwellschen Gleichung gefordert wird, oder die Konstanz von c, ausschließlich für einen Beobachter, der bei der Lichtquelle sitzt. Ich entschied mich für das erstere, weil ich der Überzeugung war, daß jedes Licht durch Frequenz und Intensität allein definiert sei, ganz unabhängig davon, ob es von einer bewegten oder von einer ruhenden Lichtquelle kommt. Es kam mir ferner nicht in den Sinn, daran zu denken, daß die abgebeugte Strahlung sich in puncto Fortpflanzung anders verhalten können als die in dem betreffenden Punkte neu emittierte Strahlung. Derartige Komplikationen schienen mir viel unberechtigter als jene, welche der neue Zeitbegriff mit sich bringt.[30]

In a letter drafted on the back of a letter from Mr. A. Rippenbein in 1952, Einstein states:[31]

> Ihr Versuch, die spezielle Relativität zu ersetzen durch die Annahme, daß die Lichtgeschwindigkeit *relativ zur Lichtquelle* konstant sei, ist zuerst von Ritz vertreten worden. Diese Annahme ist mit dem Michelson-Versuche und mit der Aberration vereinbar. [Einstein then refers to De Sitter's refutation of Ritz's theory][32] . . . Außerdem verlangt diese Theorie, daß überall und in jeder bestimmten Richtung Lichtwellen verschiedener Fortpflanzungsgeschwindigkeit möglich sein sollen. Es dürfte unmöglich sein, eine irgendwie vernünftige elektromagnetische Theorie aufzustellen, die solches leistet. Dies ist der hauptsächliche Grund, aus dem ich schon vor der Aufstellung der speziellen Relativitätstheorie diesen an sich denkbaren Ausweg verworfen habe.[33]

The Context of Discovery

Now I am ready to discuss what light my examination of Einstein's use of the Michelson–Morley experiment in the "context of justification" may shed upon its role in the "context of discovery" of the special theory of relativity. On the basis of the evidence so far presented, the following account seems a reasonable conjecture.[34]

There are two major choices involved in Einstein's development of the special theory of relativity; each of his choices is embedded in one of the two postulates of the theory. The first was the choice between acceptance of the relativity principle as valid for all phenomena, and in particular for electrodynamic phenomena; or acceptance of the Maxwell–Lorentz theory, with its unique ether frame of reference. The second was the choice between an emission theory of light, for which its velocity would be constant only with respect to a frame of reference in which its source was at rest; or adoption of the principle of light constancy, for which the velocity of light would be constant, regardless of the state of motion of its source with respect to (initially) one inertial frame of reference.

The Michelson–Morley experiment played a role in the first choice. Einstein was certainly aware of the result before formulating the special theory, and it was one piece of evidence in favor of adoption of the relativity principle. Just how important a piece of evidence is not clear. In the Kyoto lecture (1922) Einstein says just after the passage quoted earlier:

> When I was still a student, and still playing with this idea, I learned of the strange result of Michelson's experiment, and I realized that if one accepts his result as correct, it would probably be wrong to consider the earth as moving relative to the ether. This, then, was the first step which led me to what I today call the principle of special relativity. After that, I began to think that, even though the earth revolves around the sun, this motion cannot be detected by experiments with light.

It is hard to accept the statement that he learned of the Michelson–Morley experiment while still a student, in view of the 1901 letter to Grossmann, written after his graduation from the ETH. Many later statements by Einstein deny the primary influence of the Michelson–Morley experiment on his adoption of the relativity principle.[35] Whatever the exact role it played in inclining him toward extension of the relativity principle to electrodynamics, the Michelson–Morley experiment was only one among a number of such influences. One other major influence (electromagnetic induction) has been briefly discussed above.

On the other hand, Einstein found the Maxwell–Lorentz theory, as expounded in the *Versuch*, extremely satisfactory in almost all other respects. If he was *a priori* inclined to adopt the relativity principle, he was enough of a physicist to have at least considered stifling these inclinations, and the doubts they engendered about the Maxwell–Lorentz theory, in the face of its undoubted successes. Lorentz had shown that all first order (v/c) effects of the earth's motion through the ether must be absent on the basis of his theory, which also offered explanations of Fresnel's dragging coefficient, and abberation. If Einstein had not found the additional hypothesis invoked in the explanation offered for the Michelson–Morley experiment (contraction hypothesis) to be artificial,[36] he might have been inclined to accept the Maxwell–Lorentz theory even though it seemed to conflict with the relativity principle. This was probably the major contribution of the Michelson–Morley experiment to his reasoning.

Once he had opted for the relativity principle, he was faced with the problem of squaring this principle with all of electrodynamics, especially optics. Here, the Michelson–Morley experiment was of *no help* at all. By the relativity principle, any optical experiment in which source, apparatus and detectors are at rest in the same inertial frame *must* give a null result. That is, the result must be the same regardless of the state of motion of this inertial frame—and *also* regardless of the theory of light propagation adopted. So any hypothesis about light propagation consistent with the relativity principle was consistent with the null result of the Michelson–Morley experiment. Einstein considered two alternative hypotheses, as the letter to Ehrenfest states: (see footnote 28)

> I well knew that the principle of constancy of the velocity of light was something quite independent of the relativity principle; and I weighted which was more probable: the principle of the constancy of c, as required by Maxwell's equations, or the constancy of c exclusively for an observer at rest with respect to the source of light.

It was his inability to "set up any sort of reasonable electromagnetic theory" on the basis of the latter hypothesis, which helped to convince him[37]

> . . . daß eine prinzipiell von der Lorentzschen verschiedene Theorie, welche auf einfachen und anschaulichen Voraussetzungen beruht und dasselbe leistet, nicht aufgestellt werden konnte.

Now he faced the final, critical stage of his work: How to reconcile his two principles. On the emission hypothesis, the Galilean velocity composition law was no problem: if the velocity of light with respect to its source was c, and the velocity of the source with respect to a laboratory frame of reference was v, then the velocity of light with respect to the laboratory frame of reference was $c + v$. This constituted no violation of the relativity principle, any more than a similar argument for a bullet fired from a moving gun. The velocity of the source introduced a distinction between otherwise equivalent inertial frames. But on the hypothesis of the *independence* of the velocity of light from the velocity of its source in (initially) one inertial frame, this way out was not open. If the Galilean velocity composition law was correct, there would be one and only one inertial frame in which the velocity of light was constant and independent of direction—and the velocity of the source could not be invoked to explain the unique position of that frame. The relativity principle would thus be violated. In his Kyoto talk (1922), Einstein described this last stage of his efforts:

> Why are these two things inconsistent with each other? I felt that I was facing an extremely difficult problem. I suspected that Lorentz's ideas had to be modified somehow, but spent almost a year on fruitless thoughts. And I felt that was a puzzle not to be easily solved.
>
> But a friend of mine living in Bern (Switzerland)[38] helped me by chance. One beautiful day, I visited him and said to him: "I presently have a problem that I have been totally unable to solve. Today I have brought this 'struggle' with me." We then had extensive discussions, and suddenly I realized the solution. The very next day, I visited him again and immediately said to him: "Thanks to you, I have completely solved my problem."
>
> My solution actually concerned the concept of time. Namely, time cannot be absolutely defined by itself, and there is an unbreakable connection between time and signal velocity. Using this idea, I could now resolve the great difficulty that I previously felt.
>
> After I had this inspiration, it took only five weeks to complete what is now known as the special theory of relativity.

When Lorentz's theory was reexamined in the light of this new concept of time, and the consequent modification of the concept of space, it was found to require no formal changes but only reinterpretation. The conflict between the relativity principle and Maxwell–Lorentz electrodynamics had indeed been only "apparent."

Naturally this account is conjectural. It is also much too "smooth" and logical, particularly in its use of terms like "postulate" and "principle": Einstein surely formulated his conclusions in these terms at a very late stage of their development. But I hope it gives a sketch of the sort of path that could have led to the final formulation of the theory,[39] is consistent with the known evidence, and shows the role of the Michelson–Morley experiment.

I thank Prof. Gerald Holton for his helpful comments on earlier versions of this paper.

Notes

[1] Hans Reichenbach, "The Philosophical Significance of the Theory of Relativity", in P. A. Schilpp (ed), *Albert Einstein: Philosopher-Scientist* (Open Court, La Salle, 1969). The quotations are from pp. 292–3. The first part of this passage is quoted by Gerald Holton in his essay "On the Origins of the Special Theory of Relativity", in G. Holton, *Thematic Origins of Scientific Thought* (Harvard University Press, Cambridge, 1973). Holton's papers on the origins of special relativity, collected in this volume, are fundamental to all further work on this topic.

[2] A. Einstein to Dr. H. L. Gordon, May 3, 1949. This is item Number 58 – 217 in the Control Index to the Einstein Archive, which may be consulted at Mudd Library, Princeton University. Unpublished items in the Einstein Archives will be referred to by this Control

Index Number. All quotations, except those from Japanese, will be given in the original language or the language of first publication, as well as English translation if necessary.

[3] There is at least one major unpublished exposition of the subject which is not available. This is a manuscript dating from before World War I, prepared for Marx's Handbuch der Radiologie, but not published due to the war. A description of this seventy-three page manuscript, together with a table of contents and reproduction of one page, appeared in an auction catalogue of Hauswedell and Nolte, Auktion 191, 1973. [Now published as "Manuscript on the Special Theory of Relativity," in *The Collected Papers of Albert Einstein*, vol. 4, *The Swiss Years: Writings 1912–1914*, Doc. 1 (Princeton University Press, Princeton 1995), pp. 9–108].

[4] The talk was delivered December 14, 1922. Einstein's translator on the Japanese trip was Jun Ishiwara, a Japanese physicist who had visited Einstein in 1912, and written extensively on relativity theory in both scientific and popular journals. He published his "Record of Professor Einstein's Lectures" in J. Ishiwara, *Einstein Koen-roku* (Tokyo, Tosho 1971), reprint of *Einstein Kyozyu Koen-roku* (Kaizō-sha, Tokyo 1923). I shall cite from an unpublished translation by Akira Ukawa, revised by me. Excerpts, in another translation, may be found in T. Ogawa, "Japanese Evidence for Einstein's Knowledge of the Michelson–Morley Experiment," *Jap. Stud. Hist. Sci.* 18, 73 (1979).

[5] Einstein described the scene in a short poem he wrote to accompany a drawing of the scene:

"Gedrängt das Volk, gespitzt die Ohren
Sie sitzen alle wie verloren,
In Sinnen tief, verzückt der Blick
Ergeben in ein hart' Geschick."
Der Einstein an der Tafel steht
Die Predigt rasch von Stapel geht,
Und Ishiwara flink und fein
Schreibt alles in sein Büchlein ein.

(For drawing and text, see reference 4). This might be rather freely translated as:

Compressed the crowd, unpraised their ears
All sitting as if lost for years
In thinking deep, with raptured gaze
Given over quite to fate's hard ways.
Old Einstein at the blackboard stands
His sermon flowing with mouth and hands
While Ishiwara with sharp, quick look
Writes it all down in his little book.

[6] In a future paper I hope to examine all of these expositions, in a discussion of the origins of the special theory not centered on the role of the Michelson–Morley experiment.

[7] Following the title and bibliographical data on each paper is its number in the Readex Bibliography of Einstein's writings: N. Boni, M. Russ and D. H. Laurence, *A Bibliographical Checklist and Index to the Collected Writings of Albert Einstein*, (Readex Microprint, New York, 1960). Future references to these articles will be Readex Number and year. [These papers may now be consulted in *The Collected Papers of Albert Einstein*, being published by Princeton University Press].

[8] This contrasts with the treatment of the Michelson–Morley experiment in many places in the secondary literature. Loyd Swenson, in his *The Ethereal Aether* (University of Texas, Austin, 1972) states on p. 163:

> Although the Michelson–Morley results had provided only ready-made evidence for the credibility of Einstein's so-called light principle (i.e., the postulate that the speed of light is the same regardless of direction and does not vary with motion of the source of the observer)...

Robert Shankland in "Conversations with Albert Einstein II," *Am. J. Phys.* **41**, 895 (1973), says on p. 896:

> ... it is evident that the importance of the Michelson–Morley experiment for Einstein was that it gave positive confirmation to his belief that the speed of light is invariant in all inertial frames, independent of the motion of source, apparatus or observer.

Shankland starts this sentence with the words: "As clearly reported by Max Wertheimer, who in 1916 discussed with Einstein the development of his ideas in special relativity in great detail ...;" but I can find no support for his conclusion in Wertheimer's references to the Michelson–Morley experiment. See M. Wertheimer, *Productive Thinking, Enlarged Edition* (Harper, New York, 1959), Chapter Ten. Since the book was written long after the period reviewed in this paper, I have not included its reports of interviews with Einstein in 1916.

Just at random, I cite several other sources treating the Michelson–Morley experiment as evidence for the light constancy principle: A. Shadowitz, *Special Relativity*, (W.B. Saunders, Philadelphia, 1968), p. 7; W. H. McCrea, *Relativity Physics* (Methuen, London, 1935, revised fourth edition 1954), pp. 4–7; H. Melcher, *Relativitätstheorie in elementarer Darstellung, mit Aufgaben und Lösungen*, (VEB Deutscher Verlag der Wissenschaften, fifth edition, Berlin 1978), p. 78; M. Schwartz, *Principles of Electrodynamics* (McGraw-Hill, New York 1972), p. 110. Later sections of this paper will make it clear why I believe this to be incorrect.

[9] The Michelson–Morley experiment showed that phenomena were in accord with the relativity principle, even where this was not to be understood on the basis of Lorentz's theory. It looked, therefore, as if the Lorentz theory had to be given up again and replaced by a theory, the foundations of which were in accord with the relativity principle, for such a theory would predict the outcome of the Michelson–Morley experiment at once.

[10] The Michelson experiment suggested the assumption that all phenomena relative to a coordinate system moving with the earth, or more generally relative to any acceleration-free moving system, would take place according to the same laws. In the sequel, we shall call this assumption "the principle of relativity" for short.

[11] "... the equal legitimacy of all inertial systems (special principle of relativity), which was proved in a particularly incisive manner by Michelson's famous experiment." ("On the Theory of Relativity," 1921) Cited from A. Einstein, *Ideas and Opinions* (Crown Publishers, New York, 1954), pp. 246–249.

[12] But the negative result of the Michelson–Morley experiment shows that in one definite case even an effect of second order (proportional to v^2/c^2) was not present, in spite of the fact that on the basis of Lorentz's theory it would have had to be perceptible.

[13] The Lorentz theory arouses our mistrust from the circumstances that it appears to violate the principle of relativity.

14 The successes of the Lorentz theory were so significant that physicists would have unhesitatingly dropped the principle of relativity, if an important experimental result had not existed, of which we must now speak, namely the Michelson experiment.

15 His earliest preserved essay, "Über die Untersuchung des Ätherzustandes im magnetischen Feld," makes this clear. See J. Mehra, "Albert Einstein's erste Wissenschaftliche Arbeit," *Phys. Bl.* **27**, 386 (1971), which reproduces part of the manuscript and transcribes the text. The manuscript was dated "1894 oder 1895" by Einstein in his own hand in 1955.

16 Corroboration of this account occurs in the biography of Einstein, written with his cooperation by his son-in-law Rudolf Kayser. This was only published in English, with a preface by Einstein, under the pseudonym A. Reiser: *Albert Einstein, A Biographical Portrait* (A. & C. Boni, New York, 1930). Mention of Einstein's interest in the problem of the earth's motion through the light ether, and of his projected experiment, is on p. 52. Reiser-Kayser adds: "But there was no chance to build this apparatus. The skepticism of his teachers was too great, the spirit of enterprise too small."

17 Albert Einstein to Marcel Grossmann, Control Index Number 11-485. "A considerably simpler method for the investigation of the relative motion of matter with respect to the light ether has again occurred to me, which is based on ordinary interference experiments. If only inexorable destiny gives me the time and peace necessary to carry it out. When we meet again, I will tell you about it."

18 H. Fizeau, *Pogg. Ann.* **92**, 652 (1854).

19 H. Lorentz, *Proc. Acad. Sci. Amsterdam* **4**, 678 (1902).

20 P. Nordmeyer, *Ann. Physik* **11**, 284 (1902). Nordmeyer was a student of Bucherer's and did the work for his Thesis. An article by Bucherer preceded Nordmeyer's in the *Annalen*, in which Bucherer showed that a negative result was to be anticipated even on the assumption of a moving ether: A. H. Bucherer, *Ann. Physik* **11**, 270 (1902).

21 J. Laub, *Jahrb. d. Radioaktivtät* **7**, 405 (1910). Laub lists it among the optical experiments to first order in (v/c) which gave negative results.

22 P. Speziali (ed.), *A. Einstein-M. Besso Correspondance 1902–1955* (Hermann, Paris, 1972), p. 4. "In the immediate future I will occupy myself with [studies of] molecular forces in gases, and then carry out comprehensive studies in electron theory."

23 H. A. Lorentz, *Versuch einer Theorie der elektrischen und optischen Erscheinungen in bewegten Körpern* (E. J. Brill, Leiden, 1895). Evidence for Einstein's study of the Versuch included his testimony in the Kyoto lecture, as well as later comments.

24 See M. Flückiger, *Albert Einstein in Bern* (Haupt, Bern, 1974), pp. 71–80, for Einstein's relations with the Bern Scientific Society.

25 G. Holton, "On Trying to Understand Scientific Genius," American Scholar 41, 95 (1971). Reprinted as Chapter 10 of his *Thematic Origins of Scientific Thought* (Harvard, Cambridge, 1973).

26 A. Einstein, "The Fundamental Idea of General Relativity in Its Original Form," Control Index Number 2-070 (original in Morgan Library, New York City). I quote from the translation given in footnote 25: In the construction of special relativity theory, the following, not-yet-mentioned thought concerning the Faraday [experiment] on electromagnetic induction, played for me a leading role.

27 The thought that one is dealing here with two fundamentally different cases was for me unbearable [war mir unerträglich]. The difference between these two cases could be not a real difference but rather, in my conviction, only a difference in the choice of the

reference point... The phenomenon of the electromagnetic induction forced me to postulate the (special) relativity principle. [Footnote:] The difficulty that had to be overcome was in the constancy of the velocity of light in vacuum which I had first thought I would have to give up. Only after groping for years did I notice that the difficulty rests on the arbitrariness of the kinematical fundamental concepts...

[28] A. Einstein to Paul Ehrenfest, June 1912. Control Index Number 9-333. Other letters in which Einstein discusses his consideration of an emission hypothesis for light include: A. Einstein to M. Viscardini, April 28, 1922, Control Index Number 25-301 and A. Einstein, draft letter to C. O. Hines, February 1952, Control Index Number 12-251.

[29] P. Ehrenfest, "Zur Frage der Entbehrlichkeit des Lichtäthers," *Phys. Zeitschr.* **13**, 317 (1912).

[30] I was not in the least annoyed by your article. On the contrary. Such considerations are quite familiar to me from the pre-relativistic period. I well knew that the principle of the constancy of the velocity of light was something quite independent of the relativity principle; and I weighed which was more probable: the principle of the constancy of c, as required by Maxwell's equations, or the constancy of c exclusively for an observer at rest with respect to the source of light. I decided for the former, because I was convinced that any light is completely defined by frequency and intensity, quite independently of whether it comes from a moving light source or one at rest. It did not even enter my mind to imagine that a deflected radiation propagated through a point could behave differently than radiation newly emitted at the point in question. Such complications seemed to me much more unreasonable than those which the new concept of time involves.

[31] Control Index Number 20-040.

[32] For a discussion of the Ritz theory, and other emission theories, including De Sitter's refutation, see W. Pauli, *The Theory of Relativity* (Pergamon Press, London, 1958). This is the translation of his 1921 article: "Relativitätstheorie," *Encykl. d. Math. Wiss.* V 2, 539 (1921).

[33] Your attempt to replace special relativity with the assumption that the velocity of light is constant *relative to the source of light* was first advocated by Ritz. This assumption is compatible with the Michelson experiment and with abberation ... [Einstein then refers to De Sitter's refutation of Ritz's theory]. In addition this theory requires that everywhere and in each definite direction light waves with different velocities of propagation shall be possible. It must be impossible to set up any sort of reasonable electromagnetic theory which accomplishes this. This is the principal reason why, even before setting up the special theory of relativity, I rejected this way out, in itself conceivable.

[34] "What song the *Syrens* sang, or what name *Achilles* asumed when he hid himself among women, though puzzling questions, are not beyond all conjecture": Sir Thomas Browne, "Hydriotaphia, Urne-Buriall, or, A Discourse of the Sepulchrall Urnes lately found in Norfolk," in this *Religio Medici and Other Writings* (E. P. Dutton, New York, 1951), p. 177. I am grateful to Mr. E. A. Poe for drawing my attention to this quotation.

[35] See numerous quotations in G. Holton, "Einstein, Michelson and the 'Crucial' Experiment," *Isis* 60, 133 (1969), reprinted as Chapter 9 of his *Thematic Origins of Scientific Thought* (Harvard, Cambridge, 1973).

[36] In the manuscript referenced in footnote 26, Einstein says: Durch diese [Lorentz-Fitzgerald contraction] Hypothese wird uns formal dem Tatbestande genügt, aber der Geist bleibt bei alledem unbefriedigt. Sollte die Natur uns wirklich in einen Äthersturm gesetzt haben und sollte sie dabei andererseits die Naturgesetze accurat so eingereiht haben, daß

wir von diesem Sturme nichts bemerken können?
This [Lorentz-Fitzgerald contraction] hypothesis formally suffices for the facts of the situation, but in spite of that the mind remains dissatisfied. Is nature really supposed to have placed us in an ether gale, and on the other hand exactly so arranged the laws of nature that we can notice nothing of this gale?

[37] Readex 40 (1911), p. 4: . . . that a theory differing in principle from Lorentz's, based upon simple and intuitive assumptions and accomplishing as much, could not be set up.

[38] The friend, of course, was Michele Besso, by then his co-worker at the Patent Office, and the only person he thanks for help in his 1905 paper (Readex 9).

[39] For a very clear and insightful account of the logical structure of Einstein's argument in the 1905 paper, which properly evaluates the role of the light constancy postulate, see R. B. Williamson, "Logical Economy in Einstein's, On the Electrodynamics of Moving Bodies," *Stud. Hist. Phil. Sci.* 8, 49 (1977).

Einstein on the Theory of Relativity

John Stachel

I

Einstein was the first physicist to formulate clearly the new kinematical foundation for all of physics inherent in Lorentz's electron theory. This kinematics emerged in 1905 from his critical examination of the physical significance of the concepts of spatial and temporal intervals. The examination, based on a careful definition of the simultaneity of distant events, showed that the concept of a universal or absolute time, on which Newtonian kinematics is based, has to be abandoned; and that the Galilean transformation between the coordinates of two inertial frames of reference has to to be replaced by a set of spatial and temporal transformations that agree formally with a set that Lorentz had introduced earlier with a quite different interpretation. Through its interpretation of these transformations as elements of a space-time symmetry group corresponding to the new kinematics, the special theory of relativity, as it later came to be called, provided physicists with a powerful guide in the search for new dynamical theories of fields and particles, and gradually led to a deeper appreciation of the role of symmetry criteria in physics. The special theory of relativity also provided philosophers with abundant material for reflection on the new views of space and time. The special theory, like Newtonian mechanics, still assigns a privileged status to the class of inertial frames of reference. The attempt to generalize the theory to include gravitation led Einstein to formulate the equivalence principle in 1907. This was the first step in his search for a new theory of gravitation denying a privileged role to inertial frames, a theory that is now known as the general theory of relativity.

Einstein presented the special theory in *Einstein 1905r* (Doc. 23), a paper which is a landmark in the development of modern physics. In the first part of this paper, Einstein presented the new kinematics, basing it on two postulates, the relativity principle and the principle of the constancy of the velocity of light. In

The Collected Papers of Albert Einstein,
Vol. 2, *The Swiss Years: Writings 1901–1909* pp. 253–274

the second part, he applied his kinematical results to the solution of a number of problems in the optics and electrodynamics of moving bodies. This volume includes a number of other papers on relativity. Three of these, *Einstein 1905s* (Doc. 24), *1906e* (Doc. 35), and *1907h* (Doc. 45), present arguments for one of the most important consequences of the theory, the equivalence of mass and energy. Two papers, *Einstein 1906g* (Doc. 36) and *1907e* (Doc. 41), suggest new experimental tests of the theory. In his reply to a paper by Ehrenfest, *Einstein 1907g* (Doc. 44), Einstein clarified the kinematical nature of the special theory. *Einstein 1907j* (Doc. 47) is the first major review of the foundations of the theory, as well as of its applications to date (corrections appeared in *Einstein 1908b* [Doc. 49]). The review also contains discussions of several topics Einstein had not previously treated. Particularly notable are Einstein's comments on the equivalence principle and its relationship to the problem of gravitation. A brief discussion comment, *Einstein et al. 1909b* (Doc. 59), concerns an objection to the theory. Two papers on the relativistic electrodynamics of moving media were written in collaboration with Jakob Laub, *Einstein and Laub 1908a* (Doc. 51), *1908b* (Doc. 52). These papers and corrections in *Einstein and Laub 1908c*, (Docs. 53 and 54) are discussed in the editorial note, "Einstein and Laub on the Electrodynamics of Moving Media," pp. 503–507. Einstein's Salzburg talk, *Einstein 1909c* (Doc. 60), stating the relation between the theory of relativity and Einstein's views on the structure of radiation, is also discussed in the editorial note, "Einstein's Early Work on the Quantum Hypothesis," pp. 147–148, and in the Introduction, pp. xvii–xviii. In 1913, *Einstein 1905r* (Doc. 23) and *Einstein 1905s* (Doc. 24) were reprinted in *Blumenthal 1913*.[1]

Strictly speaking, it is anachronistic to use the term "the theory of relativity" in discussing Einstein's first papers on the subject.[2] In them, he referred to the "principle of relativity" ("Prinzip der Relativität" or "Relativitätsprinzip").[3] Max Planck used the term "Relativtheorie" in 1906 to describe the Lorentz–Einstein equations of motion for the electron, and this expression continued to be used from time to time for several years.[4] Bucherer seems to have been the first person to use the term "Relativitätstheorie" in the discussion following Planck's lecture.[5] The term was used in an article by Ehrenfest[6] and adopted by Einstein in 1907 in his reply.[7] Although Einstein used the term from time to time thereafter, for several years he continued to employ "Relativitätsprinzip" in the titles of his articles.[8] In 1910 the mathematician Felix Klein suggested the name "Invariantentheorie," but this suggestion does not seem to have been adopted by any physicist.[9] In 1915 Einstein started to refer to his earlier work as "the special theory of relativity" ("die spezielle Relativitätstheorie") to contrast it with his later "general theory" ("allgemeine Theorie").[10] In *Einstein 1907j* (Doc. 47) he does refer to the need for generalizing the "principle of relativity" in order to include gravitation in the theory, but throughout the present volume the phrase "the theory of relativity" is used to denote the special theory.

II

In his 1905 paper, as well as in his 1907 and 1909 reviews of the theory, Einstein described the theory of relativity as arising from a specific problem: the apparent conflict between the principle of relativity and the Maxwell–Lorentz theory of electrodynamics.[11] While the relativity principle asserts the physical equivalence of all inertial frames of reference, the Maxwell–Lorentz theory implies the existence of a privileged inertial frame.

The principle of relativity originated in classical mechanics.[12] Assuming Newton's laws of motion and central force interactions, it can be demonstrated that it is impossible to determine the state of motion of an inertial frame by means of mechanical experiments carried out within a closed system with center of mass at rest in this frame. This conclusion, well known and empirically well confirmed by the end of the nineteenth century, was sometimes called the principle of relative motion, or principle of relativity.[13]

The introduction of velocity-dependent forces between charged particles led to doubts about the validity of the relativity principle for magnetic interactions. The wave theory of light appeared to invalidate the principle for optical phenomena. The theory seemed to require an all-pervading medium, the so-called luminiferous ether, to explain the propagation of light in the absence of ordinary matter. The assumption that the ether moves together with matter seems excluded by the phenomenon of aberration and by Fizeau's results on the velocity of light in moving media. If the ether is not dragged with matter, it should be possible to detect motion relative to a reference frame fixed in the ether by means of optical experiments. However, all attempts to detect the motion of the earth through the ether by optical experiments failed.[14]

Maxwell's electromagnetic theory was intended to provide a unified explanation of electric, magnetic, and optical phenomena. With its advent, the question arose of the status of the principle of relativity for such phenomena. Does the principle follow from the fundamental equations of electrodynamics?[15] The answer to this question depends on the form of Maxwell's equations postulated for bodies in motion. Hertz developed an electrodynamics of moving bodies based on the assumption that the ether moves with matter, in which the relativity principle holds.[16] In addition to its inability to account for the optical phenomena mentioned above, Hertz's theory was unable to explain several new electromagnetic phenomena, and it soon fell out of favor.[17]

By the turn of the century, when Einstein started working on the electrodynamics of moving bodies, Lorentz's very successful version of Maxwell's theory had gained wide acceptance.[18] Lorentz's electrodynamics is based on a microscopic theory that came to be known as the electron theory.[19] The theory makes a sharp distinction between ordinary, ponderable matter and the ether. Ordinary matter is composed of finite-sized material particles, at least some of which are electrically charged. All of space, even those regions occupied by material particles, is pervaded by the ether, a medium with no mechanical properties, such as mass. The ether is the seat of all electric and magnetic fields. The ether only acts on matter through the electric and magnetic forces that the fields exert on charged particles. By assuming such atoms of electricity ("electrons"), the theory incorporates an important element of the pre-Maxwellian continental tradition into Maxwell's theory, from which it took the field equations.

The parts of the ether are assumed to be immobile relative to each other. Hence, Lorentz's ether defines a rigid reference frame, which is assumed to be inertial. It is in this frame that Maxwell's equations are valid; in other frames, the Galilei-transformed form of these equations hold. Hence it should be possible to detect the motion of the earth through the ether by suitably designed terrestrial electromagnetic or optical experiments. Lorentz was well aware of the failure of all attempts to detect the motion of the earth through the ether, in particular such sensitive optical attempts as the Michelson-Morley experiment,[20] and attempted to explain this failure on the basis of his theory.

His basic approach to this problem in 1895 was to use the theorem of "corresponding states" ("correspondierende Zustände") in combination with the well-known contraction hypothesis.[21] The theorem is essentially a calculation tool that sets up a correspondence between phenomena in moving systems and those in stationary systems by introducing transformed coordinates and fields. On this basis, Lorentz was able to account for the failure of most electromagnetic experiments to detect the motion of the earth through the ether. In 1904 he showed how to explain the failure of all such experiments by a generalization of his theory. He introduced a set of transformations for the spatial and temporal coordinates (soon named the Lorentz transformation by Poincaré)[22] and for the electric and magnetic field components, such that under these transformations Maxwell's equations in the absence of charges take the same form in all inertial frames.[23] Lorentz's approach to the explanation of failure of attempts to detect motion through the ether, thus, was to show that the basic equations of the electron theory, in spite of the fact that they single out the ether rest frame, can still be used to explain the failure of all optical and electromagnetic attempts to detect the earth's motion through the ether.

Einstein's work was based on a new outlook on the problem. Instead of regarding the failure of electromagnetic and optical experiments to detect the earth's motion through the ether as something to be deduced from the electrodynamical equations, he took this failure as empirical evidence for the validity of the principle of relativity in electrodynamics and optics. Indeed, he asserted the universal validity of the principle, making it a criterion for the acceptability of any physical law. In this respect he gave the principle of relativity a role similar to that of the principles of thermodynamics, an example that he later stated helped to guide him.[24] Rather than being deductions from other theories, such principles are taken as postulates for deductive chains of reasoning, resulting in a theory formulating general criteria that more specialized theories must satisfy.[25]

Einstein now confronted the problem of making Maxwell–Lorentz electrodynamics compatible with the principle of relativity. He did so by means of a principle drawn from electrodynamics, the principle of the constancy of the velocity of light. That the velocity of light is independent of that of its source, and has a constant value in the ether rest frame, can be deduced from the Maxwell–Lorentz theory. Einstein dropped the ether from consideration, and took the constancy of the velocity of light as a second postulate, supported by all the empirical evidence in favor of the Maxwell–Lorentz theory. When combined with the relativity principle, this leads to an apparently paradoxical conclusion: the velocity of light must be the same in all inertial frames. This result conflicts with the Newtonian law of addition of velocities, forcing a revision of the kinematical foundations of electrodynamics. Einstein showed that the simultaneity of distant events is only defined physically relative to a particular inertial frame, leading to kinematical transformations between the spatial and temporal coordinates of two inertial frames that agree formally with the transformations introduced by Lorentz in 1904.

Einstein next considered the implications of the new kinematics for electrodynamics and mechanics. By eliminating the concept of the ether, he in effect asserted that electromagnetic fields do not require an underlying substratum.[26] He showed that the Maxwell–Lorentz equations for empty space remain invariant in form under the new kinematical transformations when the transformation laws for the electric and magnetic fields are appropriately defined. He deduced appropriate transformation laws for charge densities and velocities from the requirement that Maxwell's equations remain invariant when convection currents are added. Finally, by assuming that Newton's equations hold for a charged particle at rest, he was able to use a kinematical transformation to deduce the equations of motion of a charged particle ("electron") with arbitrary velocity.

The problems connected with the formulation of an electrodynamics of moving bodies consistent with all the experimental evidence were discussed frequently during the years Einstein was working on his theory. Statements similar to many of the individual points made in *Einstein 1905r* (Doc. 23) occur in the contemporary literature, and Einstein may well have been familiar with some of the books and articles in which they do.[27] But his approach to the problem, leading to the peculiar combination of these ideas in his paper, is unique—particularly the recognition that a new kinematics of universal applicability is needed as the basis for a consistent approach to the electrodynamics of moving bodies.

III

Einstein's work on relativity grew out of his long-standing interest in the electrodynamics and optics of moving bodies.[28] His first scientific essay, written in 1895, discussed the propagation of light through the ether.[29] The next year, as he later recalled, the following problem started to puzzle him:

> If one were to pursue a light wave with the velocity of light, one would be confronted with a time-independent wave field. Such a thing doesn't seem to exist, however! This was the first childlike thought-experiment concerned with the special theory of relativity.
>
> Wenn man einer Lichtwelle mit Lichtgeschwindigkeit nachläuft, so würde man ein zeitunabhängiges Wellenfeld vor sich haben. So etwas scheint es aber doch nicht

zu geben! Dies war das erste kindliche Gedanken-Experiment das mit der speziellen Relativitätstheorie zu tun hat.[30]

By this time Einstein presumably was familiar with the principle of relativity in classical mechanics. While preparing for the ETH entrance examination in 1895, he had studied the German edition of Violle's textbook.[31] Violle actually based his treatment of dynamics on the "principle of relative motions" ("Prinzip der relativen Bewegungen") together with the principle of inertia.[32]

About 1898, Einstein started to study Maxwell's electromagnetic theory, apparently with the help of Drude's textbook.[33] By 1899, after studying Hertz's papers on the subject, he was at work on the electrodynamics of moving bodies.[34] He discussed this topic a number of times in letters to Mileva Marić between 1899 and 1901,[35] once refering to "our work on relative motion" ("unsere Arbeit über die Relativbewegung").[36] In December 1901, Einstein also explained his ideas on the subject to Professor Alfred Kleiner of the University of Zurich, who encouraged him to publish them;[37] but there is no evidence that Kleiner played a further role in the development of these ideas.

Einstein's comments show that in 1899 his viewpoint on electrodynamics was similar to that of Lorentz;[38] but, aside from this similarity, there is no evidence that Einstein had yet read anything by Lorentz.[39] Shortly afterward, Einstein designed an experiment to test the effect of the motion of bodies relative to the ether on the propagation of light; in 1901 he designed a second such experiment, but was unable to carry out either.[40] Near the end of 1901, he was at work on what he described as "a capital paper" ("eine kapitale Abhandlung") on the electrodynamics of moving bodies, asserting his renewed conviction of the correctness of his "ideas on relative motion" ("Ideen über die Relativbewegung").[41] His words may indicate that he already doubted whether motion with respect to the ether is experimentally detectable.[42] Soon afterward he wrote that he intended to study Lorentz's theory in earnest.[43]

There is direct or strong indirect contemporary evidence that, by 1902, Einstein had read or was reading the following works on electrodynamics or optics: *Drude 1894, 1900c*; *Helmholtz 1892, 1897*; *Hertz, H. 1890a, 1890b*; *Lorentz 1895*; *Voigt 1896*; and *Wien 1898*.[44] There is a later report that he read *Föppl 1894* as a student.[45] Comments in his letters on articles published in the *Annalen der Physik* between 1898 and 1901 indicate that during those years he looked at that journal regularly, and studied a number of articles in it.[46] It is reasonable to suppose that he continued to do so between 1902 and 1905.[47] During these years a number of significant articles on the electrodynamics and optics of moving bodies appeared in the *Annalen*.[48] He cited several works published before 1905 in his later articles on relativity, and it is possible that he read one or more of these before 1905.[49] Einstein's readings on the foundations of science during these years, as reported by his friend Maurice Solovine, are described in the Introduction. He later attributed great significance in the development of the theory of relativity to his reading of Hume, Mach, and Poincaré.[50]

Belief in the reality of the ether was widespread at the turn of the century.[51] However, Einstein was familiar with several works that questioned the certainty of its existence. Mill, in the course of a discussion of "the Hypothetical Method" in his *Logic*, gives a number of reasons for skepticism concerning "the prevailing hypothesis of a luminiferous ether."[52] Poincaré, in *La science et l'hypothèse*, raised the question of the existence of the ether, even if he offered no clear answer.[53] Ostwald, in his *Lehrbuch der allgemeinen Chemie*, suggested that the ether hypothesis could be replaced by a purely energetic treatment of radiation.[54]

Few contemporary documents throw any light on Einstein's work on electrodynamics between 1902 and 1905. On 22 January 1903, he wrote Michele Besso: "In the near future I want to deal with molecular forces in gases, and then make a comprehensive study of electron theory" ("In der nächsten Zeit will ich mich mit den Molekularkräften in Gasen abgeben, und dann umfassende Studien in Elektronentheorie machen").[55] On 5 December 1903, Einstein gave a talk to the Naturforschende Gesellschaft Bern on "The Theory of Electromagnetic Waves" ("die Theorie der elektromagnetischen Wellen").[56] By the time Einstein wrote his friend Conrad Habicht early in 1905, the theory was practically complete:

> The ... paper exists only as a sketch and is an electrodynamics of moving bodies that utilizes a modification of the theory of space and time.
>
> Die ... Arbeit liegt erst im Konzept vor und ist eine Elektrodynamik bewegter Körper unter Benützung einer Modifikation der Lehre von Raum and Zeit.[57]

Later statements by Einstein suggest several important elements in the development of his ideas on relativity before *Einstein 1905r* (Doc. 23) was written that are not recorded in any known contemporary documents. In 1932 he gave a general characterization of "the situation that led to setting up the theory of special relativity" ("die Situation, die zur Aufstellung der speziellen Relativitätstheorie geführt hat"):

> Mechanically all inertial systems are equivalent. In accordance with experience, this equivalence also extends to optics and electrodynamics. However, it did not appear that this equivalence could be attained in the theory of the latter. I soon reached the conviction that this had its basis in a deep incompleteness of the theoretical system. The desire to discover and overcome this generated a state of psychic tension in me that, after seven years of vain searching, was resolved by relativizing the concepts of time and length.
>
> Mechanisch sind alle Inertialsysteme gleichwertig. Nach den Erfahrungen erstreckt sich diese Gleichwertigkeit auch auf die Optik, bewz. Elektrodynamik. In der Theorie der letzteren erschien aber diese Gleichwertigkeit unerreichbar. Ich gewann früh die Ueberzeugung, das dies in einer tiefen Unvollkommenheit des theoretischen Systems seinen Grund habe. Der Wunsch diese aufzufinden und zu beheben, erzeugte einen Zustand psychischer Spannung in mir, der nach sieben Jahren vergeblichen Suchens durch Relativierung des Begriffe Zeit und Länge gelöst wurde.[58]

In 1952 he wrote:

> My direct path to the special theory of relativity was mainly determined by the conviction that the electromotive force induced in a conductor moving in a magnetic field

> is nothing other than an electric field. But the result of Fizeau's experiment and the phenomenon of aberration also guided me.
>
> Mein direkter Weg zur speziellen Relativitäts-Theorie wurde haupsächlich durch die Überzeugung bestimmt, dass die in einem im Magnetfeld bewegten Leiter induzierte elektromotorische Kraft nichts anderes sei als ein elektrisches Feld. Aber auch das Ergebnis des Fizeau'schen Versuches und das Phänomen der Aberration führten mich.[59]

Beyond their well-known role as evidence against the assumption that the ether is completely carried along by moving matter, it is not clear what role the result of Fizeau's experiment and the phenomenon of aberration played in Einstein's thinking.[60] Possibly its role depended on the fact that, in both cases, the observed effect only depends on the motion of matter (water in the first case, a star in the second) relative to the earth, and not on the presumed motion of the earth with respect to the ether.

In the case of electromagnetic induction, Einstein earlier gave a more detailed account of its role. In 1920, Einstein wrote:

> In setting up the special theory of relativity, the following ... idea about Faraday's electromagnetic induction played a guiding role. According to Faraday, relative motion of a magnet and a closed electric circuit induces an electric current in the latter. Whether the magnet is moved or the conductor doesn't matter; only the relative motion is significant ... The phenomena of electromagnetic induction ... compelled me to postulate the principle of (special) relativity.
>
> Bei der Aufstellung der Speziellen Relativitätstheorie hat für mich der folgende ... Gedanke über die Faraday'sche magnet-elektrische Induktion eine führende Rolle gespielt. Bei der Relativbewegung eines Magneten gegenüber einem elektrischen Stromkreise, wird nach Faraday in letzterem ein elektrischer Strom induziert. Ob der Magnet bewegt wird oder der Leiter, ist gleichgültig; es kommt nur auf die Relativ-Bewegung an ... Die Erscheinungen der magnetelektrischen Induktion zwang mich dazu, das (spezielle) Relativitätsprinzip zu postulieren.[61]

In a footnote he added:

> The difficulty to be overcome then lay in the constancy of the velocity of light in vacuum, which I first thought would have to be abandoned. Only after groping for years did I realize that the difficulty lay in the arbitrariness of the fundamental concepts of kinematics.
>
> Die zu überwindende Schwierigkeit lag dann in der Konstanz der Vakuum-Lichtgeschwindigkeit, die ich zunächst aufgeben zu müssen glaubte. Erst nach jahrelangem Tasten bemerkte ich, dass die Schwierigkeit auf der Willkür der kinematischen Grundbegriffe beruhte.

His stong belief in the relativity principle and abandonment of "the constancy of the velocity of light in vacuum" led Einstein to explore the possibility of an emission theory of light. In such a theory, the velocity of light is only fixed relative to that of its source, so it is clearly consistent with the relativity principle. Newton's corpuscular theory of light is an emission theory, and Einstein's search for such a theory may have been one source of his light quantum hypothesis. In 1912, commenting on Ritz's emission theory,[62] Einstein referred to "Ritz's conception, which before the theory of relativity was also mine" ("die Ritz'sche Auffassung, die ... vor der Rel.-Theorie auch die meine war").[63] Elsewhere, he expanded on this theme:

> I knew that the principle of the constancy of the velocity of light was something quite independent of the relativity postulate, and I weighted which was more probable, the principle of the constancy of c, as required by Maxwell's equations, or the constancy of c exclusively for an observer located at the light source, I decided in favor of the former...
>
> Ich wusste wohl, dass das Prinzip von der Konstanz der Lichtgeschwindigkeit etwas von dem Relativitätspostulat ganz unabhängiges ist, und ich erwog, was wahrscheinlicher sei, das Prinzip von der Konstanz von c, wie es von den Maxwell'schen Gleichungen gefordert wird, oder die Konstanz von c, ausschliesslich für einen Beobachter, der bei der Lichtquelle sitzt. Ich entschied mich für das erstere ...[64]

In 1924, Einstein described the sudden resolution of his dilemma:

> After seven years of reflection in vain (1898–1905), the solution came to me suddenly with the thought that our concepts and laws of space and time can only claim validity insofar as they stand in a clear relation to our experiences; and that experience could very well lead to the alteration of these concepts and laws. By a revision of the concept of simultaneity into a more malleable form, I thus arrived at the special theory of relativity.
>
> Nach siebenjährigem vergeblichem Nachdenken (1898–1905) kam mir plötzlich die Lösung mit dem Gedanken, daß unsere Begriffe und Gesetze über Raum und Zeit nur insofern Geltung beanspruchen dürfen, als sie mit den Erlebnissen in klaren Beziehungen stehen, und daß die Erfahrung sehr wohl dazu führen könne, daß wir diese Begriffe und Gesetze abändern. Durch eine Revision des Begriffes der Gleichzeitigkeit unter gestaltbarer Form gelangte ich so zur speziellen Relativitätstheorie.[65]

In a talk at Kyoto University in 1922, Einstein is reported to have said that, after a year of struggle with the problem of how to reconcile Lorentz's theory with his ideas on relativity, he visited a friend one day to discuss the problem in detail with him. The next day Einstein said to his friend: "Thanks to you, I have completely solved my problem."[66] The friend was presumably Michele Besso, then his colleague at the Swiss Patent Office and the only person whose help is acknowledged in Einstein's first paper on relativity.[67]

Work on this paper was apparently completed very rapidly after this. In 1952 Einstein wrote that "between the conception of the idea for the special theory of relativity and the completion of the relevant publication five or six weeks elapsed" ("Zwischen der Konzeption der Idee der speziellen Relativitätstheorie und der Beendigung der betreffenden Publikation sind fünf oder sechs Wochen vergangen").[68]

Einstein's comments suggest the following stages in his work on the theory of relativity.

(1) He became convinced that, as is the case for mechanical phenomena, only the relative motions of ponderable bodies are significant in determining electromagnetic and optical phenomena; at some point, this conviction led him to abandon the concept of the ether.

(2) He temporarily abandoned Lorentz's theory of electrodynamics, which appears to attach physical significance to absolute motion (i.e., motion with respect to empty space or the ether).

(3) He explored the possibility of an alternative electrodynamical theory, which would justify the emission hypothesis about the velocity of light relative to its source.

(4) Abandoning such attempts, he reexamined Lorentz's theory, at some point focusing his concern on the conflict of his ideas on relative motion with a particular consequence of Lorentz's theory: the independence of velocity of light of the velocity of its source.

(5) He recognized that this conflict involves previously tacitly accepted kinematical assumptions about temporal and spatial intervals, leading him to examine the meaning of the concept of the simultaneity of distant events.

(6) He defined simultaneity physically, and constructed a new kinematical theory based on the relativity principle and the light principle, thus resolving the apparent conflict between them.

There have been a number of attempts at a detailed reconstruction of Einstein's development of the theory of relativity, attempts which often differ significantly in their conclusions.[69] Such a reconstruction has to take into account other strands in Einstein's work at this time. In particular, by the time he wrote the relativity paper, he no longer regarded Maxwell's electromagnetic theory as universally valid, and had already proposed his light quantum hypothesis.[70] He had also shown that the equipartition theorem, which his work on the foundations of thermodynamics convinced him is valid for the most general classical-mechanical system, combined with Maxwell's theory, leads to an incorrect law for black-body radiation. Thus, he already challenged the unlimited validity of both classical mechanics and of Maxwell's theory. For further discussion of his doubts about classical theories of matter and radiation, see the Introduction, pp. xvi–xxix, and the editorial note, "Einstein's Early Work on the Quantum Hypothesis," pp. 134–148.

Einstein later recalled that, uncertain how to proceed in the search for better theories of the structure of matter and radiation, he became convinced that "only the discovery of a universal formal principle could lead ... to assured results" ("nur die Auffindung eines allgemeinen formalen Prinzips ... zu gesicherten Ergebnissen führen könnte").[71] Such principles play a role analogous in this respect to the role played by the principles of thermodynamics. The theory of relativity is based on just such principles: even though suggested originally by specific mechanical and electromagnetic theories, the principles of relativity and of the constancy of the velocity of light are supported by empirical evidence that is independent of the validity of these theories. For further discussion of the role of such principles in Einstein's thought, see the Introduction, pp. xxi–xxii, xxvi.

IV

According to his sister's memoir, Einstein was anxious about whether his relativity paper would be accepted by the *Annalen der Physik*.[72] After it was accepted, he eagerly anticipated an immediate response to its publication, even though he expected it to be critical. He was greatly disappointed when his paper was not even mentioned in the following issues of the *Annalen*. Sometime afterward, she recounts, he received a letter from Max Planck, requesting explanations of a few obscure points in the work.[73]

> After the long period of waiting, this was the first sign that his paper was being read at all. The happiness of the young scholar was all the greater, since acknowledgement of his accomplishment came from one of the greatest contemporary physicists ... At that time Planck's interest signified infinitely much for the morale of the young physicist.
>
> Nach der langen Wartezeit war dies das erste Zeichen, dass seine Arbeit überhaupt gelesen worden war. Die Freude des jungen Gelehrten war um so grösser, da die Anerkennung seiner Leistung von einem der grössten Physiker der Gegenwart herrührte ... In jenem Zeitpunkt bedeutete das Interesse Plancks in moralischer Beziehung unendlich viel für den jungen Physiker.

Planck and Einstein continued to correspond, and Planck discussed Einstein's paper in the University of Berlin's physics colloquium during the fall of 1905.[74] During the next few years, Planck wrote several papers developing further consequences of the relativity principle,[75] and interested his assistant, Max Laue,[76] and one of his students, Kurd von Mosengeil,[77] in working on related problems. A few years later, Einstein paid tribute to Planck's role in promoting the theory of relativity:

> The attention that this theory so quickly received from colleagues is surely to be ascribed in large part to the resoluteness and warmth with which he intervened for this theory.
>
> Der Entschiedenheit und Wärme, mit der er für diese Theorie eingetreten ist, ist wohl zum großen Teil die Beachtung zuzuschreiben, die diese Theorie bei den Fachgenossen so schnell gefunden hat.[78]

Other physicists also started to discuss Einstein's work in 1905 and 1906. Two months after it appeared, Kaufmann cited it in a preliminary report of his recent experiments on the mass of electrons in β-rays.[79] The following year, in a fuller discussion of his results, while noting that the two theories yield the same equations of motion for the electron, he gave the first clear account of the basic theoretical difference between Lorentz's and Einstein's views.[80] Drude, the editor of the *Annalen*, cited Einstein's paper in the second edition of his standard text on optics,[81] as well as in his article on optics in the *Handbuch der Physik*.[82] Röntgen wrote to Einstein asking for copies of his papers on electrodynamics,[83] presumably in connection with a talk Röntgen was to give on the equations of motion of the electron.[84] Sommerfeld, who heard the talk, soon read Einstein's work and was so impressed that he decided to give a colloquium on it.[85] During 1907, Einstein was in correspondence about the theory with Planck, Laue, Wien,[86] and

Minkowski.[87] In the same year, he was asked to write a review article on relativity, which appeared in Stark's *Jahrbuch der Radioaktivität* at the end of the year,[88] and a major publishing house inquired about the possibility of a book on his research.[89] A reference by Ehrenfest in 1907 to Einstein's theory as a "closed system" ("abgeschlossenes System")[90] led Einstein to clarify his view of the nature of the theory.[91] By 1908, the theory of relativity, though still controversial, and often not clearly distinguished from Lorentz's electron theory, was a major topic of discussion among leading German-speaking physicists.[92]

V

The topic "electrodynamics of moving bodies," as understood at the time Einstein wrote his 1905 paper, usually included not only the microscopic electron theory discussed in that paper, but also the macroscopic theory, involving conduction currents in polarizable and magnetizable moving media. The field equations of such a theory could either be postulated phenomenologically, as Cohn did;[93] or they could be derived from an underlying microscopic theory, as in Lorentz's electron theory.[94] Einstein did not turn to the macroscopic theory until 1908. His work with Laub on that subject is discussed in the Editorial Note, "Einstein and Laub on the Electrodynamics of Moving Media," pp. 503–507.

Since the theory of relativity grew out of Einstein's long-standing concern with electrodynamics, and his applications of the theory were primarily in this field, the theory was often looked upon as essentially another version of Lorentz's electron theory (see Section II, pp. 191–193). Einstein soon felt the need to make clear the distinction between the kinematical results of the theory, deduced from the two principles of the theory,[95] and the application of such kinematical results to the solution of problems in the optics and electrodynamics of moving bodies, to the derivation of the equations of motion of a charged particle[96]—or indeed to any physical theory. He pointed out that the postulates of the theory do not constitute a "closed system" ("abgeschlossenes System"), but only a "heuristic principle, which considered by itself alone only contains assertions about rigid bodies, clocks, and light signals" ("heuristisches Prinzip, welches für sich allein betrachtet nur Aussagen über starre Körper, Uhren und Lichtsignale enthält"). Beyond such assertions, the theory could only establish "relations between otherwise apparently independent laws" ("Beziehungen zwischen sonst voneinander unabhängig erscheinenden Gesetzmäßigkeiten") of physics.[97]

A few months after first publishing the theory of relativity, Einstein discovered a relation that particularly intrigued him, the relation between inertial mass and energy. He wrote Conrad Habicht:

> One more consequence of the electrodynamical paper has also occurred to me. The principle of relativity, together with Maxwell's equations, requires that mass be a direct measure of the energy contained in a body; light transfers mass. A noticeable decrease of mass should occur in the case of radium. The argument is amusing and attractive; but I can't tell whether the Lord isn't laughing about it and playing a trick on me.

Eine Konsequenz der elektrodynamischen Arbeit ist mir noch in den Sinn gekommen. Das Relativitätsprinzip im Zusammenhang mit den Maxwellschen Grundgleichungen verlangt nämlich, daß die Masse direkt ein Maß für die im Körper enthaltene Energie ist; das Licht überträgt Masse. Eine merkliche Abnahme der Masse müßte beim Radium erfolgen. Die Überlegung ist lustig und bestechend; aber ob der Herrgott nicht darüber lacht und mich an der Nase herumgeführt hat, das kann ich nicht wissen.[98]

The ideas that inertial mass is associated with electromagnetic energy was often discussed before 1905. Around the turn of the century, it was suggested that all mechanical concepts could be derived from those of electromagnetism.[99] In particular, there were attempts to derive the entire inertial mass of the electron from the energy associated with its electromagnetic field.[100] It was also proved that a radiation-filled container manifests an apparent inertial mass, which (if the mass of the container is neglected) is proportional to the energy of the enclosed radiation.[101]

In his first paper on this subject, Einstein argued that, as a consequence of the relativity principle, inertial mass is associated with *all* forms of energy.[102] He was only able to establish the association for a process involving the emission of electromagnetic radiation by a system, but argued that the result is independent of the mechanism by which the system loses energy. In addition, he was able to show that changes in energy are associated with changes in inertial mass equal to the changes in energy divided by c^2. His argument was criticized in 1907 by Planck, who presented his own argument to show that a transfer of heat is associated with a similarly related transfer of inertial mass.[103]

Soon afterward, Stark attributed the discovery of the relationship between mass and energy to Planck.[104] Einstein wrote Stark:

I was rather disturbed that you do not acknowledge my priority with regard to the connection between inertial mass and energy.

Es hat mich etwas befremdet, daß Sie bezüglich des Zusammenhanges zwischen träger Masse u. Energie meine Priorität nicht anerkennen.[105]

After receiving a conciliatory reply from Stark, acknowledging his priority,[106] Einstein replied, regretting his original testy reaction:

People, to whom it is granted to contribute something to the progress of science, should not allow pleasure in the fruits of their common work to be clouded by such matters.

Die Leute, denen es vergönnt ist, zum Fortschritt der Wissenschaft etwas beizutragen, sollten sich die Freude über die Früchte gemeinsamer Arbeit nicht durch solche Dinge trüben lassen.[107]

Einstein returned to the relation between inertial mass and energy in 1906 and in 1907, giving more general arguments for their complete equivalence,[108] but he did not achieve the complete generality to which he aspired.[109] In his Salzburg talk, Einstein strongly emphasized that inertial mass is a property of all forms of energy, and therefore electromagnetic radiation must have mass. This conclusion strengthened Einstein's belief in the hypothesis that light quanta manifest particle-like properties.[110]

In 1905, Einstein proposed a number of other experimentally testable consequences of his theory, in particular the equations of motion of the electron.[111] His use of the terms "transverse mass" and "longitudinal mass" indicates that he was familiar with some of the earlier work on this topic.[112] The following year he suggested another experimental test of these equations employing cathode rays.[113]

In this paper, Einstein first mentioned Kaufmann's experimental investigations of the motion of electrons in β-rays.[114] Starting in 1901, Kaufmann had carried out a series of experiments on the deflection of β-rays by electric and magnetic fields. In 1905 he asserted that his recent experiments yielded data for the dependence of the transverse mass on velocity that were incompatible with the (identical) predictions of the Lorentz and Einstein theories, but were compatible with those of the Abraham and Bucherer electron models.[115] Kaufmann's work occasioned considerable discussion. Lorentz was disheartened by the apparent refutation of his theory.[116] Planck subjected the experiment to a careful analysis, and concluded that Kaufmann's work could not be regarded as a definitive refutation of the Lorentz-Einstein predictions.[117] Röntgen, one of the leading German experimentalists, is reported also to have felt that Kaufmann's results were not decisive, because his observations were not that accurate.[118] In his 1907 review, Einstein discussed Kaufmann's results at some length, especially their apparent irreconcilability with the Lorentz-Einstein predictions.[119] Commenting on a figure showing Kaufmann's results and the relativistic predictions, Einstein wrote: "Considering the difficulty of the experiment, one might be inclined to regard the agreement as satisfactory" ("In Anbetracht der Schwierigkeit der Untersuchung möchte man geneigt sein, die Übereinstimmung als eine genügende anzusehen"). However, he noted that the deviations are systematic and well outside Kaufmann's error limits.

> Whether the systematic deviations are based upon a source of error not yet considered, or on lack of correspondence between the foundations of the theory of relativity and the facts, can only be decided with certainty when more manifold observational data are at hand.
>
> Ob die systematischen Abweichungen in einer noch nicht gewürdigten Fehlerquelle oder darin ihren Grund haben, daß die Grundlagen der Relativitätstheorie nicht den Tatsachen entsprechen, kann wohl erst dann mit Sicherheit entschieden werden, wenn ein mannigfaltigeres Beobachtungsmaterial vorliegen wird.

Although Einstein evidently accepted experiment as the ultimate arbiter of the fate of any theory, he was cautious about accepting Kaufmann's results as definitive, perhaps because of his familiarity with Planck's critical analyses of the experiments.[120] What he found even more difficult to accept were alternative equations of motion for the electron that are based on what he regarded as arbitrary dynamical assumptions about the shape of a moving electron. While conceding that Kaufmann's data seemed to favor the theories of Abraham and Bucherer,[121] Einstein concluded:

In my opinion, however, a rather small probability should be ascribed to these theories, since their fundamental assumptions about the mass of a moving electron are not supported by theoretical systems that embrace wider complexes of phenomena.

Jenen Theorien kommt aber nach meiner Meinung eine ziemlich geringe Wahrscheinlichkeit zu, weil ihre die Maße des bewegten Elektrons betreffenden Grundannahmen nicht nahe gelegt werden durch theoretische Systeme, welche größere Komplexe von Erscheinungen umfassen.[122]

This cautious attitude toward Kaufmann's results proved justified. During the following years, controversies over the interpretation of the experimental results prevented investigations of this topic from playing a decisive role in contemporary evaluations of the theory of relativity.[123] Bestelmeyer carried out β-ray experiments generally regarded as inconclusive, while Bucherer's results favoring the Lorentz-Einstein equations were seriously questioned.[124] Experiments using cathode rays, reported by several investigators starting in 1910, proved similarly inconclusive. Almost a decade elapsed until generally accepted data on the variation of electron mass with velocity were at hand; these data supported the relativistic predictions.[125]

In 1907, Einstein agreed to write a review article on relativity for Johannes Stark's *Jahrbuch der Radioaktivität und Elektronik*.[126] He indicated what he had read on the subject by this time:

Besides my own, I know of a paper by H. A. Lorentz (1904),[127] one by E. Kohn,[128] one by Mosengeil,[129] as well as two by Planck.[130] I am not familiar with other relevant theoretical papers.

Ausser meinen eigenen Arbeiten ist mir eine Arbeit von H. A. Lorentz (1904), eine von E. Kohn, eine von Mosengeil sowie zwei von Planck bekannt. Andere die Sache betreffende theoretische Arbeiten sind mir nicht bekannt geworden.[131]

Einstein indicated the care he had taken to make the review pedagogically useful:

I have taken great pains to clarify the assumptions employed, by—as far as possible—introducing these assumptions one by one and pursuing their consequences in sequence.

An die Klarlegung der benutzten Annahmen habe ich viel Sorgfalt verwendet, indem ich—soweit es anging—jene Annahmen einzeln einführte und der Reihe nach ihre Konsequenzen verfolgte.[132]

The review covers relativistic kinematics, optics, electromagnetic theory, the relativistic dynamics of a particle, and the relativistic dynamics and thermodynamics of an extended system. He summarized the results of a number of his earlier papers on the theory of relativity, sometimes simplifying his earlier proofs. He adopted Planck's approach to relativistic dynamics and thermodynamics, although often giving his own proofs of Planck's results.[133]

In his Salzburg lecture, Einstein reviewed the historical background of the theory of relativity in order to stress some consequences of the theory for the problem of the structure of the radiation field.[134] By discarding the concept of the ether and by showing light consists of "structures" ("Gebilde") that carry inertial mass from an emitter to an absorber, he stressed that the theory of relativity opens the way for a theory of light that includes both corpuscular and undulatory features. He went on to argue the necessity of such a theory. For further discussion of the Salzburg lecture, see the Introduction, pp. xvii–xviii, and the editorial note, "Einstein's Early Work on the Quantum Hypothesis," pp. 147–148.

Einstein wrote three additional reviews of relativity between 1910 and 1912.[135] *Einstein 1910a* and *Einstein 1911e*, written for nonspecialized physics audiences, discuss historical and foundational questions. The technically most detailed of all his reviews is an unpublished manuscript written about 1912.[136] After 1912, Einstein's active research interests no longer included what soon came to be known as the special theory of relativity. He occasionally wrote pedagogical articles on aspects of the special theory,[137] and continued to include reviews of the subject in many expositions of the theory of relativity as a whole.[138]

VI

The 1907 review concludes with Einstein's first published discussion of gravitation. The final section entitled "Principle of Relativity and Gravitation" ("Relativitätsprinzip und Gravitation"),[139] takes the equality of gravitational and inertial mass as its starting point.[140] It follows from this equality that there is no observable difference between the behavior of a mechanical system in a gravitation-free uniformly accelerating reference frame and the behavior of the same system in an inertial frame in which there is a uniform gravitational field. Generalizing this mechanical equivalence, Einstein postulated "the complete physical equivalence of a gravitational field and a corresponding acceleration of the reference system" ("die völlige physikalische Gleichwertigkeit von Gravitationsfeld und entsprechender Beschleunigung des Bezugssystems"), which allows one "to replace a homogeneous gravitational field by a uniformly accelerated reference system" ("ein homogenes Gravitationsfeld durch ein gleichförmig beschleunigtes Bezugssystem zu ersetzen").[141] This equivalence enabled him to draw conclusions about the effects of a uniform gravitational field on physical processes from the analysis of such processes in a uniformly accelerated reference system. He thus deduced that spectral lines from the sun show a shift toward longer wave lengths when compared to the corresponding terrestrial lines (an effect now called the gravitational red shift) and that a light ray passing through a gravitational field is deflected from a straight-line path; but he noted that the effect is too small to be observed in the earth's gravitational field. Although his correspondence during the intervening years contains several references to the problem of gravitation, Einstein did not publish again on this topic until 1911.[142] (See Volume 4 [full reference in note 136], the Editorial Note, "Einstein on Gravitation and Relativity: The Static Field," pp. 122–128).

NOTES

[1] See Otto Blumenthal's foreword to *Blumenthal 1913*. Notes added to *Einstein 1905r* (Doc. 23) in this reprint are printed in editorial footnotes to that paper. For a discussion of the authorship of these notes, see *Einstein 1905r* (Doc. 23), note 8.

[2] See *Einstein 1905r* (Doc. 23), *1905s* (Doc. 24), *1906e* (Doc. 35), *1906g* (Doc. 36), and *1907e* (Doc. 41).

[3] He also referred to "die Relativitätselektrodynamik" in *Einstein 1907h* (Doc. 45) on pp. 372, 381, and to "das Relativitätssystem" in *Einstein 1907g* (Doc. 44) on p. 207.

[4] See *Planck 1906b*, p. 424. In a critical review of Kaufmann's experiments at the 1906 meeting of the Gesellschaft Deutscher Naturforscher und Ärzte, Planck compared the "Relativtheorie" of Lorentz and Einstein with the "Kugeltheorie" ("sphere theory") of Abraham. The term "Relativtheorie" was still in use in 1910 (see *Noether 1910*). Planck continued to use "Prinzip der Relativität" or "Relativitätsprinzip" to describe the approaches of Lorentz and Einstein, between which he did not distinguish. See *Planck 1906a, 1907a*.

[5] For Bucherer's comment, see *Discussion/Planck 1906*, p. 760.

[6] *Ehrenfest 1907*, p. 205.

[7] See *Einstein 1907g* (Doc. 44), pp. 206, 207; see also *Einstein 1907j* (Doc. 47), p. 439.

[8] He did not use "Relativitätstheorie" in a title until *Einstein 1911e*.

[9] See *Klein, F. 1910*, p. 287.

[10] See *Einstein 1915b*, p. 778.

[11] See *Einstein 1905r* (Doc. 23), pp. 891–892; *Einstein 1907j* (Doc. 47), pp. 411–413; and *Einstein 1909c* (Doc. 60), pp. 485–487. See *Miller 1981b* for a detailed study and an English translation of *Einstein 1905* (Doc. 23).

[12] See *Torretti 1983* for a discussion of the Newtonian principle of relativity, with some historical references.

[13] See, e.g., *Violle 1892*, p. 90, and *Poincaré 1902*, pp. 135–137, 281, to name two sources that Einstein had read by 1905 (see *The Collected Papers of Albert Einstein* Vol. 1, *The Early Years, 1879–1902* (Princeton University Press, 1987) "Albert Einstein—Beitrag für sein Lebensbild," p. lxiv, for Violle; and *Solovine 1956*, p. viii, for Poincaré).

[14] For a survey of ether theories, centering on attempts to detect the motion of the earth through the ether, see *Hirosige 1976*; for a survey focusing on the Michelson-Morley experiment, see *Swenson 1972*. For a survey of work on the optics of moving bodies in the nineteenth century, see *Sesmat 1937*.

[15] For discussions of the problem of relative motion in current theories of electrodynamics, see *Cohn 1900, 1902*, and *Abraham/Föppl 1904*, pp. 430–436, which speaks of the "axiom of relative motion" ("Axiom der Relativbewegung").

[16] See *Hertz, H. 1890b*. For a review of Hertz's theory, see *Hirosige 1966*.

[17] It appears that Hertz was aware of the difficulties optical phenomena present for his theory. He explicitly stated that it was meant to explain electromagnetic phenomena in the narrower sense (see *Hertz, H. 1890b*). However, even here it yields incorrect predictions. For its rejection see, e.g., *Lorentz 1892c*, p. 6 (reprint edition); *Cohn 1900*, p. 518; *Cohn 1902*, p. 29; and *Abraham/Föppl 1904*, pp. 427–428, 435.

[18] Theories similar to Lorentz's electron theory were proposed by Wiechert (*Wiechert 1896*) and Larmor (*Larmor 1894, 1895, 1897*). Emil Cohn proposed a macroscopic electrodynamics of moving media (*Cohn 1900, 1902*), which received some attention. For a review of these theories, see *Hirosige 1966*. Einstein mentioned Cohn's theory in *Einstein 1907j* (Doc. 47), p. 413, and a copy of *Cohn 1904a* is in Einstein's reprint collection (now in Wix Library, Weizmann Institute, Rehovoth, Israel).

[19] For Lorentz's program see, e.g., *Lorentz 1892c*, pp. 70–71. For reviews of his theory; see *Lorentz 1904c, 1909b*. For discussions of the theory; see *McCormmach 1970a, 1970b*.

[20] *Michelson 1881*; *Michelson and Morley 1887*. For Lorentz's first discussion of Michelson's experiment, see *Lorentz 1886*.

[21] See *Lorentz 1985*.

[22] See *Poincaré 1905b*.

[23] See *Lorentz 1904a*. Lorentz did not have transformation laws for the charge density and velocity that would make the equations with sources fully invariant. These laws were supplied by Poincaré, who recognized that Lorentz's transformations form a group, under which the Maxwell–Lorentz equations remain invariant in form. Poincaré regarded these results as an explanation of the apparent universal validity of the principle of relativity (see *Poincaré 1905b, 1906*).

[24] Einstein first made the comparison between the relativity principle and the principles of thermodynamics in *Einstein 1907g* (Doc. 44). He stated that these principles had served as his "example" ("Vorbild") in *Einstein 1979*, p. 48.

[25] Einstein later formulated the distinction between "theories of principle" ("Prinziptheorien") and "constructive theories" ("konstruktive Theorien") to explain the relationship between the two types of theories (see *Einstein 1919*). See the Introduction, pp. xxi–xxii, for further discussion of Einstein's use of such principles.

[26] This concept of electromagnetic fields is first stated explicitly in *Einstein 1907j* (Doc. 47), p. 413; but it is implicit in *Einstein 1905i* (Doc. 14), as well as in *Einstein 1905r* (Doc. 23).

[27] Some of these are cited in the editorial notes to *Einstein 1905r* (Doc. 23) (see notes 5, 6, 9, 10, 12). Two papers by Emil Cohn, *Cohn 1904a, 1904b*, published in response to *Lorentz 1904a*, deserve special mention. They are "remarkable ... because in some respects they remind us of the point of view taken in the theory of relativity" (*Hirosige 1966*, p. 35). For evidence of Einstein's later familiarity with Cohn's work, see Section V. However, there is still no evidence indicating when he first read Cohn.

[28] See Vol. 1 [full reference in note 13], the editorial note, "Einstein on the Electrodynamics of Moving Bodies," pp. 223–225.

[29] See "On the Investigation of the State of the Ether in a Magnetic Field" (Vol. 1 [full reference in note 13], Doc. 5).

[30] *Einstein 1955*, p. 146. See also *Einstein 1979*, p. 48, 50.

[31] *Violle 1892, 1893*. For a reference to the evidence about when Einstein studied Violle's textbook, see note 13.

[32] See *Violle 1892*, pp. 90–91, for the statement of the two principles, and pp. 92–94 for the derivation of Newton's second law (see p. 90 for Violle's discussion of the role of such principles in physics). A copy of *Violle 1892* is in Einstein's personal library. A marginal annotation in Einstein's hand on p. 94 indicates that he read this section of the book.

[33] *Drude 1894*. For evidence that Einstein read Drude, see Einstein to Mileva Marić, 16 April–8 November 1898 and after 16 April 1898 (Vol. 1 [full reference in note 13], Docs. 40 and 41), especially the descriptive notes to both letters.

[34] *Hertz, H. 1890a, 1890b*. For Einstein's discussion of Hertz's work, see Einstein to Mileva Marić, 10 August 1899 (Vol. 1 [full reference in note 13], Doc. 52).

[35] See Vol. 1 [full reference in note 13], the editorial note, "Einstein on the Electrodynamics of Moving Bodies" pp. 223–225, for a list and discussion of these letters.

[36] See Einstein to Mileva Marić, 27 March 1901 (Vol 1 [full reference in note 13], Doc. 94). There is no indication in their letters of the nature of her collaboration, if any, in his research.

[37] See Einstein to Mileva Marić, 19 December 1901 (Vol. 1 [full reference in note 13], Doc. 130).

[38] See Einstein to Mileva Marić, 10 August 1899 (Vol. 1 [full reference in note 13], Doc. 52).

[39] Einstein may have already read *Wien 1898*, which includes a review of Lorentz's theory. For evidence that Einstein read this article, see Einstein to Mileva Marić, 28 September 1899 (Vol. 1 [full reference in note 13], Doc. 57).

[40] See Einstein to Mileva Marić, 10 September 1899 (Vol. 1 [full reference in note 13], Doc. 54), and Einstein to Marcel Grossmann, 6 September 1901 (Vol. 1, Doc. 122).

[41] Einstein to Mileva Marić, 17 December 1901 (Vol. 1 [full reference in note 13], Doc. 128). *Earman et al. 1982* suggests that Einstein may have incorporated elements of his earlier work on electrodynamics into *Einstein 1905r* (Doc. 23).

[42] In 1899 he read *Wien 1898* (see note 39), which briefly describes a number of experimental attempts to detect the motion of the earth through the ether. The Michelson–Morley experiment is included among those "with negative result" ("mit negativem Ergebniss").

[43] See Einstein to Mileva Marić, 28 December 1901 (Vol. 1 [full reference in note 13], Doc. 131). Einstein later stated that *Lorentz 1895* is the only work by Lorentz that he knew in 1905 (Einstein to Carl Seelig, 19 February 1955). It is possible, however, that he learned about subsequent developments in Lorentz's work through the comments of others on Lorentz's widely discussed theory.

[44] For *Drude 1894*, see Einstein to Mileva Marić, after 16 April 1898 (Vol. 1 [full reference in note 13], Doc. 41). For *Helmholtz 1892* and *Hertz, H. 1890a, 1890b*, see Einstein to Marić, 10 August 1899 (Vol. 1, Doc. 52). For *Helmholtz 1897* and *Wien 1898*, see Einstein to Marić, 28 September 1899 (Vol. 1, Doc. 57). For *Voigt 1896*, see Einstein to Marić, 28 November 1901 (Vol. 1, Doc. 126). For *Drude 1900c* and *Lorentz 1895*, see Einstein to Marić, 28 December 1901 (Vol. 1, Doc. 131).

[45] See *Kayser 1930*, p. 49 and *Frank 1979*, p. 38. It is possible that Föppl's book was brought to his attention by Joseph Sauter, H. F. Weber's former *Assistent*, who was already at the Swiss Patent Office when Einstein started work there. Sauter cited *Föppl 1894* in an article attempting a mechanical explanation of Maxwell's equations (see *Sauter 1901*, p. 332).

[46] See Einstein to Mileva Marić, 28 May 1901 (Vol. 1 [full reference in note 13], Doc. 111), for references to his reading of the *Annalen der Physik*, and to his study of papers in the *Annalen* by Lenard and Reinganum. See Einstein to Marić, 28 September 1899 (Vol. 1, Doc. 57), for a reference to a paper by Wien, and Einstein to Marić, 4 April 1901 (Vol. 1, Doc. 96), for references to papers by Drude and Planck, all in the *Annalen*.

[47] In spite of his statement that the libraries were closed during his free time (Einstein to Johannes Stark, 25 September 1907), he must have found a way to follow some of the periodical literature. The same letter cites several recent journal articles, and his writings of this period refer to a number of others.

[48] See, e.g., *Abraham 1903, 1904b*; *Bucherer 1903*; *Cohn 1902*; *Gans 1905*; *Hasenöhrl 1904*; *Nordmeyer 1903*; *Oppolzer 1902*; *Wien 1904*. For a discussion of Einstein's readings in electrodynamics before 1905, see *Miller 1981b*, pp. 87–92.

[49] For example, "the relevant works" ("die einschlägigen Arbeiten") of Emil Cohn are mentioned in *Einstein 1907j* (Doc. 47), p. 413. *Poincaré 1900* is cited in *Einstein 1906e* (Doc. 35).

[50] See the Introduction, pp. xxiii–xxiv.

[51] In 1909, Einstein cited a statement in *Khvolson 1902* as typical: "The probability of the hypothesis of the existence [of the ether] ... approaches extremely near to certainty"("Die Wahrscheinlichkeit der Hypothese von der Existenz ... grenzt außerordentlich nahe an Gewißheit") (*Einstein 1909c* [Doc. 60], p. 482).

[52] See *Mill 1872*, vol. 2, pp. 12, 20 (see *Solovine 1956*, p. viii, for evidence that Einstein read Mill). Mill also noted several phenomena which appear to favor the emission theory (ibid., p. 23).

[53] See *Poincaré 1902*, pp. 199–202. For a reference to the evidence about when Einstein read this book, see note 13.

[54] See *Ostwald 1893*, part I, pp. 1014–1016. For evidence that Einstein read Ostwald's *Lehrbuch*, see Einstein to Wilhelm Ostwald, 19 March 1901 (Vol. 1 [full reference in note 13], Doc. 92), and Einstein to Mileva Marić, 10 April 1901 (Vol. 1, Doc. 97).

[55] By "Elektronentheorie," Einstein presumably here meant Lorentz's "Elektronentheorie der Elektrodynamik" (*Abraham 1903*, p. 105). There is also an "Elektronentheorie der Metalle," in which Einstein had earlier shown great interest (see Vol. 1 [full reference in note 13], the editorial note, "Einstein on Thermal, Electrical, and Radiation Phenomena," pp. 236–237).

[56] See the Minutes of the Society for that date in the Stadt-und Universitätsbibliothek, Bern, Switzerland, and printed in *Verhandlungen 1904*, p. 328.

[57] Einstein to Conrad Habicht, 18 May–8 June 1905.

[58] Einstein to Erika Oppenheimer, 13 September 1932.

[59] Message by Einstein, prepared for R. S. Shankland to read at a celebration of the centennial of Michelson's birth, 19 December 1952, at Case Institute.

[60] Einstein cited Fizeau's experiment in this connection in several early papers (see *Einstein 1909c, 1910a, 1911e*). Aberration was often similarly cited at the time (see, e.g., *Abraham 1905*, p. 342). *Einstein 1918* refutes the claim that aberration is compatible with an ether-drag theory.

[61] Unpublished manuscript, entitled "Fundamental Ideas and Methods of the Theory of Relativity, Presented as it Developed" ("Grundgedanken und Methoden der Relativitätstheorie, in ihrer Entwicklung dargestellt"). The undated manuscript was finished early in 1920 (see Einstein to Robert W. Lawson, 26 December 1919, and 22 January 1920). [It will appear in Vol. 7 of *The Collected Papers of Albert Einstein*].

[62] See, e.g., *Ritz 1908a, 1908b*.

[63] Einstein to Paul Ehrenfest, 25 April 1912. Ehrenfest had discussed Ritz's emission theory in *Ehrenfest 1912*.

[64] Einstein to Paul Ehrenfest, 20 June 1912. The letter explains in some detail why he had rejected the emission theory. For additional accounts of his early interest in an

emission theory, see Einstein to M. Viscardini, 18 April 1922; Einstein to C. O. Hines, February 1952; and Einstein to A. Rippenbein, after 25 August 1952.

[65] This recording is transcribed in *Herneck 1966a*; the quotation is from p. 134. *Herneck 1976*, p. 349, describes the provenance of the recording: "The discographic document of February 6, 1924, is registered in the catalogue of the former 'Institute for Sound Research' of the University of Berlin under the file number 'Autophon Nr. 56'. I have transcribed the text from a copy on magnetic tape" ("Das diskographische Dokument vom 6.2.1924 ist in der Kartei des ehemaligen 'Instituts für Lautforschung' der Berliner Universität unter der Signatur 'Autophon Nr. 56' registriert. Den Text habe ich nach einer Magnetbandkopie aufgezeichnet").

[66] See the report of Einstein's talk, given on 14 December 1922, in *Ishiwara 1971*, pp. 78–88.

[67] See *Einstein 1905r* (Doc. 23), p. 921.

[68] Einstein to Carl Seelig, 11 March 1952.

[69] See, e.g., *Earman et al. 1982*; *Goldberg 1983*; *Hirosige 1976*; *Holton 1973*, part II; *Miller 1981b*; *Pais 1982*; *Schaffner 1982*.

[70] See *Einstein 1905i* (Doc. 14). In *Einstein 1907h* (Doc. 45), pp. 372–373, Einstein discussed why Maxwell's theory could nevertheless be used with confidence within the limits of its validity.

[71] *Einstein 1979*, p. 48; English translation, p. 49.

[72] See *Winteler–Einstein 1924*, pp. 23–24. Winteler-Einstein also reports that Einstein originally intended to submit the work on special relativity as his doctoral dissertation (for her account of this claim, see the editorial note, "Einstein's Dissertation on the Determination of Molecular Dimensions," p. 175).

[73] It is not clear just when Planck first wrote to Einstein. His first known reference to Planck occurs in a letter of 3 May 1906 to Maurice Solovine: "My papers are well received and are giving rise to further research. Prof. Planck (Berlin) wrote me recently about this" ("Meine Arbeiten finden viel Würdigung und geben Anlass zu weiteren Untersuchungen. Prof. Planck (Berlin) schrieb mir neulich darüber"). If Winteler-Einstein's account is correct, the letter from Planck referred to cannot be his first. The earliest surviving letter from Planck is dated 6 July 1907.

[74] According to Max Laue, who was then Planck's *Assistent*, Planck discussed Einstein's paper at one of the first, if not *the* first, of the physics colloquia during the winter semester. See *Laue 1952*.

[75] See *Planck 1906a, 1907a*. He also analyzed Kaufmann's experiments during this period (see Section V).

[76] See *Laue 1907*.

[77] See *Mosengeil 1907*.

[78] *Einstein 1913*, p. 1079.

[79] See *Kaufmann 1905*.

[80] See *Kaufmann 1906a*, pp. 491–493. See also *Kaufmann 1906b*, which contrasts Cohn's theory to those of Lorentz and Einstein.

[81] See *Drude 1906a*, p. 467.

[82] See *Drude 1906b*, p. 1387.

[83] See Wilhelm Röntgen to Einstein, 18 September 1906.

[84] The talk is discussed in a letter of Arnold Sommerfeld to Wilhelm Wien, 23 November 1906 (Deutsches Museum, Munich, Wien Nachlaß, Mappe Sommerfeld).

[85] See Arnold Sommerfeld to Wilhelm Wien, 23 November 1906 (Deutsches Museum, Munich, Wien Nachlaß, Mappe Sommerfeld). The letter indicates that Wien had previously commented to Sommerfeld on Einstein's paper. By early 1908, Sommerfeld was corresponding with Einstein (see *Eckert and Pricha 1984*). During the winter semester of 1908–1909 Sommerfeld gave a course on the theory of relativity at the University of Munich, which he believed to be the first such course (see *Jungnickel and McCormmach 1986b*, p. 283).

[86] Einstein had a lengthy correspondence with Wien on the question of whether superluminal signal velocities are compatible with Maxwell's theory. See *The Collected Papers of Albert Einstein* Vol. 5 *The Swiss Years: Correspondance, 1902–1914* (Princeton University Press, 1993), the Editorial Note, "Einstein on Superluminal Signal Velocities," pp. 56–60.

[87] Minkowski wrote Einstein to request a copy of his paper for discussion at a Göttingen seminar (see Hermann Minkowski to Einstein, 9 October 1907). The seminar on electrodynamics during the 1907 winter semester was conducted by Hilbert and Minkowski (see *Pyenson 1985*, p. 83).

[88] For correspondence with Stark about this review, see *Hermann 1966*. For Einstein's review paper, see *Einstein 1907j* (Doc. 47). The paper is discussed in this volume in Section V.

[89] See B. G. Teubner to Einstein, 3 October 1907.

[90] *Ehrenfest 1907.*

[91] See *Einstein 1907g* (Doc. 44). For a discussion of this paper, see Section V below.

[92] For the reception of the theory of relativity in Germany and several other countries, see *Goldberg 1984*, part II; and *Glick 1987.*

[93] See *Cohn 1900, 1902, 1904a, 1904b.*

[94] See, e.g., *Lorentz 1904c.*

[95] See *Einstein 1905r* (Doc. 23), "Kinematischer Teil," pp. 892–907.

[96] See *Einstein 1905r* (Doc. 23), "Elektrodynamischer Teil," pp. 907–921.

[97] *Einstein 1907g* (Doc. 44), pp. 206–207. The article is a reply to *Ehrenfest 1907.*

[98] Einstein to Conrad Habicht, end of June–end of September 1905.

[99] *Wien 1900* explicitly states the program of an electromagnetic foundation of mechanics. For a survey of the so-called electromagnetic world view, see *McCormmach 1970b.*

[100] See *Abraham 1902a, 1902b, 1903.* For a contemporary review of such attempts, see *Bucherer 1904*, pp. 51–68.

[101] See *Hasenöhrl 1904, 1905*, and *Mosengeil 1907.* For a contemporary review of work on this topic that does not use the theory of relativity, see *Hasenöhrl 1909b.*

[102] See *Einstein 1905v* (Doc. 27).

[103] See *Planck 1907a.* Section 17. The critical comment on Einstein's argument is in a footnote on p. 565. For a discussion of Planck's argument, and later criticisms of Einstein's derivation, see *Stachel and Torretti 1982.* [See this volume, pp. 215–222].

[104] See *Stark 1907.*

[105] Einstein to Johannes Stark, 17 February 1908.

[106] See Johannes Stark to Einstein, 19 February 1908.

[107] Einstein to Johannes Stark, 22 February 1908.

[108] See *Einstein 1906e* (Doc. 35) and *Einstein 1907h* (Doc. 45).

[109] See *Einstein 1907h* (Doc. 45), pp. 371–372, for Einstein's dissatisfaction with arguments based on special cases.

[110] See *Einstein 1909c* (Doc. 60), p. 490.

[111] See *Einstein 1905r* (Doc. 23), Section 10. Einstein restricted himself to slowly accelerated electrons.

[112] The concepts were introduced by Lorentz (see *Lorentz 1900*). The terms were apparently introduced by Abraham (see *Abraham 1902a*), and commonly employed thereafter in discussions of proposed equations of motion for the electron (see, e.g., *Abraham 1903, Bucherer 1904*).

[113] See *Einstein 1906g* (Doc. 37).

[114] See *Kaufmann 1905, 1906a*, which include references to Kaufmann's earlier work. For citation of contemporary reviews of attempts to determine the variation of electron mass with velocity, see note 123. For recent accounts of Kaufmann's experiments and the subsequent discussion of them, see *Miller 1981b*, pp. 334–352; *Cushing 1981.*

[115] See *Kaufmann 1905.* For Lorentz's theory, see *Lorentz 1904a.* For Abraham's model, see *Abraham 1902a, 1902b, 1903.* For Bucherer's model see *Bucherer 1904*, pp. 57–58. Langevin had independently proposed the same hypothesis as Bucherer about the shape of a moving electron. See *Langevin 1905c.*

[116] See the letter of Henrik Lorentz to Henri Poincaré, 8 March 1906, reproduced in *Miller 1980*, pp. 83–84; and Lorentz's comment in his 1906 lectures at Columbia University, printed in *Lorentz 1909b*, p. 213.

[117] See *Planck 1906b*, and the discussion following his paper, reported in *Discussion/Planck 1906*; see also *Planck 1907b.*

[118] Röntgen's views were expressed in a talk to the Bavarian Academy of Sciences, reported in a letter of Arnold Sommerfeld to Wien, 23 November 1906 (Deutsches Museum, Munich, Wien Nachlaß, Mappe Sommerfeld).

[119] See *Einstein 1907j* (Doc. 47), pp. 437–439.

[120] On 1 November 1907, Einstein thanked Stark for calling his attention to Planck's work on Kaufmann's experiments (*Planck 1906b*). Einstein evidently wrote to Planck requesting a copy, for Planck replied by postcard on 9 November, stating that he had sent copies of his two papers on the subject (*Planck 1906b, 1907b*) to Einstein, and adding further comments on the experiments.

[121] See note 115.

[122] *Einstein 1907j* (Doc. 47), p. 439. For an earlier statement of Einstein's attitude toward the unifying power of theories, see Einstein to Marcel Grossmann, 14 April 1901 (Vol. 1 [full reference in note 13], Doc. 100).

[123] For reviews of the early experiments on the variation of the mass of the electron with its velocity; see *Laub 1910*; *Laue 1911b*, pp. 16–18; *Guye and Lavanchy 1916*, pp. 288–292; *Pauli 1921*, p. 636; and *Lorentz 1922*, chap. VII.

[124] Einstein corresponded with Bucherer in 1908, shortly after the latter carried out his experiments.

[125] See *Guye and Lavanchy 1916*, and *Glitscher 1917*.

[126] Einstein to Johannes Stark, 25 September 1907. See also note 47.

[127] *Lorentz 1904a*.

[128] That is, Emil Cohn. See *Cohn 1900, 1902, 1904a*, and *1904b*. A copy of *Cohn 1904a* is in Einstein's collection of reprints, now in Wix Library, Weizmann Institute, Rehovoth, Israel. See also note 27.

[129] *Mosengeil 1907*.

[130] Presumably a reference to *Planck 1906a, 1907a*. Einstein did not yet know of *Planck 1906b, 1907b* (see note 120).

[131] Einstein to Johannes Stark, 25 September 1907. In his reply (Stark to Einstein, 4 October 1907), Stark cited *Planck 1907a, Laue 1907*. Stark later told Einstein about Planck's studies of Kaufmann's experiments.

[132] Einstein to Johannes Stark, 1 November 1907.

[133] Although Einstein did not publish anything further on relativistic thermodynamics, it appears that later in his life he had doubts about the validity of Planck's approach (see, e.g., Einstein to Max von Laue, 29 January 1952).

[134] See *Einstein 1909c* (Doc. 60), pp. 482–490.

[135] After 1912, Einstein's reviews of relativity include discussions of the generalized theory, with primary emphasis on the problem of gravitation.

[136] The untitled manuscript was written in 1912 for *Marx 1924*, publication of which was delayed due to the First World War (see Erich Marx to Einstein, 2 January 1922, and Einstein's updated reply to Marx's letter of 3 March 1922). See *The Collected Papers of Albert Einstein*, Vol. 4 *The Swiss Years: 1912–1914* (Princeton University Press, 1995), pp. 3–108, for this manuscript.

[137] See, e.g., *Einstein 1935*.

[138] See *Einstein 1917a, Einstein 1921b*, for early examples.

[139] See *Einstein 1907j* (Doc. 47), Section V, pp. 454–462.

[140] In *Einstein 1934* (first published in English in *Einstein 1933*), Einstein stated that he was convinced of this equality even though not yet aware of the experiments of Eötvös (see *Eötvös 1890*).

[141] *Einstein 1907j* (Doc. 47), p. 454.

[142] See *Einstein 1911g*.

Einstein's First Derivation of Mass–Energy Equivalence

John Stachel and Roberto Torretti

There is a persistent tradition, apparently starting with a paper by Ives, that Einstein's original derivation of the mass–energy equivalence was fallacious.[1] It has probably gained most currency through the account in Jammer's well-known book *Concepts of Mass*.[2] Jammer states:[3]

> It is a curious incident in the history of scientific thought that Einstein's own derivation of the formula $E = mc^2$, as published in his article in the *Annalen der Physik*, was basically fallacious. In fact, what for the layman is known as "the most famous mathematical formula ever projected"[4] in science was but the result of a *petitio principii*, the conclusion of begging the question.

Arzeliès, in his treatise on relativity,[5] also adopts Ives's discussion. He states:[6]

> It might appear rather piquant that a relation of this importance was introduced into physics by this expedient. In fact it is just one example among others of the slight importance of logic in physical research.

We leave a decision about the importance of logic in writing treatises on relativity to the reader after finishing this article.

Arthur Miller, in his recent book on the emergence and early interpretation of special relativity,[7] endorses Jammer's verdict.

Before considering the quality of the arguments on which Ives, Jammer, and Arzeliès base their criticism, let us repeat Einstein's derivation. As we shall see, far from being "basically fallacious," it is quite sound.

Einstein[8] considers a body at rest in an inertial frame which emits electromagnetic radiation (Einstein speaks of a light wave) of total energy L in two equal but oppositely directed amounts. Since the emissions are symmetrical, the body loses energy but not momentum, so it remains at rest in the original inertial system.

American Journal of Physics
Vol. 50, pp. 760–763
August 1982

Einstein assumes that its initial energy with respect to this frame was E_0; and that energy is conserved in this frame, so that its final energy E_1 with respect to the frame differs from E_0 by the amount L:

$$E_0 = E_1 + L. \tag{1}$$

Now we consider the same act of emission from another inertial frame, with respect to which the body is moving with a velocity $\mathbf{v}$. Note that $\mathbf{v}$ is arbitrary, and that its magnitude v can be chosen as small as we desire. If energy conservation is assumed in the first inertial frame, the relativity principle requires it to hold in the second as well. If H_0 and H_1 are the initial and final energies with respect to the second frame, Einstein shows that the initial and final energies differ by the amount $L/\sqrt{(1 - v^2/c^2)}$:

$$H_0 = H_1 + L\sqrt{(1 - v^2/c^2)}. \tag{2}$$

This result is established by using the law for the transformation of the energy of a plane light wave from one inertial frame to the other, derived in his first paper on the special theory of relativity.[9] Now Einstein subtracts Eq. (1) from Eq. (2) to obtain:

$$(H_0 - E_0) - (H_1 - E_1) = L[1/\sqrt{(1 - v^2/c^2)} - 1]. \tag{3}$$

Einstein proceeds:

> H and E are energy values of the same body, referred to two coordinate systems in motion relative to each other, in one of which [...] the body is at rest. *It is therefore clear* that the difference $H - E$ can differ from the kinetic energy K of the body with respect to the other system [...] only by an additive constant C which depends on the choice of the arbitrary additive constants in the energies H and E.[10]

We may therefore set

$$\begin{aligned} H_0 - E_0 &= K_0 + C, \\ H_1 - E_1 &= K_1 + C, \end{aligned} \tag{4}$$

for the constant C will not be altered by the emission of radiation. Substituting from Eq. (4) into Eq. (3), we obtain

$$K_0 - K_1 = L[1/\sqrt{(1 - v^2/c^2)} - 1], \tag{5}$$

where K_0 and K_1 are the initial and final kinetic energies of the body with respect to the inertial frame in which its velocity is $\mathbf{v}$. This, the decisive step of Einstein's derivation, depends on the acceptance of Eqs. (4). Though "clear" to Einstein, their validity has not been quite so evident to others. The following relativistic argument will, we hope, contribute to make it so.

Consider an isolated body in equilibrium, and therefore at rest in some inertial frame, which we call its rest frame. Its internal state will be characterized by a set of parameters whose values (with respect to its rest frame) we symbolize collectively by S. By the relativity principle, it must be possible for the body to have the same internal state S when in uniform motion with velocity $\mathbf{v}$. For otherwise one could determine the absolute state of motion of the body by determining its internal state. Its energy in its rest frame (in the absence of any external fields, of course) is a function of its internal state only. Then its energy in an inertial frame with respect to which it is moving can only be a function of S and its velocity. Because of the isotropy and homogeneity of space, the body's energy can depend only on the magnitude of its velocity—as well as on its internal state:

$$E = E(v, S).$$

The kinetic energy of the body, by definition, is equal to the work necessary to bring the body from the state of rest to uniform motion with velocity $\mathbf{v}$. But, by conservation of energy, this must be equal to the difference between its energy for the state S and speed v and its energy for the same internal state when at rest:

$$K = E(v, S) - E(0, S).$$

K depends on v and S:

$$K = K(v, S).$$

We can now rewrite Eqs. (1)–(5) in our new notation, by setting

$$\begin{aligned} E_0 &= E(0, S_0), \quad H_0 = E(v, S_0), \quad K_0 = K(v, S_0); \\ E_1 &= E(0, S_1), \quad H_1 = E(v, S_1), \quad K_1 = K(v, S_1), \end{aligned}$$

where S_0 and S_1 denote, respectively, the initial and the final internal state of the body, assumed to settle into a new equilibrium state S_1 after emitting the radiation. Obviously,

$$[E(v, S_0) - E(0, S_0)] - [E(v, S_1) - E(0, S_1)] = K(v, S_0) - K(v, S_1),$$

whence Eq. (5), that is, in our notation

$$K(v, S_0) - K(v, S_1) = L[1/\sqrt{(1 - v^2/c^2)} - 1], \tag{5'}$$

follows at once.

Note how our argument rests on the relativity principle. We have proceeded from a consideration of the same body with respect to two inertial frames to a comparison of the body in two states of motion with respect to the same inertial frame; in other words, from a passive to an active interpretation of the Lorentz transformation. But the equivalence of the two viewpoints is just what the relativity principle (indeed any symmetry principle) asserts: the active interpretation is always possible.[11]

Einstein remarks that Eq. (5) means that "the kinetic energy of the body [...] decreases as a consequence of the emission of light, and indeed by an amount independent of the qualities of the body." This is not strictly accurate, since the amount does depend somehow on the relationship between S_0 and S_1; but what

Einstein presumably had in mind is expressed in our notation by the fact that the right-hand side of the equation is independent of S_0 and S_1. He immediately notes that "The difference [of the two kinetic energies] depends on the velocity in the same way as the kinetic energy of the electron," which he had previously calculated in his first paper on special relativity.[12] Note that this is a *result* of his argument, not an assumption or a trivial consequence of his having assumed that $K(v, S)$ for any body was of the same form as the kinetic energy of the electron.

Finally, Einstein draws his conclusion from Eq. (5) by passing to the Newtonian limit:

> Neglecting magnitudes of fourth and higher order [in v/c] we obtain
>
> $$K(v, S_0) - K(v, S_1) = (1/2)(L/c^2)v^2. \tag{6}$$
>
> From *this* equation it immediately follows: If a body emits energy L in the form of radiation, then its mass diminishes by L/c^2.[13]

This terse and subtle text demands some comment. Passage to the Newtonian limit is justified and does not impair the exact validity of the conclusion,[14] because, as we have already noted, the moving frame, which is introduced only for the sake of comparison, can be arbitrarily slow. Einstein is arguing that a body *at rest*, which emits energy in such a way as to remain at rest, thereby decreases its inertial mass. He is not directly concerned with how the energy of the body varies with its velocity. The reader should also note how carefully Einstein avoids any assumption about how the relativistic kinetic energy depends on velocity, mass, or any other parameters of the body's internal state.

We can define the inertial mass[15] for a body in translational motion (in keeping with the requirement that special relativistic dynamics have a Newtonian limit as $v \to 0$) by

$$m = \lim_{v \to 0} \frac{K}{v^2/2}\,.$$

It follows from Eq. (6) that

$$\lim_{v \to 0} \frac{(K_0 - K_1)}{v^2/2} = \frac{L}{c^2}\,.$$

With the above definition of inertial mass, this can be rewritten

$$m_0 - m_1 = L/c^2,$$

which is clearly an exact, not an approximate, result.

We are therefore entitled to the inference that a body at rest emitting electromagnetic radiation must lose inertial mass equal in amount to the energy lost divided by c^2.

Einstein then remarks that "it is obviously inessential that the energy subtracted from the body should turn precisely into energy of radiation." He then jumps to the conclusion that "the mass of a body is a measure of its energy content," which clearly suggests that the rest mass of a body might be totally convertible into other forms of energy. Soon after, Einstein sought to substantiate the above remark by proving the mass–energy equation for other cases of energy exchange due to electromagnetic interaction, but not involving the emission or absorption of radiation.[16] However, it was not until much later that he gave a purely dynamic derivation of the mass–energy equivalence. [17] The final conclusion: that the entire mass of a body is in effect a measure of its energy, is of course unwarranted by Einstein's premises. This is what he wrote in defense of it in 1907:

> A mass μ is equivalent—insofar as its inertia is concerned—to an energy content of the magnitude μc^2. Since we can arbitrarily fix the zero of [the rest energy], we are not even able to distinguish, without arbitrariness, between a "true" and an "apparent" mass of the system. It appears much more natural to regard all inertial mass as a store of energy.[18]

Later events have obviously borne out this presumption.

We turn now to the consideration of Ives's, Jammer's, and Arzeliès's criticism of Einstein's argument.[19] None of the critics disputes that up to Eq. (3) it is correct, but they all take exception to Einstein's next step. To Ives "it is by no means 'clear that, etc.' " in the Einstein passage quoted above right after Eq. (3). Jammer voices an even stronger disapproval. According to him, after "correctly" proving Eq. (3), Einstein "mistakenly put" $H - E$ [or, in our notation, $E(v, S) - E(0, S)$] equal to the kinetic energy. And yet it is hard to see what else one could mean by the kinetic energy of a body with internal state S and speed v. Ives, Jammer, and Arzeliès do not tell us why they believe that Eqs. (4) are doubtful or erroneous, but immediately embark on the following remarkable exercise in logical analysis, for which Jammer and Arzeliès give all due credit to Ives. They assume that the kinetic energy of a body of rest mass m, with respect to an inertial frame in which it is moving with speed v is equal to

$$K = mc^2[1/\sqrt{(1 - v^2/c^2)} - 1]. \tag{7}$$

Einstein had proved Eq. (7) for the kinetic energy of an "electron," i.e., a charged structureless particle, in his first relativity paper;[20] but he studiously avoided using it in the derivation of the mass–energy equivalence, even though, as we shall now see, it would have simplified his task. (Had he used it he could indeed have been justly accused of question begging, for he had as yet no grounds for assuming that the dependence of the kinetic energy on internal parameters can be summed up in a rest mass term. Only if we know the mass–energy equivalence relation will this seem reasonable.) From Eqs. (7) and (3), Ives, Jammer and Arzeliès infer that

$$(H_0 - E_0) - (H_1 - E_1) = [L/(m_0 - m_1)c^2](K_0 - K_1), \tag{8}$$

where we denote the vacuum speed of light by c, the initial and final rest mass of the body by m_0 and m_1, respectively, and otherwise use Einstein's notation. Comparing Eq. (8) and Eqs. (3) and (5), we verify that, if Eq. (7) is assumed, Eq. (5) holds good if and only if

$$m_0 - m_1 = L/c^2. \tag{9}$$

Thereupon one would normally conclude that, if Eq. (7) is added to Einstein's premises, the mass–energy relation (9) follows directly from them, without having to go to the Newtonian limit. For surely, if Eq. (5) is true only if Eq. (9) is true, and Einstein's premises entail that Eq. (5) is true, they also entail the truth of Eq. (9), as a single application of *modus ponendo ponens* will make clear. At this point, however, Ives, Jammer, and Arzeliès veer off the trodden path of ordinary propositional logic and pronounce their indictment. According to Ives, "what Einstein did by setting down [Eq. (4)] as 'clear' was to *introduce* the relation [Eq. (9)]. *Now this is the very relation the derivation was supposed to yield.*" According to Jammer, by comparing Eq. (8) with Einstein's assumption (4), "we see that Einstein unwittingly assumed" Eq. (9), "which is exactly the contention to be proved." All three authors agree that Einstein has thereby committed a "fallacy," which Jammer, as we saw, diagnoses as a *petitio principii.* Arzeliès similarly speaks of a "vicious circle." Now it is plain that Eqs. (7) and (8) jointly imply that Eq. (3) entails Eq. (5). If Einstein had used this method for proving Eq. (5) and had then proceeded to infer Eq. (9) from it, he would indeed have perpetrated a *petitio.* The method, however, does not occur in his work, but is a "rational reconstruction" which we have just devised in order to make sense out of the claims of his critics. (It is the method that he ought to have followed to earn their reprimand.) Nowhere in his paper does he invoke Eq. (7), let alone Eq. (9), to prove Eq. (5), which, as we saw, he derives from Eqs. (3) and (4). Jammer asserts without proof that Eqs. (4) are "mistaken," and if he were right Einstein's derivation would indeed break down. Not, be it noted, because if Eqs. (4) are wrong Einstein's argument would, by the accepted standards, be fallacious; but rather because, though logically impeccable, it would then rest on a falsehood. In all likelihood, Ives, Jammer, and Arzeliès are innocent of the gross misrepresentation of Einstein's method of proof which we have put forward on their behalf. If that is the case, all they can possibly be objecting to is the fact that the goal of Einstein's argument, i.e., Eq. (9), is a necessary condition of the joint assertibility of Einstein's premises—so that the latter cannot be true unless the former is true as well. But this is a relation in which the conclusion of an argument should stand to its premises whenever the argument is valid by standard—and by every known nonstandard—logic. Thus if we are not willing to countenance some mind-boggling metalogical innovation, we have to declare that Ives, Jammer, and Arzeliès—not Einstein—are guilty of a logical error.

Ives seems to have been inspired in his criticism by a remark of Planck's, which he quotes. Commenting on Einstein's derivation in a 1907 paper, Planck[21] said that it was based on an "assumption permissible only as a first approximation." To understand Planck's point, made in a footnote near the end of the paper, we must go back to its Introduction. He states there:

> We usually conceive the total *energy* of a moving ponderable body as additively composed of a term which is independent of the internal state of the body and only varies with its velocity: the translational energy; and a second term which is independent of the velocity and only depends on the internal state [...]: the internal energy of the body.[22]

He then points out that any ponderable body contains a contribution to its energy from thermal radiation, and that the thermal radiation energy does not fall into two such terms. Therefore "a division of the energy into internal and translational energy" is not possible.

Thus after he has established the mass–energy equivalence for thermal energy, he remarks in a footnote:

> Essentially the same conclusion was drawn by A. Einstein [. . .] from the application of the relativity principle to a particular radiation process, however, with the assumption, permissible only as a first approximation, that the total energy of a moving body is additively composed of its kinetic energy and its energy with respect to a system in which it is at rest.[23]

But, as we have seen above, the definition of the kinetic energy implicit in Einstein's derivation, and which we have made explicit, does not at all imply that the kinetic energy is "independent of the internal state of the body." All that is implied by our definition is that the net energy necessary to bring the body from an initial state of rest in internal state S to a final state of motion with velocity $\mathbf{v}$ and the same internal state S must be independent of the particular process by which it was brought to that final state. But this is just the principle of energy conservation.

To summarize: Einstein uses the following principles in his derivation: the principle of relativity, the law of conservation of energy, the existence of a Newtonian limit for relativistic dynamics, and the relativistic law of transformation of the energy of an electromagnetic wave. These premises are certainly strong enough to derive the mass–energy equivalence relation; and in that sense, of course, the conclusion is required by the premises, as indeed it must be in any logically correct argument. If one of his assumptions turned out to be wrong, that would make his argument inconclusive, but not fallacious.

NOTES

[1] H. I. Ives, *J. Opt. Soc. Am.* **42**, 540 (1952). Ives's criticism of Einstein's derivation was criticized in turn by J. Riseman and I. G. Young, *J. Opt. Soc. Am.* **43**, 618 (1953), to whom Ives replied in the same issue, p. 619. This exchange did not touch on the issues discussed in the present article.

[2] M. Jammer, *Concepts of Mass in Classical and Modern Physics* (Harvard University, Cambridge, MA, 1961).

[3] See footnote 2, p. 177.

[4] Jammer here quotes William Cahn, *Einstein* (Citadel, New York, 1955).

[5] H. Arzeliès, "Études Relativistes;" the discussion of Einstein's various proofs of $E = mc^2$ is in the volume *Rayonnement et dynamique du corpuscule chargé fortement accéléré* (Gauthier-Villars, Paris, 1966), pp. 74–79.

[6] See footnote 5, p. 76.

[7] A. I. Miller, *Albert Einstein's Special Theory of Relativity: Emergence (1905) and Early Interpretation (1905–1911)* (Addison-Wesley, Reading, MA, 1981). On p. 377 Miller states "Ives's [footnote 1] analysis of Einstein's [footnote 8] derivation of the mass–energy equivalence revealed a logical inconsistency—Einstein had assumed the result to be proven at the outset."

[8] A. Einstein, *Ann. Phys. Ser. 4* **18**, 639 (1905). An English translation may be consulted in *The Principle of Relativity* (Methuen, London, 1923—subsequently reprinted by Dover), pp. 67–71; but, since the translation is sometimes not quite accurate, we have translated directly from the German. We use the standard symbol c, instead of Einstein's V, to designate the speed of light *in vacuo*.

[9] A. Einstein, *Ann. Phys. Ser. 4* **17**, 891 (1905). The proof is given in Sec. 8. See foonote 8 for reference to one English translation in *The Principle of Relativity*. A better translation is contained in the Appendix to Miller, footnote 7.

[10] Eistein, footnote 8, pp. 640–641, emphasis added.

[11] For those to whom the talk of active and passive transformations sounds esoteric, the following consideration may be helpful. Let the body with internal state S be initially at rest in an inertial frame F and moving with velocity $\mathbf{v}$ in an inertia frame F'. Its energy with respect to F is then $E(0, S)$. Now accelerate the body to velocity—$\mathbf{v}$ in F. According to what we said above, its energy with respect to F increases thereby to $E(v, S) = E(0, S) + K(v, S)$—where $v = |\mathbf{v}|$. Since the body with internal state S is now at rest in F', its energy with respect to F' will be, by the relativity principle, equal to $E(0, S)$. Returning to the initial situation we see that, by symmetry, when the body had energy $E(0, S)$ with respect to F, it must have had—*modulo* an arbitrary constant independent of v and S—the energy $E(v, S) = E(0, S) + K(v, S)$ with respect to F'.

[12] Einstein, footnote 9, p. 920 (Sec. 10).

[13] Einstein, footnote 8, p. 641. Emphasis added. We have also translated Einstein's notation into ours.

[14] We cannot agree with Whittaker's statement that the proof "was put forward only as approximate." See E. T. Whittaker, *A History of the Theories of Aether and Electricity, Vol. II, The Modern Theories 1900–1926* (Nelson, London, 1953), p. 52. Whittaker also notes without comment that Einstein's "reasoning has been criticised" by Ives.

[15] Although a variable inertial mass, and even transverse and longitudinal masses were used in the early days of SRT, Einstein came to the view that only one (rest) mass should be defined; and that otherwise one should rather speak about the law of variation of energy or momentum with velocity.

[16] A. Einstein, *Ann. Phys. Ser. 4* **23**, 371 (1907); *Jahrb. Radioaktivität* **4**, 411 (1907); **4**, 440ff (1907).

[17] A. Einstein, *Am. Math. Soc. Bull.* **41**, 223 (1935). As is well known, relativistic dynamics, free from electromagnetic presuppositions, was founded by G. N. Lewis and R. C. Tolman, *Philos. Mag.* **18**, 510 (1909).

[18] Einstein, footnote 16, second article, p. 442.

[19] Ives, footnote 1, pp. 542–543; Jammer, footnote 2, pp. 177–179.

[20] Einstein, footnote 9, pp. 919ff.

[21] M. Planck, *K. Preuss Akad. Wiss. Sitzungsber.* (2), 542 (1907).

[22] Planck, foonote 21, p. 542.

[23] Planck, footnote 21, p. 566.

Part V

General Relativity

Einstein's Odyssey: His Journey from Special to General Relativity

John Stachel

In May 1915, Albert Einstein wrote to a former student who had asked for advice on the choice of a thesis topic: "It is quite singular in scientific endeavors: often nothing is of greater importance than seeing where it is not advisable to apply one's time and effort. On the other hand, one must not strive for goals whose attainment is easy. One must acquire an instinct for what is just attainable with the utmost exertions. This magnetic work, for example [Einstein here refers to his recent work on the Einstein–De Haas effect], could have been done by any scalawag [*Lump*, in German]. But General Relativity is quite another matter. Now really to have attained this goal is the highest satisfaction of my life, even if no colleague has yet recognized the depth and necessity of this journey."

Curiously enough, the last stretch—one of the most intensely difficult—still lay ahead of Einstein during the months *after* he wrote this letter. Yet the passage correctly conveys his feelings about his work on General Relativity and its significance, feelings which deepened and strengthened in later years.

What were the major steps on this journey, which profoundly affected the shape of contemporary physics? How did this journey affect Einstein's own views on the nature of physics, indeed his world-view? (Before trying to answer these questions, let me stress that, in spite of periods of intense and exclusive concentration on the gravitational problem from 1907 through 1915, Einstein continued important work on quantum theory and statistical mechanics—work which placed him in the forefront of these fields.)

In 1905 Einstein formulated what is now called the Special Theory of Relativity. Later he described it as a theory of "principle." What he meant was that it established a set of rather general rules that any physical theory purporting to

Reprinted from *The Sciences*, March 1979

explain a particular realm of phenomena (such as mechanics or electromagnetism) would need to satisfy. In this respect, Einstein often compared Special Relativity to thermodynamics. What were these principles?

The Relativity Principle. Newton's law of inertia states that bodies acted on by no external forces either stay at rest or move in a straight line at constant speed. But rest or motion can only be described with respect to some frame of reference. Clearly this law cannot hold for all observers, regardless of their own state of motion. The frame of reference of any observer for whom the law of inertia holds is called an "inertial frame of reference."

The Earth, to a certain approximation, is an inertial frame of reference. Now consider a train moving at constant speed along a straight track. An observer on the train will see an object on the ground near the track moving backward at constant speed. In this example, we see that the law of inertia holds for both the train's and the Earth's frame of reference. More generally, any frame of reference moving in a straight line at constant speed with respect to an inertial frame is also an inertial frame. Newton's laws of mechanics say that the laws of motion with respect to an inertial frame are the same for *all* bodies, even if they are acted upon by external forces. Thus, all inertial frames are mechanically equivalent.

But what Einstein did was to extend this principle. He postulated that *all* the laws of physics—mechanical or otherwise—*must* take the same form in any inertial frame.

Constant Speed of Light. Einstein also postulated that the speed of light with respect to an inertial frame is independent of the motion of the light source. Together with the relativity principle, this implies that the speed of light must be the *same* constant in all inertial frames.

This was absurd if one accepted Newton's ideas about the existence of "absolute" time. The two principles seemed irreconcilable. But Einstein saw that the absurdity disappeared if only we could give up our notion of "absolute" time. What Einstein recognized was that even time is relative. Time, it turned out, was relative to one's choice of inertial frame of reference.

A few other physicists, notably Max Planck, almost immediately grasped the significance of Einstein's theory and they, together with Einstein, set out to see how to reconcile the various fields of physics with the principles of special relativity. But Einstein was troubled by another question: What was so special about inertial frames? Could the relativity principle be extended beyond inertial frames of reference? Could the laws of physics take the same form in some more general class of reference frames? But here Newton's law of inertia seemed to bar the way.

In an inertial frame of reference, when a body was not subject to external forces it remained unaccelerated. Yet someone in an accelerated frame of reference would observe such a body to be accelerated. (Think of how objects at rest on the ground appear to move as you ride on a merry-go-round.) If one were to apply Newton's laws of motion in the accelerated reference frame to such objects, it would have to be concluded that they were subject to some curious forces, which an observer in an inertial frame would regard as non-existent. These "forces" have come to be called "inertial forces" (although "noninertial forces" might be a better name, since they arise when we describe motion with respect to noninertial frames of reference). The laws of physics, then, really do appear simpler in an inertial frame (no inertial forces), and thus this class of frames seems to be singled out. It was essentially this observation on which Newton based his ideas about absolute space.

Early in his journey, Einstein was much influenced by the late nineteenth century Austrian physicist Ernst Mach's critique of Newtonian mechanics. Mach had said that only bodies and their relative motions should enter the laws of physics. They must be freed from any reference to absolute space. The inertial properties of a body, such as its mass, should then somehow result from its interaction with all the other bodies in the universe. Much later, after Einstein had set himself the task of explaining ordinary bodies as structures within a unified field, he completely repudiated this idea of Mach's. But at the time he was developing the General Theory of Relativity, Mach was a great source of inspiration to him.

It was Einstein's attempt to fit Newton's theory of gravitation into the framework of Special Relativity that convinced him that he *must* generalize the relativity principle to accelerated frames of reference. He was led to this conclusion by what he came to regard as the most fundamental fact about gravitation: All bodies fall with the same acceleration in a gravitational field. In Newton's theory of gravity this is a simple consequence of a curious coincidence: The mass that measures a body's resistance to acceleration (inertial mass) is numerically exactly equal to the mass that produces and responds to gravitational forces (gravitational mass). Einstein seized upon this "coincidence" as the vital clue to a new theory of gravitation. He held on to it throughout his search for that new theory, which lasted from 1907 to 1915. Einstein showed that, if we accept the equality of gravitational and inertial mass as fundamental, the relativity principle must be extended to accelerated frames of reference too.

Gravitational Equivalence. The first step toward this extension came in 1907 when he discussed a constant gravitational field. Consider the gravitational field in a room. As Galileo had demonstrated, if we drop objects in the room, they will all fall down with the same acceleration. But now let us consider a room out in space, far from the Earth's (or any other) gravitational pull. Suppose this room is being accelerated *upward*, like a rocket, with the same numerical acceleration as the objects falling in the room on the surface of the Earth. To an observer in the accelerated room out in space, an object he or she releases would appear to be accelerating *downward*, just as in the room on the Earth. Because of the equality of inertial and gravitational mass, there is no way we can tell the difference by mechanical experiments between what goes on in the two rooms. Einstein guessed that not only mechanical laws but *all* the laws of physics were the same in the two rooms, whether on Earth or out in space. He postulated that an accelerated frame of reference in which there was *no* gravitational field was completely equivalent to an inertial frame of reference in which there was a *constant* gravitational field. He called this postulate the principle of equivalence.

Einstein drew some novel physical consequences from his principle of gravitational equivalence. A light ray observed from an inertial frame—in which there is no gravitational field—always travels in a straight line (in empty space, at least). But an observer looking at such a light ray from an accelerated frame of reference would see its path as curved. (Just think of the paths of raindrops on an automobile window when the car is accelerating.) The equivalence principle led Einstein to the conclusion that a gravitational field must curve the path of a light ray. In 1907 he did not see any way of confirming this prediction, but we shall see that he soon came up with the right sort of experiment to test it.

Later, he both restricted and extended the equivalence principle. He recognized that only in sufficiently small regions of space and time can one completely imitate the effects of a gravitational field by an accelerated frame of reference. (The gravitational field in a room can be so replaced for a limited period of time; but the field over the *entire* surface of the Earth cannot.) He came to regard any accelerating frame of reference as just as acceptable as an inertial frame. The previously dubious inertial forces needed in such a frame were to be regarded as nothing but a special kind of gravitational force.

Between 1907 and 1911 the gravitational problem receded into the background of Einstein's activities. One reason, no doubt, was that from 1908 onwards he was finally accepted into the academic world, and had to cope with a heavy teaching load as well as training research students. During this period he was also deeply occupied with trying to find a fundamental solution to the quantum puzzle, a solution which ultimately eluded him.

When he returned to intensive work on gravitation in 1911, he saw that one could test the theory by experiment: Why not observe starlight which had passed near the sun on its way to the Earth? Would the light rays be curved by the sun's gravitational field? The effect would be very small and the light ray would have to pass as close to the sun as possible to make the effect detectible. It could only be detected during a solar eclipse when it was dark enough to observe such starlight. Indeed, a German scientific expedition set out to observe the 1914 solar eclipse from the Ukraine. But they were interned by the Russian authorities (luckily only briefly) after World War I broke out. Since Einstein's argument of 1911 only yielded half the predicted light deflection of his final 1915 General Relativity Theory, it is interesting to speculate what might have happened if the expedition *had* been able to complete its work. Of course, the full light deflection prediction of General Relativity was confirmed by a British team in 1919. It was the announcement of this result that made Einstein world famous.

Einstein drew another important conclusion in his 1911 paper. According to his Special Theory of Relativity, the speed of light was constant with respect to inertial frames. Einstein showed that it would not appear constant as seen by an accelerating observer if he or she measured time in a reasonable way. Drawing on his equivalence principle once again, Einstein recognized that the speed of light must be influenced by gravity. Later, he gave a formula that related the speed of light to the strength of gravity at any point. In two papers written early in 1912, he tried to extend his ideas about gravitation to certain nonconstant gravitational fields, discussing how these fields are produced by masses and how they exert forces on them. In so doing, he described the strength of gravity by this variable speed of light.

Four Dimensions. During the years when Einstein had been less occupied with gravitation, an important step in the mathematical treatment of Special Relativity had taken place. It was to prove crucial for Einstein's next step, which constituted the major conceptual breakthrough in his search for a General Relativity Theory.

The mathematician Hermann Minkowski (who had been one of Einstein's teachers) recognized that the best way to bring out an inherent symmetry between the roles of space and time in Special Relativity Theory was to unite the two mathematically into a single space-time structure, now called "Minkowski space." Minkowski's flat four-dimensional geometry included our every-day three-dimensional Euclidean geometry of space, as well as time. Einstein, who then had little use for abstract mathematics, was not impressed at first by the elegance of Minkowski's four dimensions. He felt that his new gravitational theory, with its variable speed of light, showed that the space-time symmetry, which lay at the basis of Minkowski's idea, would have to be given up.

But Einstein soon realized that it was essential not to abandon Minkowski's four dimensions. Rather, he had to adapt it to his own purposes. Some time during the latter half of 1912, he concluded that the way to treat arbitrary gravitational fields was to use a four-dimensional space-time, but drop the assumption that its geometry was flat. Instead, he concluded that the geometry of space-time was curved, which was remarkable enough. But even *more* remarkable, since the geometry of this space-time reflects the nature of the gravitational field (which in turn depends on the distribution of matter in the universe), he was led to conclude that the geometry of space-time was not something fixed and given once and for all. Instead of thinking of space and time as a stage, on which the drama of matter unfolds, we have to imagine some ultra-modern theater in which the stage itself becomes one of the actors, changing as the drama unfolds. Prior to this remarkable conclusion, space and time were thought to be absolute: While they influenced the course of all physical phenomena, they were themselves uninfluenced by them. Now, by uniting them into a four-dimensional space-time geometry, Einstein made them an integral part of physics, constituting its gravitational aspect, both influencing and being influenced by all other physical processes. Einstein, it has been said, physicalized geometry.

What does a non-flat geometry mean? Let's take a look at a two dimensional example: How could we tell the difference between a flat plane and a curved two-dimensional surface? We are not interested in the so-called extrinsic curvature of the surface: From our point of view a plane stays a plane, even if we roll it up into a cylinder. We are interested in the so-called *intrinsic* curvature: How can we tell that a plane is different from the surface of a sphere, even if we don't know anything except relationships on each surface? Suppose we measure the distances *on the surface* between all the points in some neighborhood of a point on the plane, and all the points in some neighborhood of a point on the sphere. (By distance on the surface we mean the shortest distance between the two points along a curve which lies entirely on the surface. This will be a segment of a straight line on the plane, and an arc of a great circle on the sphere.) We shall find systematic differences between these two sets of distance measurements—which is why a flat map can never give a perfectly satisfactory picture of the relations between distances on the Earth, while a globe can. If we are clever enough, we can even calculate the radius of the sphere, which is a measure of the curvature of its surface, from such sets of distance measurements. Now on a surface of variable curvature, such as a hand, the set of distance measurements in the neighborhood of any point will show different characteristics at each point of the surface, because at different points the surface is like bits of spheres of different radii. It turns out that, by giving three functions of the two coordinates of each point on a surface, one can sum up all this information about the relations between distances of points on the surface, and thus about its varying curvature from point to point. These three functions have been given the mathematical name of the metric (or measuring) tensor. In the four-dimensional space-time that Einstein was considering, a set of ten such functions are needed to characterize the intrinsic geometry—the generalized curvature properties—of space-time. Newton's gravitational theory had been able to get along with one function to describe the strength of the gravitational field as had Einstein in 1912 with the variable speed of light. Now ten components of the metric tensor of space-time were needed to describe the gravitational field.

Having arrived at this point, Einstein turned to his old friend and schoolmate, Marcel Grossmann, for help with the formidable mathematical problems of translating his ideas into a specific physical theory. Grossmann found that a group of late nineteenth-century and early twentieth-century mathematicians had already forged the tools needed. The language of tensors proved to be the one in which the rest of our story was written. Tensor analysis provides a natural mathematical device, in terms of which the behavior of a system of equations, used to characterize some physical processes, may be examined as we pass from the frame of reference in which the equations were written down originally to any other frame of reference. Since Einstein hoped that his gravitational theory would not single out any frame of reference (or class of frames) as special, he was looking for a system of equations whose form was the same in all frames of reference. Such a system of equations is called "generally covariant" in tensor analysis. Einstein and Grossmann soon succeeded in writing the equations describing the effects of gravitation on other physical processes—such as the equations of motion of a body in a gravitational field—in a generally covariant form. And Einstein found that a certain tensor, called the stress-energy tensor, was the correct relativistic generalization of the Newtonian mass in its role as the source of the gravitational field.

The stress-energy tensor describes the distribution of matter and energy in space-time. It acts to curve up the space-time structure into some non-flat four-dimensional geometry. This must be described mathematically by field equations that relate the stress-energy tensor to the metric tensor. The effect of this non-flat geometry on all physical processes is what we call gravitational—and inertial—forces.

The other parts of the problem were solved quite rapidly in late 1912 and early 1913, but there remained the question of the correct field equations to describe just how the sources generated the gravitational field. There was a tensor, formed from the metric tensor, which described certain aspects of the curvature of space-time. Called the Ricci tensor, it seemed to be just about (but not quite—as later proved to be important) the only possible candidate to be used for generally covariant gravitational field equations. But Einstein and Grossmann convinced themselves that these equations could not be correct since they did not seem to give Newton's theory of gravity as a limiting case. (Since Newton's theory gave us an extremeley accurate description of most gravitational effects, it was an important test for any new theory that it give Newton's results as a first approximation.) Einstein then reluctantly abandoned the search for generally covariant equations, and set up non-generally covariant equations for the gravitational field. This meant restricting the relativity principle: More frames of reference than just inertial ones were equivalent; but not *all* frames of reference were equivalent. Trying to make a virtue of necessity, Einstein constructed a more general argument purporting to show that the gravitational field equations could not be generally covariant.

As it turned out, both these arguments against general covariance were wrong; but it took more than two years, from early 1913 to late 1915, until Einstein could see this clearly. In the meantime there was a series of tragi-comic ups and downs, as Einstein alternately reconciled himself to non-generally covariant field equations, and tried somehow to save the principle of general covariance, which he felt *should* be valid.

In late 1915, recognizing his earlier mistakes, Einstein returned definitively to the principle of general covariance, and to the Ricci tensor, abandoned with such a heavy heart in 1913. However, things still did not come out quite right. The Ricci tensor field equations were fine for the gravitational field *outside* of matter. Indeed, Einstein also succeeded in uncovering the answer to a long-unexplained astronomical anomaly, the advance of the perihelion of Mercury. (This refers to the observation showing that Mercury's elliptical orbit around the sun is itself slowly revolving.) Einstein had been aware of this anomaly since 1907, but none of his previous attempts to come up with the right solution had been successful.

But *inside* matter, his use of the Ricci tensor resulted in a paradoxical conclusion. At first Einstein accepted the paradox and tried to turn it into a virtue, by developing a new theory of matter. But then he quickly realized that the trouble lay in the form of the field equations themselves. That little bit of freedom, which made the Ricci tensor *almost* the unique generally covariant candidate, allowed for a slight modification of the field equations. He replaced the Ricci tensor by what we now call the "Einstein tensor." With the new field equations, the paradox disappeared, to be replaced by a beautiful new result: The gravitational field equations now had as a consequence that the laws of conservation of energy and momentum for the sources of the gravitational field *must* hold. Over the next two decades it was to be realized gradually that, as a consequence of this, the field equations of General Relativity had the remarkable property of implying equations of motion for their sources—but that is another story!

Thus, by the end of 1915, the General Theory of Relativity as we know it today had been worked out by Einstein, in the course of an intellectual odyssey which had included shipwrecks and heartbreaks reminiscent of the Greek hero's journey. Einstein was not too surprised when his colleagues, who had not been fellow crew members on the voyage, were slow to accept the magnitude of what he had achieved.

Einstein himself said of this journey: "... the years of anxious searching in the dark, with their intense longing, their alternations of confidence and exhaustion and the final emergence into the light—only those who have experienced it can understand it."

Einstein had also hoped that his theory might be of some help in understanding the structure of matter, particularly in helping to solve the quantum problem. But he soon realized that General Relativity was of no help. Instead, he started on the quest for a unified field theory, which would explain the structure of ordinary matter—a quest which was to occupy him for the rest of his life.

In the course of revolutionizing theoretical physics by the development of the General Theory of Relativity, Einstein also altered his own philosophy of physics. It did not happen overnight, and in the beginning he was sometimes inconsistent in his expression of his changing outlook. Finally he came to realize how completely he had changed. From a young man, rather positivistically inclined in his view of science, he came more and more to adopt a "realist" outlook—although he never adhered dogmatically to any one philosophical credo. He came to feel that our concepts and theories could never be arrived at by induction from experience, but were free creations of the human mind. From skepticism about the importance of higher mathematics, he came to view the role of mathematics in setting up new physical theories as fundamental: Choice of the correct mathematical structure to describe physical concepts, and postulation of the simplest mathematical equations compatible with that structure were the key elements in searching for new theories. The search for a unified field theory, which was to be general-relativistic in approach, and to give an objective description of the universe, came to dominate his outlook. This approach made him skeptical of the acceptance of quantum mechanics by most of his contemporaries as a complete theory of the physical world.

In 1917, writing to Felix Klein, Einstein expressed the conviction that no theory he or anyone else had arrived at was ever to be final. "No matter how we may single out a complex from nature ... its theoretical treatment will never prove to be ultimately conclusive ... I believe that this process of deepening of theory has no limit." Perhaps that is the greatest lesson of his odyssey.

The Genesis of General Relativity

John Stachel

Responding to a letter inquiring about what influences had led him to his discoveries, Einstein replied that he could not indicate anything which had served as an external motive. He characterized his proper life's work as the search to answer three questions:

1) how the representation of a light ray depends on the state of motion of the coordinate system to which it is referred;

2) the basis for the equality of inertial and gravitational mass;

3) whether the gravitational and electromagnetic fields can be comprehended in a unified scheme.

The attempt to answer the first question led Einstein to the special theory of relativity. The search had its beginning in his teens (when he first wondered about what would be observed if one succeded in running alongside a light ray), but started in earnest during his final years at the ETH (The Swiss Federal Polytechnical Institute, where he conceived the idea of measuring the motion of the earth through the ether by an optical experiment). It continued for seven years until 1905, when he was able to rapidly complete the theory after realizing that simultaneity must be a relative concept.

The second question was the starting point of the general theory of relativity, which occupied Einstein, with varying degrees of intensity, between 1907 and 1915, when the final version of the theory was completed.

The third question was already briefly raised in the 1916 review paper on general relativity. Einstein's search for a satisfactory unified field theory started in earnest a few years later, and was to continue until the very last days of his life.

H. Nelkowski et al., eds.
Einstein Symposion Berlin,
Lecture Notes in Physics **100**, pp. 428–442

In another letter, Einstein elaborated on the nature of his motivation in attempting to answer such questions. He described it as the striving for a logically simple interpretation of empirically known connections. He described the state of psychic tension generated by what he felt to be fundamental incompleteness in the theoretical system of physics, and the striving to overcome this incompleteness, spurred on by his intense conviction of the existence of a logically simple solution. He compared his motivation to that of someone trying to solve a puzzle or chess problem, spurred on by the knowledge that the solution must exist. These comments shed some light on the significance of Einstein's cosmic religiosity: it was a deep emotional conviction of the lawful nature of the universe, which supplied the psychic energy needed to keep working on such "puzzles" over a decade or more. Even in the case of the unified field theory, where Einstein felt far from certain of his own ultimate success, he felt that some more fundamental solution to the "puzzle" must exist than that provided by quantum mechanics.

Both at the time of its creation and looking back in later years, Einstein saw his work on general relativity as something quite unique in his life. He felt that, if he had not worked out the special theory of relativity, someone else—perhaps Paul Langevin—would have done so. His approach to the general theory was entirely his own (apart from mathematical help from Marcel Grossmann), carried through with enormous efforts in the face of skepticism if not active hostility from physicists he respected, such as Planck and Abraham. He characterized his efforts on special relativity as mere child's play compared to what was needed to complete the general theory. His feeling of satisfaction at its completion marked the high point of his career. His whole later outlook on science was colored by the experience of the search; and the terms of his quest for a unified field theory were set by the nature of the generally covariant theory of gravitation that was its outcome.

In this paper I shall try to briefly outline the steps in this search to solve the second "puzzle"—the central episode in Einstein's scientific activity. I hope to publish a larger study, with full references, at a later date.

Einstein himself, in material he prepared for Erwin Freudlich's book—the first popular book on general relativity, published in 1916—divided the story of his search into three parts (I attach no mystical significance to this recurrence of triads in our story):

1) In 1907, he had the basic idea for a generalized theory of relativity, based upon the attempt to find a fundamental explanation for the equality of gravitational and inertial mass.

2) In 1912 came the recognition of the non-Euclidean nature of the space-time metric, and its physical determination by gravitation.

3) In 1915 Einstein arrived at the correct field equations for gravitation, and the explanation of the hitherto anomalous precession motion of the perihelion of Mercury.

The Equivalence Principle

In 1907, Einstein was asked by Johannes Stark to contribute a review paper on the principle of relativity to Stark's *Jahrbuch*. In this paper, Einstein first published his reflections on the relationship between the relativity principle and gravitation. From later comments, however, we know that this was not his first attempt at a relativistic theory of gravitation. He had apparently tried to set up a special-relativistic gravitational theory by some such obvious generalization of Newton's theory as transforming Poisson's equation for the Newtonian gravitational potential into d'Alembert's equation. But then he had been struck by the fact that such a theory violated the equivalence between gravitational and inertial mass. This feature of the gravitational interaction had been known for centuries, but appeared to have no deeper explanation within Newtonian theory. Einstein became convinced that this "coincidence" must hold the key to any fundamental explanation of gravitation: gravitation and inertia must be two aspects of the same phenomenon. He held to this conviction throughout the eight years it took him to arrive at his final formulation of the field equations of general relativity—years during which other colleagues tried to set up theories of gravitation that abandoned the strict equivalence between gravitational and inertial mass. Luckily, it was not until his 1914 discussion of Nordström's theory that Einstein realized that a scalar theory of gravitation was compatible with the equivalence principle. By that time, he was firmly convinced of the need for a (metrical) tensor theory of gravitation. Since it is sometimes implied that Einstein was unaware of the anomalous precession of the perihelion of Mercury at the time, let me note that in a letter of 1907 he states that he is working on a new theory of gravitation and hopes to be able to explain the anomalous precession with its help. He added that unfortunately it did not seem to want to come out right—as indeed it did not until 1915! This comment suggests, however, that he might have arrived at a gravitational analogue of the Sommerfeld special-relativistic orbit precession as early as 1907.

At any rate, he abandoned such special-relativistic attempts to generalize the Newtonian theory in 1907. From then on, he pinned his hopes for a relativistic theory of gravitation on the attempt to generalize the relativity principle from inertial frames to some wider class of accelerated frames of reference. The 1907 review paper gives us the first fruits of this attempt. Einstein had apparently been troubled even earlier by the question of why inertial frames should be singled out by the (special) relativity principle. One major source of his doubts came from his reading of Mach's *Mechanics*, with its critique of Newton's concepts of inertia and absolute space—a topic to which we shall return near the end of this paper. Of course, if one accepted the existence of one privileged frame of reference (ether frame) there was no problem of relativity. But once one gave that up, there seemed no logically compelling reason why one should stop at the relativity of inertial frames. The conventional answer to such doubts was that the laws of nature took a simpler form in the inertial frames of reference, because in these frames one did not have to introduce the fictitious so-called "inertial forces" (which would be better called "non-inertial," for our purposes) when writing down

Newton's law of motion. But Einstein saw that this argument really did not hold, at least in the simplest case, if one took into account the equivalence between gravitational and inertial mass. Consider a uniformly accelerated (with respect to any inertial frame) frame of reference, without any gravitational field, on the one hand; and an inertial frame of reference with a uniform gravitational field of the same magnitude and opposite direction, on the other. Because of the equivalence of gravitational and inertial mass, all bodies fall with the same acceleration in a uniform gravitational field (Galileo effect), so there is no way by a mechanical experiment to tell the two situations apart. Einstein made the bold assumption that this equivalence between the two frames—one inertial with uniform gravitational field, the other accelerated without gravitational field—held for all physical phenomena, in particular for electromagnetic effects (the parallel with his extension of the mechanical relativity principle in special relativity is obvious). Using this as a heuristic guide, he was able to draw several quantitative conclusions, which he regarded as at least a first approximation to results that any complete theory of gravitation should yield. Among these was the prediction that a light ray must be deflected by a gravitational field, but at that time he could suggest no experimental test.

For several years after that, the gravitational problem receded into the background of his efforts, although never dropped entirely. For example, in a letter to Sommerfeld in 1909, he remarks on the importance of generalizing his work on uniformly accelerated translation to uniformly rotating rigid bodies, but he does not seem to have worked on this problem until 1912. One can suggest several reasons for this:

1) Although absolutely convinced of the correctness of his basic approach, he was unsure of just how to proceed to develop his ideas further. He was much troubled that, whereas giving the space and time coordinates in inertial reference frames direct physical significance had played a crucial role in the development of special relativity, the coordinates seemed to lose their direct physical interpretation in accelerated frames of reference.

2) During these years he became deeply involved in the attempt to find a fundamental explanation of the quantum effects he had been so instrumental in bringing to light. Indeed, his first attempt at a "unified theory" came at about this time: he was looking for a unified quantum theory of electrons and electromagnetic fields. However, after several cycles of elation and defeat as one attempt after another proved a failure, he decided a fundamental theory was beyond his efforts. He returned to his earlier approach of taking the existence of the quantum (provisionally) for granted, and elucidating the consequences of its existence.

3) An even more prosaic explanation may be that during this period he was inducted into the academic world, first as a part-time *Privatdozent* in Bern, and then as Associate Professor at Zürich. His bibliography shows a definite

fall-off in his (always extraordinary) rate of publication in 1908–1909, quite consistent with the acclimatization period of a new junior faculty member.

At any rate, it was only when he moved to Prague in 1911 as a full Professor that he returned to the problem of gravitation with full vigor. In a preface to the Czech edition of his popular book on *Relativity*, he remarks that it was within the quiet halls of the Prague Institute of Theoretical Physics that he found the necessary equanimity to slowly give his earlier ideas a more definite form. The first product of Einstein's Prague ruminations was his 1911 paper "On the Influence of Gravitation on the Propagation of Light." This he regarded as basically an improved version of his earlier considerations on the equivalence principle. He stressed two conclusions, one experimental, the other theoretical, which proved important for the later history of general relativity.

The experimental conclusion was new: it would actually be possible to test the idea that light was deflected by a gravitational field. One should look for a shift in the apparent position of a star near the edge of the sun during an eclipse, compared with its position on an ordinary night. A Prague colleague put Einstein in contact with the Berlin astronomer Erwin Freundlich, who became extremely enthusiastic about the possibility of testing this and other astronomical consequences of Einstein's ideas about gravitation (e.g., gravitational red shift), and devoted much of the next decade to such efforts. Bad weather and World War conspired to twice keep Freundlich from getting the needed eclipse photographs. Since Einstein's 1911 paper predicted only half the deflection of the final 1915 theory, it is interesting to speculate on what might have been the reception of Einstein's final theory if it had appeared *after* an experimental test had shown the deflection was twice what Einstein had earlier predicted.

The theoretical conclusion that was to prove important for Einstein's further work was his treatment of the velocity of light—actually an improved derivation of a result obtained in the 1907 paper. Measured with the global time coordinate of a uniformly accelerated reference frame, c is no longer a constant, but depends on the acceleration. By Einstein's equivalence principle, this meant that the velocity of light in a gravitational field must be variable; and Einstein gave a formula showing how it varied with the gravitational potential. Using this, it was easy to derive the deflection of a light ray in a gravitational field by analogy with that of a light ray in a medium with variable index of refraction.

It is clear from some comments at the beginning of his next paper in 1912 that Einstein had been thinking about the application of his equivalence principle to the uniformly rotating frame of reference; but this was not the problem to which he turned his attention in the body of the paper. The reason perhaps lies in the fact that he could hardly accept his own conclusions, as indicated by the tentative nature of his comments about the rotating disk, and did not know what to do with the paradoxical result to which he was led—a result to which we shall return shortly. In line with his step-by-step approach to the gravitation problem (as he characterized it in a letter to Ehrenfest), he devoted two papers to the problem of the static gravitational field, a problem which he felt able to handle on the basis of his concept of the variable speed of light. If the speed of light depends on the gravitational potential in a static gravitational field, then we may replace the gravitational potential by the (variable) speed of light. Gravitational field equations for the speed of light are set up in the first 1912 paper, and modified in the second to accord with the conservation laws when gravitational momentum and energy are taken into account. Naturally, this leads to a non-linear theory of the gravitational field. But the establishment of equations of motion for a particle in the gravitational field proved to be the most important result of these papers. In an "Addendum" to the second of the papers, added in proof, Einstein notes that the equations of motion can be derived from a variational principle, which is formally the same as Planck's special-relativistic variational principle for particle motion, except that the speed of light is now a function of position (not time, of course, since only the static case is treated).

The Metric Tensor

Before turning to the next and crucial step in the story, it is important to recall that, during the years between 1907 and 1911, Minkowski had presented his four-dimensional formulation of special relativity theory, and it had been brought to the attention of the theoretical physics community largely by the expository efforts of Sommerfeld and Laue. At first, Einstein's reaction had been rather negative. From his student years on, he later remembered, he had felt that the physicist's mathematical requirements were quite modest; and that any toying with advanced mathematics was pure luxury—or worse, needless pedantry. Jests are reported, in which Einstein spoke of the Göttingen mathematicians as making relativity so hard that the physicists wouldn't be able to understand it. This skeptical attitude was only reinforced by his 1912 work on the static gravitational field. The concept of a variable speed of light when gravity was taken into account seemed to undercut the mathematical basis for the symmetry between space and time that underlay Minkowski's approach to special relativity; there are a number of comments to this effect in Einstein's 1912 correspondence. However, his notebooks from this period seem to indicate that he was studying Minkowski's approach, possibly via Sommerfeld's influential exposition.

Now we have the elements that went into the next step in Einstein's search, the crucial step: his adoption of a non-flat four-dimensional metric tensor as the correct mathematical representation of the gravitational field. An attempt to understand how Einstein arrived at this truly revolutionary viewpoint is complicated by the fact that there is not a hint of it in his 1912 papers on gravitation; while the very next paper in 1913 starts off from this viewpoint. Therefore, one can only speculate on exactly what happened in the half-year between. Here is my (incomplete) outline of the story:

After completing his treatment of the static case, Einstein planned to treat stationary gravitational fields, as he informed Ehrenfest. This may have turned his attention back to the rigidly rotating disk, the simplest stationary problem. What had troubled Einstein about his earlier analysis was that it seemed to show that measuring rods at rest on such a rotating disk would show a ratio of its circumference to its diameter different from π. Einstein seems to have confirmed this earlier, tentative conclusion. He had good reason to believe the result because it could be based on reasoning from the viewpoint of a non-rotating inertial frame (i.e., purely special-relativistic viewpoint), plus the assumption that small measuring rods at rest on the disk are unaffected by their acceleration. A non-Euclidean geometry must hold in the disk frame of reference. But then, by the equivalent principle, this rotating frame of reference should be equivalent to an inertial frame of reference plus a certain gravitational field. Einstein concluded that, in the presence of a gravitational field, measuring rods will map out a non-Euclidean geometry.

Einstein already knew something about non-Euclidean geometry. About the only mathematical lectures that stood out favorably in his remembrances of his years at the ETH were those by the geometer Karl Friedrich Geiser, who had lectured on infinitesimal geometry—another name for differential geometry. From the notebooks of Einstein's classmate and later collaborator Marcel Grossmann, it is known that these lectures included at least some elements of the Gaussian theory of surfaces, including the use of arbitrary curvilinear coordinates on a two-dimensional surface.

Einstein seems to have put this idea of curvilinear coordinates for two-dimensional non-Euclidean geometries together with his earlier result on the variational principle for a particle in a static gravitational field, to conclude that his variational integrand could be regarded as the line element of a non-flat four-dimensional space-time, invariant under arbitrary transformations of the space-time coordinates. What was required by the break down of the invariance of special relativity under the Lorentz group when gravitation was taken into consideration was not to drop Minkowski's four-dimensional point of view. Rather, one had to generalize the group of coordinate transformations and apply them to a non-flat space-time. From this point of view, it became obvious that the correct mathematical representation of the gravitational field was the array of ten coefficients of the coordinate differentials in the expression for the line element—the metric tensor.

Two problems remained: to reformulate the non-gravitational parts of physics in the light of this new, non-Euclidean metrical approach; and to find the correct field equations for the gravitational field itself. But the crucial step had been taken: the recognition of the dual role of the metric tensor as the representation of both the space-time structure and of the gravitational field. Once this mathematical structure had been identified, it was only a matter of time until the two problems just mentioned were solved. Einstein was not certain just how wide the new group of coordinate transformations should be, under which the resulting theory would be invariant. But clearly, any answer short of the most general coordinate transformations would only raise the question again: why stop here? So he seems to have felt strongly that the final theory must be invariant under arbitrary coordinate transformations.

The Field Equations

Shortly after his return to Zurich, in October 1912, as Professor at the ETH, Einstein had more or less clearly formulated this program. He must have felt that his mathematical background was not adequate to work out these problems alone. He turned for help once more to his former ETH schoolmate Marcel Grossmann. The latter had placed his careful notebooks at Einstein's disposal when final examination time came around, and had urged his father to intervene with the head of the Swiss Patent Office when Einstein was desperate for a job after graduation. Now Grossmann was a Professor of Mathematics at the ETH. Perhaps remembering Grossmann's earlier work on non-Euclidean geometry, Einstein asked Grossmann whether mathematical methods existed for formulating equations that were invariant under arbitrary coordinate transformations. Grossmann found that the work of Riemann, Christoffel, Ricci and Levi-Civita, from the mid-nineteenth century to the early years of the twentieth, had already provided the tools needed. Tensor analysis proved to be the mathematical formalism adapted to setting up generally covariant equations. Einstein and Grossmann soon succeeded in writing generally covariant equations incorporating the effect of the gravitation field on other physical systems; actually, much of this work had already been done by others and only had to be adapted to the new point of view. It also quickly became clear that the ten-component stress-energy tensor was the correct relativistic generalization of the Newtonian mass as the source of the metric-gravitational field.

But the search for the correct field equations for the gravitational field, relating this stress energy tensor to the metric tensor, proved a much longer story. As mentioned above, Einstein had hoped that the entire theory would be generally covariant. Grossmann pointed out that there was a certain tensor, formed from the metric tensor and its first and second derivatives and now known as the Ricci tensor, which was practically the unique candidate for generally covariant field equations. But calculation seemed to show that a theory based on these field equations did not yield Newton's theory of gravitation as a first approximation for weak gravitational fields. Such a "correspondence limit" was correctly seen as a criterion for any new theory of gravitation, but not until 1914 did Einstein realize that the calculation had been incorrectly interpreted. But, Einstein being Einstein, he had not remained satisfied with rejecting general covariance on the basis of a mere calculation. He had developed a simple meta-argument to show that the equations for the gravitational field *could* not be generally covariant. By the time he realized the Newtonian limit argument was in error, the meta-argument held him in its sway. What was that argument? Einstein stated that the causality principle demanded that a given distribution of matter and energy—a given stress-energy tensor mathematically speaking—should result (modulo appropriate boundary conditions) in a unique gravitational field. So far so good; but he felt that a unique gravitational field should correspond to a unique metric tensor. Now whether this is correct or not depends on how we interpret "unique." If we mean physically unique, that is fine (there are subtle mathematical problems even here, which we need not discuss). But Einstein interpreted it to mean unique as a mathematical function of the coordinates. Clearly no generally covariant field equations can have that property, precisely because their form as a function of the coordinates must change under coordinate transformation. Just because a theory was generally covariant, mathematically different-looking solutions corresponding to the same sources could be physically identical. But it was apparently only late in 1915 that Einstein realized this flaw in his meta-argument.

Under the sway of first the erroneous Newtonian limit argument, and then the erroneous meta-argument, Einstein developed a non-covariant set of equations for the gravitational field. Yet he was never really comfortable with giving up general covariance. His correspondence from early 1913, when he did the original work with Grossmann, until late 1915 is filled with justifications of the non-covariant equations and statements of how satisfied he was with them—probably as much to convince himself as his correspondents—and equally passionate statements that somehow the final theory must be generally covariant.

During this period he was also engaged in discussions—and sometimes polemical interchanges—with the authors of other theories of gravitation, notably Abraham, Mie and Nordström. Einstein, working with Fokker, was happy to find that one of Nordström's scalar theories of gravitation could be put into generally covariant form; they remark at the end of their 1914 paper on this subject that one should perhaps reexamine the question of general covariance for the Einstein–Grossmann theory (it is here that Einstein acknowledges the error in his earlier work). But somehow this did not stop him from publishing the meta-argument against general covariance in another 1914 paper. Since Nordström's theory did not predict any deflection of light by a gravitational field, Einstein looked upon future eclipse results as providing the critical test between the two theories. Abraham, after his efforts to create a special-relativistic theory of gravitation with a variable speed of light had been criticized by Einstein, abandoned the relativity concept altogether (at least verbally) and delivered himself of stinging attacks on Einstein's efforts. He did not fail to note that Einstein had first abandoned Lorentz invariance and then general covariance. Einstein rather welcomed these polemics—at least someone was publicly paying attention to his work on gravitation. Although his public comments were much more restrained in tone than Abraham's, in his private correspondence he gave vent to such gems as a reference to one of Abraham's theories as a stately horse, but lacking three legs. He referred privately to Mie's special-relativistic gravitational theory as fantastic and having a vanishingly small inner probablity.

When Einstein joined the Prussian Academy of Sciences in 1914, he included a brief discussion of the basis of his hopes for a generalized relativity theory in his Inaugural Address. Planck, in his reply for the Academy, took the rather unusual step of criticizing these hopes, probably because Einstein's remarks aligned the latter with Mach's views (Planck and Mach were then engaged in a bitter controversy over the philosophical foundations of their views on physics) on the relativity of inertia.

Difficulties with his non-covariant theory continued to pile up. He found that a mathematical argument, developed to show that his Lagrangian was unique, actually permitted almost any Lagrangian. He found that the rotating disk metric was not a solution to his field equations. He found that the value calculated for the anomalous precession of the perihelion of Mercury was not in agreement with the observed value. So in mid-1915 he started a fundamental reexamination of the whole problem, in the course of which he must have found the flaw in his meta-argument, and returned happily to the generally covariant approach.

In speculating on the reasons for the long delay between the initial formulation of the covariant program in late 1912–early 1913 and its final vindication in late 1915, one must bear in mind, of course, the revolutionary nature of what Einstein was trying to do. What seems obvious, from a contemporary viewpoint, about mathematical properties of covariant equations was clearly not so obvious to one working his way towards such equations from primarily physical viewpoints and criteria. In addition, one may speculate on the influence of various non-scientific events on Einstein's work during this period. It was the period of his move to Berlin; of his final separation from his first wife, who returned to Zurich with the children, to whom he was deeply attached, shortly after the move to Berlin; and of the outbreak of the first World War, which so deeply affected Einstein's whole outlook. In any case, he did succeed in overcoming most of his confusions about general covariance by late 1915, and the development of the theory proceeded quite rapidly from that point on.

Quite rapidly, but not without another mis-step that Einstein later regretted, he had immortalized in print. He first readopted the 1913 generally covariant field equations, in which the Ricci tensor was set equal to the gravitational coupling constant times the stress-energy tensor. This results in field equations that are satisfactory outside of matter, where the stress-energy tensor vanishes. Indeed, Einstein was extremely pleased when he was able finally to derive a value for the Mercury perihelion precession in good agreement with the observed one. But where matter did not vanish, his field equations were not mathematically consistent, he found, unless the trace of the stress-energy tensor vanished. His first response was to make a virtue of necessity and to argue that, although the trace of the phenomenological stress-energy tensor of ordinary matter might be non-vanishing, the trace of the electromagnetics stress-energy tensor did vanish. Perhaps a fundamental theory of ordinary matter would show that it was basically electromagnetic in nature (such speculations were not uncommon at the time, notably in the work of Mie), so that at a fundamental level the trace of the stress-energy tensor did vanish.

Shortly after publishing this justification of his equations, Einstein realized that they could be slightly modified so that they remained generally covariant without requiring any condition on the trace of the stress-energy tensor. Thus, Einstein finally arrived at the field equations that we know today as the Einstein equations for the gravitational field. In 1916 he published a review article on the general theory of relativity, in which he pointed out that the principle of general covariance did not force any definite assumptions about the nature of matter (i.e., non-gravitational sources of the gravitational field). He left it as an open question whether a combined theory of gravitation and electromagnetism could shed further light on the nature of matter. Here we may see an oblique comment on his earlier error, as well as on Hilbert's continuation of the Mie program; and the germ of Einstein's interest in a unified field theory.

Cosmology and Epistemology

But Einstein did not regard his work as complete for another, and to him more pressing, reason. We have mentioned that one of the strong motivations for Einstein's search for a generalization of the relativity principle was his reading of the critique of Newton in Mach's *Mechanics*. There have been many varying interpretations of Mach's text; but let it suffice here to say that Einstein interpreted Mach as demanding a theory in which the inertial properties of a body would not come from its resistance to motion with respect to absolute space, but rather from its interaction with all the rest of the matter in the universe. On such an interpretation, the difference in shape between a rotating and non-rotating liquid (Newton's bucket experiment) must be due to the former liquid's rotating relative to the distant matter in the universe, while the latter does not (the first argument for an extension of the relativity principle in Einstein's 1916 review paper is a version of this Machian epistemological argument). During the long years of his search for the general theory, Einstein had often consoled himself with the strength of this argument—several times in direct correspondence with Mach, in which he defended the latter against Planck, at that time the best-known opponent among physicists of Mach's epistemological credo. Without wishing to imply that Einstein was a full-fledged Machian—which he was not; or that he was an opponent of Planck—to whom he was otherwise very close; he certainly drew much of his inner conviction in the search for a generalized relativity principle from this aspect of Mach's doctrine.

Yet when his theory was finally completed, it did not embody Einstein's version of Mach's principle. He had been able to show to his satisfaction that the inertia of a body was influenced by the presence of other bodies in its neighborhood; but it was by no means the case that the total inertial of one body was determined by the mutual interactions between all the bodies in the universe. Indeed, from Einstein's point of view, it was a scandal that a solution to his field equations should exist which corresponds to the presence of a single body in an otherwise "empty" universe (i.e., no other bodies present). He made this clear in a number of letters in 1916–17 to Schwarzchild, Klein, de Sitter and others. The way out for him was to construct a cosmological model, in which the matter entirely determined the metric. This was the basic motivation for his 1917 static closed cosmological model, which he looked upon as the completion of his general-relativistic program. He was even willing to modify his field equations once more, by introducing the cosmological constant term, in order to achieve this end. De Sitter soon found an "empty" solution to the new field equations, but it was apparently not static. Einstein was disturbed by the existence of this solution, and found various reasons for rejecting it.

In later years, of course, he was forced to give up his static model when it became clear that a non-static expanding model gave the most natural explanation for the observations of galactic red shifts. By then he was also no longer wedded to Mach's principle. He realized that it was ultimately tied to a matter (in the conventional, particulate sense of the word) ontology, and that Mach's arguments were irrelevant in the context of a field ontology, basic to his approach to the unified field problem. But it took some time until these observational and ontological arguments finally convinced Einstein to give up his static model. Regardless of the original motivation for (and the fate of) this particular model, it must be recognized that Einstein's basic approach to cosmology—setting up a metrical model and attempting to correlate its features with observations—has dominated theoretical work in the field for over half a century.

The development of general relativity not only revolutionized theoretical physics, it also changed Einstein's views on epistemology. As he stated in a letter to Lanczos, his work on gravitation turned him from a skeptical empiricist into a believing rationalist. Neither of these extremes should be taken literally as a completely adequate characterization of Einstein's early or late philosophical views. But they do correctly indicate the trend of development of his views, and the crucial role that his work on general relativity played in this evolution. He more and more came to see the search for mathematical simplicity as the source of progress in theoretical physics—even if it was in constant need of control by experience.

On the other hand, it would not be correct to say that Einstein regarded what he had done as a reduction of physics—even the gravitational part of physics—to geometry. He explicitly repudiated such an interpretation of his work in his comments on Myerson's book on relativity in 1927, as well as on numerous other occasions. He sought a generally covariant field description for physical reality, but did not regard such a description as any more or less of a geometrization of physics than the fact that distance occurs in Newton's law of gravitation, or that Maxwell's equations can be written in terms of vectors and tensors.

Einstein reacted sharply against any attempt to extract a philosophy from his theories. He stated that a mathematical-scientific theory such as relativity might well be the *object* of philosophical investigation, but could never form the *basis* of a philosophy.

He was aware, even during the earliest days of his triumph, of the limitations of what he had done. As we have seen, he realized almost at once that general relativity did not solve the problem of the structure of matter. He discussed the possibility, as early as 1917, that no continuum theory might be able to do justice to the quantum nature of matter. Yet he also felt that the possibility that it *could* do so was not excluded, even by the triumph of quantum mechanics; and that this possibility should be explored to the end, particularly in view of the lack of any adequate non-continuum mathematical model. He often acknowledged the mortality of all physical theories, including his own.

The Rigidly Rotating Disk as the "Missing Link" in the History of General Relativity

John Stachel

1. Introduction

Working with the Einstein Archive at the Institute for Advanced Study has given me a chance to become familiar with some of the material in this most extensive repository of documents on the life and activities of Albert Einstein, collected indefatigably over the last quarter of a century by the Trustees of his Estate, Dr. Otto Nathan and Miss Helen Dukas; and organized by Miss Dukas, the Archivist of the collection, whose memory is undoubtedly its most important single resource. It has also made me realize how much the material held in the Archive can contribute towards the study of many problems in the history of modern physics, to say nothing of many cultural, social, and political topics.

As a small example, I shall discuss the question of the relativistic rigidly rotating disk, a topic that has been the subject of extensive—and intensive—discussion from the early days of the special theory of relativity to the present.[1] An examination of Einstein's treatment of this problem is of interest not only because it shows his way of treating the issues involved, but because it seems to provide a "missing link" in the chain of reasoning that led him to the crucial idea that a nonflat metric was needed for a relativistic treatment of the gravitational field.

A. Held, ed.
General Relativity and Gravitation
One Hundred Years After the Birth of Albert Einstein
Vol. 1, pp. 1–15

2. Einstein's Treatment of the Rotating Disk

Einstein's first mention of rigidly rotating bodies that I have located in the Archive is in a letter of September 29, 1909, to Arnold Sommerfeld:

> The treatment of the uniformly rotating rigid body seems to me to be of great importance on account of an extension of the relativity principle to uniformly rotating systems along lines of thought analogous to those that I tried to carry out for uniformly accelerated translation in the last section of my paper published in the *Zeitschrift für Radioaktivität* (EA 21–377).[2]

Einstein is referring here to his first published attempt in 1907 to develop a relativistic treatment of gravitation based on the equivalence principle applied to a spatially uniform static gravitational field (Einstein 1907).

The occasion for Einstein's comment was probably the discussion of Max Born's paper (1909) at the Salzburg meeting of the German Society of Scientists and Physicians, which took place September 21–25, 1909. Born had presented his definition of rigid motions (and thus of rigid bodies, insofar as they are capable of existing) in special relativity. Sommerfeld (1909) had commented on Born's talk and Born, in a later paper (1910), noted that he and Einstein had discussed the rigid-body problem at Salzburg, and were puzzled "that a [rigid] body at rest can never be brought into uniform rotation;" this problem was discussed almost simulaneously by Paul Ehrenfest in a paper (1909) submitted September 29—the same day as Einstein's letter to Sommerfeld—and soon became known as "Ehrenfest's paradox." In spite of the importance he attached to the problem, and the intense discussion occasioned by Ehrenfest's paper,[3] Einstein published nothing directly on this question during the next few years. His only contribution to the discussion was an answer to one of the points raised by Varičak (1911) in a comment on Ehrenfest's paradox.[4] Einstein's note (1911) made no reference to the rotating-disk problem but confined itself to rebutting Varičak's aspersions on the "reality" of the Lorentz contraction.

Einstein's first published reference to the rigidly rotating disk is a hesitant one. It occurs in the first of two papers on static gravitational fields, written in 1912 during his stay in Prague (Einstein 1912a, 1912b). It dates from February 1912 and begins by reviewing his previous work on the uniformly accelerating coordinate system, pointing out that

> Such a system K, according to the equivalence principle, is strictly equivalent to a system at rest in which a matter-free static gravitational field of a certain kind exists. Let spatial measurements in K be made with measuring rods, which—when compared with each other at rest at some point of K—have the same length; assume that the theorems of [Euclidean] geometry are valid for lengths measured in this way, and thus also for the relationship between the coordinates x, y, z and other lengths. This stipulation is not automatically permissible, but contains physical assumptions that ultimately could prove to be invalid. For example, they most probably do not hold in a uniformly rotating system, in which, on account of the Lorentz contraction, the ratio of the circumference of a cirle to its diameter would have to differ from π using our definition of length. The measuring rod, as well as the coordinate axes, are to be treated as rigid bodies. This is permissible in spite of the fact that, according

> to [special] relativity theory, rigid bodies cannot really exist. For one can imagine the rigid measuring body replaced by a large number of small non-rigid bodies, so aligned alongside each other that they do not exert any pressure forces on each other since each is separately held in place.

The tentative nature of his conclusions reflects Einstein's puzzlement during this period over the problem of the relationship between coordinates and measurements with rods and clocks, a point to which we shall later return.[5]

There are references to rotating frames of reference in several of Einstein's papers on the developing general theory of relativity,[6] as well as in the correspondence;[7] but the context is Einstein's interpretation of the equivalence principle: the explanation of the inertial forces occurring in such frames as equivalent to gravitational forces, and there is no direct reference to the rotating-disk problem. The next time that the rotating-disk argument occurs in Einstein's writings—this time without the tentative note—is in the 1916 review paper, in which he presented the final version of the general theory, together with various arguments for it (Einstein 1916). Although the paper is well known and easily accessible, I quote the paragraph in full for the sake of completeness:

> In a space which is free of gravitational fields we introduce a Galilean system of reference $K(x, y, z, t)$, and also a system of co-ordinates $K'(x', y', z', t')$ in uniform rotation relatively to K. Let the origins of both systems, as well as their axes of Z, permanently coincide. We shall show that for a space-time measurement in the system K' the above [special relativistic] definition of the physical meaning of lengths and times cannot be maintained. For reasons of symmetry it is clear that a circle around the origin in the X, Y plane of K may at the same time be regarded as a circle in the X', Y' plane of K'. We suppose that the circumference and diameter of this circle have been measured with a unit measure infinitely small compared with the radius, and that we have the quotient of the two results. If this experiment were performed with a measuring-rod at rest relatively to the Galilean system K, the quotient would be π. With a measuring rod at rest relatively to K', the quotient would be greater than π. This is readily understood if we envisage the whole process of measuring from the "stationary" system K, and take into consideration that the measuring rod applied to the periphery undergoes a Lorentzian contraction, while the one applied along the radius does not. Hence Euclidean geometry does not apply to K'. The notion of co-ordinates defined above, which presupposes the validity of Euclidean geometry, therefore breaks down in relation to the system K'. So, too, we are unable to introduce a time corresponding to physical requirements in K', indicated by clocks at rest relatively to K'. To convince ourselves of this impossibility, let us imagine two clocks of identical constitution placed, one at the origin of co-ordinates, and the other at the circumference of the circle, and both envisaged from the "stationary" system K. By a familiar result of the special theory of relativity, the clock at the circumference—judged from K—goes more slowly than the other, because the former is in motion and the latter at rest. An observer at the common origin of co-ordinates, capable of observing the clock at the circumference by means of light, would therefore see it lagging behind the clock beside him. As he will not make up his mind to let the velocity of light along the path in question depend explicitly on the time, he will interpret his observations as showing that the clock at the circumference "really" goes

more slowly than the clock at the origin. So he will be obliged to define time in such a way that the rate of a clock depends upon where the clock may be.

We therefore reach this result:—In the general theory of relativity, space and time cannot be defined in such a way that the differences of the spatial co-ordinates be directly measured by the unit measuring-rod, or differences in the time co-ordinate by a standard clock (Einstein 1916, pp. 115–117).

Note that the "rigidly rotating disk" is not actually referred to in these considerations. However, that Einstein had it in mind—or at least was not averse to its consideration in this context—is made clear by the expanded discussion of the topic that he gave in his *Relativity* book of 1916 (Einstein 1917), Chapter 23 of which is devoted to the topic "Behavior of Clocks and Measuring-Rods on a Rotating Body of Reference." In this amplified discussion, he adds: "In order to fix our ideas, we shall imagine K' to be in the form of a plane circular disk, which rotates uniformly in its own plane about its centre," and thereafter phrases his discussion in terms of the disk. He also adds an element not made explicit in his previous discussion, asserting that an observer at rest on the disk is entitled to regard "the force acting on himself, and in fact on all other bodies which are at rest relative to the disk ... as the effect of a gravitational field." Thus, he links up his treatment of the rotating disk with his earlier treatment of rotating reference frames; but we shall continue to confine ourselves to the disk aspect of his discussion.

He draws the conclusion "that the propositions of Euclidean geometry cannot hold exactly on the rotating disk, nor in general in a gravitational field, at least if we attribute the length 1 to the [measuring] rod in all positions and in every orientation." He immediately follows this discussion with a discussion in Chapter 24 of the "Euclidean and Non-Euclidean Continuum," and in Chapter 25 with a discussion of the use of "Gaussian Co-ordinates" to treat non-Euclidean continua mathematically. Since the book is still in print and easily accessible, I shall omit this long discussion.

In *The Meaning of Relativity*, based upon his 1921 Princeton lectures, Einstein again gives a similar discussion of the rotating disk. This time I shall quote his conclusions, since they briefly summarize the material in Chapters 23, 24, and 25 just mentioned:

Space and time, therefore, cannot be defined with respect to K' as they were in the special theory of relativity with respect to inertial systems. But, according to the principle of equivalence, K' may also be considered as a system at rest, with respect to which there is a gravitational field (field of centrifugal force, and force of Coriolis). We therefore arrive at the result: the gravitational field influences and even determines the metrical laws of the space-time continuum. If the laws of configuration of ideal rigid bodies are to be expressed geometrically, then in the presence of a gravitational field the geometry is not Euclidean.

The case that we have been considering is analogous to that which is presented in the two-dimensional treatment of surfaces. It is impossible in the latter case also to introduce co-ordinates on a surface (e.g., the surface of an ellipsoid) which have a simple metrical significance, while on a plane the Cartesian co-ordinates x_1, x_2, signify

> directly lengths measured by a unit measuring rod. Gauss overcame this difficulty, in his theory of surfaces, by introducing curvilinear co-ordinates which, apart from satisfying conditions of continuity, were wholly arbitrary, and only afterwards these co-ordinates were related to the metrical properties of the surface. In an analogous way we shall introduce in the general theory of relativity arbitrary co-ordinates x_1, x_2, x_3, x_4, which shall number uniquely the space-time points, so neighboring events are associated with neighboring values of the co-ordinates; otherwise, the choice of co-cordinates is arbitrary. We shall be true to the principle of relativity in its broadest sense if we give such a form to the laws that they are valid in every such four-dimensional system of co-ordinates, that is, if the equations expressing the laws are co-variant with respect to arbitrary transformations.
>
> The most important point of contact between Gauss's theory of surfaces and the general theory of relativity lies in the metrical properties upon which the concepts of both theories, in the main, are based (Einstein 1922a, pp. 60–61).

The topics are presented in the same order as in the *Relativity* book (Einstein 1917); first the discussion of the disk, leading to the conclusion that a non-Euclidean geometry holds on the disk, and therefore space (and time) coordinates cannot be given a direct physical meaning as in the special theory. The two-dimensional Gaussian theory of curved surfaces is recalled, based upon the possibility of introducing entirely arbitrary coordinate systems, and then using the metric tensor to describe the metrical properties of the surface. The analogy is then drawn with the use of arbitrary space-time coordinates and the metric tensor to characterize the gravitational field mathematically. Later on, I shall comment upon the possible historical significance of this order of presentation. But first I shall finish the account of Einstein's discussions of the disk.

In *The Evolution of Physics*, his popular book written with Leopold Infeld, Einstein again reverts to the example of the rotating disk to show the necessity to introduce non-Euclidean geometry if one wants to generalize the principle of relativity so that it applies to noninertial frames of reference (Einstein and Infeld 1938, pp. 226–234).

I have not found any other discussion of the rotating disk in Einstein's published writings (in a far from exhaustive search, I must add); but there are a number of letters in which Einstein refers to the subject, giving a much more detailed discussion than in any of the printed sources, and explicitly replying to some of the objections that were offered to his treatment. He also makes several comments on the importance of the problem for the development of the general theory of relativity, which will be of great value for my later historical discussion.

The first such letter I have come across is one to Joseph Petzoldt, the well-known positivist philosopher, and the author of several early essays claiming special relativity theory as a triumph of the positivistic approach (see, for example Petzoldt 1912). On July 26, 1919, Petzoldt wrote Einstein a letter (EA 19-055) in which he raised the objection to Einstein's treatment of the rotating disk (an objection that will not be unknown to the connoisseur of the literature on this subject) that the Lorentz contraction of the rotating rods on the circumference implies that the circumference of the rotating disk should be *shorter* than 2π times the radius.

Einstein replied at length in a letter of August 19, 1919, which has been published in German; I translate it here:

> As concerns the rotating disk, I cannot agree with you at all. It is well to remark that a rigid circular disk at rest must break up if it is set into rotation, on account of the Lorentz contraction of the tangential fibers and the noncontraction of the radial ones. Similarly, a rigid disk in rotation (produced by casting) must explode as a consequence of the inverse changes in length, if one attempts to bring it to the rest state. If you fully take into account this state of affairs, your paradox vanishes.
>
> Now you believe that a rigidly rotating circular line must have a circumference that is less than $2r\pi$ because of the Lorentz contraction. The basic error here is that you instinctively set the radius r of the rotating circular line equal to the radius r_0 that the circular line has in the case when it is at rest. This however, is not correct; because of the Lorentz contraction rather $2\pi r = 2\pi r_0\sqrt{1-(v^2/c^2)}$.
>
> The treatment of the metric of the circular disk runs as follows in detail. Let U_0 be the circumference, r_0 the radius of the rotating disk, considered from the standpoint of K_0 [that is, the rest frame]; then, on account of ordinary Euclidean geometry,
>
> $$U_0 = 2\pi r_0 \tag{1}$$
>
> U_0 and r_0 naturally are to be thought of as measured with nonrotating measuring rods, i.e., at rest relative to K_0.
>
> Now let me imagine corotating measuring rods of rest length 1 laid out on the rotating disk, both along a radius as well as the circumference. How long are these, considered from K_0? Let us imagine, in order to make this clearer to ourselves, a "snapshot" taken from K_0 (definite time t_0). On this snapshot the radial measuring rods have the length 1, the tangential ones, however, the length $\sqrt{1-(v^2/c^2)}$. The "circumference" of the circular disk (considered from K) is nothing but the number of tangential measuring rods that are present in the snapshot along the circumference, whose length considered from K_0 is U_0. Therefore,
>
> $$U = U_0/\sqrt{1-(v^2/c^2)}\,. \tag{2}$$
>
> On the other hand, obviously
>
> $$r = r_0 \tag{3}$$
>
> (since the snapshot of the radial unit measuring rod is just as long as that of a measuring rod at rest relative to K_0).
>
> Therefore, from (2), (3), $U/r = U_0/r_0(1/\sqrt{1-(v^2/c^2)})$, or on account of (1) $= 2\pi/\sqrt{1-(v^2/c^2)}$.
>
> You make the analogous error for clocks as for measuring rods. The *rotating observer notes very well that, of his two equivalent clocks, that placed on the circumference runs slower than that placed at the center.* We again prove this by considering the entire process from K_0. Let U_z be the clock at the center, U_p the one at the periphery. Considered from K_0, U_p goes slower than U_z; a corotating observer placed next to U_z therefore also sees U_p as going slower than U_z. For it is clear that—judged from K_0—the time between the occurrence of a position of the hands of the clock and its perception by our observer is constant (independent of the time). I hope this explanation will suffice (EA 19-069; Thiele 1971, pp. 71–73).

Apparently, Petzoldt did not find this explanation fully satisfactory, and in a missing letter must have objected to the introduction of rigid bodies into the argument (again, an objection not unknown to the connoisseur). Einstein replied in a second letter, of August 23, 1919, which has also been published in the original German:

> I also think that only a personal discussion can produce real clarity. I request that you therefore visit me soon (after making an appointment by telephone). In the meantime, the following on the matter: I know quite well, naturally, that rigid bodies cannot exist according to relativity theory. But one can proceed with advantage as if such did exist; i.e., it is a question of an idealization that can be applied in certain considerations without any contradiction. The considerations of my letter are to be understood in this sense.
>
> You have incorrectly set the radius of the rotating "rigid" circular line equal to r_0. Because the circumference, *thought of as materialized by itself*, contracts because of the Lorentz transformation. It would be otherwise, if only the *radii* were thought of as materialized, *but not the tangential connections of their endpoints.*
>
> What you say about peripheral measuring rods and clocks is quite untenable. It is a question of the unjustified taking over of results of special relativity to accelerated reference systems (relative to the inertial system). Freundlich and Schlick are absolutely correct here. By your sort of reasoning one could just as well conclude that every light ray must propagate rectilinearly with respect to an arbitrary rotating system, etc. Your misunderstanding is quite fundamental (EA 19-072; Thiele 1971, p. 73).

As late as 1951, Einstein again gave a detailed discussion of the disk in his draft reply (EA 25-482) to a letter from an Australian medical student named Leonard Champion, who had been teaching himself general relativity but could not find anyone on the staff of Melbourne University able to answer all of his questions.[8] In particular, he had run across the account of the rotating disk by Whittaker (1949), who mentioned that Lorentz and Eddington regarded the geometry of the disk as Euclidean, while others, including Einstein, took it as non-Euclidean. Attempting to resolve the conflict between the sources, Champion made a calculation that amounted essentially to taking the metric of the disk to be given by the line element orthogonal to the world lines of the disk; i.e.,

$$d\sigma^2 = g_{ij} - g_{0i}g_{0j}/g_{00},$$

where the world lines of the points of the disk are given by $x^i = \text{const}$ ($i = 1, 2, 3$). Einstein's reply agreed fully with Champion's calculation. He stated that one had to assume the existence of rigid infinitesimal rods, which implies that if two such rods once agreed in length, when compared, they would always do so, no matter what sort of gravitational field each might afterwards have been subjected to; and a similar assumption must be made for clocks. That is, physical objects that measure the metrical interval are assumed to exist. (Einstein points out that his assumption could be wrong, even though the gravitational field equations were correct.) It follows that the length of an elementary measuring rod is the orthogonal interval between the world lines of the endpoints of the rod.

He then states that he does not know what Eddington meant by claiming the geometry on the rotating disk is flat. While four-dimensional Minkowski space is naturally flat, no matter what coordinates are used, this is not the case for the geometry of the disk as measured with measuring rods rotating with the disk.

He points out that, to set up a rigidly rotating disk, one would first have to melt a disk at rest, then set the molten disk into rotation and solidify it while it rotates. He admits that there are not really any completely rigid bodies, since if there were one could signal with superluminal velocities; but he maintains that the use made of rigid bodies in his argument seems justified. He states that this example of the disk was of "decisive importance" to him in setting up the general theory of relativity because it showed that a gravitational field (here equivalent to the centrifugal field) causes non-Euclidean arrangements of measuring rods, and thus compelled a generalization of Euclidean space. He emphasizes that the behavior of the rotating measuring rods can be obtained from special-relativistic considerations, since everything is considered from the nonrotating frame of reference.

In this letter Einstein thus brings together in summary form all of his considerations on the rigidly rotating disk, together with his answers to many objections to his treatment.

Although this paper is primarily historical in its aim, it is perhaps worth noting one epistemological feature of Einstein's argument, because it is of some importance for the historical discussion. Einstein sees the argument for the necessity of non-Euclidean metrical relations on the rigidly rotating disk to be based upon three premises:

1. Special relativity holds in a global inertial frame, in which no gravitational field is present.

2. Any coordinate system may be used, and indeed not only mathematically, but may be interpreted as a physical frame of reference, provided that the appropriate gravitational (cum-inertial) field is introduced.

3. A small measuring rod does not change its length in any gravitational field.

The first assumption represents Einstein's conviction that special relativity theory retained its validity within any gravitational theory as the important limiting case in which no noninertial gravitational field occurs. The second assumption is finally embodied in general relativity in the postulate that the metric tensor is the appropriate mathematical representation of the gravitational field potentials. The third assumption, in the context of the metric interpretation of the second, gives physical significance to the metrical interval. That the third assumption really is an independent one is a point that Einstein emphasized a number of times.

In addition, he makes a most significant remark for the history of the development of general relativity, about the importance of his considerations in convincing him of the need to go over to non-Euclidean geometries in his treatment of the gravitational field. While I have not found any discussion of such a role for the rotating disk in Einstein's published writings on the origins of the general theory of relativity (Einstein 1921, 1933, 1949), it is not the first time that the claim is made in his correspondence. In a letter (EA 26-351) presumed to date from the winter of 1939–40 to Hyman Levy, the English Marxist mathematician and author of a number of books on modern science and philosophy for popular audiences, Einstein comments on Levy's latest book (Levy 1939). After stating how pleased he was with much of the book, he recommends that Levy correct one glaring error in later editions. Levy had stated on page 595 that observers on the disk would verify Euclidean geometry. Einstein points out that just the opposite is the case, and adds that it was just the recognition that non-Euclidean geometry holds on the rotating disk which convinced him, at the time he was working on his gravitation theory, that Euclidean geometry could not hold for rigid bodies in the presence of a gravitational field.

There are other references to the rotating disk problem in the correspondence; but we now have essentially all of Einstein's basic ideas connected with the problem. I will try to use these ideas to help solve a problem in the history of general relativity.

3. The Rotating Disk as a "Missing Link"

In his discussion of the development of general relativity in the "Autobiographical Notes," Einstein points out that the significance of the equivalence principle in requiring a generalization of the special theory was clear to him in 1908. He then adds

> Why were another seven years required for the construction of the general theory of relativity? The main reason lies in the fact that it is not so easy to free oneself from the idea that co-ordinates must have an immediate physical meaning (Einstein 1949, p. 67).

In the context of the development of the general theory of relativity as a theory of gravitation (leaving aside the question of possible generalized unified field theories), I think it is clear that what is meant is that only the coordinates-cum-metric tensor in some coordinate system have a physical meaning.

In trying to trace Einstein's journey from the special to the general theory, the following difficulty presents itself.[9] In the papers up to and including those published in 1912, there is no mention of the need for a nonflat space-time, much less of the metric tensor as mathematical representation of the gravitational field. Yet the first paper of 1913 presents us with a full-fledged argument for the representation of the gravitational field by $g_{\mu\nu}$, together with the development of four-dimensional tensor analysis on a Riemannian manifold, the Riemann tensor, etc. (Einstein and Grossmann 1913). Of course, the problem of the correct field equations for the metric tensor was not resolved until late in 1915; but once the crucial step of the appropriate mathematical description of the gravitational field had been taken, it was only a matter of time until the right field equations were found. I shall argue that the consideration of the rotating disk is a "missing link" in the crucial developments that must have taken place in late 1912.

By the end of March 1912, Einstein had completed his work on the static gravitational field, which he treated by introducing the concept of a variable speed of light, which took over the role of gravitational potential (Einstein 1912a, 1912b). At this point he even felt compelled to give up the symmetry between space and time that had characterized the special theory of relativity, especially in Minkowski's formulation of space-time.[10] He had also arrived at the conclusion that his previous formulations of the equivalence principle only held locally. Rather than quote his papers on this question, we shall summarize his account in a letter to Ehrenfest, since this letter also takes us an important step forward beyond the papers. In this letter (EA 9-333; undated, but marked by Ehrenfest as received July 7, 1912), he states that his papers on the static gravitational field (Einstein 1912a, 1912b) indicate that the equivalence hypothesis can only hold for infinitesimally small fields. He notes that his discussion of static gravitational fields corresponds to the electrostatic case in electromagnetic theory; while what he calls "the general static case" would include the analogue of magnetostatic fields. He mentions the "rotating ring" as an example of a system that will generate such a nonstatic but time-independent field.

Thus, Einstein, pursuing his step-by-step approach to the problem, was ready by July 1912 to attack what he called the "general static case"—what we would today call the case of stationary gravitational fields.[11] This presumably led him to look again at the rotating-disk problem—the simplest case of a stationary gravitational field—which he had already tentatively discussed in early 1912, as mentioned above. There exist some notes, unfortunately undated, in Einstein's notebooks from roughly this period that may preserve evidence of this study. At this point, the argument quoted herein could have occurred to him: from special-relativistic considerations (which he certainly did not doubt hold in the absence of a gravitational field) plus the hypothesis that any coordinate system may be used, provided it is treated as a frame of reference with a corresponding gravitational field (which was a leitmotiv of his search for a general theory of relativity), and the assumption that a unit measuring rod always measures the same length in any gravitational field, he would have concluded that "in a gravitational field Euclidean geometry could not hold with respect to the arrangement of rigid bodies," as he put it in the letter to Levy. The results obtained could also have helped to shake him free from any lingering idea that "co-ordinates must have an immediate metrical meaning." As we have seen, this was one of the points implicit in his earliest printed discussion of the rotating disk. Together with the idea that the equivalence principle only holds infinitesimally, this may have reminded him of the use of Gaussian coordinates to describe the line element of curved surfaces, in which the idea that Euclidean geometry holds infinitesimally plays such a major role. We have evidence that Einstein, in his studies at the ETH, had become familiar with the Gaussian formula for the line element through the lectures on infinitesimal geometry given by Professor Geiser,[12] lectures which stood out in Einstein's memory fifty years later, when he described them as "true masterpieces of pedagogical art that later helped me very much in wrestling with general relativity" (Einstein 1955). At any rate, we have Einstein's words to assure us that

> I first had the decisive idea of the analogy of mathematical problems connected with the theory and Gauss's theory of surfaces in 1912 after my return to Zurich [which took place in August 1912], without knowing at that time Riemann's and Ricci's or Levi-Civita's work (Einstein 1922b).[13]

Minkowski's four-dimensional formulation played an important role in Einstein's considerations at this point, as he tells us (Einstein 1955), and he soon saw that what he needed was a four-dimensional generalization of Gauss's two-dimensional surface theory, and that the flat metric tensor of Minkowski's formulation of special relativity had to be generalized to a nonflat metric. At this point, with the mathematical problem already well formulated, he approached Marcel Grossmann for help,[14] with the well-known results, which must have been largely attained by October 29, 1912, when he wrote Sommerfeld

> I am now occupying myself exclusively with the problem of gravitation and believe that, with the aid of a local mathematician who is a friend of mine [Grossmann], I'll now be able to master all difficulties. But one thing is certain, that in all my life I have never struggled as hard and that I have been infused with great respect for mathematics, the subtler parts of which, in my simple-mindedness, I had considered pure luxury up to now! Compared to this problem, the original relativity theory [i.e., special relativity] is child's play (EA 13-082; Hermann 1968, p. 26).[15]

Actually, three years were to elapse before Einstein was truly able to "master all the difficulties" of the general theory—but that is another story! Meanwhile, if my reconstruction is approximately accurate, we can see that Einstein's recapitulations of the rotating-disk story in the 1916 paper and the popular book (Einstein 1916, 1917), as well as in the Princeton lectures (Einstein 1922a) and the Einstein-Infeld book (Einstein and Infeld 1938), would not only represent a certain logical order of presentation of the material leading up to the recognition of the need for a nonflat space-time structure to describe the gravitational field, but would also represent a fairly accurate historical reconstruction of Einstein's own journey. Since Einstein notes in the preface of the 1916 book that he has attempted to present his ideas "on the whole, in the sequence and connection in which they actually originated" (Einstein 1917), this is perhaps not so surprising.[16] What is more surprising, if our reconstruction is more or less correct, is the lack of mention of the rotating-disk problem in any of his papers on gravitational theory from 1907 through 1915. We can hazard the guess that the reason for this is the amazingly brief period—some time between mid-July and mid-October 1912—when the problem played its role; and the fact that the next critical step, generalizing the result to the four-dimensional metric tensor, was taken almost immediately afterwards. It was this latter great leap forward which provided the starting point for the Einstein-Grossmann investigations, and Einstein started his section of their paper (1913) with an account of this step. Thus the disk problem, having played its role of a "missing link," modestly stayed in the background; but that role was never forgotten by Einstein, as his later letters show.

Supplementary Note

Since completing this article, I have come across an exchange of letters between Einstein and von Laue on the rotating disc, dating from 1950 when von Laue was revising his book on general relativity. von Laue objected to Einstein's treatment of this problem in *The Meaning of Relativity*, noting among other things that the Riemann–Christoffel tensor vanishes even if one transforms to a rotating coordinate system. Einstein replied:

> It is true that in that case the R_{iklm} [the components of the Riemann tensor] vanish, so that one might say: "There is no gravitational field present." However, what characterizes the existence of a gravitational field from the empirical standpoint is the non-vanishing of the Γ^l_{ik} [the components of the affine connection], not the non-vanishing of the R_{iklm}. If one does not think in such intuitive (*anschaulich*) ways, one cannot comprehend (*begreifen*) why something like curvature should have anything to do with gravitation in the first place. In any case no reasonable man would have hit upon something in that way. The key to the understanding of the equality of inertial and gravitational mass would have been missing.

This comment seems to me not only to emphasize again the importance of the rotating disc problem in Einstein's thinking; but to make more precise the nature of his thought process.

Notes

[1] Bibliographies of the literature on the rotating disk may be found in Arzeliès 1966 and Grøn 1975. A recent paper by Grünbaum and Janis (1977) gives additional references.

[2] "Die Behandlung des gleichförmig rotierenden starren Körpers scheint mir von grosser Wichtigkeit wegen einer Ausdehnung des Relativitätsprinzips auf gleichförmig rotierende Systeme nach analogen Gedankengängen, wie ich sie im letzten meiner in der *Zeitschr. f. Radioktivit.* publizierten Abhandlung für gleichförmig beschleunigte Translation durchzuführen versucht habe." This letter (EA 21-377) is not included in the published volume of the Einstein-Sommerfeld correspondence (Hermann 1968).

[3] See Klein 1970, pp. 152–154, for an account of this discussion, with references.

[4] In a letter of April 12, 1911, to Ehrenfest (EA 9-316), Einstein suggested that Ehrenfest reply, noting that a short reply was needed to avoid confusion; however, Einstein took on the job himself.

[5] In a reply to an attack by Abraham a few months later, Einstein notes "One sees already from the previously treated highly special case of the gravitation of masses at rest that the space-time coordinates lose their simple physical interpretation; and it still cannot be foreseen what form the general space-time transformation equations may take. I should like to ask all colleagues to have a try at this important problem!" (Einstein 1912c, pp. 1063–1064). Curiously enough, a paper using a definition of spatial distances on the disk equivalent to Einstein's and actually deriving the metric of the rotating disk had been published two years earlier by Theodor Kaluza (1910). The paper was to have been delivered by Kaluza at the 1910 Naturforscherversammulung in Königsberg, where he was then working; but he took sick and only the published version appeared under the title "Zur Relativitätstheorie," which gave no idea of its contents. I have found no evidence that Einstein—or anyone else in the long history of the rotating-disk problem for that matter—was aware of the existence of Kaluza's work.

[6] See, for example, pp. 1031–1032 of Einstein's summary survey of the state of general relativity theory (Einstein 1914). The fact that rotating reference frames did not satisfy the field equations of the Einstein-Grossmann theory, while they satisfy the generally covariant field equations of general relativity, played an important role in motivating Einstein's abandonment of the former in favor of the latter when he discovered this in 1915, as has been pointed out in Earman and Glymour 1978b.

[7] See, for example, Einstein's letter to Mach (EA 17-454), published in Herneck 1966.

[8] No record exists in the Archive to indicate that a reply was actually sent.

[9] I have consulted on this topic Illy 1977; Lanczos 1972; Guth 1970; and Earman and Glymour 1978a, 1978b. I am extremely grateful to Dr. Illy, of the Institute of Isotopes of the Hungarian Academy of Sciences, for making his work available to me.

[10] For example, he states in a letter to Smoluchowski of March 24, 1912; "The simple schema of the equivalence of the four dimensions does not hold here in the way it does with Minkowski" (EA 20-597; Teske 1969).

[11] In another letter to Ehrenfest, undated but surely from a little earlier in 1912, Einstein speaks of his work on the static case as finished, and states that he is considering the "dynamic case" now, "again proceeding from the special to the more general case" (EA 9-321).

[12] Marcel Grossmann's notes of Geiser's lectures have been preserved and are now in the ETH Library HS 421:15. They may have been used by Einstein to study for his examinations (see Einstein 1955). They contain a discussion of curvilinear coordinates and the Gaussian line element for the plane (private communication from Professor Res Jost).

[13] "Den entscheidenden Gedanken von der Analogie des mit der Theorie verbundenen mathematischen Probleme mit der Gauss'schen Flächentheorie hatte ich allerdings erst 1912 nach meiner Rückkehr nach Zurich, ohne zunächst Riemanns und Riccis, sowie Levy-Civitas Forschungen zu kennen."

[14] Both Einstein 1955 and Einstein 1922b state explicitly that the mathematical problems to be solved were formulated by Einstein before he approached Grossmann for help in their solution. On the other hand, neither of them makes any reference to Georg Pick or any mathematical help received from him in Prague in the formulation of the problem. Indeed, I have quoted the passage from the Preface to the Czech edition of Einstein 1917 to the effect that the "decisive idea" occurred to Einstein after his return to Zurich. This stands in contrast to Philip Frank's account in his biography of Einstein (Frank 1947). Since Frank is usually very careful in his account, and moreover was Einstein's successor in Prague and presumably had a chance to speak to Pick, the discrepancy is puzzling.

[15] For all but the last sentence, I have used the translation in McCormmach 1976, p. xxviii.

[16] Max Wertheimer, in his discussion of the origins of special relativity based on discussions with Einstein, states: "In the course of one of his books he did report some steps in the process" (Wertheimer 1945, p. 168), and later makes it clear that the book was Einstein 1917. On page 174 he says: "For what now followed in Einstein's thinking we can fortunately report paragraphs from his own writing [then follows a reference to pages 14 ff. of the German edition of Einstein 1917]. He wrote them in the form of a discussion with the reader. What Einstein says here is similar to the way his thinking proceeded..." Wertheimer's account of the development of the special theory has been attacked recently as unreliable, but not on this point; see Miller 1975.

References

Arzeliès, Henri (1966). *Relativistic Kinematics*. Oxford and New York: Pergamon.

Born, Max (1909). "Über die Dynamik des Elektrons in der Kinematik des Relativiitätsprinzips." *Physikalische Zeitschrift* 10:814–817.

—— (1910). "Über die Definition des starren Körpers in der Kinematik des Relativitätsprinzips." *Physikalische Zeitschift* 11:233–234.

Earman, John and Glymour, Clark (1978a). "Lost in the Tensors: Einstein's Struggles with Covariance Principles 1912–1916." *Studies in History and Philosophy of Science* 9:251–278.

—— (1978b). "Einstein and Hilbert: Two Months in the History of General Relativity." *Archive for History of Exact Sciences* 19:291–308.

Ehrenfest, Paul (1909). "Gleichförmige Rotation starrer Körper und Relativitätstheorie." *Physikalische Zeitschrift* 10:918.

Einstein, Albert (1907). "Über das Relativitätsprinzip und die aus demselben gezogenen Folgerungen." *Jahrbuch der Radioakitivität und Elektronik* 4:411–462.

—— (1911). "Zum Ehrenfestschen Paradoxon. Bemerkung zu V. Varičaks Aufsatz." *Physikalische Zeitschrift* 12:509–510.

—— (1912a). "Lichtgeschwindigkeit und Statik des Gravitationsfeldes." *Annalen der Physik* 38:443–458.

—— (1912b). "Zur Theorie des statischen Gravitationsfeldes." *Annalen der Physik* 38:443–458.

—— (1912c). "Relativität und Gravitation. Erwiderung auf eine Bemerkung von M. Abraham." *Annalen der Physik* 38:1059–1064.

—— (1914). "Die formale Grundlage der allgemeinen Relativitätstheorie." *Königlich Preussische Akademie der Wissenschaften* (Berlin). *Sitzungsberichte*: 1030–1085.

—— (1916). "Die Grundlage der allgemeinen Relativitätstheorie." *Annalen der Physik* 49:769–822. Reprinted in translation in Lorentz, Hendrik Antoon et al. *The Principle of Relativity: A Collection of Original Memoirs on the Special and General Theory of Relativity*, W. Perrett and G. B. Jeffery, trans. London: Methuen, 1923; reprint New York: Dover, 1952, pp. 109–164.

—— (1917). *Über die spezielle und die allgemeine Relativitätstheorie. (Gemeinverständlich)*. Braunschweig: Friedrich Vieweg & Sohn. The preface is dated December 1916. Page numbers are cited from the English translation *Relativity, the Special and General Theory: A Popular Exposition*. Robert W. Lawson, trans. London: Methuen, 1920; New York: Holt, 1921.

—— (1921). "A Brief Outline of the Development of the Theory of Relativity." *Nature* 106:782–784.

—— (1922a). *The Meaning of Relativity: Four Lectures Delivered at Princeton University*. Princeton: Princeton University Press.

—— (1922b). "Vorwort des Autors zur Tschechischen Ausgabe." In *Einstein a Praha*, J. Bičak, ed. Prague: Prometheus, 1979, p. 42; written in 1922 for a Czech edition of Einstein 1917.

—— (1933). "Notes on the Origin of the General Theory of Relativity." In *Ideas and Opinions*, pp. 285–290. New York: Crown, 1954.

——— (1949). "Autobiographical Notes." In *Albert Einstein: Philosopher-Scientist.* Paul Author Schilpp, ed. Evanston, Illinois: The Library of Living Philosophers, pp. 1–95. A corrected text and translation has been issued as *Albert Einstein: Autobiographical Notes*, Paul Arthur Schilpp, ed. La Salle, Illinois and Chicago: Open Court, 1979. Page numbers are cited from the 1949 edition.

——— (1955). "Erinnergungen—Souvenirs." *Schweizerische Hochschulzeitung* 28 (*Sonderheft*): 145–148, 151–153. Reprinted as "Autobiographische Skizze." In *Helle Zeit-Dunkle Zeit*, Carl Seelig, ed. Zurich: Europa Verlag, 1956, pp. 9–17.

Einstein, Albert and Grossman, Marcel (1913). *Entwurf einer verallgemeinerten Relativitätstheorie und einer Theorie der Gravitation. I. Physikalischer Teil von Albert Einstein II. Mathematischer Teil von Marcel Grossmann.* Leipzig and Berlin: B.G. Teubner. Reprinted with added "Bemerkungen," *Zeitschrift für Mathematik und Physik* 62 (1914): 225–261.

Einstein, Albert and Infeld, Leopold (1938). *The Evolution of Physics: The Growth of Ideas from Early Concepts to Relativity and Quanta.* New York: Simon & Schuster.

Frank, Philipp (1947). *Albert Einstein: His Life and Times*, George Rosen, trans., Shuichi Kusaka, ed. New York: Alfred Knopf.

Gron, Ø. (1975). "Relativistic Description on a Rotating Disk." *American Journal of Physics* 43: 869–876.

Grünbaum, Adolf and Janis, Allen I. (1977). "The Geometry of the Rotating Disk in the Special Theory of Relativity." *Synthese* 34: 281–299.

Guth, Eugene (1970). "Contribution to the History of Einstein's Geometry as a Branch of Physics." In *Relativity, Proceedings of the Relativity Conference in the Midwest, held at Cincinnati, Ohio, June 2–6, 1969*, Moshe Carmeli, Stuart I. Fickler, and Louis Witten, eds. New York and London: Plenum Press, pp. 161–207.

Hermann, Armin, ed. (1968). *Albert Einstein–Arnold Sommerfeld Briefwechsel.* Basel and Stuttgart: Schwabe.

Herneck, Friedrich (1966). "Die Beziehungen zwischen Einstein und Mach, dokumentarisch dargestellt." *Wissenschaftliche Zeitschrift der Friedrich-Schiller-Universität Jena. Mathematisch-Naturwissenschaftliche Reihe* 15:1–14.

Illy, Józsek (1977). "The Birth of Einstein's Theory of Relativity." Typescript.

Kaluza, Theodor (1910). "Zur Relativitätstheorie." *Physikalische Zeitschrift* 11:977–978.

Klein, Martin (1970). *Paul Ehrenfest: The Making of a Theoretical Physicist.* Vol. 1. Amsterdam: North-Holland.

Lanczos, Cornelius (1972). "Einstein's Path From Special to General Relativity." In *General Relativity: Papers in Honour of J.L. Synge*, L. O'Raifeartaigh, ed. Oxford: Clarendon Press, pp. 5–19.

Levy, Hyman (1939). *Modern Science: A Study of Physical Science in the World Today.* New York: Alfred Knopf.

McCormmach, Russell (1976). "Editor's Forward." *Historical Studies in the Physical Sciences* 7:xi–xxxv.

Miller, Arthur I. (1975). "Albert Einstein and Max Wertheimer: A Gestalt Psychologist's View of the Genesis of Special Relativity Theory." *History of Science* 13:75–103.

Petzoldt, Joseph (1912). "Die Relativitätstheorie im erkenntnistheorischen Zusammenhange des relativistischen Positivismus." *Deutsche Physikalische Gesellschaft. Verhandlungen* 14:1055–1064.

Sommerfeld, Arnold (1909). "Diskussion." Following Born 1909. *Physikalische Zeitschrift* 10:826–829.

Teske, Armin (1969): "Einstein und Smoluchowski. Zur Geschichte der Brownschen Bewegung." *Sudhoffs Archiv. Zeitschrift für Wissenschaftsgeschichte* 53:292–305.

Thiele, Joachim (1971). "Briefe Albert Einstein an Joseph Petzoldt." *NTM—Schriftenreihe für Geschichte der Naturwissenschaften, Technik und Medizin* 8:70–74.

Varičak, V. (1911). "Zum Ehrenfestschen Paradoxon." *Physikalische Zeitschrift* 12:169–170.

Wertheimer, Max (1945). *Productive Thinking*. New York: Harper Brothers.

Whittaker, Edmund (1949): *From Euclid to Eddington: A Study of Conceptions of the External World*. Cambridge: Cambridge University Press; reprint New York: Dover, 1958.

The First Two Acts

John Stachel

Prologue: The Development of General Relativity, A Drama in Three Acts

In 1920, Einstein wrote a short list of "my most important scientific ideas."[1] The final three items on the list are:

> 1907 Basic idea for the general theory of relativity
>
> 1912 Recognition of the non-Euclidean nature of the metric and its physical determination by gravitation
>
> 1915 Field equations of gravitation. Explanation of the perihelion motion of Mercury.

Einstein's words provide the warrant for comparing the development of general relativity to a three-act drama:

Act I (1907): The formulation of the "basic idea," which he soon referred to as the equivalence principle.

Act II (1912): The mathematical representation of the gravitational field by a symmetric second rank tensor field, which enters into the line element of a four-dimensional space-time; hence this tensor is usually referred to as the (pseudo-)metric of space-time.[2]

Act III (1915): The formulation of the now-standard Einstein field equations for the metric field, and use of its spherically-symmetric solution to explain the anomalous precession of the perihelion of Mercury.

Act III certainly does not represent the end of general relativity, but a certain point of closure in its development, signaled by Einstein in his 1916 review paper:[3]

To appear in: *The Genesis of General Relativity*,
Vol. 1, *General Relativity in the Making*:
Einstein's Zurich Notebook. Part I
Jürgen Renn, Tilman Sauer, Michel Janssen, John Norton and John Stachel

> According to the general theory of relativity, gravitation plays an exceptional role as opposed to the other forces, in particular the electromagnetic... (p. 779);

consequently, he concluded his exposition with a "Theory of the Gravitational Field" (pp. 801–822).

Up to this point, the story had been essentially an account of Einstein's struggles.[4] Now that the final form of the gravitational field equations had been achieved and one of its predictions validated, the theory became the property of the physics and astronomy communities. Its further development and interpretation became a subject of discussion among many participants, among whom Einstein's voice did not always carry the day.[5]

This book tells the story from the opening curtain of Act I in 1907 until the curtain goes down on Act III at the end of 1915. The great bulk of it is devoted to Act III, embracing the events between 1912 and 1915. In particular, it centers on the understanding of the Zurich Notebook, which opens Act III and has made a signal contribution to our understanding of subsequent events. As in most plays, the final act contains the *dénouement*; but the crucial events that lead up to it take place in the first two acts. It was his formulation of the equivalence principle in Act I, and constant adherence to it as the guiding thread in his search for a relativistic theory of gravitation that set Einstein apart from other physicists who were working on the problem of fitting gravitation within the framework of the (special) theory of relativity. And as usual, the high point of the drama comes in Act II, with Einstein's remarkable decision to represent gravitation, not by a scalar field, but by the ten components of a tensor field that also describes the chrono-geometry of a non-flat four-dimensional space-time.

It is at this point, with an expression for the line element in terms of the metric tensor field, that the Zurich Notebook opens. Clearly, it cannot be fully understood or evaluated without prior knowledge of what happened during the first two acts. Unfortunately, no equivalent of the Zürich Notebook has been found for this period from 1907–1912. So this chapter attempts to present what can be learned—or surmised—about what happened on the basis of Einstein's published papers and correspondence, as well as his later reminiscences.[6] In keeping with the documentary character of this book, rather than attempting to summarize them I shall often cite Einstein's words *in extenso*.

Act I: The Equivalence Principle— "The Most Fortunate Thought of My Life"

In 1920,[7] Einstein recalled how he first arrived at the ideas behind the equivalence principle:

> While I was occupied (in 1907) with a comprehensive survey of the special theory[8] for the "Yearbook for Radioactivity and Electronics," I also had to attempt to modify Newton's theory of gravitation in such a way that its laws fitted into the theory. Attempts along these lines showed the feasibility of this enterprise, but did not satisfy me, because they had to be based on physical hypotheses that were not well-founded. Then there came to me the most fortunate thought of my life in the following form:

> Like the electric field generated by electromagnetic induction, . . . the gravitational field only has a relative existence. *Because, for an observer freely falling from the roof of a house, during his fall there exists*—at least in his immediate neighborhood—*no gravitational field.* Indeed, if the observer lets go of any objects, relative to him they remain in a state of rest or uniform motion, independently of their particular chemical or physical composition [note by AE: air resistance is naturally ignored in this argument]. The observer is thus justified in interpreting his state as being at rest.
>
> Through these considerations, the unusually extraordinary experimental law, that all bodies fall with equal acceleration in the same gravitational field, immediately obtains a deep physical significance. For if there were just one single thing that fell differently from the others in the gravitational field, then with its help the observer could recognize that he was falling in a gravitational field. If such a thing does not exist—which experiment has shown with great precision—then there is no objective basis for the observer to regard himself as falling in a gravitational field. Rather, he has the right to regard his state as one of rest and, with respect to a gravitational field, his neighborhood as field free. The experimental fact of the material-independence of the acceleration due to gravity is thus a powerful argument for the extension of the relativity postulate to coordinate systems in non-uniform relative motion with respect to each other .... The generalization of the relativity principle thus indicates a speculative path towards the investigation of the properties of the gravitational field (pp. 24–25).

Einstein alludes here to his initial attempts to set up a special-relativistic theory of gravitation, but gives no details. In 1933 he gave the fullest account of how he "arrived at the equivalence principle by a detour [*Umweg*]"[9] through such attempts.[10] After mentioning his doubts after 1905 about the privileged dynamical role of inertial systems, and his early fascination by Mach's idea that the acceleration of a body is not absolute, but relative to the rest of the bodies in the universe, he turns to the events of 1907:

> I first came a step closer to the solution of the problem when I attempted to treat the law of gravitation within the framework of special relativity. Like most authors at the time, I attempted to establish a field law for gravitation, since the introduction of an unmediated action at a distance was no longer possible, at least in any sort of natural way, on account of the abolition of the concept of absolute simultaneity.
>
> The simplest thing naturally was to preserve the Laplacian scalar gravitational potential and to supplement Poisson's equation in the obvious way by a term involving time derivatives, so that the special theory of relativity was satisfactorily taken into account. The equation of motion of a particle also had to be modified to accord with the special theory. The way to do so was less uniquely prescribed, since the inertial mass of a body might well depend on its gravitational potential. This was even to be expected on the basis of the law of the inertia of energy.
>
> However, such investigations led to a result that made me highly suspicious. For according to classical mechanics, the vertical acceleration of a body in a vertical gravitational field is independent of the horizontal component of its velocity. This is connected with the fact that the vertical acceleration of a mechanical system, or rather of its center of mass, in such a gravitational field turns out to be independent of its internal kinetic energy. According to the theory I was pursuing, however, such an independence of the gravitational acceleration from the horizontal velocity, or from the internal energy of a system, did not occur.[11]

> This did not accord with an old fact of experience, that all bodies experience the same acceleration in a gravitational field. This law, which can also be formulated as the law of equality of inertial and gravitational mass, now appeared to me in its deep significance. I was most highly amazed by it and guessed that in it must lie the key to the deeper understanding of inertia and gravitation (pp. 135–136).

Turning from later reminiscences, let us see how Einstein presented his approach to gravitation in 1907:[12]

> Up to now we have only applied the principle of relativity, i.e., the presupposition that the laws of nature are independent of the state of motion of the reference system, to *acceleration-free* reference systems. Is it conceivable that the principle of relativity also holds for systems that are accelerated relative to each other?
>
> This is not the place for an exhaustive treatment of this question. Since, however, it is bound to occur to anyone who has followed the previous applications of the relativity principle, I shall not avoid taking a position on the question here.
>
> Consider two systems in motion Σ_1 and Σ_2. Let Σ_1 be accelerated in the direction of its X-axis, and let γ be the magnitude (constant in time) of this acceleration. Let Σ_2 be at rest, but in a homogeneous gravitational field that imparts an acceleration $-\gamma$ in the direction of the X-axis to all objects. As far as we know, the laws of physics with respect to Σ_1 do not differ from those with respect to Σ_2; this is due to the circumstance that all bodies in a gravitational field are equally accelerated. So we have no basis in the current state of our experience for the assumption that the systems Σ_1 and Σ_2 differ from each other in any respect; and therefore in what follows shall assume the complete physical equivalence of a gravitational field and the corresponding acceleration of a reference system.
>
> This assumption extends the principle of relativity to the case of uniformly-accelerated translational motion of the reference system. The heuristic value of this assumption lies in the circumstance that it allows the replacement of a homogeneous gravitational field by a uniformly accelerated reference system, which to a certain extent is amenable to theoretical treatment.[13]

Some further comments on this equivalence in his next paper on gravitation in 1911[14] are illuminating. He notes that in both systems, objects subject to no other forces fall with constant acceleration:

> For the accelerated system K' [corresponding to the 1907 Σ_1 -JS], this follows directly from the Galileian principle [of inertia-JS]; for the system K at rest in a homogeneous gravitational field [corresponding to the 1907 Σ_2 -JS], however, it follows from the experimental fact that in such a field all bodies are equally strongly uniformly accelerated. This experience of the equal falling of all bodies in a gravitational field is the most universal with which the observation of nature has provided us; in spite of that, this law has not found any place in the foundations of our physical picture of the world.... From this standpoint one can as little speak of the *absolute acceleration* of a reference system, as one can of the *absolute velocity* of a system according to the usual [special-JS] theory of relativity. [note by AE: Naturally, one cannot replace an *arbitrary* gravitational field by a state of motion of the system without a gravitational field; just as little as one can transform all points of an arbitrarily moving medium to rest by a relativity transformation.] From this standpoint the equal falling of all bodies in a gravitational field is obvious.

> As long as we confine ourselves to purely mechanical processes within the realm of validity of Newtonian mechanics, we are certain of the equivalence of the systems K and K'. Our point of view will only have a deeper significance, however, if the systems K and K' are equivalent with respect to all physical processes, i.e., if the laws of nature with respect to K agree completely with those with respect to K'. By assuming this, we obtain a principle that, if it really is correct, possesses a great heuristic significance. For, by means of theoretical consideration of processes that take place relative to a uniformly accelerated reference system, we obtain conclusions about the course of processes in a homogeneous gravitational field.[15]

With hindsight, one can see that Einstein's attempt to find the best way to implement mathematically the physical insights about gravitation incorporated in the equivalence principle was hampered significantly by the absence of the appropriate mathematical concepts. His insight, as he put it a few years later, that gravitation and inertia are "essentially the same" ["*Wesensgleich*"],[16] cries out for implementation by their incorporation into a single inertio-gravitational field, represented mathematically by a non-flat affine connection on a four-dimensional manifold. But the concept of such a connection was only developed *after*, and largely in response to, the formulation of the general theory. So Einstein had to make do with what was available: Riemannian geometry and the tensor calculus as developed by the turn of the century, i.e., based on the concept of the metric tensor, without a geometrical interpretation of the covariant derivative. As I have suggested elsewhere, this absence is largely responsible for the almost three-year lapse between the end of Act I and the close of the play.[17]

Act Two—The Metric Tensor: "Just what are coordinates actually supposed to mean in physics?"

In 1949, Einstein himself raised the question of what was responsible for this long delay:

> This [recognition that the relativity principle had to be extended to non-linear transformations-JS] took place in 1908. Why were a further seven years required for setting up the general theory of relativity? The principal reason is that one does not free oneself so easily from the conception that an immediate physical significance must be attributed to the coordinates.[18]

Both the question and answer thus concern the entire period between 1907 (or 1908) and 1915. In 1933 Einstein made the answer more precise, and confined it to a shorter period of time:

> I soon saw that, according to the point of view about non-linear transformations required by the equivalence principle, the simple physical interpretation of the coordinates had to be abandoned; i.e., one could no longer require that coordinate differences be interpreted as signifying the immediate results of measurements with ideal measuring rods and clocks. This recognition tormented me a great deal because for a long time I was not able to see just what *are* coordinates actually supposed to mean in physics? The resolution of this dilemma was reached around 1912.[19]

Einstein's reference to 1912 is a clear allusion to the introduction of the metric tensor. But, as his reference to "a further seven years" after 1908 in the previous quotation suggests, the problem of the meaning of coordinates in general relativity was by no means completely resolved with the introduction of the metric tensor. Only with the resolution in 1915 of the "hole argument" ["*Lochbetrachtung*"] against general covariance that Einstein developed in 1913, did Einstein fully solve this problem, but discussion of the post-1912 aspects of the question will be found in later chapters.[20]

Now let me return to the problem of coordinates as Einstein saw it in 1907–1908. It is worth emphasizing that Einstein attributed his success in formulating the special theory in 1905 in no small measure to his insistence on defining physically coordinate systems that allow one to attach direct physical significance to coordinate differences:

> The theory to be developed—like every other electrodynamics—rests upon the kinematics of rigid bodies, since the assertions of each such theory concerns relations between rigid bodies (coordinate systems), clocks and electromagnetic processes. Taking this into account insufficiently is the root of the difficulties, with which the electrodynamics of moving bodies currently has to contend.[21]

Little wonder that Einstein was "tormented" by the problem of "just what coordinates are actually supposed to mean in physics" once they lose their direct physical significance!

This problem arose in the course of the application of the equivalence principle to linearly accelerated frames of reference, and the attempt to apply it to uniformly rotating frames, both considered within the confines of Minkowski space.[22] Its resolution came out of Einstein's work on a theory of the static gravitational field, in particular on the equations of motion of a particle in this field; and his attempt to generalize this static theory to non-static fields.

Both problem areas, accelerated system of reference in Minkowski space and static gravitational fields, ultimately led Einstein beyond the confines of Minkowski space to the consideration of non-flat Riemannian space-times. For convenience of exposition, I shall discuss these two strands of the story as if they were the subject of two separate scenes of Act Two, culminating in a third scene that ends the act. While it is broadly true that events in Scene One precede those in Scene Two, and certainly true that they both precede the events in Scene Three, to the extent that this division suggests a strict chronological separation between events in the First and Second Scenes, it does a certain violence to the actual course of events. However, it seems preferable to run this risk rather than attempt to jump back and forth between events in each of the intertwined strands of the story.[23]

Scene One—"To Interpret Rotation as Rest"

I shall start by again citing Einstein's 1949 comments on coordinates:

> The change [from the viewpoint that coordinates must have an immediate metrical significance] came about in more-or-less the following way.
>
> We start with an empty, field-free space as it appears with respect to an inertial system in accord with the special theory of relativity, as the simplest of all conceivable physical situations. Now if we imagine a non-inertial system introduced in such a way that the new system (described in three-dimensional language) is uniformly accelerated in a (suitably defined) direction with respect to the inertial system; then, with respect to this system, there exists a static parallel gravitational field. In this case, the reference system may be chosen as a rigid one, in which three-dimensional Euclidean metric relations hold. But that time [coordinate-JS], in which the field appears static, is *not* measured by *equally constituted* clocks at rest [in that system-JS]. From this special example, one already recognizes that, when one allows non-linear transformations of any sort, the immediate metrical significance of the coordinates is lost. One *must* introduce such transformations, however, if one wants to justify the equality of gravitational and inertial mass by the foundations of the theory, and if one wants to overcome Mach's paradox concerning inertial systems.[24]

Examination of Einstein's 1907 paper[25] shows that this account correctly reflects its contents. Einstein first demonstrates that—at least to first order in the acceleration—the spatial coordinates in a uniformly accelerating frame of reference retain their direct physical significance in terms of measuring rods; and thus, by the principle of equivalence, they still do so in the equivalent gravitational field. He then goes on to show that what he calls "the local time σ" [he uses both "*Ortszeit*" and "*Lokalzeit*" as names], which is essentially the proper time as measured by an ideal clock at a fixed point of the frame, differs from the "time τ," which he later called the "universal" ["*universelle*"] time,[26] which must be used to define simultaneity of distant events if one wants a time coordinate expressing the static nature of the gravitational field that is equivalent to the uniformly-accelerated one.

Thus, by the end of 1907, Einstein knew that differences between the "universal" time coordinate of events in a uniform gravitational field do not correspond to differences in the readings of ideal clocks in that field. It is true that he had shown that, at least to first order in the field strength, spatial coordinate differences still correspond to the results of measurements with rigid rods. But the fact that he felt compelled to demonstrate this for uniform gravitational fields suggests that he anticipated the possibility that similar problems might arise for the spatial coordinates in more complicated gravitational fields.

Einstein did not publish anything on gravitation between 1908 and 1911, but he continued to think about the subject:

> Between 1909–1912 while I had to teach theoretical physics at the Zürich and Prague Universities I pondered ceaselessly on the problem.[27]

The earliest surviving indication that Einstein contemplated an extension of the relativity principle beyond linearly accelerated systems dates from 1909:

> The treatment of the uniformly rotating rigid body seems to me to be of great importance on account of an extension of the relativity principle to uniformly rotating systems along lines of thought analogous to those that I attempted to carry out for

> uniformly accelerated translation in the last paragraph of my paper published in the *Zeitschrift für Radioaktivität*.[28]

What he had in mind is made more explicit in 1912 in a letter to his friend Michele Besso. After a rather full account of his new static theory (to be discussed below), he concludes: "You see that I am still far from being able to interpret [*auffassen*] rotation as rest. Every step is devilishly difficult..."[29]

As his reference to a "uniformly rotating rigid body" suggests, a solution to the problem of "interpreting rotation as rest" seemed to him to depend on developing a theory of rigid bodies in special relativity. In 1910 he wrote of this

> child of sorrow [*Schmerzenskind*], the rigid body. .... one should attempt to devise hypotheses about the behavior of rigid bodies that would permit a uniform rotation.[30]

Born had provided a definition of a relativistic rigid body in 1909, but he discussed only the case of linearly accelerated motion in any detail.[31] Further clarification soon came:

> The latest relativity-theoretical investigations of Born and Herglotz interest me very much. It really seems that in the theory of relativity there does not exist a "rigid" body with 6 degrees of freedom."[32]

This was disturbing, but brought new hope: If rigid *bodies* are incompatible with the special theory, rigid *motions* are not. In 1911, Laue summarized the situation concisely:

> The limiting concept of a body that is rigid under all circumstances, which is so useful everywhere in classical mechanics, in my opinion cannot be taken over [to the special theory-JS] on account of the impossibility of indefinitely large velocities for the propagation of elastic deformations. However this does not exclude a body moving at times like a rigid one; even according to classical mechanics, under certain circumstances a drop of fluid can move as if it were rigid.[33]

In short, the problem had been transformed from a dynamical one (what is a relativistic "rigid body" and how does it behave when accelerated?) to a kinematical one (given Born's relativistic definition, what types of "rigid motion" are possible?). Progress on the kinematical problem was much easier.[34] Indeed, in the paper cited by Einstein, Herglotz showed:

> that, as soon as one of the points of a [rigid-JS] body in Mr. Born's sense is fixed, it can only rotate uniformly about an axis passing through this point, like the usual rigid body.[35]

In the course of classifying all solutions of Born's rigidity condition, Herglotz gave the explicit form of the solution for rigid rotation about the z-axis.[36] So Einstein could discuss the kinematics of such an ideal rigid rotation, and the gravitational field which is equivalent to the inertial forces in such a rotating frame, without having to solve the dynamical problem of what types of physical system could actually undergo such a motion.

But for Einstein, there remained a second, Machian type of question: What sort of body could *induce* the gravitational field in a frame at rest that is equivalent to the inertial forces in an accelerated frame. Einstein first considered this question for linear acceleration, so we shall discuss it before returning to the problem of rotation. In a 1912 paper on gravitational induction,[37] Einstein showed that, as a consequence of the inertia of energy and the equivalence principle, a spherical shell of matter K accelerated linearly relative to an unaccelerated frame exerts such an inductive accelerating gravitational effect on a particle P enclosed in the shell.[38] He also showed that:

> the presence of the massive shell K increases the inertial mass of the particle P within it. This makes likely the assumption that the entire inertia of a massive particle is an effect of the presence of all the other masses, based on a sort of interaction with the latter. This is completely the same standpoint that E. Mach had upheld in his acute investigations on the subject (E. Mach, The Development of the Principles of Dynamics. Chapter Two. Newton's Views on Time, Space and Motion).[39] How far this conception is justified will be seen when we are in the happy possession of a usable dynamics of gravitation (p. 177).

Presumably, Einstein already had in mind the application of this induction idea to rotational acceleration. In 1921, discussing the development of the general theory, Einstein wrote:[40]

> Can gravitation and inertia be identical [*wesensgleich*]? The posing of this question leads directly to the General Theory of Relativity. Is it not possible for me to regard the earth as free from rotation, if I conceive of the centrifugal force, which acts on all bodies at rest relatively to the earth, as being a "real" field of gravitation (or part of such a field)? If this idea can be carried out, then we shall have proved in very truth the identity of gravitation and inertia. For the same effect [*Wirkung*] that is regarded as *inertia* from the point of view of the system not taking part in the rotation can be interpreted as *gravitation* when considered with respect to the system that shares the rotation.

I believe that the phrase "or part of such a field" makes clear what Einstein had in mind as his ultimate goal. The total gravitational field of the earth in a frame in which it is at rest (i.e., a co-rotating frame) consists of two parts: a gravito-static term, which would be present even if the earth were not in rotation (this is the Newtonian gravitational field), and a gravito-stationary term. The latter is usually interpreted as an inertial field, consisting of centrifugal and Coriolis terms, which would exist even if the earth were massless, i.e., in *any* rotating frame of reference. But, in accord with the principle of equivalence, these terms may be interpreted as a gravito-stationary field in a non-rotating frame of reference.

In summary: Because of his attraction to Mach's program, from the beginning of his search for a theory of gravitation based on the equivalence principle, the aim of interpreting rotation as rest-plus-a-gravitational-field appears to have loomed large in Einstein's motivation. This motive led him to consider uniformly rotating systems of reference soon after his 1907 treatment of uniformly linearly accelerated systems. But only after the clarification of the question of rigid motions do we find any signs of progress on the rotation problem.

The study of uniformly rotating reference systems then led him to the conclusion that, in this case, the spatial coordinates cannot be given a direct physical meaning. He announced this result in February 1912, in an uncharacteristically tentative tone, in the course of a discussion of the spatial coordinates in a linearly accelerated frame of reference K:

> The spatial measurement of K is done with measuring rods that—when compared with each other at rest at the same place in K—possess the same length; the theorems of [Euclidean-JS] geometry are assumed to hold for lengths measured in this way, and thus also for the relations between the coordinates x, y, z and other lengths. That this stipulation is allowed is not obvious; rather it contains physical assumptions that eventually could prove incorrect. For example, it is highly probable that they do not hold in a uniformly rotating system , in which, on account of the Lorentz contraction, the ratio of the circumference to the diameter, using our definition of lengths, must be different from π.[41]

There is evidence suggesting that he had this rotating disk argument before 1912, but it is indirect and suggestive rather than conclusive. Einstein's letter of 1909 to Sommerfeld, cited above (see note 18), was written just a few days after the Salzburg meeting of the Society of German Natural Scientists and Physicians [*Deutscher Naturforscher und Ärzte*], at which Einstein had spoken. As the editors of the Einstein Papers note:[42]

> At the Salzburg meeting Max Born had presented a paper on rigid body motion in special relativity . . ., on which Sommerfeld had commented in the discussion following the paper. Einstein and Born had discussed the subject and had discovered that setting a rigid disk into rotation would give rise to a paradox: the rim becomes Lorentz-contracted, whereas the radius remains invariant (see *Born 1910*, p. 233[43]). The existence of this paradox was first pointed out in print by Paul Ehrenfest (1880–1933) in a paper that was received on the date of this letter (see *Ehrenfest 1909*[44]).

While the line of argument about setting a rigid disk into rotation (which has come to be called "Ehrenfest's Paradox") is not the same as that in Einstein's treatment of an already rigidly-rotating disk,[45] the basic idea in both arguments is the same: Relative to an inertial frame, measuring rods at rest in a uniformly rotating frame of reference *do not* contract if aligned in a radial direction, but *do* contract if aligned orthogonally to a radial direction.

So it is reasonable to suppose that Einstein, already alerted to the possibility that coordinate differences in an accelerating frame might not be directly interpretable in terms of physical measurements, and having read Herglotz's 1910 paper, realized that this was indeed the case for the spatial coordinates in a rigidly rotating frame of reference.[46]

Indeed there is evidence that, by the end of 1911 at the latest, Einstein saw an analogy between the gravitational field that, according to the equivalence principle (conceiving rotation as rest), is equivalent to the inertial field in a uniformly rotating frame of reference and the magnetostatic (or electro-stationary) field due to a stationary circular current distribution.[47] In the letter cited earlier,[48] Max Laue alludes to:

> your [i.e., Einstein's] question whether the gravitational field strength should be represented by a four-vector or a six-vector.

We shall return to this letter at some length below, but for the moment, consider the implications of Einstein's question. The most natural generalization of the Newtonian gravitational field strength would be a four-vector, since the force exerted by the Newtonian field depends only on the position and not the velocity of a mass in that field. On the other hand, the electric and magnetic field strengths together constitute a six-vector, and the force exerted on a charge in an electromagnetic field depends on the position (electric force) and velocity (magnetic force) of the charge. Around the turn of the century, H. A. Lorentz had suggested a gravitational theory modeled on electromagnetism, in which there were gravitational analogues of the electric and magnetic forces.[49]

Thus, Einstein's question to Laue suggests that, by the end of 1911, he had reason to believe that the force exerted by the most general gravitational field might also be velocity dependent. In his 1911 paper, he had considered the gravitational analogue of a constant electrostatic field, and he was soon to consider the analogue of the general electrostatic field.[50] By 18 February 1912, Einstein was already communicating some of his results (see Einstein to Hendrik Antoon Lorentz, *Collected Papers*, vol. 5, pp. 411–413; reference to gravitation on p. 413).

There is also evidence that, by February 1912, Einstein was already considering the gravitational analogue of a magnetostatic field. Paul Ehrenfest visited Einstein in Prague during the last week of February, and Ehrenfest's diary entry for 24 February 1912 contains the following lines:

> Einstein told me about his gravitational work. [I omit some equations referring to the static case] Centrifuging of radiation.[51]

A subsequent letter from Ehrenfest makes clear the meaning of the final, rather cryptic phrase. He reports that a Russian colleague, Michael Frank, has "put me in a very uncomfortable situation" by asking Ehrenfest to translate into German a work concerning "the geometry of light rays in a uniformly rotating laboratory." After describing Frank's work, Ehrenfest adds:

> If one wanted to transform the acceleration field of uniform rotation into a corresponding force field at rest, as you do in your paper 'On the Influence of Gravitation ...' for uniform linear acceleration, then this substitute force-field would also have to give the proper Coriolis deflection for light rays.—That is the content of [Frank's] note. The thing is embarrassing [*peinlich*] for me since you had already communicated this argument to me. ... I told him that you had already told me about this (I remembered it naturally just at the moment when "Coriolis" was recognizable).[52]

So "centrifuging of radiation" refers to "the geometry of light rays in a uniformly rotating laboratory," a problem on which Einstein had evidently worked before February 1912 when he presented his results to Ehrenfest. Einstein's mention of "Coriolis" indicates that he had in mind a velocity-dependent gravitational force. It was presumably the publication of Frank's paper[53] that decided Einstein against publishing his own version of the results on light rays. He wrote Ehrenfest:

> Translate that work [of Frank-JS] in tranquility. I do not arrogate to myself any relativity-monopoly! Everything that is good, is also welcome. You needn't send the proofs.[54]

So by February 1912, on the basis of his own work (as seems probable), and if not by April, on the basis of Ehrenfest's letter, Einstein was aware that the gravitational field equivalent to a rotating frame of reference would have to exert a force on a light ray that depends not only on its position but on (at least the direction of) its velocity—something that is incompatible with a scalar theory of gravitation. So, even when writing his first paper on the static gravitational field, he was aware that a scalar theory was not possible for more general gravitational fields. I shall return to this point below.

Ehrenfest continued to work on the kinematic aspects of the gravito-stationary problem, In a postcard, he writes that he has solved the "Problem: To determine the *most general* field of world-lines that is equivalent to a *stationary* gravitational field." He mentions two "special cases": "hyperbolic motion" (i.e., constant linear acceleration) and "uniform rotation."[55] Subsequent letters outline the proof his solution.[56] In reply to one of these letters, Einstein says that he does not understand Ehrenfest's result,[57] but adds some comments on the problem:

> A rotating ring does not generate a static field in this sense [the sense of Papers I and II-JS], although it is a time-independent field. In such a field the reversibility of light paths does not hold.[58] My case corresponds to the electrostatic field in electromagnetic theory, while the more general static case would also include the analogue of the static magnetic field. I am not yet that far along. The equations I have found only relate to the static case of masses at rest.[59]

I have now given reasons for believing that, at the earliest by 1909 and the latest by the end of 1911, Einstein was aware of problems with the interpretation of both temporal and spatial coordinates in accelerating frame of reference in Minkowski space; and that, by February 1912 at the latest, had every reason to expect that the full theory of gravitation would have to pass beyond the limits of a four-dimensional scalar theory, on the one hand; and, at least spatially, beyond the limits of Euclidean geometry. Now I shall turn to Einstein's investigation of static gravitational fields, which ultimately led to the resolution of these problems.

Scene Two—"The Speed of Light is no Longer Constant"

It may well have been Max Laue who directed Einstein's attention to the crucial importance of the gravitational potential, and the possibility of its replacement by the variable speed of light. The letter Laue sent Einstein at the end of 1911 was cited above, but I must now quote from it at greater length. (Unfortunately we do not have Einstein's letter, if there was one, to which this is a reply.)[60] Discussing Einstein's 1911 paper (see note 14)], he writes:

> It seems extraordinarily characteristic to me that the gravitational potential thereby acquires a physical significance, which is completely lacking for the electrostatic potential. One could, in principle, immediately determine the former by measurement of the velocity of light.

This comment may well have been the cue that prompted Einstein's replacement of the gravitational potential by the variable speed of light in his gravito-static theory.

But, as indicated in the last section, the letter contains more significant clues to the direction in which Einstein was heading. Laue continues:

> Your question, whether the gravitational field strength should be represented by a four-vector or a six-vector, is thereby settled. Not it [the field strength-JS] but rather the potential accordingly seems to me to be the primary concept, the four-dimensional representation of which must be investigated.[61]

A little background is helpful in assessing the full significance of this comment of Laue's. After initially slighting the significance of Minkowski's four-dimensional reformulation of the special theory, Einstein had started to study it in earnest around 1910,[62] probably at least in part in response to Sommerfeld's exposition of a four-dimensional vector algebra and analysis,[63] and its incorporation and further development in Laue's textbook—the first on special relativity.[64]

A further motive for this study was Einstein's decision to include the four-dimensional approach in a major review article that he agreed to write in 1911, and started work on by 1912.[65] In this review he notes that an anti-symmetric second rank tensor "is, following Sommerfeld, usually designated as a six-vector" ["*Sechservektor*"],[66] and notes that the electromagnetic field strength, the components of which are the electric (see p. 9) and magnetic (see p. 10) field strengths, is a six-vector (see p. 81).

So it is reasonable to assume that Einstein had already mastered the four-dimensional formalism by the time that he raised the question of whether the gravitational field strength is a six-vector (as in the electromagnetic case) or a four-vector (as would be the case if it were the gradient of a scalar field). Laue's reply assumes knowledge of the fact that the electromagnetic six-vector is the curl of the electromagnetic potential four-vector;[67] which breaks up into the electric potential (a three-scalar) and the magnetic (three-)vector potential with respect to any inertial frame. If we look only at the electrostatic field strength, it can be written as the three-gradient of the electric potential. But, Laue points out, the electrostatic potential is of no physical significance (because of the possibility of what are now called gauge transformations of the electromagnetic potentials);[68] while Einstein's 1911 work showed that the gravitational potential has an immediate physical significance because of its influence on the speed of light. Therefore, Laue suggests, the important question is: What is the four-dimensional representation of the gravitational *potential*? The fact that, in the static case, it reduces to a single quantity that behaves as a scalar under three-dimensional spatial transformations is *not* decisive for answering this question. The same is true of the electrostatic potential; yet the latter is known to be the fourth (i.e., timelike) component of a four-vector.

Thus, Laue's comment could have served to draw Einstein's attention away from the representation of the gravitational field strength, and toward the question that, within a few months, was to occupy him: What is the four-dimensional representation of the gravitational potential in the non-static case?

Laue's comment, like all of his and Sommerfeld's work on the four-dimensional formalism, is situated within the context of the special theory of relativity. But, as we have seen, with his interpretation of the equivalence principle as implying an enlargement of the relativity group, Einstein had already moved beyond that context; and he soon moved into the context of non-flat space-times. Indeed, there is a comment by Einstein himself, dating from mid-1912, on the question of the four-versus six-vector representation of the gravitational field strength, that suggests the need for this shift of context:

> If the gravitational field can be interpreted within our present [i.e., special-JS] theory of relativity [*sich ... im Sinne unserer heutigen Relativitätsthorie deuten läßt*], then this can only happen in two ways. One can consider [*auffassen*] the gravitational vector either as a four-vector or as a six-vector. [In either case] one arrives at results that contradict the ... consequences of the law of the gravitational mass of energy, [namely]... that gravitation acts more strongly on a moving body than on the same body, in case it is at rest.... It must be a task of the immediate future to create a relativistic-theoretical schema in which the equivalence of gravitational and inertial mass finds expression.[69]

In a similar vein, he wrote to Wien that:

> the [special-JS] relativity theory imperatively demands a further development... since the gravitational vector cannot be fitted into the relativity theory with constant c if one demands the *gravitational* mass of energy.[70]

As it turned out such a theory would not only pass beyond the bounds of the special theory; it would involve space-times with non-flat line elements, as we shall now see.

To summarize: There are good reasons to suggest that, by the beginning of 1912, Einstein already realized that he would ultimately have to go beyond a scalar theory of gravitation. His strategy was to proceed in a step-by-step fashion towards a full dynamical theory. The first step in the program was to consider what I have called above the gravito-static case, the gravitational analogue of electrostatics; but he was already thinking about the next step, the gravito-stationary case, the gravitational analogue of magnetostatics. His ultimate goal was to develop a theory for time-dependent gravitational fields.

Let us look at the first step, gravito-statics.. By March 1912 he was able to write Paul Ehrenfest:

> The investigations of gravitational statics (point mechanics electromagnetism gravitostatics) are complete and satisfy me very much. I really believe that I have found a part of the truth. Now I am considering the dynamical case, again also proceeding from the more special to the more general [case-JS].[71]

Einstein was referring to his two papers on the static gravitational field (hereafter cited as "Paper I" and "Paper II"), completed in February and March 1912 respectively.[72] These papers center on the gravitational potential, as Laue had suggested, but effect a crucial transformation of the problem, in line with Laue's comment that the gravitational potential could "in principle be determined by measurement of the speed of light." In his 1911 paper, Einstein had already shown that, with a certain definition of the universal time:[73]

> in a static gravitational field a relation between c [the speed of light-JS] and the gravitational potential exists, or in other words, that the field is determined by c (Paper I, p. 360).

In Papers I and II $c(x, y, z)$, the spatially variable but temporally constant and direction-independent speed of light, completely replaces the gravitational potential. Most of Paper I is concerned with establishing the gravitational field equation that c obeys, and the equations of motion of a (test) particle in the static gravitational field described by c. Much of Paper II is concerned with a revision of the field equation of Paper I (Section 4), and a crucial mathematical reformulation of the equations of motion (The "Supplement" to the Proofs).[74]

Einstein's introduction of a variable speed of light brought down much scorn upon him at the time,[75] but it was absolutely crucial in initiating the sequence of steps that lead to the culmination of Act II: Einstein's jump from a scalar to a tensorial gravitational potential, in which $c(x, y, z)$ becomes one of the ten components of the metric tensor used to construct the line element of a non-flat space-time.

Paper I also contains another step in the process. Einstein shows that, if one uses a light clock, for example, to measure:

> the local time, which Abraham denotes by l, then this stands to the universal time [t] in the relation
>
> $$dl = c\,dt. \qquad \text{(Paper I, p. 141).}$$

In retrospect (remembering that here c is non-constant), we recognize in this equation the relation between the differential element of the proper time dl between two events at the same place (i.e., $x, y, z =$ const) in a static field, and the coordinate differential dt between the times of the two events, using the preferred static time coordinate t. This equation begins to answer the question of the relation between coordinates and physical measurements in a gravitational field that had been puzzling Einstein for almost five years.

But before expanding on this point, let me turn a further step in the process, contained in Paper II. The equations of motion of a particle in a static gravitational field, developed in Paper I, were rewritten in Lagrangian form in a "Supplement to the Proofs" of Part II. This step proved to be so significant that it soon led to the final resolution of the problem of the representation of the gravitional potentials.

> It is noteworthy that the equations of motion of a particle [*materielle Punkt*] in the gravitational field, take a very simple form when they are given the form of Lagrange's equations. Namely, if one takes
>
> $$H = -m\sqrt{c^2 - q^2},$$

then ... For a particle moving in a static gravitational field without the action of external forces, there holds accordingly

$$\delta\left\{\int H\,dt\right\} = 0,$$

or

$$\delta\left\{\int \sqrt{c^2\,dt^2 - dx^2 - dy^2 - dz^2}\right\} = 0.$$

Here too—as was proved by Planck for the usual [i.e., special] theory of relativity—it is seen that the equations of analytical mechanics possess a significance that extends far beyond Newtonian mechanics. Hamilton's equation as finally written down let us anticipate [*ahnen*] the structure of the equations of motion of a particle in a dynamical gravitational field.[76]

Einstein's lecture notes on mechanics, which he had been teaching since 1909,[77] show his familiarity with the use of variational techniques to derive the Lagrangian equations of motion (see pp. 91–95, 116–117). Even more important for present purposes, he stressed the coordinate-invariant nature of the resulting equations of motion:

The Cartesian coordinates of the particle no longer enter into the [variational-JS] principle. It holds independently of whatever coordinates we use to determine the position of the particles of the system (p. 93).

Now we have in hand all the strands, the interweaving of which finally allowed Einstein to take the great leap forward to a metric theory of gravitation. But let me emphasize that, however much they may help us in retrospect to understand the process, there remains something almost uncanny in how Einstein made the choices that led him so far from the path trodden by other physicists in the search for a relativistic theory of gravitation. He was about to enter an entirely new land, in which the space-time structure becomes a dynamical field.

Scene 3—"Ten Space-Time Functions"

Just what did the variational reformulation of the equations lead Einstein to "anticipate"? I propose to answer this question with the help of the first three sections of his following paper on the topic, the "*Entwurf*" paper, published early in 1913.[78] After introductory comments discussing the equivalence principle, Section 1 treats "The equations of motion of a particle in a static gravitational field." Except for one small but significant detail, it is just an expanded version of the "Supplement." The detail is notational: ds is an abbreviation for $\sqrt{c^2\,dt^2 - dx^2 - dy^2 - dz^2}$ for both the case of constant c ("the usual theory of relativity") and spatially variable but static $c(x, y, z)$.

The significance of this notation does not emerge until Section 2, which treats "Equations for the motion of a particle in an arbitrary gravitational field. Characterization of the latter":

With the introduction of a spatial variability of the quantity c we have passed beyond the framework of the theory that is now designated as "the theory of relativity;" for the

> expression designated by ds no longer behaves as an invariant under linear orthogonal transformations of the coordinates....
>
> If we introduce a new space-time system $K'(x', y', z')$ by means of an arbitrary substitution.
>
> $$\begin{aligned} x' &= x'(x, y, z, t) \\ y' &= y'(x, y, z, t) \\ z' &= z'(x, y, z, t) \\ t' &= t'(x, y, z, t) \end{aligned}$$
>
> and if the gravitational field in the original system K is static, then under this substitution equation (1) goes over into an equation of the form
>
> $$\delta\left\{\int ds'\right\} = 0,$$
>
> where
>
> $$ds'^2 = g_{11}dx'^2 + g_{22}dy'^2 + \ldots + 2g_{12}dx'\,dy' + \ldots$$
>
> and the quantities $g_{\mu\nu}$ are functions of x', y', z', t'. ...
>
> Thus we arrive at the interpretation that, in the general case, the gravitational field is characterized by ten space-time functions ...[79]

Einstein proceeds to derive the equations of motion from the generalized Hamiltonian function (which we would now call the Lagrangian):

$$\begin{aligned} H &= -m\, ds/dt \\ &= -m\sqrt{g_{11}\, dx_1^2 + \ldots + 2g_{12}\, dx_1 dx_2 + \ldots 2g_{14}\, dx_1 dx_4 + \ldots g_{44} dx_4^2}. \end{aligned}$$

He writes the three Lagrange equations for the three spatial coordinates x_i, derives expressions for the momentum of the particle and the force exerted on it by the gravitational field from them; and then derives the energy of the particle by performing the usual Legendre transformation on H. He closes this discussion by noting that:

> In the usual relativity theory only linear orthogonal substitutions are permitted. It will be shown that we are able to set up equations for the influence of the gravitational field on material processes that behave covariantly under arbitrary substitutions.[80]

Up to this point, there is nothing in Einstein's results that depends on the interpretation of the $g_{\mu\nu}$ as anything more than a generalization of his 1912 results for the static gravitational field to their form in an arbitrary reference frame. (Remember that, for Einstein, a change of coordinates amounts to a change of space-time reference frame.) In such a frame, the single static gravitational potential c is transformed into ten functions $g_{\mu\nu}$.

It is only in the next section of the paper, "Significance of the Fundamental Tensor $g_{\mu\nu}$ For the Measurement of Space and Time," that he proceeds to the geometrical interpretation of these functions in space-time. After recalling that the time coordinate had already lost its immediate physical significance in a static gravitational field, he continued:

> In this connection, we remark that *ds* is to be understood as an invariant measure for the interval [*Abstand*] between two neighboring space-time points.[81]

Presumably, this is what his results in the "Supplement" suggested to him almost immediately: if the integrand is interpreted as the interval *ds* between neighboring points, then the variational principle can be interpreted as giving rise to the equation for a geodesic in the resulting non-flat space time. In a later reminiscence,[82] Einstein stated:

> The equivalence principle allows us ... to introduce non-linear coordinate transformations in such a [four-dimensional] space [with (pseudo)-Euclidean metric]; that is, non-Cartesian ("curvilinear") coordinates. The pseudo-Euclidean metric then takes the general form:
>
> $$ds^2 = \sum g_{ik}\, dx_i\, dx_k$$
>
> summed over the indices i and k (from 1–4). These g_{ik} are then functions of the four coordinates, which according to the equivalence principle describe not only the metric but also the "gravitational field." ... This formulation so far applies only to the case of pseudo-Euclidean space. It indicates clearly, however, how to attain the transition to gravitational fields of a more general type. Here too the gravitational field is to be described by a type of metric that is a symmetric tensor field g_{ik}. ... The problem of gravitation was thereby reduced to a purely mathematical one. Do differential equations exist for the g_{ik} that are invariant under non-linear coordinate transformations? Such differential equations and *only* such could be considered as field equations for the gravitational field. The equations of motion of a particle were then given by the equation for a geodesic line.
>
> With this task in mind, I turned to my old student friend Marcel Großmann, who had in the meantime become Professor of Mathematics at the ETH (pp. 14–15).

Einstein here makes a rather precise claim about what he had accomplished before turning to Grossmann upon his move back to Zurich at the end of July 1912. Earlier, in 1923, he had made a similar claim, but with a significant addition—a reference to Gauss:

> I first had the decisive idea of the analogy between the mathematical problems connected with the theory and the Gaussian theory of surfaces in 1912 after my return to Zürich, initially without knowing Riemann's and Ricci's or Levi-Civita's investigations.[83]

In the 1955 reminiscence, Einstein noted that Carl Friedrich Geiser's course on differential geometry at the ETH played an important role in his thinking; it was in that course that he learned about Gauss's theory of surfaces, based on analysis of the distance *ds* between neighboring points on a surface, expressed in terms of an arbitrary coordinate system on the surface, thereafter often called Gaussian coordinates.[84]

The space-time interval *ds* Einstein introduced represents a generalization of what was often referred to in differential geometry as the "line element." The term "element" appears to go back to Monge, who speaks of the "elements" [*élémens*] of a curve in space. Coolidge comments: "An élémen is an infinitesimally short chord."[85] Gauss[86] makes the concept of what he calls a "line element" [he uses both "*Linienelement*" and "*Linearelement*"—see his "Anzeige," pp. 344–345] connecting a pair of points on a two-dimensional surface, central to his theory of surfaces, and shows that it may be used to define the intrinsic properties of the surface, such as its curvature. Bianchi-Lukat[87] defines ds as the "element of a curve" ["*Bogenelement*"], and comments:

> Since the expression for ds given by [$ds^2 = E\,du^2 + 2F\,du\,dv + G\,dv^2$, the right hand side being the square of Gauss's expression for the line element] holds for any arbitrary curve on the surface, it is designated the *line element* of the surface. Gauss's ideas were generalized to n-dimensional manifolds by Riemann.[88]

The basic ideas of Gauss's theory of surfaces are reproduced in Grossmann's notes on Geiser's course on differential geometry [*Infinitesimalgeometrie*].[89] Reich indicates some of the high points: "Geiser treats the line element and its special form in different coordinate systems especially intensively," indicating that he used the notation ds^2 for it. "The curvature of surfaces and especially the Gaussian measure of curvature, which Geiser derives in Gauss's fashion and with Gauss's notation, is a particular theme of the semester. The result reads: 'if $ds^2 = E\,dp^2 + 2F\,dp\,dq + G\,dq^2$, then the measure of curvature depends solely on E, F, G and their derivatives.' ... A further important point are geodesic lines. After a longish introduction, Geiser goes into the 'differential equation for geodesic lines.' ... It appears important to me that Geiser offers not only the geometrical aspect but also argues invariant-theoretically, as in the case of the metric, the measure of curvature and of geodesic lines (pp. 164–165)" Geiser included a derivation of the equation for a geodesic on a surface by variation of the integral $\int ds$, where

$$ds = \sqrt{E\,dp^2 + 2F\,dp\,dq + G\,dq^2},$$

along some curve $q = \psi(p)$ to find its minimum.[90] Comparison of this with Einstein's variational principle for the equations of motion of a particle in a static gravitational field could have suggested the analogy between Gauss's theory of surfaces and Einstein's theory of the static gravitational field.

So much for the mathematical literature. In the physics literature, the equivalent proper time element $d\tau$ had been introduced by Minkowski for timelike lines.[91] But neither the concept of, nor the notation for, the line element ds or the proper time element $d\tau$ occurs in the works of Sommerfeld or Laue developing special-relativistic vector and tensor analysis (which, as we have seen Einstein studied) until after 1912.[92] Nor does it occur in Einstein's summary of vector and tensor analysis in his unpublished review of the special theory (see above).[93]

However, the concept and even the term, were beginning to appear in the physics literature in connection with the discussion of rigid bodies.[94] Herglotz seems to have introduced them.[95] After pointing out that Minkowski had introduced the idea of representing the spatial and temporal coordinates "as the four coordinates of a point in a fourfold-extended manifold $R_4(x, y, z, t)$," he goes on:

> Similarly a measure relation [*Massbestimmung*] is introduced in this R_4, according to which (the velocity of light being set equal to 1) the square of the distance of two

infinitely neighboring points is:

$$ds^2 = dx^2 + dy^2 + dz^2 - dt^2.$$

Line elements of real length ($ds^2 > 0$) are called spatial, however those of purely imaginary length ($ds^2 < 0$) are called timelike.

Born's next paper on the definition of a rigid body[96] speaks of "a four-dimensional space $x\ y\ z\ t\ \ldots$ in which a measure relation [*Maßbestimung*] with the line element $dx^2 + dy^2 + dz^2 - c^2\,dt^2$ is introduced" (p. 233).

As noted above, Einstein refers to these papers in a 1910 letter, so we may assume that he was familiar with them.[97] And, it was in response to a criticism by Einstein that Abraham wrote:[98]

> On lines 16, 17 of my note "On the Theory of Gravitation," an oversight is to be corrected, of which I became aware through a friendly communication of Mr. A. Einstein, One should read "let us consider dx, dy, dz and $du = i\,dl = i\,cdt$ as components of a displacement ds in four-dimensional space." Thus,
>
> $$ds^2 = dx^2 + dy^2 + dz^2 - c^2\,dt^2$$
>
> is the square of the four-dimensional line element, in which the velocity of light c is determined by equation (6) [Abraham's relation between the gravitional potential and the speed of light-JS].[99]

Indeed, as noted above, in Paper I on the static field Einstein had written that, in using a light clock:

> we operate with a sort of local time, which Abraham designates with l. This stands in the relation
>
> $$dl = c\,dt$$
>
> to the universal time.

This is the earliest indication (the end of February 1912) that Einstein realized the need to use differentials of the two quantities in order to relate a coordinate time t to a physically measured time l (in this case, the proper time between two events at the same spatial point).

To summarize: on the basis of the mathematical and physical resources at his command, at some point in mid-1912, after generalizing the single gravitational potential c to the array of ten gravitational potentials g_{ik}, Einstein realized that they formed the coefficients of a quadratic form $\sum g_{ik}\,dx_i\,dx_k$, which could be regarded as the square of the invariant line element ($ds^2 = \sum g_{ik}\,dx_i\,dx_k$) of a four-dimensional space-time manifold; and that the interval ds represents a physically measurable quantity—the proper time if the interval between two events were time-like, the proper length if it were space-like (of course it would vanish for null intervals).

I suggest that it was at this point that he turned to Grossman. Continuing the quotation from the 1955 reminiscence:

> I was made aware of these [works by Ricci and Levi-Civita-JS] by my friend Großmann in Zurich, when I put to him the problem to investigate generally covariant tensors,

whose components depend only on the derivatives of the coefficients of the quadratic fundamental invariant.

He at once caught fire, although as a mathematician he had a somewhat skeptical stance towards physics. ... He went through the literature and soon discovered that the indicated mathematical problem had already been solved, in particular by Riemann, Ricci and Levi-Cività. This entire development was connected to the Gaussian theory of curved surfaces, in which for the first time systematic use was made of generalized coordinates (pp. 15, 16).

In short: While the analogy to Gaussian surface theory had occurred to Einstein *before* he consulted Grossmann, probably including the role of the line element; the connection between this theory and the later line of development from Riemann to Ricci and Levi-Civita only became clear to Einstein *after* consulting Grossmann.

Louis Kollross, another student friend of Einstein, who was also Professor of Mathematics at the ETH during this time, adds another name that belongs between those of Riemann and of Ricci and Levi-Civita:

[Einstein] spoke to Großmann about his troubles and said to him one day: "Großmann, you must help me, otherwise I'll go crazy!" And Marcel Großmann succeeded in showing him that the mathematical instrument that he needed had been created precisely in Zurich in the year 1869 by Christoffel in the paper "On the Transformation of Homogeneous Differential Expressions of the Second Degree," published in volume 70 of "Crelle's Journal" for pure and applied mathematics.[100]

A look at Grossmann's Part II of their joint paper, confirms Kollros's recollection:

The mathematical tool [*Hilfsmittel*] for the development of the vector analysis of a gravitational field that is characterized by the invariance of the line element

$$ds^2 = \sum_{\mu\nu} g_{\mu\nu}\, dx_\mu\, dx_\nu$$

goes back to the fundamental paper of Christoffel on the transformation of quadratic differential forms.[101]

So it was Marcel Grossmann, who introduced Einstein to the work of Ricci and Levi-Civita after Einstein's return to Zurich in early August 1912.[102] However, in his exposition Grossmann plays down the geometrical significance of vector and tensor analysis:

In it I have purposely left geometrical methods [*Hilfsmittel*] aside, since in my opinion they contribute little to the visualization [*Veranschaulichung*] of the concepts constructed in vector analysis (p. 325).

This distinction between tensor analysis and geometrical methods is based on the distinction Ricci and Levi-Civita make between the "fundamental quadric or form" (p. 13),which they denote by ϕ, and the line element (they never use these words), denoted by ds^2, of an n-dimensional manifold, denoted by V_n (see, e.g., pp. 128, 153). They assert that: "The methods of the absolute differential calculus depend essentially on consideration of" the fundamental form (p. 133);

but the geometrical interpretation of it "as the ds^2 of a surface" (p. 162) is merely one possibility.

Grossmann's exposition of tensor analysis is based on Chapter I, "Algorithm of the Absolute Differential Calculus" (pp. 128–144), which includes discussions of covariant differentiation and of the Riemann tensor that do not depend at all upon the geometrical interpretation of the fundamental form, but rather on the theory of algebraic and differential invariants of the fundamental form and other functions (see pp. 127 of the "Preface" and Section 1 of the first chapter, "Point transformations and systems of functions," pp. 128–130).[103] This is entirely in the spirit of Christoffel's exposition of the differential invariants of a quadratic differential form in n independent variables. Only in the last paragraph of the paper does he mention "a posthumous paper of Riemann" on "the square of the line element in a space of three dimensions."[104] And indeed, until Levi-Civita developed the concept of parallel dispacement in a manifold with metric, geometrical methods did not contribute much to the interpretation of the covariant derivative and the Riemann tensor.[105] Only starting with Chapter II of Ricci and Levi-Civita, on "Intrinsic geometry as a calculational tool," are geometrical applications to n-dimensional manifolds introduced.

Bianchi–Lukat,[106] another source that Grossmann mentions,[107] also separates the invariant-theoretical treatment of "Binary Quadratic Forms" in Chapter II from the geometrical treatment of "Curvilinear Coordinates on Surfaces," in Chapter III, which includes the introduction of "The Line Element of Surfaces" in Section 33.

—But I have already begun to encroach on the opening scene of Act III. If, so far, Einstein's intuition led him along the highway, almost without mis-step, to the search for dynamical field equations governing the behavior of the metric tensor field, the last act will show our hero's wanderings on many a curious by-way.

Notes

[1] Einstein Archives, Hebrew University of Jerusalem, Control Index No. 11 196. (Hereafter, only the number of such items will be cited.) It appears from Einstein to Robert Lawson, 22 April 1920 (No, 1-010), that it was written for the biographical note in *Relativity, the Special and General Theory: A Popular Exposition* (London: Methuen 1920). This is the English translation of the fifth German edition of Einstein's *Über die spezielle und die allgemeine Relativitätstheorie* (*Gememeinverständlich*) (Braunschweig: Vieweg 1920). I thank Dr. Josef Illy of the Einstein Papers for locating Einstein's letter to Lawson. *The Collected Papers of Albert Einstein, vol. 8, Part B, The Berlin Years: Correspondence*, Robert Schulmann et al., eds. (Princeton University Press 1998), incorrectly calendars it under 1917 (see pp. 1005–1006).

[2] Properly speaking, the term "metric" should be restricted to line elements with positive-definite signature; those with an indefinite signature are more properly termed "pseudometrics." But Einstein, and following him most physicists, referred to the four-dimensional tensor field with Minkowski signature as the metric tensor, and I shall follow that usage.

[3] "Die Grundlage der allgemeinen Relativitätstheorie," *Annalen der Physik 49* (1916): 769–822.

[4] David Hilbert played a significant role in the final moments of Act III, although not the one that is often attributed to him. See Jürgen Renn and John Stachel, "Hilbert's Foundation of Physics: From a Theory of Everything to a Constituent of General Relativity," Max-Planck-Institut für Wissenschaftsgeschichte Preprint 118 (1999).

[5] For Einstein's side of this discussion during its first years, see *The Collected Papers of Albert Einstein*, vol. 7, *The Berlin Years: Writings, 1918–1921*, Michel Janssen et al., eds. (Princeton University Press, 2001). After their first citation, the volumes of this series will be cited in the form: *Collected Papers*, vol. x.

[6] For another account, see Abraham Pais, *'Subtle is the Lord...' The Science and the Life of Albert Einstein* (Oxford: Clarendon Press/New York: Oxford University Press 1982), Part IV, pp. 177–296. For some critical comments, see John Stachel, "Einstein," *Science 218* (1982): 989–990. [See this volume, pp. 551–554].

[7] "Grundgedanken und Methoden der Relativitätstheorie, in ihrer Entwicklung dargestellt," to appear in *Collected Papers*, vol. 7. [See note 5].

[8] I shall follow the common but anachronistic practice of referring to "the special theory of relativity." During this period, Einstein initially called it "the principle of relativity" ["*das Relativitätsprinzip*"] and then, following the practice of others, "the theory of relativity" ["*die Relativitätstheorie*"]. For details, see the discussion in the Editorial Headnote, "Einstein on the Theory of Relativity," *Collected Papers*, vol. 2, p. 254. [See this volume, p. 192].

[9] "Erinnerungen-Souvenirs," *Schweizerische Hochschulzeitung 28* (*Sonderheft*) (1955), pp. 145–153; reprinted as "Autobiographische Skizze," in Carl Seelig, ed., *Helle Zeit-Dunkle Zeit In Memoriam Albert Einstein, Zürich/Stuttgart/Wien, Europa Verlag. 1955*, pp. 9–17. Citation from "Autobiographische Skizze." p. 14.

[10] "Einiges über die Entstehung der allgemeinen Relativitättheorie," the German text of a lecture given at the University of Glasgow, 20 June 1933. The German text was published in Albert Einstein, *Mein Weltbild* (Amsterdam: Querido 1934), pp. 248–256. Cited from the paperback edition, *Mein Weltbild*, Carl Seelig, ed. (Frankfurt/M, Berlin Wien: Ullstein 1981), pp. 134–138. Hereafter cited as "Einiges über die Entstehung..."

[11] As Einstein later realized, a special-relativistic theory of gravitation that does justice to the equivalence principle is possible, and indeed one was developed a little later by Gunnar Nordström. For an account of Nordström's theories, and Einstein's reaction to them, see John Norton, "Einstein, Nordström and the Early Demise of Scalar, Lorentz Covariant Theories of Gravitation," *Archives for History of Exact Sciences 45* (1992): 17–94.

[12] "Über das Relativitätsprinzip und die aus demselbem gezogenen Folgerungen," *Jahrbuch der Radioktivität und Elektronik 4* (1907): 411–462; reprinted in *The Collected Papers of Albert Einstein*, vol 2: *The Swiss Years: Writings 1900-1909*, John Stachel et al., eds. (Princeton University Press 1989), pp. 433–488; "Berichtigungen," *ibid* 5 (1908): 98–99, reprinted in *ibid*, pp. 494–495. Hereafter cited as "Über das Relativitätsprinzip."

[13] "Über das Relativitätsprinzip," p. 476; also see p. 495 for a discussion of the meaning of uniform acceleration.

[14] "Über den Einfluß der Schwerkraft auf die Ausbreitung des Lichtes," *Annalen der Physik 35* (1911): 898–908; reprinted in *The Collected Papers of Albert Einstein*, vol. 3, *The Swiss Years: Writings, 1909-1911*, Martin Klein et al, eds. (Princeton University Press 1993), pp. 486–497.

[15] *Collected Papers*, vol. 3, pp. 487–488.

[16] In 1912 Einstein regarded "the equivalence of inertial and gravitational mass" as being reducible to the "essential likeness [*Wesensgleichheit*] of both of these elementary qualities of matter and energy"; and asserts that his theory of "the static gravitational field" allows him to regard it "as physically the same in essence [*wesensgleich*] as an acceleration of the reference system." See "Relativität und Gravitation. Erwiderung auf eine Bemerkung von M. Abraham," *Annalen der Physik 38* (1912): 1059–1064, p. 1063.

[17] The entire complex of problems raised in this paragraph is discussed at length in John Stachel, "The Story of Newstein," to appear in: *Alternative Approaches to General Relativity*, Jürgen Renn et al., eds.

[18] Albert Einstein, "Autobiographical Notes," in Paul Arthur Schilpp, ed., Albert Einstein: Philosopher-Scientist (Open Court: LaSalle/Cambridge University Press, London 1949), pp. 2–94), cited from Albert Einstein, *Autobiographical Notes/A Centennial Edition* (Open Court: LaSalle and Chicago 1979), p. 63, hereafter cited as "Autobiographical Notes." Although published first in 1949, they were actually written in 1947.

[19] "Einiges über die Entstehung..." (ref. in note 10), p. 137.

[20] For a historical discussion of Einstein's hole argument, see my 1980 Jena paper, published as John Stachel, "Einstein's Search for General Covariance, 1912–1915," *Einstein Studies*, vol. 1, *Einstein and the Development of General Relativity*, Don Howard and John Stachel, eds. (Birkhäuser: Boston 1989), pp. 63–100. [See this volume, pp. 301–337].

[21] Albert Einstein, "Zur Elektrodynamik bewegter Körper," *Annalen der Physik 17* (1905): 891–921; reprinted in *Collected Papers*, vol. 2, pp. 276–306, citation from p. 277 of the latter. For further discussion of this paper see the Editorial Note "Einstein on the Theory of Relativity," ibid., pp. 253–274. [See this volume, pp. 233–244].

[22] For a discussion of the development of Einstein's concept of the equivalence principle, see John Norton, "What Was Einstein's Principle of Equivalence?," *Studies in the History and Philosophy of Science 16* (1985): 203–246; reprinted in Don Howard and John Stachel, eds., *Einstein Studies*, vol. 1: *Einstein and the History of General Relativity* (Boston: Birkhäuser 1989), pp. 5–47.

[23] Since contemporary documents are cited with dates, the chronological sequence can easily be reconstructed.

[24] "Autobiographical Notes" (see note 18), pp. 62 and 64. An idea of what he meant by "Mach's paradox concerning inertial systems" may be gathered from his citations of Mach in Einstein's article, "Ernst Mach," *Physikalische Zeitschrift 17* (1916): 101–104; reprinted in *Collected Papers*, vol. 6, A. J. Kox et al., eds. (Princeton University Press 1996), pp. 278–281. See also "Die Grundlage der allgemeinen Relativitätstheorie," *Annalen der Physik 49* (1916): 769–822; reprinted in *Collected Papers*, vol. 6, pp. 284–339, Section 2, "Über die Gründe, welche eine Erweiterung des Relativitätspostulates naheliegen," pp. 286–288.

[25] "Über das Relativitätsprinzip" (see note 12), Section 18, pp. 476–480.

[26] Einstein did not actually introduce this term until 1912, in his first paper on the static gravitational field, in which he contrasts the "local time" and the "universal time" (see below, note 41).

[27] "Autobiographische Skizze" (see note 9), p. 14.

[28] Einstein to Arnold Sommerfeld, 29 September 1909, *The Collected Papers of Albert Einstein*, vol. 5, *The Swiss Years: Correspondence 1902-1914*, Martin Klein et al., eds.

(Princeton University Press 1993), p. 210. Einstein incorrectly names the title of the journal in which his earlier paper was published (see note 12). Einstein had described the main theme of the last section of this paper in an earlier letter to Sommerfeld (Einstein to Arnold Sommerfeld, 5 January 1908, *Collected Papers*, vol. 5, p. 86.

[29] Einstein to Michele Besso, 26 March 1912, *ibid.*, pp. 435–438; citation from p. 436.

[30] Einstein to Arnold Sommerfeld, 19 January 1910, *ibid.*, pp. 228–230; citation from p. 229.

[31] See Max Born, "Die Theorie des starren Elektrons in der Kinematik des Relativitätsprinzips," *Annalen der Physik 30* (1909): 1–56. It was his report on this work at the 1909 Salzburg meeting of the Versammlung deutscher Naturforscher und Ärtzte, "Über die Dynamik des Elektrons in der Kinematik des Relativitätsprinzips," *Physikalische Zeitschrift 10* (1909): 813–817, that provoked the above-cited letter of 1909 from Einstein to Sommerfeld. See also Max Born, "Über die Definition des starren Körpers in der Kinematik des Relativitätsprinzips," *Physikalische Zeitschrift 11* (1910): 233–234, discussed below.

[32] Einstein to Jakob Laub, 16 March 1910, *ibid.*, pp. 231–233; citation from p. 232.

[33] Max Laue, *Das Relativitätsprinzip* (Braunschweig: Vieweg, 1911), hereafter cited as *Das Relativitätsprinzip*, 1911 ed.; note at bottom of p. 107.

[34] Of course, a dynamical problem remained: how to create the circumstances that would lead a non-rigid body of a particular constitution to execute a particular rigid motion. But such special dynamical problems could be attacked *after* the general kinematical problem was solved.

[35] Gustav Herglotz, "Über den vom Standpunk des Relativitätsprinzips aus als "starr" zu bezeichnenden Körper," *Annalen der Physik 31* (1910): 393–415; citation from p. 403.

[36] *Ibid.*, p. 412.

[37] Albert Einstein, "Gibt es eine Gravitationswirkung, die der elektrodynamische Induktionswirkung analog ist?," *Vierteljahrschrift für gerichtliche Medizin und öffentliches Sanitätwesen 44* (1912): 37–40; reprinted in *Collected Papers*, vol. 4, pp. 175–179. Einstein published the article in this journal because it was part of a Festschrift for his friend, Heinrich Zangger, an expert in forensic medicine.

[38] Presumably, this would be Einstein's answer to Laue's objection to the equivalence principle: "For the gravitational field in the system K [at rest, with a uniform gravitational field] there must be present a body that causes gravitation, not however for the accelerated system K'. So a search for it must immediately decide whether there is a real gravitational field or only an accelerated reference system" (Max Laue to Albert Einstein, 27 December 1911, *Collected Papers*, vol. 5, p. 384). This letter is discussed further below. Einstein evidently tried to answer Laue's objection in a footnote to his next paper, submitted two months later: "The masses that produce this field must be thought of as at infinity" ("Lichtgeschwindigkeit und Statik des Gravitationsfeldes," *Annalen der Physik 38* (1912): 355–369, citation from p. 356. The paper on gravitational induction followed almost immediately.

[39] Exactly the same words about Mach's book occur in "Einstein's Scratch Notebook," reproduced with transcription in *Collected Papers*, vol. 3, "Appendix A," pp. 564–596; see p. 592.

[40] Albert Einstein, "A Brief Outline of the Development of the Theory of Relativity," *Nature 106* (1921): 782–784; citation from p. 783. A German draft, "Kurze Skizze zur

Entwicklung der Relativitätstheorie," has been used to correct the English text. Both appear in the *Collected Papers*, vol. 7.

[41] Albert Einstein, "Lichtgeschwindigkeit und Statik des Gravitationsfeldes," *Annalen der Physik 38* (1912): 355–369, reprinted in *Collected Papers*, vol. 4, pp. 130–145; citation from p. 131. (This is his first paper on the static gravitational field, discussed at greater length below.) For a translation of a longer portion of this passage and a fuller discussion of the rotating disk problem, see John Stachel, "Einstein and the Rigidly Rotating Disk," in *General Relativity and Gravitation One Hundred Years After the Birth of Albert Einstein*, Alan Held, ed., Plenum Press, New York and London 1980, vol. 1, pp. 1–15; reprinted in *Einstein and the History of General Relativity* (see note 22), pp. 48–62. [See this volume, pp. 245–260].

[42] See *Collected Papers*, vol. 5, p. 211, note [5].

[43] Max Born, "Über die Definition des starren Körpers in der Kinematik des Relativitätsprinzips," *Physikalische Zeitschrift 11* (1910): 233–234. The reference reads: "Mr. P. Ehrenfest . . . showed in a very simple way that a body at rest can never be brought into uniform rotation; I had already discussed the same fact with Mr. A. Einstein in Salzburg."

[44] Paul Ehrenfest, "Gleichförmige Rotation Starrer Körper und Relativitätstheorie," *Physikalische Zeitschrift 10* (1909): 918.

[45] For Einstein's way of avoiding Ehrenfest's paradox, see "Einstein and the Rigidly Rotating Disk" (ref. in note 41), pp. 6–7 and 9.

[46] Curiously, neither he nor any other contemporary ever refers to a 1910 paper by Theodor Kaluza (of later five-dimensional Kaluza–Klein theory fame) solving the problem of the "proper geometry" ["*Eigengeometrie*"] of a Born rigidly-rotating body: "Zur Relativitätstheorie," *Physikalische Zeitschrift 11* (1910): 977–978. Kaluza was prevented by illness from presenting his work at the 1910 Königsberg meeting of the *Deutsche Naturforscher und Ärzte* which may help to explain its lack of impact on the rigid body discussion.

[47] However, there is also an important difference between the two: The gravitational field equivalent to the inertial forces in an accelerated reference frame does not appear to correspond to any material sources, while the analogous electromagnetic fields are produced by a charged ring—rotating or not. As noted above, this was the purport of Max Laue's criticism of Einstein's treatment of the gravitational field equivalent to the inertial forces in a uniformly accelerated frame of reference (see note 38), to which Einstein hoped to provide a Machian answer.

[48] Max Laue to Albert Einstein, 22 December 1911, *Collected Papers*, vol. 5, pp. 384–385; citation from p. 385.

[49] Hendrik Antoon Lorentz, "Beschouwingen over de zwaartekracht," *Verslagen van de Gewone Vergaderingen der Wis- en Natuurkundige Afdeeling, Koninklijke Akademie van Wetenschappen te Amsterdam 8* (1899–1900): 603–620; English translation, "Considerations on gravitation," *Proceedings of the Section of Sciences, Koninklijke Akademie van Wetenschappen te Amsterdam 2* (1899–1900): 559–574.

[50] This was probably at least in part in response to Abraham's work on the problem of gravitation, which appeared early in 1912. Abraham's first two papers on the subject, dated "December 1911," were received 14 December and published in the issue of 1 January 1912 of *Physikalische Zeitschrift* ("Zur Theorie der Gravitation," *Physikalische Zeitschrift 13*

(1912): 1–4; "Das Elementargesetz der Gravitation," *ibid.*, pp. 4–5. There is evidence that Einstein had corresponded with Abraham about his theory before publication (see below).

[51] Rijksmuseum voor de Geschiednis der Natuurwetenschappen, Leiden: Ehrenfest Collection, Ehrenfest Notebook 4-11, Microfilm number 12.

[52] Ehrenfest to Einstein, draft letter before 3 April 1912, *Collected Papers*, vol. 5, pp. 439–445, citation from p. 440.

[53] Michael L. Frank, "Bemerkung betreffs der Lichtausbreitung in Kraftfeldern," *Physikalische Zeitschrift 13* (1912): 544–545. The paper is dated 7 March 1912. Frank does not suggest that the force field equivalent to a force-free rotating frame of reference need be gravitational.

[54] Einstein to Ehrenfest, 25 April 1912, *Collected Papers*, vol. 5, pp. 450–451; citation from p. 450.

[55] Ehrenfest to Einstein, 14 May 1912, *Collected Papers*, vol. 5, pp. 460–461. In later chapters, these two special case will become very familiar to the reader since they are the two test cases that Einstein uses again and again to evaluate candidate gravitational field equations.

[56] Only Ehrenfest's drafts of his letters have been preserved: Ehrenfest to Einstein: after 16 May 1912, *Collected Papers*, vol. 5, pp. 461–464 (see pp. 462–463); 29 June 1912, *Collected Papers*, vol. 5, pp. 487–496. Einstein to Ehrenfest, 25 April 1912, *Collected Papers*, vol. 5, pp. 450–451 (see p. 451); 27 April 1912, *Collected Papers*, vol. 5, p. 455; before 20 June 1912, *Collected Papers*, vol. 5, pp. 484–486 (see pp. 485–486). Only letters containing references to gravitation are cited. Ehrenfest later published a paper on this subject: "Over Einsteins Theorie van het stationaire gravitatieveld," *Versl. Akad. Amsterdam 21* (1913): 1234–1239; English version, "On Einstein's theory of the stationary gravitation field," *Proc. Amsterdam Acad. 15* (1913): 1187–1191.

[57] Einstein to Ehrenfest, before 20 June 1912, *Collected Papers*, vol. 5, pp. 484–486; see p. 485.

[58] This assertion is analogous to the result discussed above that Frank had published, but with a subtle difference. Like Einstein, Frank had discussed Minkowski space-time as seen from a uniformly rotating frame of reference and the field (he does not specify it as gravitational) equivalent to the inertial forces present in such a frame. Here, Einstein is discussing the field generated by a rotating material ring, which he must have realized would be non-Minkowskian since this is true even for the field of a non-rotating material ring.

[59] See ref. in note 57, p. 486.

[60] Since Einstein did not raise the question of whether the gravitational field strength is a four-vector or a six-vector in his 1911 paper, I assume that the phrase "Your question" in Laue's letter refers to either a previous letter or conversation.

[61] Max Laue to Albert Einstein, 27 December 1911, *Collected Papers*, vol. 5, p. 384.

[62] In 1908, Einstein and Laub thought it worthwhile to publish a paper rederiving Minkowski's four-dimensional results on electrodynamics (see "Die Grundgleichungen für die elektromagnetischen Vorgänge im bewegten Körper," *Königliche Gesellschaft der Wissenschaften zu Göttingen. Mathematisch-physikalische Klasse. Nachrichten* (1908): 53–111) in three-dimensional form because that "work makes rather great demands mathematically on the reader" (Albert Einstein and Jakob Laub, "Über die elektromagnetischen Grundgleichungen für bewegte Körper," *Annalen der Physik 26* (1908): 532–540: reprinted

in *Collected Papers*, vol. 2, pp. 509–517). In a review talk on special relativity given in January 1911, Einstein included a brief discussion of "the highly interesting mathematical development that the theory has undergone, primarily due to Minkowski who unfortunately died so young," noting that it had led to "a very perspicacious representation of the theory, which essentially simplifies its application" (Das Relativitäts-Theorie," *Naturforschende Gesellschaft in Zürich, Vierteljahrschrift 56* (1911): 1–14; reprinted in *Collected Papers*, vol. 3, pp. 425–439). For a discussion of Minkowski's work and the varying forms of its assimilation by the physics and mathematics communities, see Scott Walter, "The Non-Euclidean Style of Minkowskian Relativity," in *The Symbolic Universe/Geometry and Physics 1890–1930*, Jeremy Gray, ed. (Oxford University Press 1999), pp. 91–127.

[63] Arnold Sommerfeld, "Zur Relativitätstheorie I. Vierdimensionale Vekoralgebra," *Annalen der Physik 32* (1910): 749–776; "Zur Relativitätstheorie II. Vierdimensionale Vekoranalysis," *Annalen der Physik 33* (1910): 649–689. He states that the formalism he presents, "is (aside from imaginary coordinates) an immediate generalization of the customary three-dimensional vector methods", and provides "a complete substitute for the matrix calculus used by Minkowski" (I, p. 749).

[64] Laue notes that he has "taken into account extensively the mathematical development of the theory that Sommerfeld has recently given;" *Das Relativitätsprinzip*, 1911 ed., p. vi. Einstein commented: "His book on relativity theory is a little masterpiece," Einstein to Alfred Kleiner, 3 April 1912, *Collected Papers*, vol. 5, pp. 445–446.

[65] This article, prepared for Erich Marx's *Handbuch der Radiologie*, was completed but has only been published recently: See "Manuscript on Relativity," *The Collected Papers of Albert Einstein*, vol. 4, *The Swiss Years: Writings, 1912–1914*, Martin Klein et al., eds. (Princeton University Press 1995), pp. 9–108. For its history see the Editorial Note, "Einstein's Manuscript on the Special Theory of Relativity," *ibid.*, pp. 3–8.

[66] See *ibid.*, p. 72. Sommerfeld had given the names "four-vector" and "six-vector" to what Hermann Minkowski had called "space-time vectors of type I and II," respectively ("Die Grundgleichungen für die elektromagnetischen Vorgänge im bewegten Körper," *Königliche Gesellschaft der Wissenschaften zu Göttingen. Mathematisch-physikalische Klasse. Nachrichten* (1908): 53–111; see pp. 65–68.

[67] See Laue's book, *Das Relativitätsprinzip*, 1911 ed., pp. 99–100, which defines the "four-potential vector" ["*Viererpotential*"] as the four-vector whose four-curl is the electromagnetic field six-vector, and notes that the four-potential vector is only determined up to the four-gradient of a scalar.

[68] This was, of course, long before discussions of the physical significance of the electromagnetic potentials, based on the Aharonov-Bohm effect, took place.

[69] "Relativität und Gravitation. Erwiderung auf eine Bemerkung von M. Abraham," *Annalen der Physik 38* (1912): 1059–1064; reprinted in *Collected Papers*, vol. 4, pp. 181–188. The paper was received on 4 July 1912. The citations are from pp. 1062–1063.

[70] Einstein to Wilhelm Wien, 17 May 1912, *Collected Papers*, vol. 5, p. 465.

[71] Einstein to Paul Ehrenfest, 10 March 1912, *Collected Papers*, vol. 5, p. 428.

[72] "Lichtgeschwindigkeit und Statik des Gravitationsfeldes," *Annalen der Physik 38* (1912): 355–369, reprinted in *Collected Papers*, vol. 4, pp. 130–145; "Zur Theorie des statischen Gravitationsfeldes," *Annalen der Physik 38* (1912): 443–458, reprinted in *ibid.*, pp. 147–164. For a discussion of these papers, see the Editorial Note "Einstein on Gravitation and Relativity: The Static Field," *ibid.*, pp. 122–128.

[73] "The time in the field [that is] defined by the stipulation that the speed of light c depends indeed upon the position but not on direction" as Einstein explained to Michele Besso, 26 March 1912, *Collected Papers*, vol. 5, p. 435. This definition of the time is given in more detail in Paper I: "We think of the time in the [uniformly accelerated-JS] system K as measured by clocks of such a nature and such a fixed arrangement at the spatial points of K that the time intervals—measured with them—that a light ray takes to go from a point A to a point B of the system K does not depend on the moment of emission of the light ray at A [static condition]. Further it turns out that simultaneity can be defined without contradiction in such a way that, with respect to the settings of the clocks, the stipulation is satisfied that all light rays passing a point A of K have the same speed of propagation, independent of their direction" [isostropy condition] (I, pp. 357–358).

[74] The revision of the gravitational field equation is discussed in some detail in Jürgen Renn and Tilman Sauer, "Pathways out of Classical Physics," *The Genesis of General Relativity*, vol. 1, *General Relativity in the Making: Einstein's Zurich Notebook*, to appear. The remainder of paper II is concerned with electromagnetism (Sections 1 and 2) and thermodynamics (Section 3) in a static gravitational field, topics I shall not consider.

[75] Abraham, for example, exulted: "Einstein ... had already given up his postulate of constancy of the velocity of light at the turn of the year, which was so essential for his earlier theory; in a recent work he abandons the requirement of the invariance of the equations of motion under Lorentz transformations, thereby delivering the coup de grace to the theory of relativity. Those who, like the author, have repeatedly had to warn against the siren song of this theory, can only greet with satisfaction the fact that its originator has now convinced himself of its untenability" (Max Abraham, "Relativität und Gravitation, Erwiderung auf eine Bemerkung des Hrn. A. Einstein," *Annalen der Physik 38* (1912): 1056–1057).

[76] Paper II, p. 458.

[77] Albert Einstein, "Lecture Notes for Introductory Course on Mechanics at the University of Zurich, Winter Semester 1909–1910," *Collected Papers*, vol, 3, pp. 11–129. See also the Editorial Note, "Einstein's Lecture Notes," *ibid.*, pp. 3–10, especially Section II, pp. 4–6. Einstein also taught mechanics in Prague during the winter semester of 1911 (see *ibid.*, "Appendix B: Einstein's Academic Courses," pp. 598–600).

[78] *Entwurf einer verallgemeinerten Relativitätstheorie und einer Theorie der Gravitation/ I. Physikalischer Teil von Albert Einstein*, B. G. Teubner, Leipzig/Berlin 1913; reprinted in *Collected Papers*, vol. 4, pp. 303–323. For a discussion of this paper, see the Editorial Note "Einstein on Gravitation and Relativity: The Collaboration with Marcel Grossmann," *ibid.*, pp.294–301. See also the Editorial Note "Einstein's Research Notes on a Generalized Theory of Relativity," *ibid.*, pp. 192–199, especially Section II, pp. 193–195.

[79] *Ibid.*, p. 6.

[80] *Ibid.*, p. 7.

[81] *Ibid.*, p. 8.

[82] "Autobiographische Skizze" (see note 9).

[83] In the Preface to the Czech edition of his popular book on relativity. The German text is in the *Collected Papers*, vol. 6, p. 535, note [4].

[84] In the 1955 "Autobiographische Skizze" (see note 9), Einstein described the lectures as "true masterpieces of pedagogical art, which later helped me very much when wrestling with general relativity" (pp. 10–11). In a letter of 24 April 1930 to Walter Leich, Einstein

wrote: "Geiser was dry only in the large lectures, otherwise I owe him the most of all." For an outline of the contents of Geiser's course on "Infinitesimalgeometrie" given in the Winter Semester of 1897/1898 and the Summer Semester of 1898, based on Grossmann's lecture notes, see *Collected Papers*, vol. 1, pp. 365–366. Grossmann's notes are preserved in the ETH Bibliothek, Hs 421: 15 & 16.

[85] See Julian Lowell Coolidge, *A History of Geometrical Methods* (Oxford: Clarendon Press 1940), "Book III Differential Geometry," Chapters I, II and III, pp. 318–387. The citations are from p. 322.

[86] Carl Friedrich Gauss, "Disquisitiones generales circa superficies curvas," *Commentationes societatis regiae scientiarum Gottingensis recentiores 6* (1828): 99–146; reprinted in *ibid.*, *Werke*, vol. 4 (Göttingen: Königliche Gesellschaft der Wissenschaften 1880), pp. 217–258. A valuable notice ["Anzeige"] by Gauss appeared the *Göttingische gelehrte Anzeigen, Stück 177* (1827): 1761–1768, reprinted in *ibid.*, pp. 341–347; hereafter cited as "Anzeige." The basic idea of using the line element to investigate the properties of a surface had already been used in Gauss's 1822 Prize Essay, "Allgemeine Auflösung der Aufgabe die Theile einer gegebenen Fläche auf einer andern gegebenen Fläche so abzubilden dass die Abbildung dem Abgebildeten in den kleinsten Theilen ähnlich wird," *Astronomische Abhandlungen. Heft 3* (Altona: Schumacher 1825); reprinted in *ibid.*, pp. 189–216. For discussions, see Coolidge, Chapter III, section 1, pp. 355–359; and P. Stäckel, "Gauß als Geometer," *Materialen für eine wissenschafliche Biographie von Gauß*, Felix Klein et al., eds. (Leipzig: Teubner 1918), Heft V, pp. 25–142, Section V, "Die allgemeine Lehre von den krummen Flächen," pp. 104–138.

[87] Luigi Bianchi, *Vorlesungen über Differentialgeometrie*, transl. by Max Lukat, 2nd ed. (Leipzig and Berlin: Teubner 1910); herafter cited as "Bianchi-Lukat." The citations are from pp. 60–61. As we shall see below, this work was consulted by Marcel Grossmann.

[88] Bernhard Riemann, "Ueber die Hypothesen, welche der Geometrie zu Grunde liegen," *Abhandlungen der königlichen Gesellschaft der Wissenschaften zu Göttingen 13* (1868): 133–150. This was the posthumous publication of Riemann's *Habilitationsschrift* of 1854.

[89] These notes are described in Karin Reich, *Die Entwcklung des Tensorkalküls* (Basel, Boston, Berlin: Birkhäuser 1994), pp. 163–166.

[90] Grossmann notes for 10 June 1898.

[91] See Section III, p. 108 of "Raum und Zeit," *Physikalische Zeitschrift 10* (1909): 104–111; it was soon reprinted as a separate booklet, *Raum und Zeit* (Leipzig: B.G. Teubner 1909); and then in Minkowski's *Gesammelte Abhandlungen*, David Hilbert, ed. (Leipzig: B.G. Teubner 1911), vol. 2, pp. 431–434.

[92] Laue only introduces the proper time and uses it to define the four-velocity in the second edition of *Das Relativitätsprinzip* (Braunschweig: Vieweg 1913); see pp. 57 and 69.

[93] Einstein does define the four-velocity vector by the following equation:

$$G_\mu = \frac{dx_\mu}{\left\{\sqrt{-\sum dx_\sigma^2}\right\}}$$

but without any explanation of the expression in the denominator, which does not occur anywhere else in the paper (p. 84).

[94] See Giulio Maltese and Lucia Orlando, "The Definition of Rigidity in the Special Theory of Relativity and the Genesis of the General Theory of Relativity," *Studies in the History and the Philosophy of Modern Physics 26B* (1995): 263–303.

[95] Gustav Herglotz, "Über den vom Standpunk des Relativitätsprinzips aus als "starr" zu bezeichnenden Körper," *Annalen der Physik 31* (1910): 393–415; citation from p. 394.

[96] Max Born, "Über die Definition des starren Körpers in der Kinematik des Relativitätsprinzips" (see note 21), p. 233.

[97] "The latest relativity-theoretical investigations of Born and of Herglotz interest me very much..." Albert Einstein to Jakob Laub, 16 March 1910, *Collected Papers*, vol. 5, pp. 231–233; citation from p. 232.

[98] Max Abraham, "Berechtigung," *Physikalische Zeitschrift 13* (1912): 176. This correction was submitted to issue no. 4, which had a closing date of 2 February 1912, but was published on 15 February 1912.

[99] Abraham reiterated this point in his next paper, "Die Erhaltung der Energie und der Materie im Schwerkraftfelde," *Physikalische Zeitschrift 13* (1912): pp. 311–314: "As a result of the variability of c, the Lorentz group only holds in the infinitely small, so that dx, dy, dz and $du = i\,cdt$ represent the components of an infinitely small displacement in a four-dimensional space" (p. 312). Note that, in contrast to Einstein, Abraham's c may be a function of all four space-time coordinates.

[100] Louis Kollross, "Erinnerungen-Souvenirs," *Schweizerische Hochschulzeitung 28* (*Sonderheft*) (1955), pp. 169–173; reprinted as "Erinnerungen eines Kommilitonen," in Carl Seelig, ed., *Helle Zeit- Dunkle Zeit In Memoriam Albert Einstein, Zürich/Stuttgart/ Wien, Europa Verlag. 1955*, pp. 17–31. Citation from latter, p. 27.

[101] Marcel Grossmann, *Entwurf einer verallgemeinerten Relativitätstheorie und einer Theorie der Gravitation/ II. Mathematischer Teil von Marcel Grossmann*, B. G. Teubner, Leipzig/Berlin 1913, pp. 23–38; reprinted in *Collected Papers*, vol. 4, pp. 324–339; citation from p. 324 of the latter. Hereafter cited as "Entwurf, Mathematischer Teil." The paper cited is Elwin Bruno Christoffel, "Ueber die Transformation der homogenen Differentialausdrücke zweiten Grades," *Journal für die reine und angewandte Mathematik 70* (1869): 46–70.

[102] Gregorio Ricci and Tullio Levi-Civita, "Methodes de calcul différentiel absolu et leurs applications," *Mathematische Annalen 54* (1901): 125–201. This paper has been tanslated into English in *Lie Groups: History, Frontiers and Applications, Series A*, vol. 2. *Ricci and Levi-Civita's Tensor Analysis Paper*, transl. and comments by Robert Hermann (Brookline: Math Sci Press).

[103] Only in Section 4 of this chapter, "Applications to vector analysis," is "the ds^2 of [Euclidean three-] space as the fundamental form ϕ" introduced (p. 135). The paper has a number of lapses: for example, the fundamental form is introduced on p. 130 without the name, and the notation ϕ for it is used on p. 132, before both are defined on p. 133. Most serious, the Christoffel symbols of the second kind are introduced and used to define the covariant derivative on p. 138, without ever being defined or related to the symbols of the first kind which are defined and then used to define the covariant form of the Riemann tensor (here called "the covariant system of Riemann") on p. 142.

[104] Ref. 44, p. 70.

[105] For a discussion of this question see "The Story of Newstein," reference in note 17.

[106] Bianchi–Lukat.

[107] See "Entwurf, Mathematischer Teil," p. 330.

How Einstein Discovered General Relativity: A Historical Tale with Some Contemporary Morals

John Stachel

Introduction

Like a classical drama, Einstein's attempt to generalize his original 1905 (special) theory of relativity is divided into three acts:

In 1907, Einstein attempted to set up a relativistically invariant theory of gravitation. He realized, however, that this theory failed to obey the equivalence principle: all bodies did not fall with the same acceleration in a gravitational field. He regarded satisfaction of the equivalence principle as the fundamental criterion for an acceptable gravitational theory and decided that the 1905 relativity theory must be generalized in order to include gravitation.

In 1912, he concluded that no scalar generalization of Newtonian gravitation theory would be adequate. He decided that a non-flat generalization of the four-dimensional metric tensor of Minkowski space-time was the correct mathematical representation of the gravitational field.

After several years during which he was convinced that the gravitational field equations could not be generally covariant, in 1915 Einstein adopted (the practically unique) generally covariant field equations for the metric tensor. He found that the spherically-symmetric solution to these equations accounted for the well-known anomaly in the precession of Mercury's orbit.

In Stachel (1979), I gave a summary of the historical aspects of this drama; in Stachel (1980) I went into more detail about what happened during the crucial second act. Here, I shall discuss the conceptual difficulties facing Einstein at each

M. A. H. MacCallum, ed.
General Relativity and Gravitation
Proceedings of the 11th International Conference on General Relativity and Gravitation, pp. 200–208

step, and how he coped with them. Since physicists usually do not recapitulate the development of a theory when learning it, they are often not aware of such difficulties, nor of how they were overcome. These historical issues are closely connected with some of the unique difficulties encountered in current attempts to quantize the gravitational field, so this story is of more than historical interest.

Finally, I shall discuss some ways in which the currently accepted formulation of the mathematical foundations of the general theory does not do full justice to Einstein's conception of the theory, and suggest some ways in which this formulation might be improved.

The Equivalence Principle

The main difficulty at this stage was to see that, in contrast to all other forces of nature, gravitation bears a special relationship to inertia. As Einstein put it, gravitation and inertia are '*wesensgleich*', essentially the same. This insight implies that the class of global inertial frames singled out in the special theory of relativity can have no place in a relativistic theory of gravitation.

> You consider the transition to special relativity as the most essential thought of relativity, not the transition to general relativity: I consider the reverse to be correct. I see the most essential thing in the overcoming of the inertial system, a thing that acts upon all processes but undergoes no reaction. This concept is, in principle, no better than that of the center of the universe in Aristotelian physics (Einstein to Georg Jaffe, 19 January 1954).

Other physicists then working on a relativistic theory of gravitation found this idea difficult if not impossible to accept, as do many physicists to this day. As a consequence of this insight, Einstein was skeptical of all attempts to start with a special-relativistic quantum field theory and make it generally covariant.

> Contemporary physicists do not see that it is hopeless to take a theory that is based on an independent rigid space (Lorentz-invariance) and later hope to make it general relativistic (in some natural way) (ibid).

Such comments apply *a fortiori* to attempts to quantize the general theory itself by treating it as if it were no more than a special-relativistic theory that happens to be non-linear and hence invariant under a particularly nasty gauge group.

The Metric Tensor

The main difficulty at this stage was to grasp the dual nature of the metric tensor: it is both the mathematical object which represents the space-time structure (chronogeometry) and the set of 'potentials,' from which representations of the gravitational field (Christoffel symbols) and of the tidal 'forces' (Riemann tensor) may be derived. The space-time structure of general relativity is generally non-flat; more important, it is itself a set of dynamical field variables. The metric tensor not only acts upon the matter fields, but is acted upon by them, via their stress-energy tensors.

Within a few years, Levi-Civita, Weyl and Cartan generalized the Christoffel symbols to the concept of affine connection. This concept served to make the relationship between the mathematical representations of various physical concepts much clearer. Just because it is not a tensor field, the connection field provides an adequate representation of the gravitational-cum-inertial field required by Einstein's interpretation of the equivalence principle. There is no (unique) decomposition of the connection field into an inertial connection plus a gravitational tensor field.

> ... what characterizes the existence of a gravitational field from the empirical standpoint is the non-vanishing of the [components of the affine connection- JS], not the non-vanishing of the [components of the Riemann tensor- JS]. If one does not think in such intuitive (*anschaulich*) ways, one cannot grasp why something like curvature should have anything at all to do with gravitation. In any case, no rational person would have hit upon anything otherwise. The key to the understanding of the equality of gravitational and inertial mass would have been missing (Einstein to Max von Laue, September 1950).

Discussions of the thermal ambience registered by accelerating particle detectors in Minkowski space-time, summarized, for example, in Sciama (1979), pp. 706–714, serve to emphasize the physical significance of the non-vanishing of the affine connection in certain frames of reference, even when the Riemann tensor vanishes. The current fashionableness of gauge theories has also made elementary particle physicists well aware of the fundamental significance of connections in physics.

Attempts to quantize the gravitational field while interpreting the resulting formalism in a 'background' space-time (usually taken to be flat) are basically efforts to impose a decomposition of the connection field into an inertial connection, which is left classical, plus a gravitational field tensor, which is quantized. From Einstein's point of view, this is a violation of the injunction: 'What God has joined together let no man put asunder.'

Of course, there is no guarantee that Einstein's approach to gravitation is correct; but it is certainly worth exploring the possibility that it is and that the attempt to reconcile gravitation and quantization involves something more fundamental than minute, observationally insignificant gravitational corrections to special-relativistically calculated effects.

General Covariance

The main difficulty at this stage was to realize that, in a theory in which the metric tensor is not given *a priori*, the points of the space-time manifold (which, in the physical interpretation, are to be correlated with events) only derive their physical individuation from the metric field.

Although he considered the possibility of generally covariant field equations in 1913, Einstein rejected them because of an argument that seemed to show that, given the distribution of matter and energy everywhere outside some region of space-time, no generally covariant field equations could uniquely determine the gravitational field inside that 'hole.' For, if the stress-energy tensor outside the region determined *one* such solution $G(x)$, then dragging that solution over the region by any point transformation that reduces to the identity outside and on the boundary of the 'hole' will result in *another* solution $G'(x)$ (in the *same* coordinate system, as Einstein realized), which satisfies the same field equations, is identical with $G(x)$ outside the 'hole,' but differs from it inside the 'hole.'

Only two years later, after he realized that its validity depends on a tacit assumption, was he able to move beyond this 'hole' argument to adopt generally covariant field equations. Norton (1984) tells the story in some detail. Einstein had been assuming that the physical identity of the points of the region, that is, the event corresponding to the argument x in $G(x)$ and $G'(x)$, was established prior to the specification of the metric. In any theory in which the metrical structure is given *a priori*, the physical identity of points of the space-time manifold is indeed established independently of any dynamical considerations, But, in general relativity the metrical structure forms part of the set of dynamical variables, which must be determined before the points of space-time have any physical properties.

In special relativity, for example, it is assumed that the individuation of the points of Minkowski space is established by a framework of rigid rods and clocks or some similar device before one begins to do electrodynamics. Hence, even though such fields are mathematically isomorphic, in Minkowski space one can distinguish physically between Coulomb fields with the same charge centered at different points in the same inertial frame, or at the origins of different inertial frames. In general relativity, on the other hand, it makes no sense to try to make a physical distinction between Schwarzchild fields with the same mass. All such Schwarzchild space-times must be identified as corresponding to a *single* gravitational field, since the points of their respective manifolds only inherit their physical individuation and properties from the metric field. Speaking more generally, a space-time (manifold with a metric) corresponds to a gravitational field; but a gravitational field corresponds to an *equivalence class* of space-times. I know of only a few textbooks of general relativity that make this point clearly, notably Hawking and Ellis (1973), p. 56, and Sachs and Wu (1977), p. 27. Without understanding this point, it is impossible, for example, to make sense of the Cauchy problem in general relativity (see Hawking and Ellis, pp. 227–228).

This is the true significance of the concept of general covariance, as Peter Bergmann among others has long known. Unfortunately, this profound insight of Einstein was originally expressed in the language of coordinate systems, rather than in the coordinate-free language of modern formulations of differential geometry. This has been responsible for a great deal of misunderstanding of the concept.

Attempts to quantize the field equations of general relativity utilizing some variant of the techniques of special-relativistic quantum field theory often fail to take into account problems peculiar to generally covariant theories. For example, the physical interpretation of the usual quantum formalism depends heavily on the presence of a given space-time metrical structure. With its help, such concepts as the position of a measuring apparatus, the timing of a measurement, etc., may be introduced. Without such a pre-existent metrical structure—indeed, when this metrical structure is included among the dynamical variables—the meaning of such concepts (or their replacement by others) becomes a major problem in interpreting the formalism.

It is often assumed that such problems of interpretation can be avoided by the introduction of a background metric, usually taken to be flat, and often also used as the starting point for an iterative approximation procedure. In addition to the problem of justifying its introduction, discussed above, there is the problem of interpreting this background metric itself. Even though the two are the same in mathematical form, Minkowski space as a particular solution to the field equations is conceptually distinct from Minkowski space as the given space-time structure of the special theory of relativity. All of the problems of physical individuation of the points of space-time discussed above apply to Minkowski space in the first sense (or to any other background solution) just as much as they do to any other solution of the field equations.

Problems with the Standard Mathematical Formulation

The standard mathematical formulation of the general theory (see the texts mentioned above, for example) starts with a bare differentiable manifold, the points of which are interpreted as physical events, and then imposes various fields on this manifold. In particular, a second-rank symmetric covariant tensor field is interpreted as the space-time metrical structure interrelating these events, and as the gravitational potentials. While this way of introducing the space-time structure is quite adequate in the case of non-generally covariant theories, in which this structure is given *a priori*, it does not do full justice either to Einstein's relational concept of space-time, or to the actual practice of relativists in formulating and solving problems.

In Einstein (1952), he wrote:

> On the basis of the general theory of relativity . . . space as opposed to 'what fills space' . . . has no separate existence . . . If we imagine the gravitational field, i.e., the functions g_{ik} to be removed, there does not remain a space of the type [of Minkowski space in special relativity- JS], but absolutely *nothing*, not even a 'topological space.' For the functions g_{ik} describe not only the field, but at the same time also the topological and metrical structural properties of the manifold . . . There is no such thing as an empty space, i.e. a space without a field. Space-time does not claim existence on its own, but only as a structural quality of the field.

In one of his last comments on the question, Einstein (1954), he noted:

> It required a severe struggle to arrive at the concept of independent and absolute space, indispensible for the development of theory. It has required no less strenuous exertions subsequently to overcome this concept—a process which is by no means as yet completed.

The standard mathematical treatment clearly does not fulfill Einstein's criterion: no metric, no anything. The manifold of space-time events is introduced prior to the introduction of any other structures.

Even more important, the standard treatment does not correctly describe the actual practice of relativists in solving the field equations. At least as early as the 1917 discussion between Einstein and Felix Klein about the topology of the spatial cross-sections of the Einstein static cosmological model, there has been some understanding by relativists that the topology of a solution to the field equations is not given *a priori*, but has to be investigated. Einstein had further experience with this question when, together with Nathan Rosen, he developed the 'bridge' interpretation of the Schwarzchild metric.

Having lived through the exciting period during which the Kruskal interpretation of the Schwarzchild solution was developed and black hole physics rose to prominence, I need hardly remind this audience that we do not start out with a fixed manifold topology, but solve the field equations on a generic patch. Once such a solution has been found, the problem of obtaining its maximal global extension is a major one. It involves a lot of hard work, which generally includes the formulation of criteria for an acceptable extension in order to arrive at a unique answer. Questions such as maximal analytic extension, geodesic completeness (timelike and/or null), and the adjunction of ideal boundary points at infinity (null and/or spatial) have generated some of the most interesting research on general relativity in recent years.

Yet we persist in formalizing what we do as if the choice of manifold topology were made at the outset. It appears to me unwise to allow this gap between official ideology and daily practice. Not only does it mislead students and researchers in other fields attempting to understand the nature of our work; the gap may conceal important new insights into the nature of the theory. The correct mathematical formulation of a problem often suggests further and unexpected avenues of progress.

Fiber Bundles and Sheaves

Recently (Stachel 1986), I discussed in some detail how use of the fibre bundle concept can resolve the first of the problems mentioned above (no metric, no anything). I shall briefly summarize the advantages of this approach to generally covariant field theories. First of all, it deals with entire classes of solutions to the field equations, rather than with individual metrics, thus focussing attention on the fact that the space-time structure itself is not given, but must be determined as a cross-section of the bundle satisfying certain criteria (generally, field equations). Second, the use of jet extensions of the bundle gives a structure on which field equations may be formulated. Third, and most relevant to the problem at hand, points of space-time may be represented, not by points of the base manifolds as is usually done, but by a mapping of points of a cross-section of the bundle into points of the base manifold. Events are then defined as sets of such mappings in the equivalence class of cross-sections corresponding to a unique gravitational field. Thus, if there is no gravitational field, there are no events.

However, the fibre bundle approach clearly does not solve the second problem discussed in the previous section. The topology of the base manifold is given *a priori*, so that a different fibre bundle must be introduced *a posteriori* for each topologically distinct class of solutions. The process of finding the global topology cannot be formalized within the fibre bundle approach. It appears that sheaf cohomology theory is the appropriate mathematical theory for dealing with the problem of going from local to global solutions, as noted by Wells (1980), p. 36. This is not the time nor the place, nor am I the person to discuss this suggestion in detail. Let me close by appealing to the mathematically more sophisticated members of our group to work on closing this outstanding gap between our creed and our deed.

References

Einstein, A. (1952). Relativity and the problem of space. In *Relativity: The Special and the General Theory*, pp. 135–137. New York: Crown.

—— (1954). Foreword. In *Concepts of Space*, Jammer, M., pp. xiii–xvi. Cambridge, MA: Harvard University Press.

Hawking, S. and Ellis, G. (1973). *The Large Scale Structure of Space-Time*. Cambridge: Cambridge University Press.

Norton, J. (1984). How Einstein Found his Field Equations: 1912–1915. In *Historical Studies in the Physical Sciences*, J. L. Heilbron, ed., vol. 14, pp. 253–316. Berkeley: University of California Press.

Sachs, R. and Wu, H. (1977). *General Relativity for Mathematicians*. New York: Springer-Verlag.

Sciama, D. (1979). Black Holes and Fluctuations of Quantum Particles: An Einstein Synthesis. In *Relativity, Quanta and Cosmology in the Development of the Scientific Thought of Albert Einstein*, F. de Finis, ed., vol. 2, pp. 681–724. New York: Johnson Reprint Corporation.

Stachel, J. (1979). The Genesis of General Relativity. In *Einstein Symposion Berlin*, H. Nelkowski et al., eds., pp. 428–442. Berlin: Springer-Verlag. [See this volume, pp. 233–244].

—— (1980). Einstein and the Rigidly Rotating Disk. In *General Relativity and Gravitation One Hundred Years After the Birth of Albert Einstein*, A. Held, ed., vol. 1, pp. 1–15. New York: Plenum Press. [See this volume, pp. 245–260].

—— (1986). What a Physicist Can Learn from the Discovery of General Relativity. In *Proceedings of the Fourth Marcel Grossmann Meeting on General Relativity*, R. Ruffini, ed., pp. 1857–1862.

Wells, R. O. (1980). *Differential Analysis on Complex Manifolds*, New York: Springer-Verlag.

Einstein's Search for General Covariance, 1912–1915

John Stachel

This paper is based on an invited talk given July 17, 1980, at the Ninth International Conference on General Relativity and Gravitation, Jena, Germany (DDR). A written version of this talk has circulated privately since the fall of 1980. Since that version has been cited in print several times since then, I have decided to publish it with only minor revisions to improve the exposition, but without attempting to update it. For a more recent account of my views see Stachel 1986b.

1. Introduction

Einstein listed the stages of his search for a generally covariant theory of gravitation in a biographical note written in 1916:

1907 Basic idea for the general theory of relativity.

1912 Recognition of the non-Euclidean nature of the metric and of the physical determination of the latter by gravitation.

1915 Field equations of gravitation. Explanation of the perihelion motion of Mercury. (EA 11-196)[1]

There were thus three key moments in Einstein's development of the general theory of relativity:

1. Adoption of the principle of equivalence as the crucial element in a relativistic theory of gravitation (1907).
2. Recognition that the gravitational field must be characterized mathematically via a four-dimensional (pseudo-) Riemannian metric tensor (1912).
3. Discovery of the final form of the field equations relating the metric tensor to the sources of the gravitational field (1915).

Reprinted from D. Howard and J. Stachel, eds.
Einstein and the History of General Relativity, Einstein Studies, Vol. 1

Elsewhere, I have discussed this story in broad outlines (Stachel 1979a, 1979b), and have tried to contribute to a more detailed understanding of the decisive second step (Stachel 1980). Most discussions of the development of general relativity have focused on the third step, in particular on the puzzling question: Why—once the decisive step of representing the gravitational field by the metric tensor had been taken—did Einstein take so long to arrive at the final form of the field equations? (See Earmann and Glymour 1978a, 1978b; Hoffmann 1972; Lanczos 1972; Mehra 1973; Vizgin and Smorodinskiî 1979.)

I shall suggest an answer that is (no doubt) still incomplete, but that differs from the existing accounts in several respects. In particular, I shall try to explain why it took Einstein over two years to return to general covariance after rejecting it in 1913:

1. without convicting Einstein or Marcel Grossmann of an elementary mathematical error in their original argument rejecting covariance (Einstein and Grossmann 1913) [see Section 1];

2. showing that Einstein's subsequent "hole" argument against general covariance involves a nontrivial *physical* question about the nature of space-time [see Sections 2 and 3]; and

3. giving a new interpretation of what seem to be some of Einstein's most "operationalistic" declarations in his 1916 exposition of the general theory (Einstein 1916) [see Section 5].

Circumstances outside of Einstein's purely scientific activities may have played some role in the length of the delay. Nineteen-fourteen was the year of his move to Berlin and his separation from his first wife and their children, who soon returned to Zurich. It was also, of course, the year of the outbreak of the first World War. It is clear that Einstein was affected deeply by these personal and political events. His turn towards political activity dates from his disillusionment with the chauvinistic reaction of most German intellectuals to the war.[2] It is also clear, however, that he found solace for his personal and political worries in his work. Leaving these issues aside, I shall confine myself here to the intellectual aspects of the problem.

In his "Autobiographical Notes," Einstein himself tried to answer the question of why it took so long from the time of his initial insight into the significance of the equivalence of inertial and gravitational mass until the general theory was completed:

> This happened in 1908 [actually it was 1907, as he noted elsewhere –JS]. Why were another seven years required for the construction of the general theory of relativity? The main reason lies in the fact that it is not so easy to free oneself from the idea that co-ordinates must have an immediate metrical significance (Einstein 1949, p. 67).

Einstein's comment is often interpreted as a reference to the need to recognize the role of the metric tensor in the mathematical representation of the gravitational field. But Einstein had taken this step by the end of 1912, as mentioned earlier. If this interpretation were complete, why the further delay?

I shall try to show that it was because Einstein still had to free himself from the idea that points of space-time (events) are physically individuated apart from their metrical properties; or, more accurately, from the idea that the points of a matter-free portion of a four-dimensional manifold are individuated as spatio-temporal events in some way that is independent of the properties they inherit, so to speak, from the presence of a metric tensor field on the manifold.

If the latter assertion is included in the interpretation of Einstein's statement about coordinates, then I believe that Einstein's answer is essentially correct. Be this as it may, my thesis is that the problem of the individuation of space-time points was the basic reason for Einstein's long delay in accepting general covariance *after* he had recognized the role of the metric tensor.

2. The Einstein–Grossmann Paper

In the fall of 1912, Einstein returned to Zurich from Prague to accept a professorship in physics at the Eidgenössische Technische Hochschule (ETH). He later recalled:

> I had the decisive idea of the analogy between the mathematical problems connected with the theory [of gravitation –JS] and the Gaussian theory of surfaces only after my return to Zurich in 1912, without then knowing Riemann's or Ricci's or Levi-Civita's investigations. I was made aware of these only by my friend Grossmann in Zurich, when I posed him the problem of finding generally covariant tensors whose components depend only on derivatives of the coefficients of the quadratic fundamental invariant [components of the metric tensor –JS]. (Einstein 1922; EA 1-014).[3]

Other reminiscences by Einstein,[4] as well as a study of Marcel Grossmann's ETH notebooks,[5] demonstrate that the lectures of Professor Carl Friedrich Geiser, which Einstein heard as a student at the ETH, had familiarized him with the Gaussian theory of two-dimensional surfaces. Thus, aside from probable anachronism in his use of the words, "generally covariant tensors," to describe what he wanted from Grossmann, there is no reason to doubt Einstein's account of what he knew, and what he wanted to learn, when in the fall of 1912 he turned for help to his old school friend, now professor of mathematics at the ETH. Grossmann "at once caught fire" ("fing sofort Feuer"), as Einstein recounts (Einstein 1955, p. 151), and promised to help Einstein with the mathematical problems involved in working out his gravitational theory, but on the condition that Grossmann bear no responsibility for the physics.

By late 1912, Einstein had thus achieved the fundamental breakthrough in his search for a relativistic theory of gravitation based upon the equivalence principle: adoption of a locally Minkowskian metric tensor on a four-dimensional manifold as a mathematical representation of the gravitational field. The preceding quotation indicates that Einstein was looking for generally covariant field equations from the start. The reason seems clear: for Einstein, a mathematical coordinate system is associated with a physical frame of reference; general covariance thus corresponds to the admissibility of arbitrary reference frames. If the entire theory of gravitation could be formulated in terms of generally covariant equations, this would constitute for Einstein an extension of the (special-relativistic) relativity principle from inertial frames of reference to arbitrary frames—a truly *general* relativity (note that the German word "allgemein," as in "allgemeine Relativität," has the connotation of "universal in application"). Any frame of reference could be interpreted as an inertial frame with a certain gravitational field present in it. Thus, the distinction between gravitational and inertial effects would not be absolute, but would depend upon the frame of reference adopted.[6]

With Grossmann's help, Einstein soon succeeded in mathematically representing the effect of gravitation on all other physical processes in a generally covariant form. But there still remained the question of the equations determining the gravitational field itself. The sources of this field are represented by the rank-two stress-energy tensor (SET), the relativistic generalization of the Newtonian mass density. So the gravitational field equations must be a set of differential equations that generalize the Newtonian equation for the gravitational potential (Poisson's equation) by relating the metric tensor and its first two derivatives to the stress-energy tensor. Grossmann found that the Ricci tensor is almost the only rank-two generally covariant tensor of second differential order that can be formed from the metric tensor and its derivatives.[7] However, Einstein and Grossmann became convinced that field equations setting the Ricci tensor equal to the SET do not lead to the correct Newtonian limit for weak static gravitational fields. They gave the reasons for this belief in a joint paper on gravitation, the "Entwurf" paper (Einstein and Grossmann 1913), that was *first* published early in 1913 as a separate booklet by B.G. Teuber Verlag, and then again early in 1914 in the *Zeitschrift für Mathematik und Physik*, with an addendum signed by Einstein. (The fact that this addendum is *not* in the original edition has caused some later confusion.)

The second, "mathematical," part of the paper was written by Grossmann. After defining the Riemann tensor Grossmann states:

> The outstanding significance of these concepts for the *differential geometry* of a manifold characterized by a line element makes it *a priori* probable that these general differential tensors would also be of significance for the problem of the differential equations of a gravitational field. Indeed one can give a covariant differential tensor of second rank and second order G_{im} [the Ricci tensor, now more commonly denoted by R_{im} –JS], which could enter into those equations, namely:
>
> $$G_{im} = \sum_{kl} \gamma_{kl}(ik, lm) = \sum_{k} \{ik, km\}.$$
>
> [In notation more common today, $R_{im} = g^{kl} R_{ik\,lm} = R^{k}_{i\;km}$, where $R_{ik\,lm}$ are the covariant components of the Riemann tensor –JS.] (Einstein and Grossmann 1913, p. 257).[8]

Then comes a fateful sentence:

> But it turns out that this tensor [the Ricci tensor –JS] does *not* reduce, in the special case of an infinitely weak static gravitational field, to the expression $\Delta\varphi$ [the Laplacian of φ –JS]. (Einstein to Grossmann 1913, p. 257).[9]

What did this mean? The usual interpretation (see, e.g., Vizgin and Smorodinskiî 1979) is that Grossmann failed to see how to extract a d'Alembertian operator from the linearized approximation to the Ricci tensor because of his inability to use suitably the general covariance of the Ricci tensor—for example, by imposing a suitable coordinate condition. Since neither Grossmann nor Einstein gave any further details of the argument, it is impossible to rule out such an interpretation, and just as impossible to prove it. However, there is a simpler interpretation of the sentence that is entirely consistent with the rest of the Einstein–Grossmann paper, and that leaves Grossmann innocent of such a gross mathematical error. This interpretation does depend on the existence of a physical misconception on the part of Einstein. However, its existence is not based on conjecture. The misconception is clearly stated in part I of the paper, written by Einstein, and it explains why Grossmann arrived at a result that led Einstein to reject the Ricci tensor. The argument hinges on the word "static" in the sentence by Grossmann quoted above. At the time he was working on the Einstein–Grossmann paper, Einstein thought that, even before having the correct field equations, he knew the correct form of the metric tensor for a static gravitational field based on his earlier work on the subject (Einstein 1912a, 1912b). On page 229 of part I, he wrote down the special-relativistic Minkowski metric, with $g_{44} = c^2$ ($c =$ the velocity of light), $g_{11} = g_{22} = g_{33} = -1$, and all other components zero; and he immediately added: "The same type of degeneration is shown by static gravitational fields of the type previously considered, except that for the latter $g_{44} = c^2$ is a function of x_1, x_2, x_3 [the spatial coordinates –JS]" (Einstein and Grossmann 1913, p. 229).[10] (The words "previously considered" refer to pages 227–228, where Einstein summarized his earlier work on static gravitational fields, in which the gravitational field is represented by a time-independent but spatially variable speed of light.)

We may well imagine Grossmann substituting this *Ansatz* for the metric tensor of a static gravitational field into the formula for the Ricci tensor, neglecting all but linear terms, since he was dealing with an "infinitely weak" field. If that is what he did, he found (in linearized approximation):[11]

$$
\begin{aligned}
R_{44} &= \frac{1}{2}\Delta g_{44}, \qquad R_{4n} = 0 \\
R_{mn} &= \frac{1}{2}\frac{\partial^2 g_{44}}{\partial x_m \partial x_n} \qquad (m, n = 1, 2, 3).
\end{aligned}
$$

At first sight this result seems promising, since R_{44} does reduce to the Laplacian of g_{44} (as it does for *any* static metric, without any coordinate conditions). A moment's thought q shows, however, that the result is really disastrous. For if the Ricci tensor vanishes (as it must outside of the sources of the gravitational field), it follows from $R_{mn} = 0$ that g_{44} can depend at most *linearly* on the coordinates. Thus, it cannot possibly represent the gravitational potential of any (finite) distribution of matter, static or otherwise. If Einstein's *Ansatz* for the static metric tensor had been correct, the vanishing of the Ricci tensor *could not* provide gravitational field equations outside of matter. Einstein's response to this dilemma (if I have correctly described it) was to keep his *Ansatz* for static metrics, and drop the Ricci tensor field equations.

The root of the trouble is Einstein's belief that the spatial metric of a static gravitational field (in coordinates adapted to the static nature of the field—in coordinates adapted to the static Killing vector, as we say now) must be flat. This belief, which is implicit in his geometrical reinterpretation of his 1912 papers on static gravitational fields (Einstein 1912a, 1912b), may well have been reinforced by the circumstance that it was the move beyond static fields to consideration of a particular stationary gravitational field—the gravitational field on a rotating disk—that first led Einstein to introduce a nonflat spatial geometry (see Stachel 1980).

He apparently remained convinced that static gravitational fields have spatially flat cross sections—at least to "first order," i.e., in a linearized approximation (see the following discussion)—until late 1915, when a study of approximate solutions to a set of generally covariant field equations finally showed that this is not the case. Indeed, it is this spatial nonflatness of the static, spherically symmetric solution that yields the additional corrections to the calculated precession of the perihelion of Mercury, whose agreement with the observed anomalous precession helped to demonstrate the validity of the covariant equations (see Einstein 1915c). In a letter to Besso, dated December 21, 1915, Einstein remarked:

> Most gratifying [about the new theory –JS] is the agreement of the perihelion motion [of Mercury –JS] and the general covariance; most curious, however, the circumstance that Newton's theory of the field, even for terms of the 1st order, is incorrect for the field (occurrence of [nonflat –JS] $g_{11} - g_{33}$). Only the circumstance that $g_{11} - g_{33}$ do not occur in the first approximation of the eq[uations] of mot[ion] of a point[-particle] causes the simplicity of Newton's theory (Speziali 1972, p. 61). [12]

Interestingly enough, in connection with the attempt to reconstruct the Einstein–Grossmann argument of 1913, this is not the first letter in which Einstein told Besso about this result. In his previous letter of December 10, 1915, Einstein told Besso about the success of his generally covariant equations, and immediately added:

> This time the most obvious thing was the correct one; but Grossmann and I believed that the conservation equations were not fulfilled, and that Newton's law did not come out in first approximation. You will be astonished by the occurrence of $g_{11} \dots g_{33}$ (Speziali 1972, p. 60). [13]

It is not difficult to surmise that Einstein was indicating his own astonishment by this comment. It is also not hard to see more than a fortuitous conjunction in the juxtaposition of the remark about the Einstein–Grossmann errors of 1913 and the result about the nonflat values for $g_{11} \dots g_{33}$. It is plausible to conjecture that Einstein thought: "If only I had realized that the static spatial metric was nonflat, I could have avoided the retreat from general covariance in 1913." I shall now return to this topic.

Faced with the check to their hopes for a generally covariant theory, Einstein and Grossmann still left open the possibility of finding, ultimately, a generally covariant set of gravitational field equations. Immediately after his previously quoted remark Grossmann continued:

> Such a connection [between the general theory of tensors and the differential equations for the gravitational field –JS] would have to be present, insofar as the gravitational equations were to permit *arbitrary* substitutions [of coordinates –JS]; but in this case finding differential equations of the *second* order seems to be excluded. On the other hand, if it were determined that the gravitational equations only allow a certain group of transformations [i.e., a more limited one –JS], then it would be understandable if one could not manage with the differential tensors offered by the general theory. As explained in the Physical Part [by Einstein –JS], we are not able to take a position on these questions (Einstein and Grossmann 1913, p. 257).[14]

In part I of the paper, after noting that it might be possible to retain general covariance without using the Ricci tensor by going to differential equations of order higher than second, Einstein stated:

> The attempt to discuss such possibilities would be premature in the present state of our knowledge of the physical properties of the gravitational field. Therefore, the restriction to second order [field equations –JS] is dictated to us and we must therefore renounce attempting to set up gravitational equations that are covariant under arbitrary transformations [of coordinates –JS]. Besides, it is to be emphasized that we have no sort of justification for general covariance of the gravitational equations (Einstein and Grossmann 1913, p. 234).[15]

After then finding what he considered satisfactory equations for the gravitational field, Einstein returned to this question:

> We have only been able to prove for the gravitational equations that they are covariant with respect to arbitrary *linear* transformations; however, we do not know if there is a general group of transformations, with respect to which the equations are covariant. The question of the existence of such a group . . . is the most important one connected with the considerations given here (Einstein and Grossmann 1913, p. 240).[16]

He again states that

> we are not justified in the present state of the theory in demanding the covariance of physical equations with respect to arbitrary substitutions [of coordinates –JS].
>
> On the other hand, we have seen that an energy-momentum balance equation for material processes can be set up . . . that allows arbitrary transformations. Therefore, it seems rather natural if we assume that all physical systems of equations with the exception of the gravitational equations are to be so formulated that they are covariant with respect to arbitrary substitutions. This exceptional position of the gravitational equations with respect to all other systems is connected, in my opinion, with their having to contain only the first two derivatives of the components of the fundamental [metric –JS] tensor (Einstein and Grossmann 1913, p. 240).[17]

How did Einstein find non-generally covariant gravitational field equations? He based his search on the criteria that the field equations should:

1. generalize Poisson's equation for the Newtonian gravitational potential (obey a correspondence principle, in modern terms);

2. be invariant at least under linear transformations (manifest no *less* relativity than in the special theory);

3. include a gravitational stress-energy complex, built from the metric tensor and its first derivatives (by analogy with Maxwell's electromagnetic theory), that is a tensor under linear transformations and enters the field equations as a source term in the same way as does the stress-energy tensor of ordinary (nongravitational) matter or fields. (See Einstein and Grossmann 1913, pp. 233–234, 239 for these criteria.)

The third requirement serves to ensure that conservation laws for the *total* gravitational and nongravitational energy and momentum follow from the gravitational field equations. Einstein was able to derive a set of field equations satisfying these three requirements in part I of the Einstein–Grossmann paper.

He also gave a sketch of how Newton's law of gravitation follows from the linear approximation to the Einstein–Grossmann field equtions for the static case, a result that he later demonstrated in detail in his 1913 lecture to the Vienna meeting of the Gesellschaft Deutscher Naturforscher and Ärzte (Einstein 1913a). As mentioned earlier, he explicitly noted that the spatial metric remains flat in this approximation: "One recognizes that coordinate lengths are at the same time natural lengths ($dt = 0$). Measuring rods thus undergo no deformation due to the 'Newtonian' gravitational field" (Einstein 1913a, p. 1260).[18]

At the time he finished the paper with Grossmann, Einstein seems to have felt that its negative results regarding general covariance could be only provisional. For example, he wrote to Paul Ehrenfest on May 28, 1913:

> I am now inwardly convinced that I have found that which is correct [the solution to the gravitational problem –JS], and, at the same time, that a murmur of disappointment will, of course, go through the ranks of our colleagues when the work appears, which will be in a few weeks... The conviction to which I have slowly struggled through is that *there are no preferred coordinate systems of any kind.* However, I have only partially succeeded, even formally, in reaching this standpoint (EA 9-340).[19]

However, by November 1913, Einstein had developed a simple "meta-argument" against all generally-covariant field equations. He wrote to Ludwig Hopf on November 2:

> I am now very content with the gravitation theory. The fact that the gravitational equations are not generally covariant, which a short time ago still disturbed me so much, has proved to be unavoidable; it is easily proved that a theory with generally covariant equations cannot exist if one demands that the field be mathematically completely determined by matter (EA 13-290).[20]

The proof in question is the notorious "hole" ("Loch") argument, alluded to in the introduction to this paper, and discussed in detail in the next section. Einstein now wrote to Ehrenfest, in an undated letter of late 1913:

> The gravitational affair has been clarified to my *complete satisfaction* (the circumstance, namely, that the equations for the gr[avitational] field are only covariant with respect to *linear* transformations). It can be proven, namely, that *generally covariant*

> equations that determine the field *completely* from the matter tensor cannot exist at all. What can be prettier than this, that the necessary specialization [of the coordinate system –JS] follows from the conservation laws? The conservation laws now determine, out of all surfaces, those which are to be chosen as preferred coordinate surfaces. We can designate these preferred surfaces as planes, since the only remaining allowed substitutions are *linear* ones (EA 9-342).[21]

He wrote to Ernst Mach in a similar vein in late 1913 or early 1914:

> For me it is absurd to attribute physical properties to "space." The totality of masses generates a $g_{\mu\nu}$-field (gravitational field), which in turn regulates the course of all processes, including the propagation of light rays and the behavior of measuring rods and clocks. Events are first referred to four *quite arbitrary* space-time variables. These must then, if the conservation laws of energy and momentum are to be satisfied, be specialized in such a way that only (fully) *linear* substitutions lead from one justified reference system to another. The reference system is, so to speak, adapted to the existing world with the help of the law of conservation of energy, and loses its nebulous, *a priori* existence (Herneck 1963; EA 17-454).[22]

At this point, then, Einstein believed that the conservation laws for the total energy and momentum, including the gravitational contribution, are valid only in a preferred set of coordinate systems, related to each other by linear transformations. He soon gave up the restriction to linearly-related coordinate systems, as we shall see; but the "hole" argument still prevented him from accepting general covariance. Let us turn to this argument.

3. The "Hole" Argument, or Covariance Lost

The argument was first published in the "Bemerkungen" ("Remarks"), which form the previously mentioned addendum to the "Entwurf" paper (Einstein and Grossmann 1913), signed by Einstein alone and not published in the original (separatum) printing of the paper. We can date the origin of the argument fairly closely. The letter to Hopf quoted earlier shows that Einstein had it by November 1913. In a lecture given to the Annual Meeting of the Swiss Naturforschende Gesellschaft on September 9, 1913, Einstein stated: "It has been possible to demonstrate by a general argument that equations that completely determine the gravitational field cannot be generally covariant with respect to arbitrary substitutions" (Einstein 1913b, p. 289).[23] And in his Vienna lecture of September 23, 1913, mentioned in the previous section, he stated: "In the last few days I have found the proof that such a generally covariant solution cannot exist at all" (Einstein 1913a, p. 1257, footnote 2).[24] Since it is always possible that both of these remarks were added when the texts were printed, all we can assert is that the "hole" argument probably first occurred to Einstein in September 1913, but surely by November of that year.

The first version of the argument runs as follows:

> Let there be a portion L [for "Loch" –JS] in our four-dimensional manifold in which a "material process" is not occurring, in which therefore the $\Theta_{\mu\nu}$ [the components of the nongravitational stress-energy tensor –JS] vanish. In accord with our assumptions, the $\gamma_{\mu\nu}$ [components of the metric tensor –JS] are completely determined everywhere,

therefore also inside L, by the $\Theta_{\mu\nu}$ given outside L. We now imagine that, instead of the original coordinates x_ν, new coordinates x'_ν are introduced of the following type. Outside of L let $x_\nu = x'_\nu$ everywhere; inside L, however, let $x_\nu \neq x'_\nu$ for at least a part of L and for at least one index ν. It is clear that by means of such a substitution it can be arranged that, at least for a part of L, $\gamma'_{\mu\nu} \neq \gamma_{\mu\nu}$. On the other hand, $\Theta'_{\mu\nu} = \Theta_{\mu\nu}$ everywhere; outside of L because, in this region, $x'_\nu = x_\nu$; inside of L, however, because, for this region, $\Theta_{\mu\nu} = 0 = \Theta'_{\mu\nu}$. It follows from this that, in the case considered, if all [coordinate –JS] substitutions are allowed as justified, more than one system of $\gamma_{\mu\nu}$ belongs to the given system of $\Theta_{\mu\nu}$.

If therefore—as is the case in this paper—one maintains the requirement that the $\gamma_{\mu\nu}$ should be completely determined by the $\Theta_{\mu\nu}$, then one is forced to restrict the choice of coordinate system (Einstein and Grossmann 1913, p. 260).[25]

Substantially the same argument is repeated in two subsequent papers, Einstein 1914a and Einstein and Grossmann 1914, both dating from early 1914. In a footnote to his paper, however, Einstein added a significant explanatory comment. To the equations $g'_{\mu\nu} = g_{\mu\nu}$ for the two metric tensors, which he notes, "surely do not all hold" ("sicherlich nicht alle erfüllt sein"), he appends a note: "The equations are to be thus understood, that on the left sides each of the independent variables x'_ν are assigned the same numerical values as the variables x_ν on the right sides" (Einstein 1914a, p. 178, column 2, footnote 1).[26]

A second version of the argument is presented in his definitive exposition of the Einstein–Grossmann theory, presented to the Berlin Academy in October, 1914 (Einstein 1914b). This version is much more explicit than the earlier one about the coordinate system used at each point in the argument. It may represent a significant evolution in Einstein's thinking about the "hole" argument, as Earman and Glymour (1978a) argue. At any rate, I shall base my discussion of the argument in the next section on this final version. After developing generally covariant formulations for the behavior of physical systems in the gravitational field, Einstein turned to the laws of the gravitational field itself. He wrote:

The epistemologically satisfying thing about the previously-developed theory lies in the fact that it fulfills the relativity principle in its most far reaching sense. Formally considered, this rests on the fact that the systems of equations are *generally* covariant, i.e., with respect to arbitrary substitutions of the x_ν.

Accordingly, this suggests the requirement that the differential laws for the $g_{\mu\nu}$ must also be *generally* covariant. However, we shall show that we must restrict this requirement if we want the law of causality to be completely valid. Namely, we prove that the laws that determine the course of events for the gravitational field cannot be *generally* covariant.

§12. *Proof of the Necessity of a Restriction of the Choice of Coordinates.*

We consider a finite portion Σ of the continuum, in which no material process occurs. What happens physically in Σ is then completely determined if the quantities $g_{\mu\nu}$ are given as functions of the coordinates x_ν with respect to a coordinate system K used for the description. The totality of these functions will be symbolically designated by $G(x)$.

Let a new coordinate system K' be introduced that, outside of Σ, coincides with K; within Σ, however, it diverges from K in such a way that the $g'_{\mu\nu}$ referred to

K', like the $g_{\mu\nu}$ (and their derivatives), are everywhere continuous. The totality of the $g'_{\mu\nu}$ will be symbolically designated by $G'(x')$. $G'(x')$ and $G(x)$ describe the same gravitational field. If we replace the coordinates x'_ν by the coordinates x_ν in the functions $g'_{\mu\nu}$, i.e., if we form $G'(x)$, then $G'(x)$ also describes a gravitational field with respect to K, which however does not correspond to the actual (i.e., originally given) gravitational field.

Now if we assume that the differential equations of the gravitational field are generally covariant, then they are satisfied by $G'(x')$ (with respect to K') if they are satisfied by $G(x)$ with respect to K. They are then also satisfied with respect to K by $G'(x)$. Relative to K there then exist the solutions $G(x)$ and $G'(x)$, which are different from each other, in spite of the fact that at the boundary of the region both solutions coincide; i.e., *what happens in the gravitational field cannot be uniquely determined by generally-covariant differential equations for the gravitational field.*

Therefore, if we demand that the course of events in the gravitational field be completely determined by the laws to be set up, then we are forced so to restrict the choice of coordinate systems that it is impossible, without abandonment of the restrictive conditions, to introduce a new coordinate system K' of the type previously characterized. The continuation of the coordinate system into the interior of a region Σ must not be arbitrary (Einstein 1914b, pp. 1066–1067).[27]

4. The Meaning of the Argument

In trying to understand these passages, we must remember that the modern terminology of differential geometry, which expresses geometrical concepts in coordinate-free language and carefully distinguishes between point and coordinate transformations, for example, was not available to Einstein (I do not know when these distinctions were first clearly enunciated by mathematicians working in differential geometry, but I have the impression it was much later).[28]

On the other hand, Einstein had an astounding geometrical intuition that usually led him, using coordinate language, to conclusions that make excellent geometrical sense. In trying to interpret these passages, then, I have proceeded on the assumption that he was trying to express something nontrivial. I also presume that, if the "hole" argument *had* been based on a trivial mathematical error, Marcel Grossmann—with whom he was still in close contact, as their second joint paper (Einstein and Grossmann 1914) proves—would have set him right. Yet, as noted in the previous section, this paper repeats the hole argument.

Thus, I am led to reject one common interpretation of the "hole" argument, which assumes that Einstein did not realize that the transformation of the components of the metric tensor under a coordinate transformation results in a redescription of the *same* gravitational field in a *different* coordinate system. This *might* hold as a critique of the first (1913) version of the argument (Einstein and Grossmann 1913, p. 260; see Section 3 of this paper).[29] But the second (1914) version *explicitly affirms* that "$G'(x')$ and $G(x)$ describe the same gravitational field" (Einstein 1914b, pp. 1066–1067; see Section 3 of this paper). So Einstein's point is *not* that $G'(x')$ and $G(x)$ are different. His point is that $G'(x)$ and $G(x)$—same x—are different: "$G'(x)$ also describes a gravitational field *with respect to*

K which however *does not correspond* to the actual (i.e., originally given) gravitational field" [emphasis added –JS].

What he is here asserting, I believe, is that (in more modern language) in a region of purely gravitational field the *dragged-along field* under a point transformation results in a gravitational field physically distinct from the original one.[30] And I further assert that the physical equivalence or nonequivalence of these two fields is *not* a purely mathematical question. Or, more accurately, it does not *become* one unless and until one introduces the additional, *nonmathematical* assumption or postulate that, in regions where no matter is present, the points of a manifold are physically differentiated only by the properties that they inherit from the metric field.

Since the explanation of this assertion does not really depend on the fact that it involves the gravitational field, nor on the fact that the latter is represented mathematically by a metric tensor field, I shall discuss the question in a wider context. I shall carry out much of the discussion in coordinate-free language, but a reader not used to dispensing with coordinates need merely interpret all statements as references to the components of the objects in question with respect to a single coordinate system.

Consider a class of theories, in which space-time is represented mathematically by a four-dimensional manifold, the points of which are associated with physical events (just *how* they represent such events will be the major question). Physical processes are represented by some set of geometrical object fields on the manifold, whose exact transformation properties under coordinate transformations need not concern us.[31] Now consider a point transformation that takes each point of the manifold (or some region of it) into another point (the transformation is assumed to be invertible). We may drag the coordinates along with the points, thus defining a new coordinate system, called the dragged-along coordinate system. We may also drag along a geometrical object field with such a point transformation; this produces a new geometrical object field, the components of which have values at each dragged-along point in the dragged-along coordinate system that are numerically equal to those of the original field at the original (undragged) point in the original coordinate system. This new field is called the dragged-along field (with respect to the point transformation). Since there is a (nondenumerable) infinity of possible point transformations on a manifold, there is a (nondenumerable) infinity of such dragged-along fields. Although all are examples of the same type of geometrical object, each is a geometrical object field mathematically distinct from the others. If there are two geometrical object fields on a manifold, one may be dragged along with a point transformation, while the other is left undragged, or left behind, as we shall say. Then we may refer to one field as dragged along relative to the other. As far as the relative effect goes, it does not really matter which field is regarded as dragged along and which as left behind. Clearly, if more than two fields are given, additional possibilities exist for dragging some and leaving others behind.

Suppose a set of geometrical object fields in some region of the manifold completely describes the physical situation in that region. Now drag along *all* of these fields with some point transformation restricted to this region. Should we regard the original and dragged-along fields as representing *distinct* physical situations in this region of space-time? The answer depends on whether we regard the points of the manifold as *physically* individuated apart from the properties they inherit from the presence of the geometrical object fields. If these fields really include *all* the information about the physical situation in the region, then such properties, which would enable us to distinguish physically one point of the region from another, cannot exist. On the other hand, if the points of the region *are* physically individuated in some independent fashion, the assertion that the original fields exhaust the physical properties of the region is incorrect. In the latter case, the point transformation dragged along all the fields, *but* left behind the distinguishing properties—and such a "relative dragging" is certainly of physical significance.

It may be the case that it is not the points directly, but other structures on the manifold that are regarded as labeled independently of all other physical processes. We shall refer to such structures as individuating structures or fields. For example, Newton's assertion of an absolute space amounts to the postulation of a preferred fibration of space-time. His assertion of an absolute time amounts to postulation of a preferred foliation of space-time. Together, they give an "absolute" individuation to the points of space-time (they provide much more of course). Relative dragging of other fields with respect to such individuating structures does have physical significance. But even the most confirmed Newtonian will have to admit that he or she does not wish to distinguish *physically* between mathematical situations in which *all* of the fields—the individuating fields as well as the other fields—are dragged along by the same point transformation.

One might be tempted to assert that coordinate-labeling systems provide such individuating fields for the points of the manifold. And so they do for *mathematical* properties, such as the differential-topological structure of the manifold. But the point is (no pun intended!) that no mathematical coordinate system is *physically* distinguished per se; and without such a distinction there is no justification for physically identifying the points of a manifold—which are mathematically homogeneous, anyway—as physical events in space-time. Thus, the mathematician will always correctly regard the original and the dragged-along fields as distinct from each other. But the physicist (or indeed anyone applying differentiable manifolds) must examine this question in a different light; and the answer will depend upon the means available for making extra-mathematical distinctions between the points of the manifold—or, more concisely upon the presence of individuating structures.

A look at the question of how the abstract or generic mathematical concept of a differentiable manifold can be reached by starting from concrete examples sheds a little more light on the problem of the (nonmathematical) individuation of the points of such a manifold. Riemann, in his *Habilitationsschrift* (Riemann 1854), introduced the concept of an "*n*-fold extended magnitude," which he also refers to as a "manifold" ("Mannigfaltigkeit"). He cited two examples of a "continuous manifold" that arise from the experiences of everyday life: "the positions of sense-objects and the colors [are –JS] indeed the only simple concepts whose modes of determination form a multiply-extended manifold" (Riemann 1854, p. 3). Of course, Riemann was not referring to "the manifold of physical colors [that] has infinitely many dimensions," as Weyl put it, but to the three-dimensional "space of the perceptively given *color-qualities* of colored light" (Weyl 1927, pp. 70–71).

Russell pointed out an important difference between these two examples: "There is no inherent quality in a single point [of physical space –JS], as there is in a single color, by which it can be quantitatively distinguished from another" (Russell 1897, pp. 66–67). Let us explore this difference. Each element within the set of colors is individuated by its qualities as a color. Three quantities can be attached to the colors in such a way that each color is uniquely characterized by a particular set of values of the three quantities. These values are taken to vary continuously, so that we may think of the colors as forming a space having three dimensions.[32] One may analyze the properties of this color space, introduce a metric on it, and so forth.[33] Color space is often referred to as the color manifold, following Riemann. As colors, the elements of this manifold are obviously not homogeneous. Only if we abstract its quality as a color from each of these elements, while retaining the continuity relations between these elements, are we left with a space of homogeneous elements, each of which may be identified with a point of a generic three-dimensional mathematical manifold.[34] Conversely, if we start from such a generic manifold and wish to turn it into a color manifold, we must introduce what I have previously called individuating structures or fields in order to restore the qualitative (and quantitative) distinctions between the elements—that is, just the qualities that make them colors and nothing else.

Now let us turn to physical space and time. Any sufficiently rich theoretical model of the world around us must include enough detail to individuate each event taking place in the model. Suppose the model can be represented mathematically as a four-dimensional differentiable manifold with certain additional structures over it.[35] In the full model, the points of the manifold represent events. It is obvious that if *all* the additional structures over the manifold are abstracted, the points of the manifold lose their character as events. How many of the additional structures can be abstracted without the points of the manifold entirely losing their individual spatio-temporal qualities? The answer to this question depends on the theoretical model. If one starts from Newton's model of absolute space and time, for example, all other physical processes may be abstracted, and one will still be left with physically individuated points of space and time. (As noted previously, absolute space and time constitute individuating structures for Newton's model of space-time.) One would have to abstract absolute space and time while retaining the continuity and differentiability properties they induce, in order to get a generic four-dimensional manifold, the elements of which are completely homogeneous.

Suppose one takes a Galilean or special-relativistic model; that is, a model that postulates the existence of a privileged group of nonaccelerated inertial frames of reference, each in uniform relative motion with respect to the others. One may still individuate the points of space and time; for example, by introducing a framework of unaccelerated rigid rods (or better, rods in rigid, unaccelerated motion), together with a set of clocks at rest at each point of the framework and then suitably synchronized.[36] Such a materialization of the abstract concept of inertial frame forms a significant part of Einstein's conception of the special theory of relativity. In 1907 he wrote:

> In order to be able to describe any physical process, we must be in a position to evaluate spatially and temporally the changes that take place at the individual points of space.
>
> For the spatial evaluation of a process of infinitely short duration taking place in an element of space (point-event) we use a Cartesian coordinate system, i.e., three perpendicular rigid rods, rigidly bound together, as well as a rigid unit rod. [In a footnote at this point, Einstein adds: "Instead of 'rigid' bodies, one ... can just as well speak of solid bodies not subject to deforming forces" –JS] ... For the temporal evaluation of a point-event we use a clock, which is at rest relative to the coordinate system and in the immediate neighborhood of which the point-event takes place...
>
> We now assume that *the clocks can be set so that the speed of propagation of any light signal in vacuum—measured with the aid of these clocks—is always equal to a universal constant c*, supposing that the coordinate system is not accelerated (Einstein 1907, p. 415).[37]

Thus, in abstracting physical structures from a Galilean or special-relativistic model of the world, one may begin by abstracting all dynamical processes and leaving this framework of rods and clocks, which constitutes an inertial reference system in Einstein's sense. The points of space and time are still physically individuated by this reference system. One would have to abstract the inertial reference system, while preserving the continuity and differentiability properties that are associated with such a system, in order to arrive at a generic four-dimensional differentiable manifold. Conversely, if we start with such a generic manifold and want to arrive at a Galilean or special-relativistic model of the world, there are now two steps involved. We first introduce a frame of reference (or some equivalent set of individuating structures or fields) in such a way that the individuation of each point of space-time is fixed. Then, all the dynamical physical processes in the model are introduced.

Einstein was well aware of the artificiality of the distinction thus set up between rods and clocks on the one hand, and all other physical processes on the other; yet he emphasized the need for such a division, given the current state of physics:

> It is striking that the theory [of special relativity –JS] (except for the four-dimensional space) introduces two kinds of physical things, i.e., (1) measuring rods and clocks, (2) all other things, e.g., the electromagnetic field, the material point, etc. This, in a certain sense, is inconsistent; strictly speaking, measuring rods and clocks should emerge as solutions of the basic equations ... not, as it were, as theoretically self-sufficient entities. The procedure justifies itself, however, because it was clear from the very beginning that the postulates of the theory are not strong enough to deduce from them equations for physical events sufficiently complete and sufficiently free from arbitrariness in order to base upon such a foundation a theory of measuring rods and clocks. If one did not wish to forego a physical interpretation of the coordinates in general (something that, in itself, would be possible), it was better to permit such

inconsistency—with the obligation, however, of eliminating it at a later stage of the theory (Einstein 1949, pp. 59, 61).[38]

So far Russell's distinction between the color manifold and physical space—which we have replaced by space-time—appears valid. Now consider a general-relativistic model. In this case, there are no nondynamical physical processes in the theory. In particular, for regions from which matter or nongravitational fields are absent, the only additional structure on the manifold is the metric tensor field. This case thus resembles the color manifold case. If we abstract the metric field while retaining the continuity and differentiability properties that it induces, we are left with a generic four-dimensional manifold that has no spatial and temporal properties associated with its points. The metric tensor field plays the role of both the dynamical and the individuating fields in the previous two examples.[39] One may still introduce a physical frame of reference, but only when the metric tensor field is present. More generally, spatio-temporal individuation of the points of the manifold in a general-relativistic model is possible only after the specification of a particular metric field, i.e., only after the field equations of the theory (which constitute its dynamical problem) have been solved. Once this is done, the points of the manifold-with-metric become full-fledged physical events, endowed with gravitational as well as spatio-temporal properties. There is no half-way house, so to speak, as there is in absolute, Galilean or special-relativistic theories, where some independent individuating structure gives the points of the manifold spatio-temporal properties (a certain point in space at a certain moment in time), before a particular solution to some set of dynamical equations (Maxwell's equations, for example) imparts additional properties to these points, making them into full-fledged physical events (the point in space where, and the moment in time when, the electric and magnetic field vectors assume certain values).

One way to obtain this individuation of the points of the manifold once the metric tensor is present is to imagine the space-time to be transversed by a number of trajectories of test particles, that is, particles of mass so small that their effect on the gravitational field in which they move may be neglected. Then the intersections of the trajectories of two such particles will constitute an event that serves to physically individuate a point of the manifold. As we shall see in the final section, this is the method that Einstein adopted at the end of 1915, once he had rejected the "hole" argument and adopted generally covariant field equations.

But now let us return to the general discussion of the argument. Suppose that the geometrical object fields are subject to generally-covariant field equations involving no additional fields. That is, no additional fields, given a priori, occur in the formulation of these equations. It follows that, if a particular solution to these field equations exists in some region of the manifold, then all the dragged-along field generated from this solution by point transformations in the region are also solutions to the field equations. Are these to be regarded as physically distinct solutions? Clearly, we are again faced with our previous question, and must give the same answer: Only if there exist some individuating fields for the points of the region. If such fields exist, then all of the solutions to the generally-covariant field equations generated from the initial one by dragging must be regarded as physically distinct, and the following version of the "hole" argument is valid.

Suppose we try to fix a unique solution to the generally-covariant field equations inside the region in question. Let us specify the sources outside the region, the values of the fields outside and on the boundary of the region, together with any finite number of normal derivatives on the boundary. We still cannot specify a unique solution inside the region, because there will always be point transformations inside the region that reduce to the identity on the boundary, together with any finite number of derivatives of the transformation, but which do not reduce to the identity inside the region. These point transformations give rise to a family of dragged-along solution fields that are identical to each other to the required differential order on the boundary, but differ inside the region.

Thus, if one requires field equations for which well-posed initial or boundary value problems can be set, with or without sources, there is only one way to escape the "hole" argument: to *identify* the class of mathematically distinct dragged-along solutions with *one* physical solution. This identification does not guarantee that a given set of generally covariant field equations has a well-posed initial or boundary value problem; but at least it allows for the possibility. As we have seen, however, the only warrant for such a physical identification of mathematically distinct fields is the denial of any a priori physical labeling of the points of the manifold. The points must inherit all of their distinguishing physical properties and relations from the set of fields that is subject to the generally covariant equations. No individuating fields can be allowed in such a theory.

We can now apply the results of this discussion to Einstein's "hole" argument. Einstein considered a region in which the metric tensor is the only field present. It obeys generally covariant field equations that do not involve any other fields in this region. Our general discussion of the "hole" argument, when applied to this situation, really has two aspects. The first is the problem of the physical identity of points of the manifold, and the consequent question of whether $G(x)$ and $G'(x)$ represent distinct gravitational fields, regardless of whether or not they satisfy certain field equations. This question is relevant whether the metric tensor is specified a priori, as, for example, in the special theory of relativity, or whether it is determined by field equations relating it to the sources of the gravitational field—the sort of theory for which Einstein was searching. The second aspect of the hole argument relates exclusively to the latter case, in which the components of the metric tensor are dynamical variables to be determined from a set of partial differential equations relating them to the components of the stress-energy tensor. (The argument is complicated, but not in any essential way, by the fact that, in many cases, the expressions for the stress-energy tensor components themselves involve the metric tensor.) Then it is indeed the case that any set of generally-covariant field equations that has $G(x)$ as a solution in some empty region of space-time will also have $G'(x)$ as a solution in that region. Nothing can prevent $G'(x)$ from being a solution to generally-covariant field equations for the metric tensor field if $G(x)$ is a solution. If the points of the manifold were in some way distinguished physically as events in space-time *before* the metric tensor field were determined, this would indeed imply that no criteria could be given to single out a *physically* unique solution to the field equations. This would be a serious problem, which might well be called a violation of the law of causality, as Einstein termed it in 1914. However, once it is stipulated that the points of the manifold have no physical properties except those that they inherit from the metric field, the problem disappears. $G(x)$ and $G'(x)$, together with all other mathematically distinct metric tensor fields that can be transformed into each other by being dragged along with a point transformation, form an *equivalence class* of solutions. This equivalence class of mathematically distinct metric tensor fields corresponds to *one* physical solution to the field equations—that is, to one gravitational field. There is no further problem of uniqueness or causality—or, more correctly, none that arises from general covariance.

Some modern treatments of general relativity discuss at least one aspect of this problem: the need to formulate the theory in such a way as to allow a unique solution to the Cauchy problem for the vacuum Einstein equations.[40] This is done by introducing the concept of isometric equivalence. Two differentiable manifolds with metric tensor fields are regarded as isometrically equivalent if there exists a sufficiently smooth invertible mapping (a diffeomorphism) between the manifolds that carries the metric tensor field of one into that of the other. Two metric fields on the *same* manifold are then isometrically equivalent if and only if one can be dragged into the other. A space-time may then be defined as a four-dimensional differentiable manifold with metric tensor field of Lorentz signature. (Additional global topological restrictions may be imposed, and the differentiability structure must be explicitly stated, but we forego discussion of such details.) A gravitational field is then defined as an equivalence class of isometrically equivalent space-times—isometric equivalence being an equivalence relation. Then the set of solutions to the field equations generated by continuing properly-specified Cauchy or initial-value data off a space-like hypersurface will all be isometrically equivalent, and thus correspond to a unique gravitational field.[41] The physical presupposition of this mathematical treatment—that the lack of independent physical individuation of the points of a manifold permits identification of isometrically equivalent space-times as one gravitational field—is not usually made explicit.

Einstein, of course, started from this second aspect of the "hole" argument: the lack of mathematically unique solutions to generally covariant field equations for the metric tensor field. But I maintain that he was initially unable to accept the idea that the points of an empty region of a manifold do not have any preexistent spatio-temporal individuality, apart from that which they inherit from the metric tensor field. It was this last remnant of his belief in the physical significance of coordinates that led Einstein to regard an important *mathematical* property of generally covariant field equations as the source of nonuniqueness in their *physical* solutions, making such equations physically unacceptable to him. He was able to accept generally covariant equations for the gravitational field only after he realized that mathematically distinct solutions to such equations could be regarded as physically equivalent if one gave up the idea of any inherent physical distinction between the points of a manifold.

Einstein had worked hard to give a physical significance to the coordinates in the special theory of relativity, and the existence of an individuation of the points of space-time by means of rods and clocks formed an integral part of his interpretation of the special theory, as we noted earlier in this paper. One presumes that it was difficult for him to unlearn this lesson, which had served him so well.

Letters that Einstein wrote to Besso and Ehrenfest after he had abandoned the "hole" argument, in which he explained to them just what was wrong with it, provide strong evidence in support of my interpretation. They are cited and discussed in the final section, to which the reader may now turn, if so minded.

5. Einstein and Fokker on Nordström's Theory

After the Einstein–Grossmann (1913) paper, Einstein's next significant references to the curvature tensor and general covariance occur in a paper that Einstein wrote early in 1914 together with a young Dutch physicist, Adriaan D. Fokker (Einstein and Fokker 1914). This paper is a discussion of a gravitational theory proposed by Gunnar Nordström, a Finnish physicist then working on the problem of finding a gravitational theory compatible with the principles of special relativity (Nordström 1913). Einstein and Fokker showed that Nordström's scalar theory can be rewritten as a metric theory of gravitation, using only generally-covariant tensor equations, if one assumes that the metric tensor is conformally related to the Minkowski metric tensor of special relativity. They phrased the latter requirement as follows: "Nordström's theory ... is based on the assumption that, by proper choice of the reference system, it is possible to satisfy the principle of the constancy of the velocity of light" (Einstein to Fokker 1914, p. 324).[42] Thus, they not only assumed the metric tensor to be conformally Minkowskian, but also adopted conformally (quasi)-Cartesian coordinates. With a few additional requirements, such as that the gravitational field equations depend only linearly on the stress-energy tensor and the second derivatives of the metric tensor, they were led to field equations of the form (in modern notation):

$$R = \kappa T,$$

where R is the Ricci scalar (the trace of the Ricci tensor, formed by contracting it with the metric), κ is the gravitational coupling constant, and T is the trace of the stress-energy tensor. They showed that this equation reduces to Nordström's gravitational field equation in conformally Cartesian coordinates.

The main interest of the paper for present purposes lies in some remarks in the concluding section. Einstein and Fokker first note that they have

> shown that, basing oneself upon the principle of the constancy of the velocity of light [i.e., the hypothesis of the conformally-Minkowskian nature of the metric tensor –JS], one can arrive at the Nordström theory by purely formal considerations, i.e., without having recourse to further physical hypotheses. It seems to us, therefore, that this theory merits preference over all others retaining this principle (Einstein and Fokker 1914, p. 328).[43]

What is responsible for this success? "We note that only the application of the theory of invariants of the absolute differential calculus [i.e., tensor analysis –JS] allowed us to give a clear insight into the formal content of Nordström's theory" (Einstein and Fokker 1914, p. 328).[44] This seems to be Einstein's first recorded acknowledgment of the impression made upon him by the power of formal mathematical methods in the search for a relativistic theory of gravitation, and later in the search for a unified field theory. (Another is cited in the final section.)

But of even more direct relevance is the final paragraph of the paper, and its closing footnote:

> Finally, the role that the Riemann–Christoffel [curvature –JS] tensor plays in the present investigation, suggests that it should also open the way to a derivation of the Einstein–Grossmann gravitational equations independent of physical assumptions. The proof of the existence or non-existence of such a connection would constitute important theoretical progress (Einstein and Fokker 1914, p. 328).[45]

This sentence, the last in the paper, is footnoted: "The argument given in §4, p. 36 [of Einstein and Grossmann 1913 –JS] ... for the non-existence of such a connection does not withstand a closer examination." (Einstein and Fokker 1914, p. 328).[46]

The reference is to the separatum version of the Einstein–Grossmann paper; it corresponds to the Ricci tensor argument on page 257 of the version in the *Zeitschrift für Mathematik und Physik*, translated in Section 2 herein. Neither here, nor anywhere else, have I found any explanation by Einstein of exactly why he and Fokker believed this argument to be invalid. I can offer an explanation consistent with my earlier interpretation of Einstein's and Grossmann's comments on the Ricci tensor, but I cannot cite any further evidence that it was Einstein's explanation. The Einstein–Fokker reformulation of the Nordström theory is based on the Ricci scalar for a conformally-Minkowskian metric. It is obvious that Nordström's theory has the correct static Newtonian limit. Einstein or Fokker may have noticed that this is also true of the Ricci scalar in the case of Einstein's *Ansatz* for the static metric tensor. Thus, a theory based on the trace of the Ricci tensor does yield the correct Newtonian limit, even with Einstein's *Ansatz*.

At any rate, Einstein did not follow up the question raised in this paper in the period immediately after his work on Nordström's theory. By November 1914 Einstein thought that he could derive the Einstein–Grossmann field equations uniquely by "purely formal considerations"; but these considerations did not involve the Riemann–Christoffel tensor. It was only the recognition of the fallaciousness of this derivation, coupled with other difficulties encountered by the Einstein–Grossmann theory, that finally led Einstein in late 1915 to reexamine the entire basis of that theory.

But before turning to the final chapter of the story, I shall discuss briefly a question that may already have occurred to the reader. If Einstein and Fokker were able to give a generally covariant formulation of Nordström's theory in 1914, why did Einstein not drop his "hole" argument immediately? This argument purports to prove the impossibility of a generally covariant theory of gravitation based exclusively on the metric tensor and satisfying Einstein's causality criterion; yet Nordström's theory appears to satisfy both of these conditions. Is there not a palpable contradiction here?

Actually, there is not. The reason is that, as Misner, Thorne, and Wheeler put it, Nordström's theory is based upon a "prior geometry," that is, to quote them, "an aspect of the geometry of space that is fixed immutably, i.e., that cannot be changed by changing the distributions of gravitating sources" (Misner et al. 1973, p. 429). In Nordström's theory, the conformal metric structure is fixed a priori to be *Minkowskian*; only the conformal factor is dynamically determined by the (trace of) the stress-energy tensor. Thus the space-time structure has a built-in "rigidity," which allows Einstein and Fokker to choose a family of quasi-Cartesian coordinates. So it is plausible to assume that Einstein realized that the "hole" argument does not apply to the Nordström theory. The existence of an additional element of "prior geometry" on the space-time manifold causes the argument to fail.

6. Covariance Regained

Einstein and Grossmann soon returned to the problem of finding all transformations under which their field equations are invariant.[47] They realized that the components of their gravitational stress energy complex only behave like the components of a tensor under linear transformations; but they now considered the requirement that it be a tensor "unjustified" for unspecified reasons (Einstein and Grossmann 1914, p. 218, footnote 1). Evidently, they regarded the invariance under nonlinear coordinate transformations of the field equations as much more important for the physical interpretation of the theory (see the quotation from Einstein and Grossmann 1914, p. 216, in endnote 6). They showed that their field equations could be derived from a variational principle, and investigated the maximal group of transformations under which this principle remains invariant. They found a condition on the coordinates that they took as a criterion for membership in the set of "adapted coordinates" ("angepasste Koordinaten"), i.e., coordinate systems in which the Einstein–Grossmann field equations hold. Transformations between such adapted coordinate systems were called "justified transformations" ("berechtigte Transformationen"). They showed that, while this group includes more than the linear transformations, it does not include all possible transformations of coordinates. (We shall refer to this set of transformations as the restricted covariance group.)

In March 1914 Einstein wrote to Besso triumphantly, explaining this new development, and adding:

> Now I am completely satisfied and no longer doubt the correctness of the entire system. ... The reasonableness of the thing is too obvious. ... The general theory of invariants only acted as a hindrance. The direct route proved to be the only accessible one. What is inconceivable is only that I had to feel my way so long before I found what lay at hand (Speziali 1972, p. 53).[48]

The problem of demonstrating the uniqueness of the field equations remained, that is, the problem of demonstrating that they are the only equations that are invariant under the restricted covariance group. Einstein thought he had provided such a proof in his next major paper on the subject, a complete review of his gravitational theory written soon after his move to Berlin (Einstein 1914b, presented to the Prussian Academy of Sciences on October 29—the paper in which he presented the final version of the "hole" argument). He gave an argument to demonstrate that the Lagrangian H, variation of which leads to the Einstein–Grossmann field equations, is uniquely fixed by the requirement that it be invariant under the restricted covariance group. He was evidently quite pleased with this accomplishment: "We have now arrived in a purely formal way, i.e., without direct appeal to our physical knowledge of gravitation, at completely determinate field equations" (Einstein 1914b, p. 1076).[49]

His triumph did not last very long. Difficulties in connection with the field equations and the variational principle began to accumulate. About a year later, shortly after these difficulties had been resolved, Einstein summarized some of them in a letter of November 28, 1915 to Sommerfeld:

> I realized ... that my previous gravitational field equations were completely untenable! This was indicated by the following points:
>
> 1) I proved that the gravitational field on a uniformly rotating system does not satisfy the field equations.
>
> 2) The motion of the perihelion of Mercury came out as $18''$ instead of $45''$ per century.
>
> 3) The covariance argument in my paper of last year [Einstein 1914b –JS] does not give the Hamiltonian function H [today usually called the Lagrangian function – JS]. When suitably generalized, it allows an arbitrary H. It followed that covariance relative to "adapted" coordinate systems was a wild goose chase.
>
> After all confidence thus had been lost in the results and methods of the earlier theory, I saw clearly that only through a connection with the general theory of covariants, i.e., with Riemann's covariant [tensor –JS], could a satisfactory solution be found. I have unfortunately immortalized the last errors in this battle in the papers of the Academy. (Hermann 1968, pp. 32–33)[50]

The last sentence is a reference to problems that Einstein encountered in the course of arriving at the generally-covariant form of the field equations first announced to the Prussian Academy of Sciences at its session of November 25, 1915 (see Einstein 1915d). This paper had been preceded during November by two papers that chronicle the tortuous path by which Einstein reached the field equations, including his errors.[51] We shall not go here into the details of this last phase of Einstein's peregrinations.

Einstein again wrote to Besso triumphantly, but in a rather different vein than in his previously-quoted letter of March 1914: "My wildest dreams have been fulfilled. *General* covariance. The perihelion motion of Mercury wonderfully exact. ... This time the most obvious thing was the correct one" (Einstein to Besso, December 10, 1915; Speziali 1972, p. 60).[52]

Starting with his paper of November 18, devoted to explaining the precession of Mercury's perihelion, a new phrase appears in Einstein's discussions of general covariance: "through which time and space are deprived of the last trace of objective reality" (Einstein 1915c, p. 831).[53] In his paper of November 25 he used a rather different formulation: "The relativity postulate in its most universal formulation, which makes the spatio-temporal coordinates into physically meaningless parameters" (Einstein 1915d, p. 847).[54] But in a letter to Schlick, he returned to words closer to his original phrasing:

> Thereby [through the general covariance of the field equations –JS] time and space lose the last remnant of physical reality. All that remains is that the world is to be conceived as a four-dimensional (hyperbolic) continuum of 4 dimensions. (Einstein to Moritz Schlick, December 14, 1915; EA 21-610)[55]

Similar wording occurs in Einstein's now-classic review and exposition of the final version of the theory, written early in 1916 (Einstein 1916). The phrase is quoted and the significance of his choice of words is discussed below.

Einstein still had to explain just what is wrong with the "hole" argument against general covariance, in which, as we have seen, he believed so strongly for over two years. He discussed this question in letters to Besso and Ehrenfest. (The reader may refer back to the translation of Section 12 of the Einstein–Grossmann "Entwurf" paper [Einstein and Grossmann 1913] in Section 3 for the version of the "hole" argument referred to in these letters.) On January 3, 1916, he wrote to Besso:

> Everything in the hole argument was correct up to the final conclusion. It has no physical content if, with respect to the *same* coordinate system K, two different solutions $G(x)$ and $G'(x)$ exist. To imagine two solutions simultaneously on the same manifold has no meaning and indeed the system K has no physical reality. The hole argument is replaced by the following consideration. Nothing is physically *real* but the totality of space-time point coincidences. If, for example, all physical happenings were to be built up from the motions of material points alone, then the meetings of these points, i.e., the points of intersection of their world lines, would be the only real things, i.e., observable in principle. These points of intersection obviously are preserved during all transformations [of coordinates –JS] (and no new ones occur) if only certain uniqueness conditions are preserved. It is therefore most natural to require of the laws that they determine no *more* than the totality of the space-time coincidences. From what has been said, this is already attained through the use of generally-covariant equations (Speziali 1972, pp. 63–64).[56]

In a letter to Ehrenfest a week earlier, Einstein had given a more detailed explanation of his new argument:

> In §12 of my paper of last year, everything is correct (in the first three paragraphs) up to the italicized part at the end of the third paragraph. Absolutely no contradiction to the uniqueness of events follows from the fact that both systems $G(x)$ and $G'(x)$ satisfy the conditions for the gravitational field with respect to the same reference system. The apparent compulsion of this argument disappears at once if one considers that 1) the reference system signifies nothing real, 2) that the (simultaneous) realization of

two different g-systems (better said, of two different grav[itational] fields) in the same region of the continuum is impossible by the nature of the theory.

The following consideration should replace §12. The physically real in what happens in the world (as opposed to what depends on the choice of the reference system) consists of *spatio-temporal coincidences*. [In a footnote, Einstein adds "and of nothing else!" –JS.] For example, the points of intersection of two world lines, or the assertion that they *do not* intersect, are real. Such assertions referring to the physically real are thus not lost because of any (single-valued) coordinate transformations. If two systems of $g_{\mu\nu}$ (or generally, of any variables applied in the description of the world) are so constituted that the second can be obtained from the first by a pure space-time transformation, then they are fully equivalent. For they have all spatio-temporal point coincidences in common, that is, everything that is observable. This argument shows at the same time how natural is the demand for general covariance (Einstein to Ehrenfest, December 26, 1915; EA 9-363).[57]

In Einstein's classic exposition of the general theory of relativity, written early in 1916, the "hole" argument naturally does not appear (Einstein 1916, submitted March 20, 1916). However the coincidence argument, developed in the two letters just quoted to replace the "hole" argument, is given early in the paper as the principal argument for general covariance:

The general laws of nature are to be expressed by equations that hold good for all systems of coordinates, that is, are covariant with respect to any substitutions whatever (generally covariant).

It is clear that a physical theory that satisfies this postulate will also be suitable for the general postulate of relativity. For the set of *all* substitutions in any case includes those which correspond to all relative motions of three-dimensional systems of co-ordinates. That this requirement of general covariance, which takes away from space and time the last remnant of physical objectivity, is a natural one, will be seen from the following reflexion. All our space-time verifications invariably amount to a determination of space-time coincidences. If, for example, events consisted only of the motion of material points, then ultimately nothing would be observable but the meetings of two or more of these points. Moreover, the results of our measurements are nothing but verifications of such meetings of the material points of our measuring instruments with other material points, or coincidences between the hands of a clock and points on the clock dial, and observed point-events happening at the same place at the same time.

The introduction of a system of reference serves no other purpose than to facilitate the description of the totality of such coincidences. We assign to the world four space-time variables x_1, x_2, x_3, x_4 in such a way that for every point-event there is a corresponding set of values of the variables $x_1 \ldots x_4$. To two coincident point-events there corresponds one set of values of the variables $x_1 \ldots x_4$, i.e., coincidence is characterized by the identity of the co-ordinates. If, in place of the variables $x_1 \ldots x_4$, we introduce functions of them x'_1, x'_2, x'_3, x'_4, as a new co-ordinate system, so that the sets of values are made to correspond to one another without ambiguity, the equality of all four co-ordinates in the new system will also serve as an expression of the space-time coincidence of the two point-events. As all our physical experience can be ultimately reduced to such coincidences, there is no immediate reason for preferring

> certain systems of co-ordinates to others, that is to say, we arrive at the requirement of general covariance. (Einstein 1916, pp. 776–777)[58]

The context in which Einstein presented the coincidence argument and referred to space and time as losing "their last remnant of physical reality" should now be clear. We see here the traces of his rejection of the "hole" argument and his concomitant acceptance of general covariance. They serve to emphasize his recognition that only some physical process can individuate the events that make up space-time, or "the world," as it was then often called, following Minkowski (1909). Removed from the context of the "hole" argument, Einstein's words seem to invite a purely operationalist interpretation. The emphasis on observability, and even the phrase "quantities observable in principle," used in his letter to Besso, were later employed in a totally different context in discussions of the quantum theory.[59] But for Einstein, I suggest, the coincidence argument was not primarily an element of a positivistic philosophical credo. Rather, it was an important link in the chain of arguments, by which he convinced himself that a manifold becomes a space-time with a certain gravitational field only after the specification of the metric tensor field; and that, prior to such a specification, there is no physical distinction between the elements of the manifold. Similarly, his comments about space and time losing their last vestiges of objective reality were not meant, I believe, to indicate that space and time have *no* physical reality, but that they no longer have any *independent* reality, apart from their significance as the spatial and temporal aspects of the metrical field. The language in which Einstein expresses his thoughts is not very precise, and—it must be admitted—certainly influenced current positivistic terminology. But as my earlier discussion indicates, his words were primarily intended to clarify physical problems connected with his work on a relativistic theory of gravitation.

Notes

[1] "1907 – Grundgedanke für die allgemeine Relativitätstheorie; 1912 – Erkenntnis der nicht-euklidischen Natur der Metrik und der physikalischen Bedingtheit derselben durch die Gravitation; 1915 – Feldgleichungen der Gravitation. Erklärung der Perihelbewegung des Merkur." The note was prepared for Erwin Freundlich's use in his book on Einstein's theory (Freundlich 1916).

[2] See Nathan and Norden 1960, section I, "The Reality of War (1914–1918)," for documentation of this period of Einstein's life.

[3] "Den entscheidenden Gedanken von der Analogie des mit der Theorie verbundenen mathematischen Problems mit der Gauss'schen Flächentheorie hatte ich allerdings erst 1912 nach meiner Rückkehr nach Zürich, ohne zunächst Riemanns und Riccis sowie Levi-Civitas Forschungen zu kennen. Auf diese wurde ich erst durch meinen Freund Grossmann in Zürich aufmerksam, als ich ihm das Problem stellte, allgemein kovariante Tensoren aufzusuchen, deren Komponenten nur von Ableitungen der Koeffizienten der quadratischen Fundamental-Invariante abhängen."

[4] See Einstein 1955, and Einstein to Walter Leich, April 24, 1950 (EA 60-252).

[5] See Reich 1979. I am also indebted to Dr. Alvin Jaeggli and Professor Res Jost, both of the ETH, for private communications about the contents of the Grossmann notebooks, which Einstein used to prepare for his examinations.

[6] A clear statement of this view occurs in Einstein and Grossmann 1914, p. 216: "The whole theory [of gravitation –JS] developed from the conviction that all physical processes in a gravitational field proceed in exactly the same way as the corresponding processes proceed without a gravitational field, if one refers them to a suitable accelerated (three-dimensional) coordinate system ("equivalence hypothesis"). This hypothesis, based upon the experimental fact of the equality of gravitational and inertial mass, is particularly convincing if the 'apparent' gravitational field, which exists with respect to the accelerated (three-dimensional) coordinate system, can be considered as a 'real' gravitational field, that is, if acceleration transformations (i.e., nonlinear transformations) are among the transformations permitted by the theory."

This point of view stands in contrast to the recent tendency to stress the nonvanishing of the Riemann tensor as the exclusive criterion for the presence of a "true" gravitational field. Einstein was aware of this viewpoint, but did not adopt it. In 1951, Max von Laue wrote to Einstein objecting to the latter's treatment of the rotating disc problem (see Stachel 1980). Von Laue noted that the Riemannian tensor vanishes on the disc. Einstein replied: "It is true that in that case the R_{iklm} [the components of the Riemann tensor –JS] vanish, so that one might say: 'There is no gravitational field present.' However, what characterizes the existence of a gravitational field from the empirical standpoint is the nonvanishing of the Γ^l_{ik} [the components of the affine connection –JS], not the nonvanishing of the R_{iklm}. If one does not think in such intuitive ways, one cannot comprehend why something like curvature should have anything to do with gravitation at all. In any case, no reasonable person would have hit upon something in that way. The key to the understanding of the equality of inertial and gravitational mass would have been missing" (Einstein to Max von Laue, September 1950; EA 16-148).

[7] "Almost" because of the difference between the Ricci tensor and what we now call the Einstein tensor, a difference that was to cause Einstein further grief when he returned to the generally covariant approach in late 1915. For discussion of this episode, see Stachel 1979a, 1979b; Earman and Glymour 1978a, 1978b; Hoffmann 1972; Lanczos 1972; Mehra 1973; Vizgin and Smorodinskiî 1979.

[8] "Die hervorragende Bedeutung dieser Begriffsbildungen für die *Differentialgeometrie* einer durch ihr Linienelement gegebenen Mannigfaltigkeit macht es a priori wahrscheinlich, daß diese allgemeinen Differentialtensoren auch für das Problem der Differentialgleichungen eines Gravitationsfeldes von Bedeutung sein dürften. Es gelingt in der Tat zunächst, einen kovarianten Differentialtensor zweiten Ranges und zweiter Ordnung G_{im} anzugeben, der in jene Gleichungen eintreten könnte, nämlich

$$G_{im} = \sum_{kl} \gamma_{kl}(ik, lm) = \sum_{k} \{ik, km\}."$$

[9] This sentence, crucial for my interpretation: "Allein es zeigt sich, dass sich dieser Tensor im Spezialfall des unendlich schwachen statischen Schwerefeldes *nicht* auf den Ausdruck $\Delta\varphi$ reduziert," is mistranslated by Mehra (1973, p. 102) as: "This in itself shows that this tensor, in the special case of an infinitely weak gravitational field, would not reduce to the expression $\Delta\varphi$," and the crucial word "static" is also omitted. This mistranslation has apparently influenced the reading of Earman and Glymour (1978a, p. 256), who remark:

"The fact that $R_{ij} = R^{k}_{ijk}$ in itself showed, according to Grossmann's reasoning, that R_{ij} cannot have the right Newtonian limit."

10 "Dieselbe Art der Degeneration zeigt sich bei dem statischen Schwerefelde der vorhin betrachteten Art, nur daß bei diesem $g_{44} = c^2$ eine Funktion von x_1, x_2, x_3 ist."

11 Hoffmann 1972, p. 159, gives the form of the Riemann and Ricci tensors for Einstein's 1912 static metric without indicating its possible relevance to understanding the Einstein–Grossmann paper.

12 "Das Erfreulichste ist das Stimmen der Perihelbewegung und die allgemeine Kovarianz, das Markwürdigste aber der Umstand, dass Newtons Theorie des Feldes schon in Gl. 1. Ordnung unrichtig ist (Auftreten der $g_{11} - g_{33}$). Nur der Umstand, das $g_{11} - g_{33}$ nicht in den ersten Näherungen der Bew. Gl. des Punktes auftreten, bedingt die Einfachheit von Newtons Theorie."

13 "Diesmal ist das Nächstliegende das Richtige gewesen; aber Grossmann u. ich glaubten die Erhaltungssätze nicht erfüllt seien, und das Newton'sche Gesetz in erster Näherung nicht herauskomme. Du wirst über das Auftreten der $g_{11} \ldots g_{33}$ überrascht sein." The reference to the conservation equations concerns another argument against general covariance that Einstein developed in 1913; it is discussed at the end of this section. Other aspects of this letter are discussed in the final section of this paper.

14 "Ein solcher Zusammenhang müßte vorhanden sein, sofern die Gravitationsgleichungen *beliebige* Substitutionen zuzulassen hätten; allein in diesem Falle scheint es ausgeschlossen zu sein, Differentialgleichungen *zweiter* Ordnung aufzufinden. Würde dagegen feststehen, daß die Gravitationsgleichungen nur eine gewisse Gruppe von Transformationen gestatten, so wäre es verständlich, wenn man mit den von der allgemeinen Theorie gelieferten Differentialtensoren nicht auskommt. Wie im physikalischen Teile ausgeführt ist, sind wir nicht imstande, zu diesen Fragen Stellung zu nehmen."

15 "Der Versuch einer Diskussion derartiger Möglichkeiten wäre aber bei dem gegenwärtigen Stande unserer Kenntnis der physikalischen Eigenschaften des Gravitationsfeldes verfrüht. Deshalb ist für uns die Beschränkung auf die zweite Ordnung geboten und wir müssen daher darauf verzichten, Gravitationsgleichungen aufzustellen, die sich beliebigen Transformationen gegenüber als kovariant erweisen. Es ist übrigens hervorzuheben, daß wir keinerlei Anhaltspunkte für eine allgemeine Kovarianz der Gravitationsgleichungen haben."

16 "Wir haben für die Gravitationsgleichungen nur beweisen können, daß sie beliebigen *linearen* Transformationen gegenüber kovariant sind; wir wissen aber nicht, ob es eine allgemeine Transformationsgruppe gibt, der gegenenüber die Gleichungen kovariant sind. Die Frage nach der Existenz einer derartigen Gruppe ... ist die wichtigste, welche sich an die hier gegebenen Ausführungen anknüpft."

17 "[wir sind] bei dem gegenwärtigen Stande der Theorie nicht berechtigt, die Kovarianz physikalischer Gleichungen beliebigen Substitutionen gegenüber zu fordern.

"Anderseits aber haben wir gesehen, daß sich eine Energie-Impuls-Bilanzgleichung für materielle Vorgänge hat aufstellen lassen ... welche beliebige Transformationen gestattet. Es scheint deshalb doch natürlich, wenn wir voraussetzen, daß alle physikalischen Gleichungssysteme mit Ausschluß der Gravitationsgleichungen so zu formulieren sind, daß sie beliebigen Substitutionen gegenüber kovariant sind. Die diesbezügliche Ausnahmestellung der Gravitationsgleichungen gegenüber allen anderen Systemen hängt nach meiner Meinung damit zusammen, daß nur erstere zweite Ableitungen der Komponenten des Fundamentaltensors enthalten dürften."

[18] "Man erkennt, daß Koordinatenlängen zugleich natürliche Längen sind ($dt = 0$). Maßstäbe erleiden also durch das 'Newtonsche' Gravitationsfeld keine Verzerrung."

[19] "Ich bin nun innerlich überzeugt, das Richtige getroffen zu haben, zugleich freilich auch, dass ein Murmeln der Entrüstung durch die Reihe der Fachgenossen gehen wird, wenn die Arbeit erscheint, was in wenigen Wochen der Fall sein wird... Die Überzeugung, zu der ich mich langsam durchgedrungen habe, ist die, *dass es bevorzugte Koordinatensysteme überhaupt nicht gibt.* Doch ist es mir nur teilweise gelungen, auch formal bis zu diesem Standpunkte vorzudringen."

[20] "Mit der Gravitationstheorie bin ich nun sehr zufrieden. Die Thatsache, dass die Gravitationsgleichungen nicht allgemein kovariant sind, welche mich vor einiger Zeit noch ungemein störte, hat sich als unumgänglich herausgestellt; es lässt sich einfach beweisen, dass eine Theorie mit allgemein kovarianten Gleichungen nicht existieren kann, falls verlangt wird, dass das Feld durch die Materie mathematisch vollständig bestimmt wird."

[21] "Die Gravitationsaffäre hat sich zu meiner *vollen Befriedigung* aufgeklärt (der Umstand nämlich, dass die Gleichungen des Gr. Feldes nur *linearen* Transformationen gegenüber kovariant sind). Es lässt sich nämlich beweisen, dass *allgemein kovariante* Gleichungen, die das Feld aus dem materiellen Tensor *vollständig* bestimmen, überhaupt nicht existieren können. Was kann es schöneres geben, als dies, das jene nötige Spezialisierung aus den Erhaltungssätzen fliesst? Nun bestimmen die Erhaltungssätze unter allen Flächen diejenigen, welche als Koordinatenflächen mit Vorzug zu wählen sind. Diese bevorzügten Flächen können wir als Ebenen bezeichnen, weil ja nur *lineare* Substitutionen als berechtigte übrig bleiben."

[22] "Für mich ist es absurd, dem 'Raum' physikalische Eigenschaften zuzuschreiben. Die Gesamtheit der Massen erzeugt ein $g_{\mu\nu}$-Feld (Gravitationsfeld), das seinerseits den Ablauf aller Vorgänge, auch die Ausbreitung der Lichtstrahlen und das Verhalten der Massstäbe und Uhren regiert. Das Geschehen wird zunächst auf vier *ganz willkürliche* raumzeitliche Variable bezogen. Diese müssen dann, wenn die Erhaltungssätzen des Impulses und der Energie genüge geleistet werden soll, derart spezialisiert werden, dass nur (ganz) *lineare* Substitutionen von einem berechtigten Bezugssystem zu einem andern führen. Das Bezugssystem ist der bestehenden Welt mit Hilfe des Energiesatzes sozusagen angemessen und verliert seine nebulose apriorische Existenz."

Herneck (1963) dated the letter to the turn of the years 1911–1912 or 1912–1913. The attribution to the turn of some year is made rather certain by the closing salutation, "With best wishes for the New Year;" but what year is left open. The years suggested by Herneck seem to be ruled out by the content of the letter, which clearly refers to the work contained in Einstein and Grossmann 1913. The point of view expressed concerning the conservation equations is closest in content to the remarks on p. 178 of Einstein 1914a, which is dated January 1914. This leads me to suggest the end of 1913 as the correct date for the letter.

[23] "Es hat sich durch eine allgemeine Überlegung zeigen lassen, dass Gleichungen, welche das Gravitationsfeld vollständig bestimmen, nicht beliebige Substitutionen gegenüber kovariant sein können."

[24] "In den letzten Tagen fand ich den Beweis dafür, daß eine derartige allgemein kovariante Lösung überhaupt nicht existieren kann."

[25] "Es gebe in unserer vierdimensionalen Mannigfaltigkeit einen Teil L, in welchem ein 'materieller Vorgang' nicht stattfinde, in welchem also die $\Theta_{\mu\nu}$ verschwinden. Durch die außerhalb L gegenbenen $\Theta_{\mu\nu}$ sind gemäß unserer Annahme überall, also auch im Innern von L die $\gamma_{\mu\nu}$ vollkommen bestimmt. Wir denken uns nun statt der ursprünglichen

Koordinaten x_ν neue Koordinaten x'_ν eingeführt von folgender Art. Außerhalb L sei überall $x_\nu = x'_\nu$; innerhalb L aber sei wenigstens für einen Teil von L und wenigstens für einen Index ν $x_\nu \neq x'_\nu$. Es ist klar, daß durch eine derartige Substitution erreicht werden kann, daß wenigstens für einen Teil von L $\gamma'_{\mu\nu} \neq \gamma_{\mu\nu}$ ist. Andererseits ist überall $\Theta'_{\mu\nu} = \Theta_{\mu\nu}$, nämlich außerhalb L, weil für dieses Gebiet $x'_\nu = x_\nu$ ist, innerhalb L aber, weil für dies Gebiet $\Theta_{\mu\nu} = 0 = \Theta'_{\mu\nu}$ ist. Hieraus folgt, daß in dem betrachteten Falle, wenn alle Substitutionen als berechtigte zugelassen werden, zu dem nämlichen System der $\Theta_{\mu\nu}$ mehr als ein System der $\gamma_{\mu\nu}$ gehört.

"Wenn also—wie dies in der Arbeit geschehen ist—an der Forderung festgehalten wird, daß durch die $\Theta_{\mu\nu}$ die $\gamma_{\mu\nu}$ vollständig bestimmt sein sollen, so ist man genötigt, die Wahl des Bezugssystems einzuschränken."

[26] "Die Gleichungen sind so zu verstehen, daß auf den linken Seiten jeweilen den unabhängigen Variabeln x'_ν, dieselben Zahlenwerten erteilt werden wie auf den rechten Seiten den Variabeln x_ν."

[27] "Das erkenntnistheoretisch Befriedigende der bisher entwickelten Theorie liegt darin, daß dieselbe dem Relativitätsprinzip in dessen weitgehendster Bedeutung Genüge leistet. Dies beruht, formal betrachtet, darauf, daß die Gleichungssysteme *allgemein*, d.h. beliebigen Substitutionen der x_ν gegunüber, kovariant sind.

"Es scheint hiernach die Forderung geboten, daß auch die Differentialgesetze für die $g_{\mu\nu}$ *allgemein* kovariant sein müssen. Wir wollen aber zeigen, daß wir diese Forderung einschränken müssen, wenn wir dem Kausalgesetz vollständig Genüge leisten wollen. Wir beweisen nämlich, daß Gesetze, welche den Ablauf des Geschehens im Gravitationsfelde bestimmen, unmöglich *allgemein* kovariant sein können.

§12. *Beweis von der Notwendigkeit einer Einschränkung der Koordinatenwahl.*

"Wir betrachten einen endlichen Teil Σ des Kontinuums, in welchem ein materieller Vorgang nicht stattfindet. Das physikalische Geschehen in Σ ist dann vollständig bestimmt, wenn in bezug auf ein zur Beschreibung benutztes Koordinatensystem K die Größen $g_{\mu\nu}$ als Funktion der x_ν gegeben werden. Die Gesamtheit dieser Funktionen werde symbolisch durch $G(x)$ bezeichnet.

"Es werde ein neues Koordinatensystem K' eingeführt, welches außerhalb Σ mit K übereinstimme, innerhalb Σ aber von K abweiche, derart, daß die auf K' bezogenen $g'_{\mu\nu}$ wie die $g_{\mu\nu}$ (nebst ihren Ableitungen) überall stetig sind. Die Gesamtheit der $g'_{\mu\nu}$ bezeichnen wir symbolisch durch $G'(x')$. $G(x')$ und $G(x)$ beschreiben das nämliche Gravitationsfeld. Ersetzen wir in den Funktionen $g'_{\mu\nu}$ die Koordinaten x'_ν durch die Koordinaten x_ν, d.h. bilden wir $G'(x)$, so beschreibt $G'(x)$ ebenfalls ein Gravitationsfeld bezüglich K; welches aber nicht übereinstimmt mit dem tatsächlichen (bzw. ursprünglich gegebenen) Gravitationsfelde.

"Setzen wir nun voraus, daß die Differentialgleichungen des Gravitationsfeldes allgemein kovariant sind, so sind sie für $G'(x')$ erfüllt (bezüglich K'), wenn sie bezüglich K für $G(x)$ erfüllt sind. Sie sind dann also auch bezüglich K für $G'(x)$ erfüllt. Bezüglich K existierten dann die voneinander verschiedenen Lösungen $G(x)$ und $G'(x)$, trotzdem an den Gebietsgrenzen beide Lösungen übereinstimmten, d.h. *durch allgemein kovariante Differentialgleichungen für das Gravitationsfeld kann das Geschehen in demselben nicht eindeutig festgelegt werden.*

"Verlangen wir daher, daß der Ablauf des Geschehens im Gravitationsfelde durch die aufzustellenden Gesetze vollständig bestimmt sei, so sind wir genötigt, die Wahl des Koordinatensystems derart einzuschränken, daß es ohne Verletzung der einschränkenden Bedingungen unmöglich ist, ein neues Koordinatensystem K' von der vorhin charakterisierten

Art einzuführen. Die Fortsetzung des Koordinatensystems ins Innere eines Gebietes Σ hinein darf nicht willkürlich sein."

[28] For a history of tensor calculus, see Reich 1979. For a history of differential geometry, see Coolidge 1940, Book III. For a modern account of tensor analysis, see, for example, Bishop and Goldberg 1968. For textbooks of relativity based on the modern approach to differential geometry, see, for example, Hawking and Ellis 1973 and Sachs and Wu 1977. In this section, I have used many concepts and definitions from Schouten 1954, although it is not entirely modern in approach.

[29] However, a reading of the essential passage in this version of the argument in the light of Einstein's footnote to the "hole" argument in Einstein 1914a, p. 178 (also cited in Section 2), makes me dubious that Einstein ever committed this trivial error.

[30] For the definition of the dragged-along field of a geometrical object, see, for example, Schouten 1954, pp. 102–103. A brief discussion is given later in this section.

[31] A geometrical object, a generalization of a tensor, may be defined informally as an object whose components are determined in every coordinate system once its components are specified in one coordinate system. For example, an affine connection is a geometrical object but not a tensor. For a formal definition, see Schouten 1954, p. 57.

[32] Each color may be characterized by hue, brightness, and saturation, for example. More commonly, three primary colors are selected, and other colors are characterized by the amounts of each that must be mixed to produce a particular color.

[33] Schrödinger carried out an extensive analysis of color space and, following Helmholtz, introduced such a metric. See Schrödinger 1920, 1926. For a more recent discussion, see Weinberg 1976.

[34] I shall ignore as irrelevant for present purposes such questions as the global topology of the color manifold, the existence of a boundary, and so forth.

[35] In this case, as in the case of color space, we leave aside all questions as to whether the events in the world, or the colors we perceive, actually vary continuously and differentiably. We confine our attention to theoretical models that have such continuity and differentiability properties, without prejudice to the question of whether models without such properties may ultimately be required.

[36] Clearly, other methods exist for such an individuation, using various combinations of test particles, light rays, clocks, and other devices.

[37] "Um irgendeinen physikalischen Vorgang beschreiben zu können; müssen wir imstande sein, die in den einzelnen Punkten des Raumes stattfindenden Veränderungen örtlich und zeitlich zu werten.

"Zur örtlichen Wertung eines in einem Raumelement stattfindenden Vorganges von unendlich kurzer Dauer (Punktereignis) bedürfen wir eines Cartesischen Koordinatensystems, d.h. dreier aufeinander senkrecht stehender, starr miteinander verbundener, starrer Stäbe, sowie eines starren Einheitsmaßstabes. [In the footnote: "Statt von 'starren' Körpern, könnte ... ebenso gut von deformierenden Kräften nicht unterworfenen festen Körpern gesprochen werden."] ... Für die zeitliche Wertung eines Punktereignisses bedienen wir einer Uhr, die relativ zum Koordinatensystem ruht und in deren unmittelbarer Nähe das Punktereignis stattfindet...

"Wir nehmen nun an, *die Uhren können so gerichtet werden, daß die Fortpflanzungsgeschwindigkeit eines jeden Lichtstrahles in Vakuum—mit Hilfe dieser Uhren gemessen—*

allenthalben gleich einer universellen Konstante c wird, vorausgesetzt, daß das Koordinatensystem nicht beschleunigt ist." See also Einstein 1905, part I, §1.

[38] "Es fällt auf, dass die Theorie (ausser dem vierdimensionalen Raum) zweierlei physikalische Dinge einführt, nämlich (1) Masstäbe und Uhren, (2) alle sonstigen Dinge, z.B. das elektromagnetische Feld, den materiellen Punkt, usw. Dies ist in gewissem Sinne inkonsequent; Masstäbe und Uhren müssten eigentlich als Lösungen der Grundgleichungen (Gegenstände bestehend aus bewegten atomistischen Gebilden) dargestellt werden, nicht als gewissermassen theoretisch selbstständige Wesen. Das Vorgehen rechtfertigt sich aber dadurch, dass von Anfang an klar war, dass die Postulate der Theorie nicht stark genug sind, um aus ihr genügend vollständige Gleichungen für das Physikalische Geschehen genügend frei von Willkür zu deduzieren, um auf eine solche Grundlage eine Theorie der Masstäbe und Uhren zu gründen. Wollte man nicht auf eine physikalische Deutung der Koordinaten überhaupt verzichten (was an sich möglich wäre), so war es besser, solche Inkonsequenz zuzulassen—allerdings mit der Verpflichtung, sie in einem späteren Stadium der Theorie zu eliminieren."

Einstein had given a less explicit formulation of the same distinction in 1921: "The concept of measuring body, as well as the concept of measuring clock coordinated to it in the theory of relativity, does not correspond exactly to any object in the real world. It is also clear that the rigid body and the clock do not play the role of irreducible elements in the conceptual structure of physics, but the role of composite structures, which should not play an independent role in the construction of theoretical physics. However, it is my conviction that these concepts, at the present stage of development of theoretical physics, still must be introduced as independent concepts" (Einstein 1921, p. 127). ["Der Begriff des Meßkörpers sowie auch der ihm in der Relativitätstheorie koordinierte Begriff der Meßuhr findet in der wirklichen Welt kein ihm exakt entsprechendes Objekt. Auch ist klar, daß der feste Körper und die Uhr nicht die Rolle von irreduzibeln Elementen im Begriffsgebäude der Physik spielen, sondern die Rolle von zusammengesetzten Gebilden, die im Aufbau der theoretischen Physik keine selbständige Rolle spielen dürfen. Aber es ist meine Überzeugung, daß diese Bergriffe beim heutigen Entwicklungsstadium der theoretischen Physik noch als selbständige Begriffe herangezogen werden müssen."]

[39] For example, in the generic case, the points of the manifold are completely individuated by the values of the four nonvanishing invariants of the Riemann tensor, as Kretschmann (1917) first pointed out. We omit consideration of complications that arise, for example, in case the metric tensor field has certain symmetries. In this case, additional structures are necessary to physically individuate points on an orbit of the symmetry group.

[40] See, for example, Hawking and Ellis 1973, p. 56 and chapter 7; or Sachs and Wu 1977, pp. 27, 262.

[41] We omit numerous details and mathematical complications in this sketch. For a complete discussion see Hawking and Ellis 1973, chapter 7.

[42] "Der Nordströmschen Theorie ... liegt die Annahme zugrunde, daß es möglich sei, durch passende Wahl des Bezugssystems dem Prinzip von der Konstanz der Lichtgeschwindigkeit zu genügen."

[43] "gezeigt ... daß man bei Zugrundelegung des Prinzips von der Konstanz der Lichtgeschwindigkeit durch rein formale Erwägungen, d.h. ohne Zuhilfenahme weiterer physikalischen Hypothesen zur Nordströmschen Theorie gelangen kann. Es scheint uns deshalb, daß diese Theorie allen anderen Gravitationstheorien gegenüber, die an diesem Prinzip festhalten, den Vorzug verdient."

[44] "Wir bemerken, daß nur die Verwendung der Invariantentheorie des absoluten Differentialkalküls uns eine klare Einsicht in den formalen Inhalt der Nordströmschen Theorie zu geben vermag."

[45] "Endlich legt die Rolle, welche bei der vorliegenden Untersuchung der Riemann–Christoffelsche Differentialtensor spielt, den Gedanken nahe, daß er auch für eine von physikalischen Annahmen unabhängige Ableitung der Einstein–Großmannschen Gravitationsgleichugen einen Weg öffnen würde. Der Beweis der Existenz oder Nichtexistenz eines derartigen Zusammenhanges würde einen wichtigen theoretischen Fortschritt bedeuten."

[46] "Die in §4, s. 36, des 'Entwurfs einer verallgemeinerten Relativitätstheorie' angegebene Begründung für die Nichtexistenz eines derartigen Zusammenhanges hält einer genaueren Überlegung nicht stand."

[47] Their work, published in Einstein and Grossmann 1914, was presumably well under way, if not completed, by the time Einstein left Zurich for Berlin in April 1914. Einstein explained the principal results of the paper in a letter written to Besso earlier in March, which is quoted below.

[48] "Nun bin ich vollkommen befriedigt und zweifle nicht mehr an der Richtigkeit des Systems. ... Die Vernunft der Sache ist zu evident ... Die allgemeine Invariantentheorie wirkte nur als Hemmnis. Der direkte Weg erwies sich als der einzig gangbare. Unbegreiflich ist nur, dass ich solange tasten musste, bevor ich das Nächstliegende fand."

[49] "Wir sind nun auf rein formalem Wege, d.h. ohne direkte Heranziehung unserer physikalischen Kentnisse von der Gravitation, zu ganz bestimmten Feldgleichungen gelangt."

[50] "Ich erkannte ... dass meine bisherigen Feldgleichungen der Gravitation gänzlich haltlos waren! Dafür ergaben sich folgende Anhaltspunkte:

1) Ich bewies, dass das Gravitationsfeld auf einem gleichförmig rotierenden System den Feldgleichungen nicht genügt.

2) Die Bewegung des Merkur-Perihels ergab sich zu $18''$ statt $45''$ pro Jahrundert.

3) Die Kovarianzbetrachtung in meiner Arbeit vom letzten Jahre liefert die Hamilton-Funktion H nicht. Sie lässt, wenn sie sachgemäss verallgemeinert wird, ein beliebiges H zu. Daraus ergab sich, dass die Kovarianz bezüglich 'angepasster' Koordinatensysteme ein Schlag ins Wasser war.

"Nachdem so jedes Vertrauen in Resultate und Methode der früheren Theorie gewichen war, sah ich klar, dass nur durch einen Anschluss an die allgemeine Kovariantentheorie, d.h. an Riemanns Kovariante, eine befriedigende Lösung gefunden werden konnte. Die letzten Irrtümer in diesem Kampfe habe ich leider in den Akademie-Arbeiten ... verewigt."

[51] See Einstein 1915a, submitted on November 4, and Einstein 1915b, submitted on November 11. Einstein 1915c, submitted on November 18, is primarily devoted to an explanation of the precession of Mercury's perihelion. In a footnote, it announces that a forthcoming communication will remove the last restriction on general covariance in his earlier papers. This is an allusion to Einstein 1915d, which was submitted a week later.

[52] "Die kühnste Träume sind nun in Erfüllung gegangen. *Allgemeine* Kovarianz. Perihelbewegung des Merkur wunderbar genau. ... Diesmal ist das Nächstliegende das richtige gewesen." This quotation has been cited more fully in Section 1.

[53] "durch welche Zeit und Raum der letzten Spur objectiver Realität beraubt werden."

[54] "Das Relativitätspostulat in seiner allgemeinsten Fassung, welches die Raumzeitkoordinaten zu physikalisch bedeutungslosen Parametern macht." I thank Professor Roberto Torretti for pointing out that Einstein used words similar to these late in 1913, just before he formulated the "hole" argument (see Einstein 1913b, p. 1257). But he did so in the course of a discussion of the effect of the gravitational field on other physical processes, which he treated by generally-covariant equations. He further noted that the gravitational problem would be satisfactorily solved if the gravitational equations were also generally covariant, but that this is not the case. In addition, there is a difference between stating that the coordinates are in themselves meaningless, and stating that space and time have lost the last trace of objective reality. The latter statement first occurs late in 1915, as noted in the text.

[55] "Dadurch verlieren Zeit und Raum den letzten Rest von physikalischer Realität. Es bleibt nur übrig, dass die Welt als vierdimensionales (hyperbolisches) Kontinuum von 4 Dimensionen aufzufassen ist."

[56] "An der Lochbetrachtung war alles richtig bis auf den letzten Schluss. Es hat keinen physikalischen Inhalt, wenn inbezug auf *dasselbe* Koordinatensystem K zwei verschiedene Lösungen $G(x)$ und $G'(x)$ existieren. Gleichzeitig zwei Lösungen in dieselbe Mannigfaltigkeit hineinzudenken, hat keinen Sinn und das System K hat ja keine physikalische Realität. Anstelle der Lochbetrachung tritt folgende Überlegung. *Real* ist physikalisch nichts als die Gesamtheit der raum-zeitlichen Punktkoinzidenzen. Wäre z.B. das physikalische Geschehen aufzubauen aus Bewegungen materieller Punkte allein, so wären die Begegnungen der Punkte, d.h. die Schnittpunkte ihrer Weltlinien das einzig Reale, d.h. prinzipiell beobachtbare. Diese Schnittpunkte bleiben natürlich bei allen Transformationen erhalten (und es kommen keine neuen hinzu), wenn nur gewisse Eindeutigkeitsbedingungen gewahrt bleiben. Es ist also das natürlichste, von den Gesetzen zu verlangen, dass sie nicht *mehr* bestimmen als die Gesamtheit der zeiträumlichen Koinzidenzen. Das wird nach dem Gesagten bereits durch allgemein kovariante Gleichungen erreicht."

[57] "In §12 meiner Arbeit von letzten Jahre ist alles richtig (in den ersten 3 Absätzen) bis auf das am Ende des dritten Absatzes gesperrt Gedruckte. Daraus, dass die beiden Systeme $G(x)$ und $G'(x)$, auf das gleiche Bezugssystem bezogen, den Bedingungen des Grav. Feldes genügen, folgt noch gar kein Widerspruch gegen die Eindeutigkeit des Geschehens. Das scheinbar Zwingende dieser Überlegung geht sofort verloren, wenn man bedenkt, dass 1) das Bezugssystem nichts Reales bedeutet 2) dass die (gleichzeitige) Realisierung zweier verschiedener g-Systeme (besser gesagt zweier verschiedener Grav. Felder) in demselben Bereich des Kontinuums der Natur der Theorie nach unmöglich ist.

"An Stelle des §12 hat folgende Überlegung zu treten. Das physikalisch Reale an dem Weltgeschehen (im Gegensatz zu dem von der Wahl des Bezugssystem Abhängigen) besteht *in raumzeitlichen Koinzidenzen.* [In the footnote: "und in nichts anderem!"] Real sind z.B. die Schnittpunkte zweier verschiedener Weltinien, bezw. die Aussage, dass sie einander *nicht* schneiden. Diejenigen Aussagen, welche sich auf das physikalisch-Reale beziehen, gehen daher durch keine (eindeutige) Koordinatentransformationen verloren. Wenn zwei Systeme der $g_{\mu\nu}$ (bezw. allg. der zur Beschreibung der Welt verwandten Variabeln) so beschaffen sind, dass man das zweite aus dem ersten durch blosse Raum-Zeit-Transformation erhalten kann, so sind sie völlig gleichbedeutend. Denn sie haben alle zeiträumlichen Punktkoinzidenzen gemeinsam, d.h. alles Beobachtbare.

"Diese Überlegung zeigt zugleich wie natürlich die Forderung der allgemeinen Kovarianz ist."

[58] *"Die allgemeinen Naturgesetze sind durch Gleichungen auszudrücken, die für alle Koordinatensysteme gelten, d.h. die beliebigen Substitutionen gegen-über kovariant (allgemein kovariant) sind.*

"Est ist klar, daß eine Physik, welche diesem Postulat genügt, dem allgemeinen Relativitätspostulat gerecht wird. Denn in *allen* Substitutionen sind jedenfalls auch diejenigen enthalten, welche allen Relativbewegungen der (dreidimensionalen) Koordinatensysteme entsprechen. Daß diese Forderung der allgemeinen Kovarianz, welche dem Raum und der Zeit den letzten Rest physikalischer Gegenständlichkeit nehmen, eine natürliche Forderung ist, geht aus folgender Überlegung hervor. Alle unsere zeiträumlichen Konstatierungen laufen stets auf die Bestimmung zeiträumlicher Koinzidenzen hinaus. Bestände beispielsweise das Geschehen nur in der Bewegung materieller Punkte, so wäre letzten Endes nichts beobachtbar als die Begegnungen zweier oder mehrerer dieser Punkte. Auch die Ergebnisse unserer Messungen sind nichts anderes als die Konstatierung derartiger Begegnungen materieller Punkte unserer Maßstäbe mit anderen materiellen Punkten bzw. Koinzidenzen zwischen Uhrzeigern, Zifferblattpunkten und ins Auge gefaßten, am gleichen Orte und zur gleichen Zeit stattfindenden Punktereignissen.

"Die Einführung eines Bezugssystems dient zu nichts anderem als zur leichteren Beschreibung der Gesamtheit solcher Koinzidenzen. Man ordnet der Welt vier zeiträumliche Variable x_1, x_2, x_3, x_4 zu, derart, daß jedem Punktereignis ein Wertesystem der Variablen $x_1 \ldots x_4$ entspricht. Zwei koinzidierenden Punktereignissen entspricht dasselbe Wertesystem der Variablen $x_1 \ldots x_4$; d.h. die Koinzidenz ist durch die Übereinstimmung der Koordinaten charakterisiert. Führt man statt der Variablen $x_1 \ldots x_4$ beliebige Funktionen derselben, x'_1, x'_2, x'_3, x'_4 als neues Koordinatensystem ein, so daß die Wertesysteme einander eindeutig zugeordnet sind, so ist die Gleichheit aller vier Koordinaten auch im neuen System der Ausdruck für die raumzeitliche Koinzidenz zweier Punktereignisse. Da sich alle unsere physikalischen Erfahrungen letzten Endes auf solche Koinzidenzen zurückführen lassen, ist zunächst kein Grund vorhanden, gewisse Koordinatensysteme vor anderen zu bevorzugen, d.h. wir gelangen zu der Forderung der allgemeinen Kovarianz."

[59] For a discussion of Einstein's use of the concept of observability in principle and its later use in quantum theory, see Stachel 1986a, pp. 369–374.

References

Bishop, Richard L., and Goldberg, Samuel I. (1968). *Tensor Analysis on Manifolds*. New York and London: Macmillan.

Coolidge, Julian Lowell (1940). *A History of Geometrical Methods*. Oxford: Clarendon Press.

Earman, John and Glymour, Clark (1978a). "Lost in the Tensors: Einstein's Struggle with Covariance Principles 1912–1916." *Studies in History and Philosophy of Science* 9: 251–278.

—— (1978b). "Einstein and Hilbert: Two Months in the History of General Relativity." *Archive for History of Exact Sciences* 19: 291–308.

Einstein, Albert (1905). "Zur Elektrodynamik bewegter Körper." *Annalen der Physik* 17: 891–921.

—— (1907). "Über das Relativitätsprinzip und die aus demselben gezogenen Folgerungen." *Jahrbuch der Radioaktivität und Elektronik* 4: 411–462.

—— (1912a). "Lichtgeschwindigkeit und Statik des Gravitationsfeldes." *Annalen der Physik* 38: 355–369.

—— (1912b). "Zur Theorie des statischen Gravitationsfeldes." *Annalen der Physik* 38: 443–458.

—— (1913a). "Zum gegenwärtigen Stande des Gravitationsproblems." *Physikalische Zeitschrift* 14: 1249–1260.

—— (1913b). "Physikalische Grundlagen einer Gravitationstheorie." *Naturforschende Gesellschaft Zürich. Vierteljahrsschrift* 58: 284–290.

—— (1914a). "Prinzipielles zur verallgemeinerten Relativitätstheorie und Gravitationstheorie." *Physikalische Zeitschrift* 15: 176–180.

—— (1914b). "Die formale Grundlage der allgemeinen Relativitätstheorie." *Königlich Preussische Akademie der Wissenschaften* (Berlin). *Sitzungsberichte*: 1030–1085.

—— (1915a). "Zur allgemeinen Relativitätstheorie." *Königlich Preussische Akademie der Wissenschaften* (Berlin). *Sitzungsberichte*: 778–786.

—— (1915b). "Zur allgemeinen Relativitätstheorie (Nachtrag)." *Königlich Preussische Akademie der Wissenschaften* (Berlin). *Sitzungsberichte*: 799–801.

—— (1915c). "Erklärung der Perihelbewegung des Merkur aus der allgemeinen Relativitätstheorie." *Königlich Preussische Akademie der Wissenschaften* (Berlin). *Sitzungsberichte*: 831–839.

—— (1915d). "Die Feldgleichungen der Gravitation." *Königlich Preussische Akademie der Wissenschaften* (Berlin). *Sitzungsberichte*: 844–847.

—— (1916). "Die Grundlage der allgemeinen Relativitätstheorie." *Annalen der Physik* 49: 769–822. Reprinted as a separatum Leipzig: Johann Ambrosius Barth, 1916. Quotations are from the English translation "The Foundation of the General Theory of Relativity." In H.A. Lorentz et al. *The Principle of Relativity: A Collection of Original Memoirs on the Special and General Theory of Relativity*. W. Perrett and G.B. Jeffery, trans. London: Methuen, 1923; reprint New York: Dover, 1952, pp. 109–164. (Translation modified.)

—— (1917). *Über die spezielle und die allgemeine Relativitätstheorie* (*Gemeinverständlich*). Braunschweig: Friedrich Vieweg und Sohn.

—— (1921). "Geometrie und Erfahrung." *Preussische Akademie der Wissenschaften* (Berlin). *Sitzungsberichte*: 123–130.

—— (1922). "Vorwort des Autors zur Tschechischen Ausgabe." In *Einstein a Praha*, J. Bičak, ed. Prague: Prometheus, 1979, p. 42; written in 1922 for a Czech edition of Einstein 1917.

—— (1949). "Autobiographical Notes." In *Albert Einstein: Philosopher-Scientist*, Paul Arthur Schilpp, ed. Evanston, Illinois: The Library of Living Philosophers, pp. 1–95. A corrected text and translation has been issued as *Albert Einstein: Autobiographical Notes*, Paul Arthur Schilpp, ed. La Salle, Illinois and Chicago: Open Court, 1979. Page numbers are cited from the 1949 edition, with translations slightly modified using the 1979 edition.

—— (1955). "Erinnerungen–Souvenirs." *Schweizerische Hochschulzeitung* 28 (*Sonderheft*): 145–148, 151–153. Reprinted as "Autobiographische Skizze." In *Helle Zeit-Dunkle Zeit*, Carl Seelig, ed. Zurich: Europa Verlag, 1956, pp. 9–17.

Einstein, Albert and Fokker, Adriaan D. (1914). "Die Nordströmsche Gravitationstheorie vom Standpunkt des absoluten Differentialkalküls." *Annalen der Physik* 44: 321–328.

Einstein, Albert and Grossmann, Marcel (1913). *Entwurf einer verallgemeinerten Relativitätstheorie und einer Theorie der Gravitation. I. Physikalischer Teil von Albert Einstein. II. Mathematischer Teil von Marcel Grossmann.* Leipzig and Berlin: B.G. Teubner. Reprinted with added "Bemerkungen," *Zeitschrift für Mathematik und Physik* 62 (1914): 225–261. Page numbers are cited from the latter version.

—— (1914). "Kovarianzeigenschaften der Feldgleichungen der auf die verallgemeinerte Relativitätstheorie gegründeten Gravitationstheorie." *Zeitschrift für Mathematik und Physik* 63: 215–225.

Freundlich, Erwin (1916). *Die Grundlagen der Einsteinschen Gravitationstheorie.* Berlin: Julius Springer.

Hawking, Stephen W. and Ellis, George F.R. (1973). *The Large-Scale Structure of Space-Time.* Cambridge: Cambridge University Press.

Hermann, Armin, ed. (1968). *Albert Einstein–Arnold Sommerfeld. Briefwechsel.* Basel and Stuttgart: Schwabe.

Herneck, Friedrich (1963). "Zum Briefwechsel Albert Einsteins mit Ernst Mach (Mit zwei unveröffentlichten Einstein–Briefe)." *Forschungen und Fortschritte* 37: 239–243.

Hoffmann, Banesh (1972). "Einstein and Tensors." *Tensor* 26: 157–162.

Kretschmann, Erich (1917). "Über den physikalischen Sinn der Relativitätspostulate, A. Einsteins neue und seine ursprügliche Relativitätstheorie." *Annalen der Physik* 53: 575–614.

Lanczos, Cornelius (1972). "Einstein's Path from Special to General Relativity." In *General Relativity: Papers in Honor of J.L. Synge*, L. O'Raifeartaigh, ed. Oxford: Claredon Press, pp. 5–19.

Mehra, Jagdish (1973). "Einstein, Hilbert, and the Theory of Gravitation." In *The Physicist's Concept of Nature*, Jagdish Mehra, ed. Boston and Dordrecht: D. Reidel, pp. 92–178.

Minkowski, Hermann (1909). "Raum und Zeit." *Physikalische Zeitschrift* 10:104–111.

Misner, Charles W., Thorne, Kip S., and Wheeler, John Archibald (1973). *Gravitation.* San Francisco: W.H. Freeman.

Nathan, Otto and Norden, Heinz, eds. (1960). *Einstein on Peace.* New York: Simon and Schuster.

Nordström, Gunnar (1913). "Zur Theorie der Gravitation von Standpunkt des Relativitätsprinzips." *Annalen der Physik*, 42: 533–554.

Reich, Karin (1979). "Einstein's mathematische Voraussetzungen," Section 4.1 of "Die Entwicklung der Tensorkalküls." Habilitationsschrift. Munich.

Riemann, Bernhard (1854). "Über die Hypothesen, welche der Geometrie zu Grunde liegen." *Königliche Gesellschaft der Wissenschaften zu Göttingen. Abhandlungen* 13 (1867): 133–152. Page numbers are cited from Bernhard Riemann, *Ueber die Hypothesen, welche der Geometrie zu Grunde liegen*, Hermann Weyl, ed. Berlin: Springer, 1919. An English translation by Henry S. White is found in *A Source Book in Mathematics*, Vol. 2. D.E. Smith, ed. New York and London: McGraw-Hill, 1929, pp. 411–425.

Russell, Bertrand (1897). *An Essay on the Foundations of Geometry*. Cambridge: Cambridge University Press; reprint New York: Dover, 1956.

Sachs, Rainer K. and Wu, Hung-Hsi (1977). *General Relativity for Mathematicians*. New York, Heidelberg, and Berlin: Springer-Verlag.

Schouten, Jan Arnoldus (1954). *Ricci-Calculus: An Introduction to Tensor Analysis*, 2nd. ed. Berlin, Göttingen, and Heidelberg: Springer-Verlag; London and New York: Lange, Maxwell and Springer.

Schrödinger, Erwin (1920). "Grundlinien einer Theorie der Farbenmetrik im Tagessehen." *Annalen der Physik* 63: 397–456, 481–520.

—— (1926). "Die Gesichtsempfindungen." In *Müller–Pouillets Lehrbuch der Physik*, 11th ed. O. Lummer, ed. Vol. 2, *Lehre von der strahlende Energie (Optik)*. Braunschweig: Friedrich Vieweg & Sohn, pp. 456–560.

Speziali, Pierre, ed. (1972). *Albert Einstein–Michele Besso. Correspondance, 1903–1955*. Paris: Hermann.

Stachel, John (1979a). "Einstein's Odyssey." *The Sciences* 19: 14–15, 32–34. [See this volume, pp. 225–232].

—— (1979b). "The Genesis of General Relativity." In *Einstein Symposion Berlin*, H. Nelkowski et al., eds. Berlin, Heidelberg, and New York: Springer-Verlag, pp. 428–442 [See this volume, pp. 233–244].

—— (1980). "Einstein and the Rigidly Rotating Disk." In *General Relativity and Gravitation One Hundred Years After the Birth of Albert Einstein*, Vol. 1. A. Held, ed. New York: Plenum, pp. 1–15. [See this volume, pp. 245–260].

—— (1986a). "Einstein and the Quantum: Fifty Years of Stuggle." In *From Quarks to Quasars: Philosophical Problems of Modern Physics*, Robert G. Colodny, ed. Pittsburgh: University of Pittsburgh Press, pp. 349–385. [See this volume, pp. 367–402].

—— (1986b). "What a Physicist Can Learn from the History of Einstein's Discovery of General Relativity." In *Proceedings of the Fourth Marcel Grossmann Meeting on General Relativity*, Remo Ruffini, ed. Amsterdam: Elsevier, pp. 1857–1862.

Vizgin, Vladimir P. and Smorodinskiĭ, Y.A. (1979). "From the Equivalence Principle to the Equations of Gravitation." *Soviet Physics. Uspekhi* 22: 489–513.

Weinberg, Joseph (1976). "The Geometry of Color." *General Relativity and Gravitation* 7: 135–169.

Weyl, Hermann (1927). *Philosophie der Mathematik und Naturwissenschaft*. Handbuch der Philosophie. Munich: R. Oldenbourg. Quotations are from the revised and augmented English edition, based on a translation by Olaf Helmer, *Philosophy of Mathematics and Natural Science*. Princeton: Princeton University Press, 1949.

Belated Decision in the Hilbert–Einstein Priority Dispute

Leo Corry, Jürgen Renn, and John Stachel

According to the commonly accepted view, David Hilbert completed the general theory of relativity at least 5 days before Albert Einstein submitted his conclusive paper on this theory on 25 November 1915. Hilbert's article, bearing the date of submission 20 November 1915 but published only on 31 March 1916, presents a generally covariant theory of gravitation, including field equations essentially equivalent to those in Einstein's paper. A close analysis of archival material reveals that Hilbert did not anticipate Einstein. The first set of proofs of Hilbert's paper shows that the theory he originally submitted is not generally covariant and does not include the explicit form of the field equations of general relativity.

It took Einstein 8 years, from 1907 to 1915, to complete the general theory of relativity, based on the field equations

$$R_{\mu\nu} = -\kappa \left(T_{\mu\nu} - \frac{1}{2} g_{\mu\nu} T \right), \tag{1}$$

where $g_{\mu\nu}$ is the metric tensor representing the gravitational potentials, $R_{\mu\nu}$ is the Ricci tensor, κ is a constant, $T_{\mu\nu}$ is the stress-energy tensor of matter, and T is its trace. The principal difficulty he had to overcome was finding the right balance between the mathematical implications of a generalized principle of relativity and physical requirements such as the existence of a Newtonian limit.(1,2) Hilbert, on the other hand, only began to work seriously on gravitation in mid-1915. Concerning physics, his interests had focused since the end of 1912 on the structure of matter, and in particular, since mid-1913, on Gustav Mie's special-relativistic electromagnetic theory of matter. Then, after Einstein's visit to Göttingen in the summer of 1915, Hilbert attempted to forge a synthesis between Mie's theory and Einstein's approach to gravitation based on $g_{\mu\nu}$.(3,4)

Science
Vol. 278, pp. 1270–1273

A recent comprehensive Einstein biography, which shows promise of becoming the standard reference, offers a succinct summary of the presently accepted account of the almost simultaneous formulation of the field equations by Einstein and Hilbert:[(5)]

> In the decisive phase [of work on general relativity] Einstein even had a congenial colleague, though this caused him more annoyance than joy, as it seemed to threaten his primacy. "Only one colleague truly understood it, and he now tries skillfully to 'nostrify' [that is, appropriate] it," he complained to [Heinrich] Zangger about what he evidently regarded as an attempt at plagiarism. This colleague was none other than David Hilbert... What must have irritated Einstein was that Hilbert had published the correct field equations first—a few days before Einstein. [...]
>
> In November, when Einstein was totally absorbed in his theory of gravitation, he essentially corresponded only with Hilbert, sending Hilbert his publications and, on November 18, thanking him for a draft of his article. Einstein must have received that article immediately before writing this letter. Could Einstein, casting his eye over Hilbert's paper, have discovered the term which was still lacking in his own equations [the trace term $-1/2g_{\mu\nu}R$],[(6)] and thus 'nostrified' Hilbert?

Fölsing is convinced, in agreement with the presently accepted view among physicists and historians of science, that Einstein's and Hilbert's achievements were actually parallel and independent, with the priority in submitting the field equations in their final form going to Hilbert.[(7)] If, however, the standard account were correct, it would seem quite possible that indeed Einstein "nostrified" from Hilbert the critical trace term, still missing from the field equations in the paper he submitted on 11 November.[(8)] Hilbert's published paper is mathematically complex and might have been difficult for Einstein to fully digest so quickly, but it does clearly display the field equations of general relativity, including the critical trace term (see reference 3, p. 404). Although this possible conclusion from the accepted view usually is not drawn, the arguments by which Einstein is exculpated are rather weak, turning on his slowness in fully grasping Hilbert's mathematics.[(5)]

In the course of a project on the history of general relativity at the Max Planck Institute for the History of Science, archival work by Corry brought to light a hitherto unnoticed set of proofs of Hilbert's paper.[(9)] Detailed analysis and comparison of these proofs with both published versions of Hilbert's paper[(10)] and with Einstein's papers on gravitation from 1913 to 1915[(2)] enabled us to construct an account of the crucial weeks in November 1915 that radically differs from the standard view, excludes the possibility that Einstein plagiarized from Hilbert the last crucial step in completing general relativity, and sheds new light on Einstein's complaint of "nostrification" by Hilbert.

Both the proofs[(9)] and the final version of Hilbert's first communication[(3)] are dated "submitted on 20 November 1915," presumably referring to the original manuscript. A copy of the proofs, preserved in his archives and marked in his own hand "First proofs of my first note," bears a printer's stamp dated 6 December 1915 (Fig. 1). However, the cover of the issue in which the heavily revised published version appeared is dated 31 March 1916. Its first note cites Einstein's conclusive paper, in which he reached the final form of his generally covariant

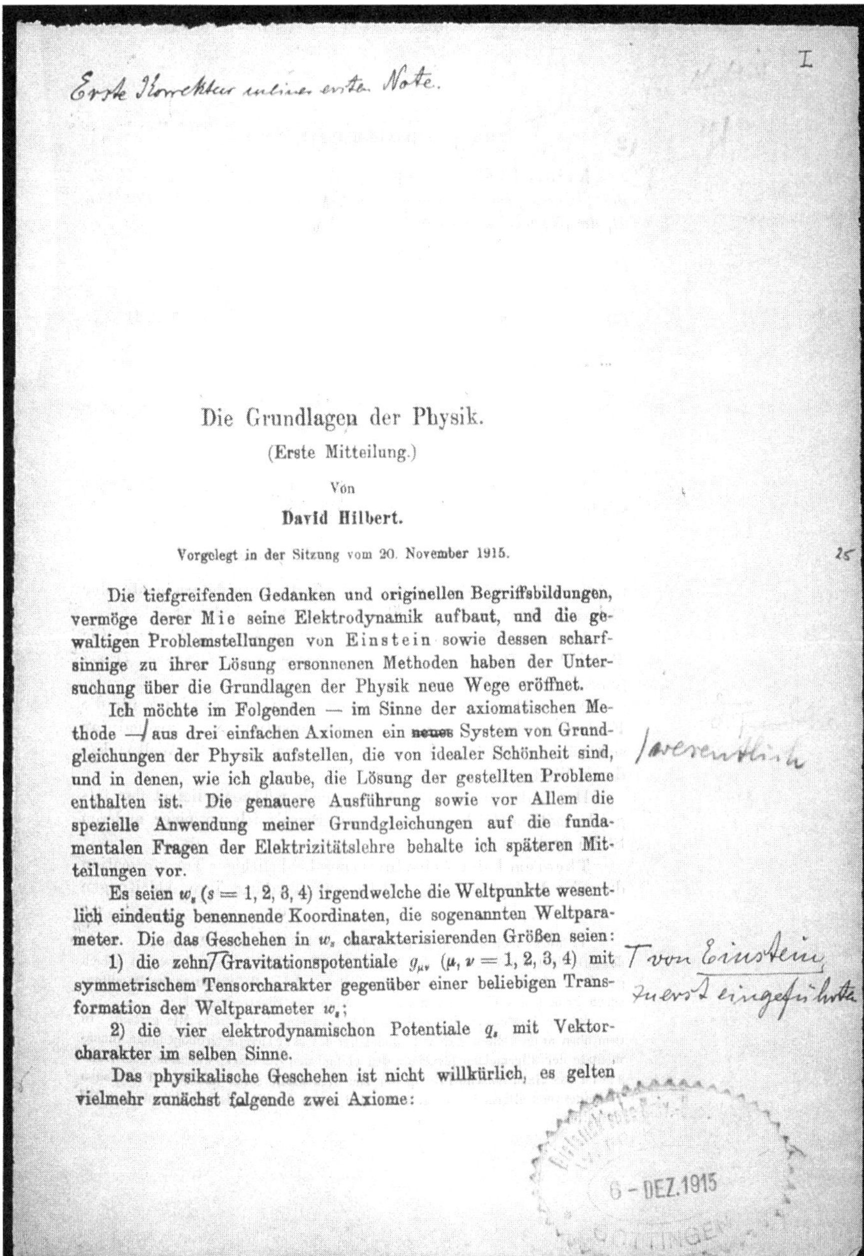

Erste Korrektur meiner ersten Note.

I

Die Grundlagen der Physik.

(Erste Mitteilung.)

Von

David Hilbert.

Vorgelegt in der Sitzung vom 20. November 1915.

Die tiefgreifenden Gedanken und originellen Begriffsbildungen, vermöge derer Mie seine Elektrodynamik aufbaut, und die gewaltigen Problemstellungen von Einstein sowie dessen scharfsinnige zu ihrer Lösung ersonnenen Methoden haben der Untersuchung über die Grundlagen der Physik neue Wege eröffnet.

Ich möchte im Folgenden — im Sinne der axiomatischen Methode — aus drei einfachen Axiomen ein ~~neues~~ System von Grundgleichungen der Physik aufstellen, die von idealer Schönheit sind, und in denen, wie ich glaube, die Lösung der gestellten Probleme enthalten ist. Die genauere Ausführung sowie vor Allem die spezielle Anwendung meiner Grundgleichungen auf die fundamentalen Fragen der Elektrizitätslehre behalte ich späteren Mitteilungen vor.

Es seien w_s $(s = 1, 2, 3, 4)$ irgendwelche die Weltpunkte wesentlich eindeutig benennende Koordinaten, die sogenannten Weltparameter. Die das Geschehen in w_s charakterisierenden Größen seien:

1) die zehn Gravitationspotentiale $g_{\mu\nu}$ $(\mu, \nu = 1, 2, 3, 4)$ mit symmetrischem Tensorcharakter gegenüber einer beliebigen Transformation der Weltparameter w_s;

2) die vier elektrodynamischen Potentiale q_s mit Vektorcharakter im selben Sinne.

Das physikalische Geschehen ist nicht willkürlich, es gelten vielmehr zunächst folgende zwei Axiome:

(Marginal handwriting: "wesentlich"; "von Einstein zuerst eingeführten"; "25")

Figure 1: The first page of a set of proofs of Hilbert's first communication, with Hilbert's handwritten corrections and a printer's stamp, dated 6 December 1915. (Reproduced with permission by the American Association for the Advancement of Science.)

theory,(11) submitted on 25 November 1915 and published on 2 December 1915. Thus, Hilbert could have revised his paper in response to Einstein's work.

Differences between the proofs and this published version of Hilbert's paper confirm this view. Two of the differences, in particular, are fundamental.

1) In the proofs, Hilbert asserts that his theory cannot be generally covariant. In addition to 10 generally covariant equations, there must be four additional noncovariant equations to guarantee causality:(12)

> Since our mathematical theorem shows that the previous axioms I and II can only provide ten essentially independent equations for the 14 potentials [of gravitation and electromagnetism]; and further, maintaining general covariance makes quite impossible more than ten essentially independent equations for the 14 potentials $g_{\mu\nu}, q_s$; then, in order to keep the deterministic character of the fundamental equations of physics, in correspondence with Cauchy's theory of differential equations [that is, to have a well-posed Cauchy problem], the requirement of four further non-invariant equations to supplement [the generally covariant gravitational equations] is unavoidable. In order to find these equations I start out by setting up a definition of the concept of energy.

After setting up an equation that he calls "the energy theorem," Hilbert introduces a third axiom:(13)

> Axiom III (axiom of space and time). The space-time coordinates are those specific world parameters for which the energy theorem . . . is valid.
>
> Using this axiom, space and time provide in fact a labelling of the world points for which the energy principle is valid.
>
> The validity of [the energy] equation . . . is a consequence of axiom III: these four differential equations . . . supplement the [generally covariant] gravitational equations . . . to yield a system of 14 equations for the 14 potentials $g^{\mu\nu}$, q_s: *the system of the fundamental equations of physics.*

Note that Hilbert distinguishes here between the world parameters, which are arbitrary, and the space-time coordinates, which are not. Here he follows Einstein's earlier argument against general covariance from his papers of 1913 to 1915. Einstein then justified his theory's lack of general covariance and the need to select "adapted coordinate systems" on the same grounds of causality and energy-momentum conservation.

Hilbert abandoned this entire argument in the published version of his first communication. In a letter to Felix Klein,(14) he remarked with regard to the role of the energy theorem in the proof version of his theory, "But I later suppressed the whole thing because the thing did not appear mature to me." In his second communication, published in 1917,(15) Hilbert gave a radically different definition of causality for a generally covariant theory, essentially the one accepted today.(16) He there critically notes the noncovariant nature of Einstein's earlier work, characteristic also of his own version in the proofs: "In his original, now abandoned theory Einstein indeed postulated four non-invariant equations for the $g_{\mu\nu}$ in order to save the causality principle in its old form."(15) Hilbert's revised definition of causality for generally covariant theories led him to explicitly reject the possibility of the coordinates being physically significant. In the 1924 version of his first communication, he described axiom II, the requirement of general covariance, for the first time as "the simplest mathematical expression for the requirement that the coordinates in themselves have no sort of physical significance ..." (see reference 10, p. 4).

2) In the proofs of his first communication, Hilbert's world function includes a gravitational term $\sqrt{g}K$[17] and indicates that the gravitational part of the field equations takes the form of the variational derivative of the gravitational term with respect to the metric. Hilbert does not, however, give the explicit form of this gravitational part of the field equations. In the published version, on the other hand, he explicitly writes down the expression for the variational derivative (see reference 3, p. 404)

$$[\sqrt{g}K]_{\mu\nu} = \sqrt{g} \cdot \left(K_{\mu\nu} - \frac{1}{2} K g_{\mu\nu} \right), \tag{2}$$

justifying his expression by the argument (see reference 3, pp. 404–405)

> ... which follows easily without calculation from the fact that, except for $g_{\mu\nu}$, $K_{\mu\nu}$ [the Ricci tensor] is the only tensor of second order and K [the trace of the Ricci tensor] is the only invariant that can be constructed from only the $g^{\mu\nu}$ and its first and second order partial derivatives...

This argument is, however, untenable because there are many other tensors of second order and many other invariants that can be constructed from the Riemann tensor; even if one requires linearity in the Ricci tensor, the crucial coefficient of the trace term remains undetermined by this argument. In the 1924 republication, he dropped this argument, replacing it with an outline of how to calculate the gravitational term (see reference 10, p. 7):

> In order to determine the expression $[\sqrt{g}K]_{\mu\nu}$ one first specializes the coordinate system in such a way that at the world point being considered all the $g_s^{\mu\nu}$ [the derivatives of the metric tensor] vanish. One finds in this way [our Eq. 2].

To summarize: Initially, Hilbert did not give the explicit form of the field equations; then, after Einstein had published his field equations, Hilbert claimed that no calculation is necessary; finally, he conceded that one is. Taken together, this sequence suggests that knowledge of Einstein's result may have been crucial to Hilbert's introduction of the trace term into his field equations.

In the light of this analysis of Hilbert's work, we can now better understand the exchange between Hilbert and Einstein in the crucial days of November 1915. On the 14th, Hilbert wrote to Einstein, inviting him to come to Göttingen 2 days later, when Hilbert intended to lecture on "my axiomatic solution of your great problem." In a postscript, he added, "Insofar as I understand your new paper, the solution given by you is completely different from mine ...."[18] Hilbert is referring to Einstein's communication of 4 November to the Prussian Academy of Sciences. On the 15th, Einstein replied,[19] excusing himself from coming on grounds of being overworked, expressing great interest in Hilbert's work and asking for a copy of Hilbert's paper as soon as possible "to satisfy my impatience."

Hilbert must have sent the requested copy or a summary of his paper immediately because, on the 18th, Einstein replied, reacting sharply to Hilbert's claim of originality. Far from thanking him for sending his communication, as Fölsing claims, Einstein began his letter by denying the novelty of Hilbert's approach: "The system given by you agrees—as far as I can see—exactly with that which I found in recent weeks and submitted to the Academy.[20] In order to claim his priority, he explained to Hilbert that he had "considered the only possible generally covariant field equations three years earlier." He also insinuated that Hilbert had not even discussed the fundamental physical problems raised by these equations:[20]

> The difficulty was not to find generally covariant equations for the $g^{\mu\nu}$; this is easy with the help of the Riemann tensor. What was difficult instead was to recognize that these equations form a generalization, and, that is, a simple and natural generalization of Newton's law.

Einstein's claim is understandable. In his 4 November paper, he had announced with a flourish his return to the Riemann tensor as the appropriate starting point for a theory of gravitation.[21] Although the theory in this paper is different from the version of Hilbert's theory that he saw, as well as from the version of the theory Einstein developed by the 18th, they are all based on the metric tensor and the only generally covariant tensor that can be built from it, the Riemann tensor, as their common foundation.

It was after this exchange with Hilbert that Einstein wrote a friend charging Hilbert with "nostrification." Einstein's letter of 18 November may have been the motive for a reputed apologetic letter (now lost) by Hilbert to Einstein (see reference 7, p. 261) and Hilbert's handwritten note, added to the proofs of 6 December, supplementing his initial reference to the gravitational potentials $g_{\mu\nu}$ with the phrase "first introduced by Einstein" (Fig. 1).

In any case, by 20 December 1915, that is, before the appearance of Hilbert's final version, Einstein's anger had subsided to the point that he offered to Hilbert a reconciliation:[22]

> There has been a certain resentment between us, the cause of which I do not want to analyze any further. I have fought against the feeling of bitterness associated with it, and with complete success. I again think of you with undiminished kindness and I ask you to attempt the same with me. It is objectively a pity if two guys that have somewhat liberated themselves from this shabby world are not giving pleasure to each other.

In the printed version of his paper, Hilbert added a reference to Einstein's conclusive paper and a concession of the latter's priority: "The differential equations of gravitation that result are, as it seems to me, in agreement with the magnificent theory of general relativity established by Einstein in his later papers" (see reference 3, p. 404). If Hilbert had only altered the dateline to read "submitted on 20 November 1915, revised on [any date after 2 December 1915, the date of publication of Einstein's conclusive paper]," no later priority question could have arisen.

References and Notes

1. The conclusive paper is A. Einstein, *Sitzungsber. Preuss. Akad. Wiss.* **1915**, 844 (1915).
2. For a historical review, based on joint work with M. Janssen, J. Norton, and J. Stachel, see J. Renn and T. Sauer, *Phys. Bl.* **52**, 865 (1996).
3. The conclusive paper is D. Hilbert, *Nach. Ges. Wiss. Göttingen* **1916**, 395 (1916).
4. For historical reviews, see (23); J. Mehra, *Einstein, Hilbert, and the Theory of Gravitation* (Reidel, Dordrecht, Netherlands, 1974); and J. Earman and C. Glymour, *Arch. Hist. Exact. Sci.* **19** 291 (1978).
5. A. Fölsing, *Albert Einstein: A Biography* (Viking, New York, 1997), pp. 375–376. The translation has been slightly modified.
6. Hilbert used the trace of the Ricci tensor in (3), whereas Einstein used the trace of the matter tensor (see Eq. 1); but these two forms are essentially equivalent.
7. See, for example, A. Pais, *"Subtle is the Lord...": The Science and the Life of Albert Einstein* (Oxford University Press, New York, 1982), pp. 257–261.
8. A. Einstein, *Sitzungsber. Preuss. Akad. Wiss.* **1915**, 799 (1915).
9. Staats-und Universitätsbibliothek Göttingen (*Handschriftenabteilung*), Cod. Ms. D. Hilbert 634.
10. Hilbert republished his paper (3) with extensive changes in *Math. Ann.* **92**, 1 (1924).
11. Einstein's paper (1) is cited on p. 395 of (3), as noted by Guth (23).
12. Pages 3 and 4 of the proofs (9); the translation of this and the following quotations is by the authors.
13. Page 7 of the proofs (9).
14. D. Hilbert, letter to F. Klein, 7 March 1918, in *Der Briefwechsel David Hilbert—Felix Klein (1886–1918)*, G. Frei, Ed. (Vandenhoeck and Ruprecht, Göttingen, 1985), p. 129.
15. D. Hilbert, *Nach. Ges. Wiss. Goettingen* **53**, 61 (1917).
16. For a discussion of Hilbert's revised definition of causality and its implications for the Cauchy problem in general relativity, see J. Stachel, in *Studies in the History of General Relativity (Einstein Studies Vol. 3)*, J. Eisenstaedt and A. J. Kox, Eds. (Birkhäuser, Boston, 1992), p. 407.
17. Hilbert denotes the Ricci tensor by $K_{\mu\nu}$ and its trace by K. He does not note the negative value of the determinant of the metric tensor g until later (10).
18. D. Hilbert, letter to A. Einstein, 14 November 1915, *Collected Papers of Albert Einstein*, vol. 8.
19. A. Einstein, letter to D. Hilbert, 15 November 1915, *ibid.*
20. —— letter to D. Hilbert, 18 November 1915, *ibid.*
21. —— *Sitzungsber. Preuss. Akad. Wiss.* **1915**, 778 (1915).
22. —— letter to D. Hilbert, 20 December 1915, *Collected Papers of Albert Einstein*, vol. 8.
23. E. Guth, in *Relativity*, M. Carmeli et al., (Plenum, New York, 1970), p. 161.

24. The authors wish to thank G. Castagnetti for his cooperation in the archival work, especially in the dating of the central documents. They also thank P. Damerow, M. Janssen, U. Majer, D. Rowe, and T. Sauer for helpful discussions. We also thank

Z. Rosenkranz and the Einstein Archives, The Hebrew University of Jerusalem, Israel, for permission to quote from Einstein's letters, and the Staats- und Universitätsbibliothek Göttingen (*Handschriftenabteilung*) for the kind permission to reproduce Fig. 1.

The Origin of Gravitational Lensing: A Postscript to Einstein's 1936 Science Paper

Jürgen Renn, Tilman Sauer, and John Stachel

Gravitational lensing, now taken as an important astrophysical consequence of the general theory of relativity, was found even before this theory was formulated but was discarded as a speculative idea without any chance of empirical confirmation. Reconstruction of some of Einstein's research notes dating back to 1912 reveals that he explored the possibility of gravitational lensing 3 years before completing his general theory of relativity. On the basis of preliminary insights into this theory, Einstein had already derived the basic features of the lensing effect. When he finally published the very same results 24 years later, it was only in response to prodding by an amateur scientist.

Sixty years ago, Einstein published a short note in *Science* entitled "Lens-Like Action of a Star by the Deviation of Light in the Gravitational Field" [1]. The note is often considered as the pioneering study of gravitational lensing, although earlier contributions have been recognized [2]. In 1920, Eddington discussed the possibility of seeing multiple images of a star if a massive object, acting as a gravitational lens, is suitably interposed between the star and an observer [3]. A few years later, Chwolson pointed out that if a star, lens, and observer are in alignment, the observer will see a ring-shaped image of the star centered on the lens [4]. Einstein's paper of 1936 deals with both effects in apparent ignorance [5] of these publications but is free of some of their shortcomings. However, it only gives the final formulas without any derivations. In view of this historical account and Chwolson's pioneering work, it has been suggested [6] that the ring-shaped images be renamed "Chwolson rings" rather than the current "Einstein rings."

In the course of a research project on the genesis of general relativity at the Max Planck Institute for the History of Science, we have identified and reconstructed calculations by Einstein on gravitational lensing closely related to his 1936 paper in notes dated to the spring of 1912. These notes show that Einstein

Science
Vol. 275, pp. 184–186

had developed the basic theory of gravitational lensing even before he completed the general theory of relativity in 1915.

Exploring consequences of a heuristic assumption about static gravitational fields, Einstein in 1911 published a paper on the deflection of light by the gravitational field of the sun [7]. The prediction of light bending was confirmed in 1919 by the famous solar eclipse expedition led by Eddington. But in 1912, Einstein was still trying to persuade astronomers to test various astrophysical consequences of his new ideas on gravitation. Erwin Freundlich was one of the few astronomers who actively engaged in such work. In spring 1912, Einstein, then a professor in Prague, visited Berlin where he met Freundlich, who was working at the Königliche Sternwarte, the Royal Observatory. It is quite possible that the issue was discussed at a meeting with Freundlich. At any rate, Einstein did the gravitational lensing calculations during his Berlin visit as evidenced by notes found in a small notebook dated to the period 1910 to 1914 (Fig. 1) [8].

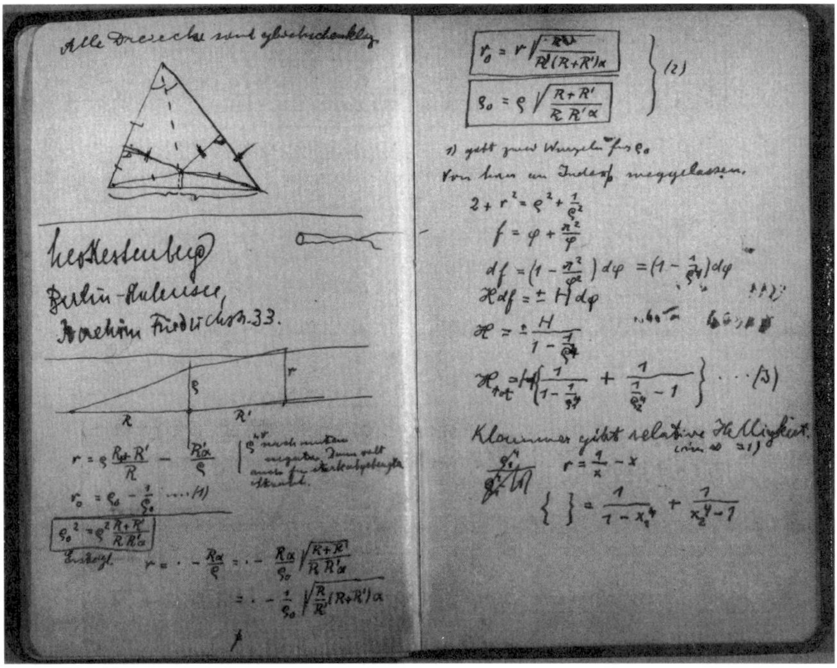

Fig. 1. Notes about gravitational lensing dated to 1912 on two pages of Einstein's scratch notebook [12]. [Reproduced with permission of the American Association for the Advancement of Science.]

These calculations appear interspersed between various notes referring to Berlin appointments and addresses during his visit of 15 to 22 April 1912 [9]. Notes on the gravitational lensing effect are contained on eight pages of the notebook and primarily deal with (i) the possibility of a double image of the source as a result of gravitational light bending and (ii) the magnification of the intensity of these images. Einstein started by sketching the geometrical constellation of gravitational lensing with a light-emitting source and a lensing star separated by a distance R (Fig. 1). An observer is located at distance R' from the lens along the axis formed by the light sources and lens and at a small distance r off of this axis. Einstein then wrote down the basic lensing equation

$$r = \rho \frac{R + R'}{R} - \frac{R'\alpha}{\rho}, \tag{1}$$

which expresses the condition that a light ray passing the lens at a distance ρ will reach the observer. According to Einstein's 1911 paper, [7] the light ray is bent by an angle α/ρ, where α depends on the mass of the lens and differs by a factor of 2 from the value following from general relativity. Introducing dimensionless units, Einstein arrived at the quadratic equation for ρ labeled (1) in his notes (Fig. 1). The two solutions to this quadratic equation correspond to the fact that a light ray can reach the observer after passing either side of the lens.

Considering the apparent brightness of the two images, Einstein also correctly calculated the magnification factor for the intensity of the deflected light [labeled (3) in Fig. 1]. On the following pages in the notebook, Einstein further discussed this magnification factor, transforming the bracketed expression of the equation labeled (3) in Fig. 1 into the form

$$\sqrt{1 + \frac{1}{r_0^2\left(1 + \frac{1}{4}r_0^2\right)}}, \tag{2}$$

with r_0 given as

$$r_0 = r\sqrt{\frac{R}{R'(R + R')\alpha}} \tag{3}$$

[labeled (2) in Fig. 1]. He also estimated the orders of magnitude of the predicted effects.

Despite this work in 1912, Einstein did not publish his results until 1936. He must have already concluded, as he put it in his 1936 paper, that "there is no great chance of observing this phenomenon" (see reference 1, p. 507) and probably for this reason did not pursue his idea any further. Even when he took up the subject again in 1936, it was only because of the initiative of an outsider to the field, a Czech electrical engineer and amateur scientist, Rudi W. Mandl. Mandl had approached Einstein and asked him to consider the possibility of a lensing effect caused by gravitational light bending. In his endeavors he was supported by "a small sum of money," granted to Mandl by the Science Service, to enable him to visit Einstein in Princeton and discuss his ideas [10]. Einstein indeed (re)did the calculations (whether he recalled that he had done the same calculation in 1912 is unknown to us) and conceded that the results might warrant a brief publication. Figure 2 shows the relevant calculations for his 1936 paper in a draft headed "Letter to Mister Mandl," dated 2 June 1936.

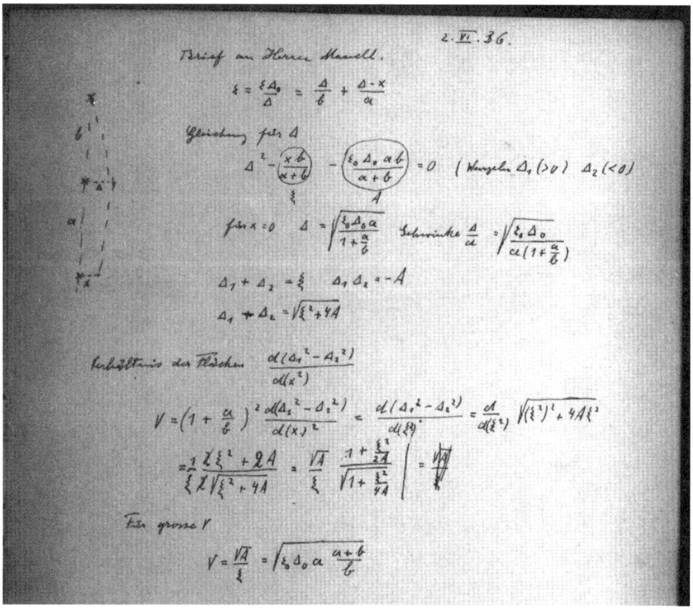

Fig. 2. Calculations on gravitational lensing made by Einstein in 1936 as background for a letter to amateur scientist R. W. Mandl. Einstein here derives the equations published in his note of the same year. The calculations are essentially equivalent to the ones of 1912 (Fig. 1). [Reproduced with permission of the American Association for the Advancement of Science.]

Closer inspection shows that this draft contains the two relevant formulas published in his 1936 note: the angle of sight for an Einstein ring and the magnification factor. What is more surprising is that the lensing equation appearing as the first equation in Fig. 2

$$\frac{\varepsilon_0 \Delta_0}{\Delta} = \frac{\Delta}{b} + \frac{\Delta - x}{a} \tag{4}$$

can be directly translated into the lensing equation that appears as the first equation in Fig. 1 by making the following identifications: $R \equiv b$, $R' \equiv a$, $\rho \equiv \Delta$, $r \equiv x$, $\alpha \equiv \varepsilon_0 \Delta_0$.

All consequences follow from the general lensing equation, and indeed, the expressions for the magnification factor obtained in 1912 and 1936 are equivalent.

The magnification obtained in 1936

$$\frac{\sqrt{A}}{\xi} \frac{1+\frac{\xi^2}{2A}}{\sqrt{1+\frac{\xi^2}{4A}}}, \tag{5}$$

where $\xi \equiv xb/(a+b)$ and $A \equiv \varepsilon_0 \Delta_0 ab/(a+b)$, as it appears in the calculation in Fig. 2, can in fact easily be transformed into Eq. 2 by noting that $r_0 \equiv \xi/\sqrt{A}$.

As a result of the persistent urging of Mandl, Einstein published these results in 1936. As Einstein acknowledged, "Some time ago, R. W. Mandl paid me a visit and asked me to publish the results of a little calculation, which I had made at his request. This note complies with his wish" (see reference 1, p. 506). In a letter to the editor of *Science*, James Cattell, Einstein expressed himself more frankly: "Let me also thank you for your cooperation with the little publication, which Mister Mandl squeezed out of me. It is of little value, but it makes the poor guy happy [11]." In spite of Einstein's pessimism, gravitational lensing was observationally confirmed in 1979 and has since become an important area of astrophysical research [2].

References and Notes

1. A. Einstein, *Science* **84**, 506 (1936).
2. P. Schneider, J. Ehlers, E. E. Falco, *Gravitational Lenses* (Springer, Berlin, 1992).
3. A. S. Eddington, *Space, Time and Gravitation* (Cambridge University Press, Cambridge, 1920), pp. 133–135 and 208–209.
4. O. Chwolson, *Astron. Nachr.* **221**, 329 (1924).
5. Einstein may well have been aware of Chwolson's short note (see reference 4) because a short note of his own (a response to W. Anderson on the problem of the theory of electron gases) appeared in the same issue of the same journal and was in fact printed on the very same page as Chwolson's note.
6. J. M. Barnothy, in *Gravitational Lenses*, J. M. Moran, J. N. Hewitt, K. Y. Lo, Eds. (Springer, Berlin, 1989), pp. 23–27.
7. A. Einstein, *Ann. Phys. (Leipzig)* **35**, 898 (1911).
8. M. J. Klein et al., Eds., *The Collected Papers of Albert Einstein* (Princeton University Press, Princeton, NJ, 1993), vol. 3, appendix A.
9. The most direct evidence is a list of appointments for his Berlin visit found on p. 36. One of the less direct hints is found in Fig. 1. Leo Kestenberg, probably an acquaintence of Elsa Löwenthal (later Einstein's second wife), was working as a piano teacher at the Klindworth-Scharwenka-Conservatory in Berlin, where Elsa Löwenthal gave a performance in February 1913.
10. R. D. Potter, letter to A. Einstein, dated 16 September 1936 (Einstein Archives, call no. 17-039).
11. "Ich danke ihnen noch sehr für ihr Entgegenkommen bei der kleinen Publikation, die Herr Mandl aus mir herauspresste. Sie ist wenig wert, aber dieser arme Kerl hat

seine Freude davon." (A. Einstein, letter to J. Cattell, dated 18 December 1936 [Einstein Archives, call no. 65-603]).

12. Figure 1 shows two facing pages of the notebook published in the facsimile reproduction of reference 8 as pages 43 and 46. Between these pages, a loose sheet containing further notes on the lensing effect was found, which was reproduced in reference 8 as pages 44 and 45.

13. We thank G. Castagnetti, J. Eisenstaedt, H. Goenner, and M. Janssen for helpful comments and Z. Rosenkranz and the Einstein Archives, Hebrew University of Jerusalem, for permission to reproduce the figures.

New Light on the Einstein–Hilbert Priority Question

John Stachel*

Introduction.[1] The development of general relativity may be compared to a traditional three-act play:

I (1907–1908): The formulation of the equivalence principle;

II (1912–1913): The choice of the metric tensor field as the appropriate relativistic generalization of the scalar Newtonian gravitational potential;

III (1913–1915): The search for the correct field equations for the metric tensor field.

The first two acts were essentially monologues starring Albert Einstein. The third act includes a dialogue between Einstein and David Hilbert, the leading Göttingen mathematician. I have told the story of all three acts from Einstein's point of view elsewhere in some detail,[2] so I shall be brief in reviewing the first two acts. But I must say more about the third act, since this is where Hilbert entered the story and the priority question about them arose.

Acts I and II. In 1907, Einstein prepared a review article on the principle of relativity and its consequences for the various branches of physics.[3] In the course of the review, he discussed the question of a relativistic theory of gravitation.[4] It was clear from the outset that the Newtonian theory would have to be modified, since it is based on the concept of a force between pairs of particles that depends on the distance between them at some moment of time. But according to the special theory of relativity, simultaneity is no longer absolute, but depends on the inertial frame of reference being considered; hence the spatial distance between two particles is now frame-dependent. The example of electromagnetism suggests how to proceed. The electrostatic Coulomb force between two charged particles

Journal of Astrophysics and Astronomy
Vol. 20, pp. 91–101

also seems to depend on their simultaneous positions; but we know how to recast the theory into a field form (Maxwell's equations) that is special-relativistically invariant, and in which interactions between charges actually propagate with a finite speed—the speed of light. The problem was to find a similar field theory of gravitation. Indeed, Newton's theory had already been cast into a field form: the gravitational potential obeys Poisson's equation with the density of matter as its source. The problem was to find a relativistic generalization of Poisson's equation. But here Einstein hit upon a fundamental feature of the gravitational field that has no parallel in electromagnetism. All objects fall with the same acceleration in a given gravitational field (Galileo's principle). Newton had explained (or better reformulated) this observation as a consequence of the equality of gravitational and inertial mass. He verified this equality to one part in a thousand, but offered no further explanation of it. It follows from this equality and Newton's laws of motion that no mechanical experiment can ever distinguish between an inertial frame of reference, in which there is a constant gravitational field $+g$, and an accelerated frame of reference with acceleration $-g$ with respect to the inertial frames, in which there is no gravitational field. Einstein went Newton one better, and proposed what he soon called the principle of equivalence:[5] No experiment, mechanical, optical, electromagnetic, or what have you, can distinguish between these two situations because there is no physical difference between them: inertia and gravitation are essentially just two sides of the same coin ("wesensgleich," as Einstein put it). But this means that, once gravitation is taken into account, the privileged role of the inertial frames—and with it the special-relativity principle—must be abandoned. Or rather, one must look for a field theory of gravitation that is invariant under some generalization of the special principle, i.e., under a group that includes transformations to (at least some) accelerated frames of reference. Einstein reached this conclusion (which forms the end of Act I) by late 1907, but for a long time he was the only physicist looking for a field theory of gravitation to accept it.[6]

At first Einstein looked for a static generalization of Newton's theory, based on the gravitational potential. By the end of 1912, he had worked out what he regarded as a satisfactory theory for the case of a static gravitational field.[7] He developed a field equation for the gravitational potential, which he identified in this case with a variable speed of light $c(x, y, z)$ instead of the constant speed of the special theory; and worked out the equations of motion for a particle in the static gravitational field. He soon realized that these equations of motion could be derived from a variational principle that implicitly involves a non-flat metric tensor field. Planck had shown that the equations of motion of a particle in Minkowski space-time (i.e., in the absence of a gravitational field) could be derived from a variational principle involving the proper time along a timelike worldline in a flat space-time with Minkowski metric tensor. Einstein found that his equations of motion for a particle in a static gravitational field could be derived from the same variational principle if he just replaced the constant c in the Minkowski metric with $c(x, y, z)$.[8] This yields a non-flat metric tensor field $g_{\mu\nu}$ of a special type; Einstein guessed that the way to proceed beyond the static case was to introduce a general non-flat metric tensor field as the generalization of the scalar Newtonian potential.[9] He presented this idea in the opening section of a paper written in early 1913,[10] concluding Act II of the play.

Act III. The rest of his 1913 paper begins the lengthy final Act III. With the help of his friend and colleague Marcel Grossmann, Einstein succeeded in showing that the effects of gravitation on all other physical processes could be taken into account by means of the non-flat metric tensor field, and that the resulting equations could be put into a generally-covariant form. That is, the group of transformations under which these equations are invariant included all invertible differentiable point transformations. (Einstein, in keeping with the mathematical treatments of the time, gave the transformations a passive interpretation as coordinate transformations; but from the modern viewpoint, it is better to give them an active interpretation as differentiable point transformations, or diffeomorphisms.) Thus the group automatically includes all possible transformations to accelerated frames of reference, which seemed to Einstein a very satisfactory solution to the problem of generalizing the relativity principle.

However, one major difficulty remained: He could not make the differential equations for the gravitational field itself generally covariant. If they were based on the Ricci tensor, just about the only generally-covariant candidate for second-order field equations constructed from the metric tensor, the equations relating the latter to its source, the stress-energy-momentum tensor describing all non-gravitational matter and fields, did not seem to allow passage to the well-verified static Newtonian limit (Poisson's equation).[11] The root of the problem lay in a misconception of what the static metric tensor field should be, and how the Newtonian limit should be taken. But this did not emerge until well after Einstein had abandoned general covariance in favor of a set of non-covariant field equations now usually called the Einstein–Grossmann field equations, first presented in the 1913 paper.

Being Einstein, he did not rest content with the Newtonian limit argument against field equations based on the Ricci tensor (indeed he seems to have become suspicious of that argument rather quickly). He soon came up with what appeared to be a devastating argument against generally-covariant gravitational field equations of any kind, which he dubbed the hole argument:[12] Consider a finite region of space-time (the hole), inside of which there is no matter. Einstein requires of any gravitational theory that the specification of all sources of the field outside the hole—together with any appropriate boundary conditions on the hole, at infinity, etc.—should uniquely determine the gravitational field inside the hole. But this seems to be impossible if the gravitational field equations are generally covariant: Suppose we have one solution to the field equations outside and inside the hole. Then let us carry out a diffeomorphism (i.e., a one-one differentiable point transformation)[13] that reduces to the identity outside and on the boundary of the hole, but is different from the identity inside the hole. For generally covariant equations, the diffeomorphic-transform of a solution is also a solution to the field equations. Thus, the same source and boundary conditions outside the hole correspond to two distinct solutions inside the hole—indeed, since the group of diffeomorphisms depends on four arbitrary functions, the number of such distinct solutions is unlimited. Therefore one must look for gravitational field equations that are not generally covariant; yet they must be invariant under a group of transformations that includes at least some non-linear ones (accelerations)—but not so many as to fall foul of the hole argument.

This is how the situation stood, as Einstein saw it, in late 1913, and so it stayed until late 1915. He remained wedded to the Einstein-Grossmann field equations and kept trying to find better and better arguments in their favor; in particular, arguments for their uniqueness and for their invariance under the maximum invariance group compatible with avoiding the hole argument.

It was not until the end of 1915, after he had returned to general covariance, that he found the flaw in his hole argument against it: In modern terms, the hole argument presupposes that some intrinsic physical significance attaches to the points of a four-dimensional manifold *before* the metric tensor field is defined on it. Once this idea is abandoned, and it is accepted that the points of the manifold inherit all their physical properties from the metric tensor field (and any other physical fields that may be present in regions outside the hole), then it is clear that two mathematical solutions inside the hole that differ only by a diffeomorphism transformation represent the *same* gravitational field; so the uniqueness requirement is satisfied. But as noted above, Einstein did not realize this until after he had returned to general covariance for other reasons.

Enter Hilbert. In the summer of 1915, while he still clung to the non-covariant theory, Einstein was invited to give a series of lectures on his gravitational theory in Göttingen (he stayed from June 29 to July 7), then the mathematical capital of Germany if not the world. In addition to Hilbert, it boasted the presence of Felix Klein, the doyen of German mathematics, Emmy Noether, and many other luminaries. Einstein, who had been having his troubles getting the physicists to accept his approach to gravitation, was delighted with the reception of his work by the Göttingers.

> In Göttingen I had the great pleasure of seeing everything understood, down to the details. I am quite enthusiastic about Hilbert. An important man. I am very curious about his opinion.[14]

For his part, Hilbert, who had been working on Mie's nonlinear electrodynamic theory, which aimed at explaining the structure of matter, was delighted with Einstein's approach to gravitation. He decided to combine Mie's four electrodynamic potentials q_μ with Einstein's gravitation potentials $g^{\mu\nu}$ in order to forge a theory of matter that would enable him to explain the structure of the electron, as well as its curious non-radiative behavior in the Bohr atom.[15] In the fall of 1915, Einstein and Hilbert entered into an intense correspondence—the only serious one either had on this topic—just as Einstein was reaching the crisis point in his efforts to shore up the Einstein-Grossmann equations, leading to their abandonment.[16] Einstein's letters to Hilbert report such milestones as: his return to general covariance and re-adoption of field equations based on equating the Ricci tensor to the stress-energy tensor, in two papers submitted on 4 and 11 November;[17] his solution of the problem of the anomalous precession of the perihelion of Mercury, submitted on 18 November;[18] and his adoption of the final form of the field equations, submitted on 25 November.[19] These equations are still based on the Ricci tensor, but now with $-1/2g_{\mu\nu}T$ (where T is the trace of the stress-energy tensor—this term is often referred to as the trace term) subtracted from the stress energy tensor on the right hand side. Meanwhile, Hilbert had not been inactive. He, too had formulated field equations—but for a combined theory of gravitation and Mie's electromagnetism, based on a generally-covariant variational principle.

The accepted account of just what happened during this period has been succinctly formulated in a recent biography of Einstein:

> In the decisive phase [of his work on general relativity] Einstein even had a congenial colleague, though this caused him more annoyance than joy, as it seemed to threaten his primacy. "Only one colleague truly understood it, and he now tries skillfully to 'nostrify' [i.e., appropriate] it," he complained to [Heinrich] Zangger about what he evidently regarded as an attempt at plagiarism. This colleague was none other than David Hilbert [. . .] What must have irritated Einstein was that Hilbert had published the correct field equations first—a few days before Einstein [. . .]
>
> In November, when Einstein was totally absorbed in his theory of gravitation, he essentially corresponded only with Hilbert, sending Hilbert his publications and, on November 18, thanking him for a draft of his article. Einstein must have received that article immediately before writing this letter. Could Einstein, casting his eye over Hilbert's paper, have discovered the term which was still lacking in his own equations [i.e., the trace term], and thus 'nostrified' Hilbert?[20]

The author argues that this is improbable, and most authorities agree; but the point is that, on the accepted account, which is based on comparison of the published version of Hilbert's paper on the foundations of physics[21] with Einstein's papers, the dating of the papers makes such an act of plagiarism by Einstein *possible*.

Recently, Leo Corry found a set of printer's proofs of Hilbert's paper, marked "First proofs of my first note" in his handwriting.[22] As a result of a study of these proofs and their dating, one can conclude that Einstein *could not* have taken anything pertaining to the field equations from Hilbert's paper, or the summary sent him in November. Indeed, the situation is quite the reverse: The question is whether Hilbert might have taken something from Einstein's 25 November paper, because it was *quite possible* for him to do so. Let me explain how this dramatic reversal of the direction of possible influence came about.

The published article by Hilbert is dated, "submitted to the session [of the Göttingen Academy of Sciences] of 20 November 1915." Einstein's conclusive paper, in which he gave the final form of his generally-covariant field equations was submitted to the session of the Prussian Academy of Sciences in Berlin of 25 November 1915. Bearing in mind that Hilbert sent a summary or draft of his paper to Einstein, to which the latter replied on November 18, it seems that Einstein had available at least the essence of Hilbert's work a week before he submitted his conclusive paper of 25 November. And so he did—but it was Hilbert's work as presented in the proofs. These proofs are dated exactly the same way as the published version—"submitted to the session of 20 November 1915"—and bear a printer's date stamp—"6 December 1915"—indicating that by then they were typeset. These dates would mean nothing if the proofs were essentially the same as the published version. But they are not: they differ in several important respects, the most crucial being that:

1) the theory that they present is *not* generally covariant; in addition to generally-covariant field equations, there are four equations whose purpose is precisely to limit the coordinate system.

2) the generally-covariant gravitational field equations are not written down explicitly but merely as the variational derivative, which is not evaluated, of a Lagrangian. In particular, there is no hint of a trace term.

Since the proofs are dated November 20, they represent the status of Hilbert's work submitted on that date to the Academy and presumably previewed in his earlier communication to Einstein. Since they bear the printer's stamp dated December 6, any changes Hilbert made before that date are incorporated in the proofs (indeed, he had probably made no changes since he marked them "first proofs"). Thus, there is no chance that Einstein could have learned about the need for a trace term in his equations from a perusal of Hilbert's work before Einstein submitted his paper of November 25, including the trace term.

On the other hand, Hilbert did not start correcting the proofs until 6 December at the earliest (he finally gave up the revision and completely rewrote the article, as we shall see); and Einstein's 25 November paper was published on 2 December. Since Hilbert's paper was not actually published until March 1916, there *was* plenty of opportunity for Hilbert to see Einstein's paper before publishing his own. Indeed, there is no doubt that he did: Hilbert's paper, which was actually published in the issue of the *Göttingen Nachrichten* dated 31 March, includes references to all of Einstein's November papers including that of 25 November. Since we now know that Hilbert drastically rewrote his original (proofs) version sometime between December 1915 and March 1916, the question is no longer what could Einstein have gotten from Hilbert, but: What could Hilbert have gotten from Einstein?

Hilbert's proofs. We shall only discuss the two major points of difference between the proofs and the published paper that are most relevant to the priority question; for more detailed comparisons one can consult the references in footnote 15. First of all, as noted above, in the proofs Hilbert asserts that the theory he is developing *cannot* be generally covariant. He bases this assertion on a slightly more sophisticated version of Einstein's hole argument (he cites Einstein's 1914 discussion of that argument), involving the Cauchy problem on an initial hypersurface rather than boundary conditions on a hole:

> Since our mathematical theorem [a version of what later became known as Noether's theorem, which guarantees the existence of four identities between generally covariant field equations] shows that the previous axioms I [the existence of a Lagrangian

> for the four electromagnetic and ten gravitational equations] and II [the general covariance of that Lagrangian] can only provide ten essentially independent equations for the 14 potentials [of gravitation and electromagnetism]; and further, maintaining general covariance makes quite impossible more than ten essentially independent equations for the 14 potentials g_{μ}, q_s; then, in order to keep the deterministic character of the fundamental equations of physics, in correspondence with Cauchy's theory of differential equations [that is, to have a well-posed Cauchy problem], the requirement of four further non-invariant equations to supplement [the generally-covariant gravitational equations] is unavoidable. In order to find these equations I start out by setting up a definition of the concept of energy (pp. 3–4 of the proofs, translation from Corry, Renn and Stachel, footnote 1).

Without going into the details of the resulting energy theorem, I shall only cite Hilbert's:

> Axiom III (axiom of space and time). The space-time coordinates are those specific world parameters [Hilbert's name for arbitrary coordinates] for which the energy theorem ... is valid.
>
> The validity of [the energy] equation ... is a consequence of Axiom III; these four differential equations ... supplement the [generally-covariant] gravitational equations ... to yield a system of 14 equations for the 14 potentials $g^{\mu\nu}$, q_s: *the system of the fundamental equations of physics* (original emphasis; p. 7 of the proofs, translation from Corry, Renn and Stachel, footnote 1).

Thus, Hilbert was adopting Einstein's line of reasoning from 1913–mid 1915 just at the time that Einstein was abandoning it, in favor of a return to general covariance. Of course, Hilbert did include ten generally covariant gravitational equations in his "system of the fundamental equations of physics," but Einstein was not impressed by this move. Immediately after receiving Hilbert's account of his new theory, Einstein wrote him on 18 November:

> The difficulty was not to find generally covariant equations for the $g^{\mu\nu}$; this is easy with the help of the Riemann tensor. What was difficult instead was to recognize that these equations form a generalization, and indeed a simple and natural generalization, of Newton's law. I only succeeded in doing so in the last few weeks (I sent you my first communication); while I had already considered the only possible generally covariant equations, which have now proven to be the correct ones, three years ago with my friend Grossmann. Only with a heavy heart did we give them up, because the physical discussion appeared to show me their incompatibility with Newton's law (my translation).[23]

That Einstein's claims about his earlier work are quite accurate is proved by an examination of his 1912 Zürich research notebook.[24]

The second major point about Hilbert's proofs concerns the form of his gravitational field equations. In the proofs, these equations do not appear explicitly. Hilbert's Lagrangian contains a gravitational term, which is probably $\sqrt{-g}\ R$ in modern terms,[25] where R is the Ricci scalar (unfortunately, the few lines where the Lagrangian is defined are missing from the proofs), and he indicates that the gravitational field equations result from taking the variational derivative of this term with respect to the metric tensor. But he does not evaluate this variational derivative at all.

Why did Hilbert decide to publish his work before he had accomplished his original aim: to explain the behavior of the electron? He submitted his work to the Göttingen Academy two days after Einstein's letter of 18 November,[26] which opened with a sentence I have not yet quoted:

> The system given by you agrees—as far as I can see—completely with what I have found in the last few weeks and sent to the Academy;

and closed with the news that:

> I have today handed in a work to the Academy, in which without any supplementary hypothesis I have quantitatively derived the perihelion motion of Mercury, discovered by Leverrier, from general relativity (my translation).[27]

As Tilman Sauer suggests,[28] Hilbert may well have felt some urgency to establish his own claims in the light of Einstein's successes.

It was shortly after receiving Hilbert's summary of his theory, which presumably contained both points discussed above, that Einstein complained about Hilbert's "nostrification" of his theory. The letter, cited earlier, continues:

> In my personal experience, I have hardly ever learned to know better the wretchedness of human beings than on the occasion of this theory and what is connected with it. But I don't give a damn (my translation).[29]

Hilbert was apparently aware of Einstein's unhappiness. When he got the proofs of his paper (presumably on or about 6 December), he added an additional complimentary reference to Einstein among several other handwritten insertions; but then he seems to have realized that he would have to completely rewrite the paper in the light of Einstein's four November papers (as noted above, they are all referenced in the published version of Hilbert's paper).[30]

The published paper. Hilbert removed all references to his argument about the need for four non-generally covariant equations to supplement the ten generally-covariant gravitational field equations. He did not discuss the subject of causality for the field equations in this paper at all. Only in 1917, in his second paper on the foundations of physics, did he return to it.[31] He simply asserted Axioms I and II, and dropped Axiom III and all other references to preferred coordinate systems. He completely changed his energy discussion to take into account the full general covariance of the present version of his theory.

He also now wrote down the explicit form of the gravitational field equations that follow from the variational derivative of his Lagrangian: the trace term $1/2g_{\mu\nu}R$, where R is the trace of the Ricci tensor, is subtracted from the Ricci tensor, giving what we now call the Einstein tensor as the left-hand side of the field equations. (His right-hand side is the stress-energy tensor for Mie's electrodynamic theory.)

Indeed, he was the first to give the field equations in this form. As noted above, Einstein, in his paper of 25 November had the Ricci tensor on the left-hand side, with the trace of the stress-energy tensor subtracted on the right-hand side. The two forms are completely equivalent, of course; but whether Hilbert or Einstein immediately recognized this is not clear; nor is it clear, if Hilbert did recognize it, whether it influenced him in any way. What is clear is that the argument he offers for the form of the left-hand side of the field equations is fallacious. Instead of evaluating the Einstein tensor, he simply says:

> ... which follows easily without calculation from the fact that, except for $g_{\mu\nu}$, $K_{\mu\nu}$ [the Ricci tensor] is the only tensor of second order and K [the Ricci scalar] is the only invariant that can be constructed from only the $g^{\mu\nu}$ and its first and second order partial derivatives... (pp. 404–405 of ref. 20, translation from Corry, Renn and Stachel, footnote 1).

The argument is fallacious, of course, because many other tensors of second order in the derivatives of the metric and many other invariants can be constructed from the Riemann tensor. Even if one requires linearity in the second derivatives, the fact that the coefficient of the trace term is exactly 1/2 remains quite undetermined by this argument. Hilbert himself seems to have recognized its untenability: When he reprinted the paper in 1924,[32] he dropped it and simply sketched a correct method of evaluating the variational derivative.

Hilbert added several clear acknowledgements of Einstein's priority in constructing a generally-covariant gravitational theory based on the metric tensor, and continued to acknowledge it on many later occasions. If he had only added a few words to the dateline of the published paper: "revised version submitted on—any date after 6 December," the whole later priority question could have been avoided.[33]

At any rate, the priority issue between Einstein and Hilbert quite soon was happily resolved. On 20 December 1915, well before Hilbert's paper appeared in print, Einstein wrote him:

> There has been a certain resentment between us, the cause of which I do not wish to analyze. I have fought against the feeling of bitterness associated with it, and indeed with complete success. I again think of you with unclouded friendliness and I ask you to attempt the same with me. It is objectively a pity if two genuine chaps, who have liberated themselves to some extent from this sorry world, do not give each other mutual pleasure (my translation).[34]

Notes

[1] This talk is based on joint work with Leo Corry and Jürgen Renn. Leo Corry, Jürgen Renn and John Stachel, "Belated Decision in the Hilbert–Einstein Priority Dispute," *Science* **278**, 1270–1273 (1997); [See this volume, pp. 339–345].

[2] See John Stachel, "The Genesis of General Relativity," in H. Nelkowski et al., eds. *Einstein Symposion Berlin, Lecture Notes in Physics* **100** (Berlin, Heidelberg, New York: Springer-Verlag, 1980), pp. 428–442; "Einstein and the Rigidly Rotating Disc," in A. Held ed. *General Relativity and Gravitation One Hundred Years After the Birth of Albert Einstein* (New York: Plenum, 1980), pp. 1–15, reprinted in D. Howard and J. Stachel, eds. *Einstein and the History of General Relativity/Einstein Studies*, vol. 1 (Boston, Basel, Stuttgart: Birkhäuser, 1989), pp. 48–62; "How Einstein discovered general relativity: A historical tale with some contemporary morals," in M. A. H. MacCallum, ed. *General Relativity and*

Gravitation/ Proceedings of the 11th International Conference on General Relativity and Gravitation (Cambridge: Cambridge University Press, 1987), pp. 200–208; "Einstein's Search for General Covariance, 1912–1915," in D. Howard and J. Stachel, eds. *Einstein and the History of General Relativity/Einstein Studies*, vol. 1 (Boston, Basel, Stuttgart: Birkhäuser, 1989), pp. 63–100. [All are included in Section V of this volume].

[3] Albert Einstein, "Über das Relativitätsprinzip und die aus demselben gezogenen Folgerungen," *Jahrbuch der Radioaktivität und Elektronik* **4** (1907), pp. 411–462; reprinted in John Stachel et al., eds. *The Collected Papers of Albert Einstein*, vol. 2, *The Swiss Years: Writing 1900–1909*, Princeton University Press, 1989, pp. 433–488. At that time he still referred to the principle, rather than the theory, of relativity, when referring to what was later called special relativity.

[4] See the reference in footnote 2, pp. 476–484.

[5] See Albert Einstein, "Lichtgeschwindigkeit und Statik des Gravitationsfeldes," *Annalen der Physik* **38**, 355–369 (1912), p. 365.

[6] Some physicists looking for a relativistic theory of gravitation continued to adhere to the special-relativistic principle (Gunnar Nordström and Gustav Mie); one physicist abandoned the relativity principle entirely in his search for a field theory of gravitation (Max Abraham).

[7] See the reference in footnote 4, and Albert Einstein, "Zur Theorie des statischen Gravitationsfeldes," *Annalen der Physik* **38**, 443–458 (1912); both papers are reprinted in Martin J. Klein et al., eds. *The Collected Papers of Albert Einstein*, vol. 4, *The Swiss Years: Writings 1912–1914*, Princeton University Press, 1995, pp. 130–144, 147–162.

[8] See the second reference in footnote 6, p. 458.

[9] The consideration of stationary gravitational fields, and in particular the gravitational field on a rotating disc, also played a role in convincing Einstein of the need to go beyond flat space to solve the gravitational problem. See John Stachel, "Einstein and the Rigidly Rotating Disc," reference in footnote 1.

[10] See Albert Einstein and Marcel Grossmann, *Entwurf einer verallgemeinerten Relativitätstheorie und einer Theorie der Gravitation*, Leipzig, Berlin, B.G. Teubner, 1913. Part I, "Physikalische Teil," was written by Einstein, Part II, "Mathematischer Teil," by Grossmann. The paper is reproduced in Martin J. Klein, et al., eds. *The Collected Papers of Albert Einstein*, vol. 4, *The Swiss Years: Writings 1912–1914*, Princeton University Press, 1995, pp. 303–339.

[11] See John Stachel, "Einstein's Search for General Covariance, 1912–1915," reference in footnote 1.

[12] For more detailed discussions, see John Stachel, "The Meaning of General Covariance: The Hole Story," in John Earman et al., eds. *Philosophical Problems of the Internal and External World/Essays on the Philosophy of Adolf Grünbaum* (Konstanz: Universitätsverlag, Pittsburgh: University of Pittsburgh Press, 1993), pp. 129–160; and "Einstein's Search for General Covariance, 1912–1915," reference in footnote 1.

[13] By speaking of diffeomorphisms instead of coordinate transformations, I am modernizing the argument, but am not being unfaithful to the essence of Einstein's argument.

[14] Einstein to Arnold Sommerfeld, 15 July 1915, in Robert Schulmann et al., eds. *The Collected Papers of Albert Einstein*, vol. 8, *The Berlin Years: Correspondence, 1914–1918 Part A: 1914–1917*, Princeton University Press, 1998, Doc. 96, p. 147.

[15] For discussions of Hilbert's work during this period, see Leo Corry, "From Mie's Electromagnetic Theory of Matter to Hilbert's Unified Foundations of Physics," *Studies in History and Philosophy of Modern Physics* 30 (1999): 159–183; Tilman Sauer, "The Relativity of Discovery: Hilbert's First Note on the Foundations of Physics," *Archive for the History of Exact Sciences* 53 (1999): 529–575; and Jürgen Renn and John Stachel, "Hilbert's Foundations of Physics: From a Theory of Everything to a Constituent of General Relativity," (Max-Planck-Institut für Wissenschaftsgeschichte Preprint 118, 1999).

[16] For discussions of the reasons for his abandonment of the Einstein-Grossmann equations and his return to general covariance, see John Stachel, "Einstein's Search for General Covariance," reference in footnote 1; John Norton, "How Einstein Found His Field Equations, 1912–1915," *Historical Studies in the Physical Sciences* **14**, 253–316 (1984); Michel Janssen, "Rotation as the Nemesis of Einstein's Entwurf Theory," in Hubert Goenner et al. (eds.), *The Expanding Worlds of General Relativity, Einstein Studies*, vol. 7, Boston, Basel, Berlin: Birkhäuser, 1999.

[17] Albert Einstein, "Zur allgemeinen Relativitätstheorie," *Königlich Preußischen Akademie der Wissenschaften* (Berlin), *Sitzungberichte*, 778–786 (1915), "Zur allgemeinen Relativitätstheorie (Nachtrag), *ibid.*, 799–801 (1915).

[18] Albert Einstein, "Erklärung der Perihelbewegung des Merkur aus der allgemeinen Relativitätstheorie," *Königlich Preußischen Akademie der Wissenschaften* (Berlin), *Sitzungberichte*, 831–839 (1915).

[19] Albert Einstein, "Die Fieldgleichungen der Gravitation," *Königlich Preußischen Akademie der Wissenschaften* (Berlin) *Sitzungsberichte*, 844–847 (1915). The papers cited in notes 17, 18 and this note are reprinted in Martin J. Klein et al., eds. *The Collected Papers of Albert Einstein*, vol. 6, *The Berlin Years: Writings 1914–1917*, (Princeton University Press, 1996), pp. 215–223, 226–228, 234–242, and 245–248.

[20] Albrecht Fölsing, *Albert Einstein: a Biography* (Viking, New York, 1997), pp. 375–376. The translation has been slightly modified and is taken from Corry, Renn, and Stachel, footnote 1.

[21] David Hilbert, "Die Grundagen der Physik (Erste Mitteilung)," *Nachrichten von der Königl. Gesellschaft der Wissenschaften zu Göttingen. Mathematisch-physikalische Klasse*, 395–407 (1915).

[22] The proofs are in the Handschriftenabteilung of the Staats- und Universitätsbibliothek Göttingen, Cod. Ms. D. Hilbert 634.

[23] Albert Einstein to David Hilbert, 18 November 1915, in Robert Schulmann et al., eds. *The Collected Papers of Albert Einstein*, vol. 8, *The Berlin Years: Correspondence, 1914–1918 Part A: 1914–1917*, (Princeton University Press, 1998), Doc. 148, pp. 201–202.

[24] Albert Einstein, "Research Notes on a Generalized Theory of Relativity," in Martin J. Klein et al., eds. *The Collected Papers of Albert Einstein*, vol. 4, *The Swiss Years: Writings 1912–1914*, (Princeton University Press, 1995), pp. 201–260 for a transcription, pp. 630–682 for a facsimile. For a brief review of its significance, see Jürgen Renn and Tilman Sauer, "Einsteins Züricher Notizbuch," *Physikalische Blätter* 52, 865–872 (1996).

[25] Hilbert does not introduce a minus sign since he is working with one imaginary coordinate, and he writes (K) for what we now usually designate by (R).

[26] It is a sad reflection on the state of the mails today that in wartime Germany, mail between Berlin and Göttingen seems to have taken only a day or so.

[27] Reference in note 23.

[28] See the reference in note 15.

[29] Albert Einstein to Heinrich Zangger, 26 November 1915, in Robert Schulmann et al., eds. *The Collected Papers of Albert Einstein*, vol. 8, *The Berlin Years: Correspondence, 1914–1918 Part A: 1914–1917*, (Princeton University Press, 1998), Doc. 152, pp. 204–205; citation from p. 205.

[30] References in notes 17, 18, and 19.

[31] David Hilbert, "Die Grundlagen der Physik (Zweite Mitteilung)," *Nachrichten von der Königl. Gesellschaft der Wissenschaften zu Göttingen. Mathematisch-physikalische Klasse*, 53–76 (1917).

[32] It was reprinted, with extensive changes, in *Mathematische Annalen* 92, 1 (1924).

[33] It was not impossible to do this in the *Göttingen Nachrichten*. Related papers by Klein and Noether, for example, carry such revised datelines (see the paper by Renn and Stachel cited in note 15).

[34] Albert Einstein to David Hilbert, 20 December 1915 in Robert Schulmann et al., eds. *The Collected Papers of Albert Einstein*, vol. 8, *The Berlin Years: Correspondence, 1914–1918 Part A: 1914–1917*, (Princeton University Press, 1998), Doc. 167, pp. 222.

Part VI

Quantum Theory

Einstein and the Quantum: Fifty Years of Struggle

John Stachel

Science is the attempt to make the chaotic diversity of our sense experience correspond to a logically uniform sytem of thought.

—ALBERT EINSTEIN
The Fundaments of Theoretical Physics

In the correspondence of his last years, Einstein returns a number of times to the theme of his half-century-long struggle with the quantum. On December 12, 1951, for example, he wrote to his old friend and patent office colleague Michele Besso: "The whole fifty years of conscious brooding [*Grübelei*] have not brought me nearer to the answer to the question 'What are light quanta?' Nowadays every scalawag [*Lump*] believes that he knows what they are, but he deceives himself."[1]

He wrote to Max von Laue, the only "Aryan" German colleague from Einstein's Berlin years to whom he still felt close, on January 17, 1952, commenting on the latest edition of von Laue's textbook on special relativity. After pointing out that he was aware of the limits of validity of Maxwell's theory before 1905, he added: "But unfortunately the fifty years that have passed since then have not brought us closer to understanding the atomistic structure of radiation. On the contrary! The De Broglie waves are obviously the counterpart of the electromagnetic waves."[2]

Again writing to Besso on December 10, 1952, Einstein said: "at any rate we are just as far from a really rational theory (of the dual nature [*Doppelnatur*] of light quanta and particles) as fifty years ago! P.S. A really rational theory would have to *deduce* the elementary structures (electrons, etc.) not *posit* them from the outset" (*Besso Correspondance*, pp. 482–3).

I shall discuss a few of the things Einstein accomplished in the course of those fifty years of struggle with the quantum; what he was trying to accomplish but felt he had failed to do; the methods that guided him both in his successes and his failures; and why he regarded what he—and others—had done as bringing us no nearer to a really rational understanding of the quantum.

R. Colodny, ed.
From Quarks to Quasars: Philosophical Problems of Modern Physics

Fortunately, I feel no obligation to give an exhaustive survey of Einstein's work on quantum theory, since there are a number of excellent recent studies. These include the work of Martin Klein, conveniently summarized with references to his earlier papers in his paper at the IAS Einstein Centennial;[3] an article and book by Abraham Pais, as well as his IAS Centennial talk;[4] and Max Jammer's book and papers at the Berlin and Jerusalem Einstein Centennials.[5] References in these sources lead into further literature on this topic, so I can proceed in good conscience with my unsystematic survey, which mixes detailed discussion of a few points with a broad, but only sketchily substantiated, characterization of Einstein's work on quantum problems.

I thank the Hebrew University of Jerusalem, which holds the copyright, for allowing me to quote from both published and unpublished letters and papers by Einstein.

1. Einstein and the Bohr Model

I shall start this paper with a minor historical mystery and end with some major physical ones. Einstein described his first published work on the quantum problem, in a letter to his friend Konrad Habicht, as "very revolutionary." This paper, received March 17, 1905, by the *Annalen der Physik*, was entitled "On a Heuristic Viewpoint Concerning the Generation and Transformation of Light." Later, I shall discuss the nature of Einstein's approach to the quantum in this and his other early papers, but I want to draw your attention to an almost unnoticed remark by Michele Besso. Besso was Einstein's sounding board for his ideas about relativity: he is the only person thanked in the 1905 relativity paper. Writing to Einstein on January 17, 1928, Besso said: "For my part, in the years 1904 and '05 I was your public. If, in connection with the drafting of your papers on the quantum problem, I deprived you of a part of your fame, yet in return I secured a friend for you in Planck."[6] No reply has been found. I shall leave aside the question of why Besso claims to have gained Planck's friendship for Einstein. The rest of the quote suggests that, at the time of Einstein's early work on the quantum theory, Besso held Einstein back from publishing some idea that would have brought Einstein even more fame. Considering the fame Einstein had achieved by 1928, Besso could have been referring to no small item! Is there any evidence that Einstein held back some potentially important speculation around 1905? Yes, there is.

It is sometimes thought that Einstein was not originally interested in atomic structure and did not concern himself with this problem in his early years. However, there are several contemporary items of evidence to the contrary. Einstein wrote to Habicht in 1905, probably between June and September: "there is not always a fully developed subject for my musings. At least none that appeals. There would of course be the subject of the spectral lines; but I believe that a simple connection of these phenomena with others already studied does not exist at all, so that the subject for the moment seems to promise very little."[7]

A letter to Philip Lenard a few months later makes it clear that Einstein had closely followed Lenard's studies of atomic spectra, and was ready to express some of his "musings":

> The experiments known to me do not exclude that possibility that the emission or absorption of each individual spectral line is connected with a definite state of the emitting or absorbing center (atom), which [state] is characteristic for it [i.e., the emission or absorption]. ... According to the indicated conception the absorption of a series by a (cold) vapor should be interpreted thus, that the absorption of light of the line ν_1 makes the absorbing center in question receptive for light of the line ν_2, etc. Then an absorption of ν_2 by the vapor would only be possible with simultaneous absorption of ν_1.[8]

This was written *after* the 1905 paper in which Einstein had proposed that emission and absorption of light of frequency ν be conceived as taking place in quanta of energy $h\nu$ (actually, Einstein at that time did not use Planck's h, but the equivalent $R/N\beta$). Perhaps it is not too far fetched to speculate, based on this letter, that around the end of 1905 Einstein had the idea of discrete energy states (levels, we would say today) of the atom in which it did not radiate. If we multiply ν_1 and ν_2 by h, we might try to interpret Einstein's remarks in the letter according to Figure 1 below.[9]

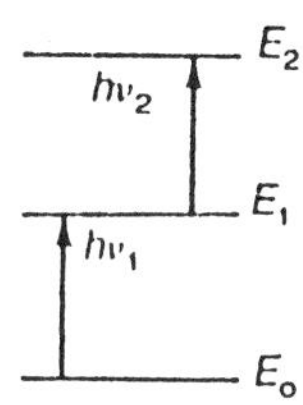

At the time, Lenard did not reply to this letter; when he did write Einstein, in 1909, he mentioned that it was still on his writing desk.

Is there any further evidence to support this wild speculation? Yes, but only a little more. In September 1913, Georg von Hevesy, a radio-chemist then working at Rutherford's laboratory, attended the Vienna meeting of the German Society of Natural Scientists and Physicians (the German AAAS) at which Einstein delivered a review of his work on general relativity up to that date. Hevesy met Einstein, and they discussed Bohr's theory of the hydrogen atom, which had been published that spring. Hevesy gave accounts of the conversation in letters to Rutherford and Bohr. Hevesy to Rutherford:

> Speaking with Einstein on different topics we came to speak of Bohr's theory, he told me that he had once similar ideas but he did not dare to publish them. "Should Bohr's theory be right, it is of the greatest importance." When I told him about the Fowler spectrum the big eyes of Einstein looked still bigger and he told me "Then it is one of the greatest discoveries."
>
> I felt very happy hearing Einstein saying so.[10]

Hevesy to Bohr:

> Then I asked him about his view on your theorie. He told me, it is a very interesting one, important one if it is right and so on and he had very similar ideas many years ago but had no pluck to develop it; I told him then that it is established now with certainty that the Pickering–Fowler spectrum belongs to He. When he heard this he

> was extremely astonished and told me: "Than the frequency of the light does not depend at all on the frequency of the electron"—(I understood him so??) And this is an *enormous achievement*. The theory of Bohr must be then wright. I can hardly tell you how pleased I have been and indeed hardly anything else could make me such a pleasure then this spontaneous judgement of Einstein.[11]

Why Einstein was so impressed by the fact that the frequency of the spectral lines was distinct from any mechanical frequency of the Bohr orbits can be gathered from a comment he made much later: "The important fact that, according to Bohr's theory, the frequency of the emitted radiation is not determined by electrical masses that undergo periodic processes of *the same* frequency, can only strengthen us in this doubt of the independent reality of the wave field [i.e., the classical Maxwell field]."[12]

There is also a reference to the relation of Bohr's work to Einstein's early work in a letter from Einstein to Besso in 1948. I cannot resist including the unrelated but delightful parenthetical remark of Einstein which immediately follows: "It is also not correct that the Bohr conditions for electron orbits follow from my quantum papers. One can only conclude that, if the mechanical picture of the atom is retained, such rules must exist. (You know quite well too, that in spite of its great practical successes, I do not consider the present statistical quantum theory a good approach. It's the same with me as the Jews with the 'Messiah')" (*Besso Correspondance*, p. 391).

There are two more very slight bits of evidence, perhaps more psycho-history than history of science, which may interest you if you have been at all swayed by my argument. The conversation with Hevesy is not the only time Einstein's reaction to Bohr's theory was recorded. In a conversation with Pais in the fifties, A. D. Fokker reminisced about working with Einstein in Zürich: "In 1913–14 he [Fokker] studied with Einstein in Zurich and there he gave the first colloquium on Bohr's theory of the hydrogen atom. Einstein, Laue and Stern were among the audience. Einstein did not react immediately, but kept a meditative silence."[13] If he was learning that some of his own unpublished ideas of almost a decade earlier had been developed and carried to such a successful conclusion by someone else, he might well have meditated.

Finally, I quote from Einstein's 1949 "Autobiographical Notes," in many ways the summing up of his scientific career. He never makes any claim to have anticipated Bohr in this document—or anywhere else, as far as I know—but there is a curious transition at one point. In talking about the early years of this century, he says:

> All of this was quite clear to me shortly after the appearance of Planck's fundamental work; so that, without having a substitute for classical mechanics, I could nevertheless see to what kind of consequences this law of temperature-radiation leads for the photoelectric effect and for other related phenomena of the transformation of radiation-energy, as well as for the specific heat of (especially) solid bodies. All my attempts, however, to adapt the theoretical foundation of physics to this [new type of] knowledge failed completely. It was as if the ground had been pulled out from under one, with no firm foundation to be seen anywhere, upon which one could have built. That this

> insecure and contradictory foundation was sufficient to enable a man of Bohr's unique instinct and sensitivity to discover the principle laws of the spectral lines and of the electron-shells of the atoms together with their significance for chemistry appeared to me as a miracle—and apears to me as a miracle even today. This is the highest form of musicality in the sphere of thought. (P. 45, translation corrected from new edition)[14]

The reference to Bohr's work—the only one in the "Autobiographical Notes"—is completely out of the chronological order that Einstein by and large follows in this essay. One may speculate that there was an association of ideas at this point: recalling his own early attempts to understand atomic spectra could have led him to pay tribute at this point to the man who later so successfully developed similar ideas.

This is all the evidence I have for my conjecture. If Besso's advice influenced Einstein to drop a train of ideas that might have led to the Bohr atom in 1905, then Besso's negative influence on the history of modern physics is at least as great as his positive influence through encouragement of Einstein's work in relativity. The lack of a response from Lenard might also have discouraged Einstein from going on with this work.

Of course, it might be argued that it was primarily Einstein's basic approach to physics which led him away from this problem. Just after the lines quoted above, Einstein continues: "My own interest in those years [after the turn of the century] was less concerned with the detailed consequences of Planck's results, however important these might be. My main question was: What general conclusions can be drawn from the radiation-formula [i.e., Planck's] concerning the structure of radiation and even more generally concerning the electromagnetic foundations of physics?" (*Autobiographical Notes*, p. 47).

2. Einstein and the Quantum Hypothesis

The tone of respect—perhaps awe would not be too strong a word—which Hevesy used in recounting Einstein's views to Rutherford and Bohr indicates how highly regarded Einstein was around 1913 (before general relativity was completed) in the physics community. Some of his renown at that time no doubt came from his work on special relativity, but probably more of it was due to his work on quantum theory. It did not come, however, from his espousal of what came to be called the photon concept. The idea that electromagnetic radiation possesses particulate characteristics, put forward by Einstein in his first quantum paper of 1905, was generally looked upon as a quirk of Einstein's, to be tolerated rather than taken seriously. In a formal recommendation of Einstein for the Berlin post he assumed in 1914, Planck, Nernst, Rubens, and Warburg—leading lights of the Berlin physics community and all deeply interested in quantum theory—said: "That he may sometimes have missed the target in his speculations, as, for example, in his theory of light quanta, cannot really be held against him." What work of Einstein's did they single out for praise? After mentioning his work on special relativity, they went on:

> Far more important for practical physics is his penetration of other questions on which, for the moment, interest is focused. Thus he was the first man to show the importance of the quantum theory for the energy of atomic and molecular movements, and from this he produced a formula for the specific heat of solids which, although not yet entirely proved in detail, has become a basis for further development of the newer atomic kinetics. He has also linked the quantum hypothesis with photoelectric and photochemical effects by the discovery of interesting new relationships capable of being checked by measurement, and he was one of the first to point out the close relationship between the constant of elasticity and those in the optical vibrations of crystals.[15]

It was this work of Einstein, starting with his 1907 paper (written late in 1906) "Planck's Theory of Radiation and the Theory of Specific Heat," that actually put quantum theory in the mainstream of physics. The theory of black body radiation was considered a rather esoteric specialty by all but a handful of physicists—but a handful that included Planck, Einstein, Lorentz, and Ehrenfest, of course. Among that handful, only Einstein and Ehrenfest accepted the idea that Maxwell's classical theory of the electromagnetic field would have to be profoundly modified to account for quantum effects. The prevailing opinion was that some modification of the theory of matter and/or the interaction between matter and radiation was all that was needed. I shall return to this question shortly, but I want to emphasize that it was largely because of Einstein's work, applying the quantum hypothesis to the most varied topics in the structure of matter and the interaction between matter and radiation, that quantum theory really caught on and became a "mainstream" topic of experimental and theoretical work in the physics community. Kuhn has called attention to this point in his recent book.[16]

Among the theoreticians, Einstein was primarily responsible for this major shift of attention. Peter Debye reminisced in 1964:

> The whole thing started with a kind of interpolation formula by Planck. Nobody wanted to accept it then because it did not appear logical. In Planck's radiation formula half of the argument was continuous and the other half was based on the concept of quanta of energy, which he set equal to $h\nu$ in order to get Wien's Radiation Law. There was much trouble at the beginning. The only man who appeared sensible was Einstein. He had the feeling that if there was anything to Planck's idea then it should also appear in other parts of physics. Well, at that time, he talked about the photoeffect, specific heats, and so forth. Then I tried to formulate the theory of specific heats in a more general way.[17]

What distinguished Einstein's approach from the outset was his firm conviction that neither classical mechanics nor classical electrodynamics—nor a combination of the two—provided a "firm foundation," as he put it in the "Autobiographical Notes," on which to build theoretical physics. Thus, he rejected not only the mechanical world view inherited from the nineteenth-century masters of physics; but the new electromagnetic world view, as well as a combination of the two approaches such as Lorentz's theory of the electron. He rejected them, of course, not as useful approximations with limited domains of applicability, but as "firm foundations" for the edifice of theoretical physics. His later recollection of this early rejection was not an artifice of Einstein's memory. In 1907 he wrote:

> We do not have for the moment a complete world-picture [*Weltbild*] consistent with the relativity principle. ... In earlier papers I have shown that our present electromechanical world-picture is not suited to explain the entropy properties of radiation nor the laws of emission and absorption of radiation and those of specific heat. Rather, it is necessary to assume, in my opinion, that the state of every periodic process is such that a transfer of energy can only take place in definite quanta of finite magnitude (light quanta); therefore, that the manifold of really possible processes is a smaller one than the manifold of possible processes in the sense of our current theoretical outlook.[18]

He went on to explain why it was nevertheless possible to use current theories under certain conditions.

For a person of Einstein's temperament, to whom the scientific *Weltbild* meant so much both intellectually and emotionally, this was indeed a situation of crisis. Einstein's often-cited 1918 Planck birthday celebration talk, entitled "Principles of Research," gives some idea of the emotional significance of the *Weltbild* for him:

> Man tries to make for himself in the fashion that suits him best a simplified and intelligible picture of the world; he then tries to some extent to substitute this cosmos of his for the world of experience, and thus to overcome it. This is what the painter, the poet, the speculative philosopher, and the natural scientist do, each in his own fashion. Each makes this cosmos and its construction the pivot of his emotional life, in order to find in this way the peace and security which he cannot find in the narrow whirlpool of personal experience.[19]

Perhaps even more revealing is an address to the students of UCLA given in February 1932:

> Science as something already in existence, already completed, is the most objective, impersonal thing that we humans know. Science as something coming into being, as a goal, however, is just as subjectively, psychologically conditioned as all other human endeavors. This is so much the case that the question of the goal and meaning of science will receive quite different answers at different times and from different personalities.
>
> To be sure, all are agreed that science must establish a connection between facts of experience such that on the basis of experienced facts we are enabled to predict other such facts. According to the view of many positivists the most complete possible solution of this task is the only goal of science.
>
> I do not believe, however, that so primitive an ideal would really permit the kindling to a high degree of that researcher's passion from which really great accomplishments arise. A stronger, but also more obscure drive lies behind the tireless exertions tied to such achievements: one wants to comprehend being, reality. ... At the basis of all such attempts lies the belief that being is completely harmonious in its structure. Today we have less ground than ever before to allow ourselves to stray from this wondrous belief (Item 2-110).[20]

The large cathexis of emotion attached to his view on quantum mechancs can also be seen in a letter to the physicist Tatiana Ehrenfest, widow of Paul Ehrenfest:

"I find the ideas that there should not be laws for being ['*das Seiende*'], but only laws for probabilities, simply monstrous [*scheusslich*] (a nauseatingly indirect description)" (Einstein to T. Ehrenfest, October 12, 1939, Item 10-296).[21]

Perhaps one can appreciate now the full significance of the words, quoted earlier, in which Einstein recalled his feelings at the turn of the century: "It was as if the ground had been pulled from under one, with no firm foundation to be seen anywhere, upon which one could have built" ("Autobiographical Notes," p. 45). A little later in the "Autobiographical Notes" he states: "Reflections of this type made it clear to me as long ago as shortly after 1900, i.e., shortly after Planck's trail-blazing work, that neither mechanics nor electrodynamics could (except in limiting cases) claim exact validity. By and by I despaired of the possibility of discovering the true laws by means of constructive efforts based on known facts. The longer and more despairingly I tried, the more I came to the conviction that only the discovery of a universal formal principle could lead us to assured results" (pp. 51–53).

Note the emotion-laden content of Einstein's wording: I do not think it is rhetorical exaggeration, but takes us to the deep well-springs of Einstein's extraordinary creativity. It would require an insightful psychobiographer to follow up these clues. Lacking such skills, I shall confine myself to explaining what Einstein meant by the statement: "I became convinced that only the discovery of a universal formal principal could lead us to assured results." For this will lead to an understanding of how Einstein approached the quantum riddle and why he felt that fifty years of pondering [*Grübelei*] had not led him closer to an answer.

3. Constructive and Principle Theories

We can begin by recalling the contrast between two types of theory Einstein drew in a 1919 article for the *Times* of London:

> We can distinguish various kinds of theories in physics. Most of them are constructive. They attempt to build up a picture of the more complex phenomena out of the materials of a relatively simple formal scheme from which they start out. Thus the kinetic theory of gases seeks to reduce mechanical, thermal, and diffusional processes to movements of molecules—i.e., to build them up out of the hypothesis of molecular motion. When we say that we have succeeded in *understanding* a group of natural processes, we invariably mean that a constructive theory has been found which covers the processes in question.
>
> Along with this most important class of theories there exists a second, which I will call "principle-theories." These employ the analytic, not the synthetic, method. The elements which form their basis and starting-point are not hypothetically constructed but empirically discovered ones, general characteristics of natural processes, principles that give rise to mathematically formulated criteria which the separate processes or the theoretical representations of them have to satisfy. Thus the science of thermodynamics seeks by analytical means to deduce necessary conditions, which separate events have to satisfy, from the universally experienced fact that perpetual motion is impossible.

> The advantages of the constructive theory are completeness, adaptability, and clearness, those of the principle theory are logical perfection and security of the foundations.
>
> The theory of relativity belongs to the latter class (*Ideas and Opinions*, p. 228).

I maintain that Einstein adopted a variant of this "principle-theory" approach in all of the quantum work for which he is so justly famous: development of the photon concept; quantum theory of specific heats; wave-particle duality; the thermodynamic derivation of Planck's law; the derivation based on transition probabilities; and numerous other contributions I shall not even mention. Yet he always maintained: "When we say that we have succeeded in *understanding* a group of natural processes, we *invariably* mean that a *constructive* theory has been found which covers the processes in question" (emphasis added); so "A really rational theory would have to *deduce* the elementary structures (electrons, etc.) not *posit* them from the outset." Never satisfied with the extraordinary successes of his "principle-theory" approach to the quantum, he continued to search for a constructive theory that would give him the *understanding* which, as indicated above, meant so much to him both intellectually and emotionally. He sought it through various unified field theories, which he attempted to construct (note the significance of this word!) from 1909 (*before* general relativity) until the end of his life.

These preliminary remarks must serve to motivate my division of Einstein's efforts to come to terms with the quantum into four constituents or strands:

1. Statistical studies of radiation and matter, which led him to the firm conviction that one would never understand Planck's radiation law, matter-radiation interactions, specific heats at low temperatures, or many other properties of solid bodies without radical revision of both classical mechanics and classical electromagnetic theory.

2. The attempt to use trustworthy general principles, such as those of thermodynamics and statistical reasoning,[22] combined with the empirically warranted "quantum hypothesis" as Einstein called it, to derive reliable consequences about the behavior of radiation and matter, alone and in interaction. This procedure was not to be confused with an *understanding* of the quantum since the "quantum hypothesis" has been posited without explanation.

In his report to the first Solvay Congress in 1911, Einstein discusses what general principles can be relied upon in deriving the consequences of the quantum hypothesis. He says:

> We are all agreed that the so-called quantum theory of today is indeed a useful tool, but no theory in the ordinary meaning of the word, at any rate not a theory which could now be developed in a coherent manner. On the other hand, it has been proved that classical mechanics, as expressed in Lagrange's and Hamilton's equations, no longer can be regarded as a usable system for the theoretical representation of all physical phenomena. . . .
>
> So the question arises, on the validity of which general principles of physics we may hope to rely in the field of concern to us [i.e., quantum phenomena]. In the first place we are all agreed that we should retain the energy principle.

> A second principle to the validity of which, in my opinion, we absolutely have to adhere is Boltzmann's definition of entropy by means of probability. The weak glimmer of theoretical light that we see today over equilibrium states of processes of an oscillatory nature we owe to this principle.[23]

Einstein goes on to discuss how to attach a meaning to Boltzmann's principle that is independent of theoretical assumptions about the nature of the system to which it is applied. But I want to emphasize Einstein's approach: Take universal formal principles on which we can rely (based on their wide range of empirical success, simplicity, etc.) and apply them to empirically validated results for some quantum system, such as Planck's law for black body radiation, in order to see what can be learned about the nature and/or structure of the system in question, such as black body radiation. Other formal principles in which Einstein placed great confidence were the relativity principle[24] and the laws of thermodynamics. The latter can be looked upon as applications of the energy principle and the Boltzmann principle, but Einstein thought it important to demonstrate that certain results could be deduced from the quantum principle by purely thermodynamic arguments.

3. The search for a constructive theory of matter and radiation, which would yield an understanding of the quantum. Such a theory should be based upon "hypothetically constructed" "elements" fitted together into a "relatively simple formal scheme" from which a "picture of the more complex phenomena" could be deduced.

In his 1909 survey paper "On the Present State of the Problem of Radiation," Einstein reported on his efforts to set up a nonlinear electromagnetic field theory: "I have not yet succeeded in finding a system of equations satisfying these conditions, from which I could see that it was suited to the construction of the electrical elementary quantum [i.e., the electron] and the light quantum. The manifold of possibilities does not seem to be so great that one need be scared away from the task."[25] Thus, at least as early as 1909 Einstein was searching for a nonlinear, "unified" field theory from which electrons and light quanta could be derived. Within a few years he had given up this ambitious program:

> I have also come to the opinion, as a result of many fruitless attempts, that by purely constructive efforts [*blosses Konstruieren*] one cannot put radiation theory back on its feet. Therefore, I have attempted to arrive at a new formulation of the question purely thermodynamically, without making use of any model [*Bild*]. ...
>
> One cannot really seriously believe in the existence of countable quanta, since the interference properties of light emitted in various directions from a luminous point are really not compatible with them. In spite of this the "honorable" quantum theory is still more preferable to me than the compromises found *up to now* as a substitute (Albert Einstein to W. Wien, May 17, 1912, Item 23-558).

In later years, after he had developed the general theory of relativity, Einstein came to doubt whether synthetic-constructive efforts starting from concepts closely linked to empirical evidence could ever lead to such a unified theory. He shifted to the search for a formal scheme which, starting from a small number of highly abstract concepts, would lead after a long chain of deductions to an explanation of quantum effects:

> I am firmly convinced that every attempt to arrive at a rational theory by *synthetic* construction will have an unsatisfactory result. Only a new basis for all of physics, from which all possible processes can be deduced with logical necessity (as for example is the case with thermodynamics) can bring a convincing solution (Albert Einstein to M. Renninger, June 11, 1953, Item 20-032).

> Now you will understand why I lapsed into my apparently Don Quixotic attempts to generalize the gravitational equations. If one cannot trust Maxwell's equations, and a representation [*Darstellung*] by means of field and differential equations is indicated on account of the principle of general relativity, and one has come to despair of arriving at deeper basis [*Tieferlegung*] of the theory by intuitive [*anschaulich*]—constructive means; then no other sort of effort seems open (Einstein to von Laue, January 17, 1952, Items 16-167).

> I do not believe in micro- and macro- laws, but only in (structure) laws of general rigorous validity. And I believe that these laws are *logically simple*, and that reliance on this logical simplicity is our best guide. Thus, it would not be necessary to start with more than a relatively small number of empirical facts. If nature is not arranged in correspondence with this belief, then we have altogether very little hope of understanding it more deeply. . . .
>
> This is not an attempt to convince you in any way. I just wanted to show you how I came to my attitude. I was especially strongly impressed with the realization that, by using a semi-empirical method, one would never have arrived at the gravitational equations of empty space (Einstein to David Bohm, November 24, 1954, Item 8-055).

The line he had earlier draw between a theory of principle and a constructive theory here becomes blurred; but the goal of *deducing* quantum phenomena from some unified theory, rather than assuming their existence from the outset, remains.

The three constituents or strands so far mentioned are not to be thought of as sequentially related. They overlap and even sometimes intertwine. For example, Einstein continued to examine and test classical theories from new angles to demonstrate their insufficiency to his colleagues. And clues provided by the "principle theory" approach to the quantum hypothesis suggested leads for a constructive theory.

4. When, in the mid-1920s, matrix mechanics was developed by Heisenberg, Born, and Jordan and wave mechanics by de Broglie and Schrödinger, a fourth constituent or strand enters the picture. At first Einstein participated eagerly, if critically, in the exploration of the new ideas. After he became convinced that neither approach—nor the unification of the two in the new quantum mechanics—was "the real thing," "the true Jacob," as he sometimes expressed it, he began a critical exploration of quantum mechanics aimed at demonstrating that it could not be regarded as a complete description of physical reality. This critical examination should always be viewed, however, in the context of his search for a complete, explanatory theory.

There is no space for even a brief survey of Einstein's work in each of these areas. I shall discuss a few characteristic highlights of Einstein's approach and a few points I feel are treated inadequately in the literature, including the nature of Einstein's critique of quantum mechanics, which is still sometimes inadequately understood.

4. The Wave-Particle Duality

Einstein's discovery of what is now called the wave-particle duality, made in late 1908 or early 1909, has been extensively and well treated (notably by Klein) so I can be brief. But I cannot forbear saying a few words about it, since it is so characteristic of Einstein's approach and was so pregnant with future significance. In his early papers on statistical mechanics, Einstein had developed general methods for analyzing the fluctuations of any mechanical system, methods which he applied first to testing the atomic hypothesis by considering observable fluctuations of small particles suspended in a fluid (Brownian motion). In his first quantum paper of 1905 Einstein boldly applied this approach to a nonmechanical system: black body radiation. He showed that, in the high-frequency limit in which Wien's law holds, this radiation behaves (with respect to its energy fluctuations, at least) like a gas of independent particles. In 1908–1909 he applied his fluctuation approach to Planck's law, which holds for radiation at all frequencies. His strategy was to forget about the origin of Planck's law for the moment (he had much to say about this both earlier and later). Taking Planck's law as a valid empirical description of the radiation spectrum, what can we learn about the structure of that radiation by calculating the average fluctuation of the energy in some given frequency range in a small volume of the radiation?

To answer this question, he applied his general formula for energy fluctuations to Planck's law. The result consists of two terms: one was linear in the energy density, the other was quadratic. The linear term, as he knew from his work on Wien's law, corresponds to the fluctuations of a gas of independent particles. The second term corresponds to the fluctuations resulting from superposition of random standing electromagnetic waves in a cavity, the energy fluctuations in a small volume arising from interference. (Such a classical wave model of black body radiation leads to the Rayleigh–Jeans law, of course). He showed that there is an analogous two-term result for the momentum fluctuations of black body radiation.

Since Einstein's early views on light quanta are still sometimes misunderstood, let me quote a letter to Lorentz: "I am not at all of the opinion that one should think of light as composed of quanta which are independent of each other, and localized in relatively small regions. For the explanation of the Wien limit of the radiation formula this would indeed be the easiest way. But even the splitting of a light ray at the surface of a refracting medium completely forbids this approach. A light ray divides, but a light quantum indeed cannot divide without an alteration of frequency" (Einstein to Lorentz, May 23, 1909, Item 16-419).

During the next few years, Einstein was not always certain that the concept of light quanta was the best way to approach the study of radiation. At one particularly low moment, after the failure of several attempts at a constructive quantum theory, he wrote: "Right now I am trying to derive the law of heat conduction in rigid dielectrics from the quantum hypothesis. Whether these quanta really exist, I don't ask any more. I am also not trying any more to construct them, because I know now that my brain is not capable of it. But I am searching as diligently as possible to learn the domain of applicability of this concept [*Vorstellung*]" (Albert Einstein to M. Besso, May 13, 1911, *Besso Correspondance*, pp. 19–20). The success of his well-known 1916–1917 derivation of the Planck law, showing that individual acts of emission and absorption of radiation by atoms must be directed processes involving momentum recoil, finally reassured him about the existence and particulate nature of light quanta.[26]

Einstein wrote to Besso: "The quantum paper sent to you has again led me back to the view of the spatial corpuscularity [*Quantenhaftigkeit*] of radiation energy. But I feel that the real jest that the eternal riddle-poser has set before us is still absolutely not understood. Shall we live to see the saving idea?" (Einstein to Besso, March 7, 1917, *Besso Correspondance*, p. 103).

His 1909 work convinced Einstein that radiation could be understood only through some sort of synthesis of wave and particle concepts. He placed great emphasis on his fluctuation results, later making the ability to explain fluctuation phenomena a challenge with which to test the new quantum mechanics. The famous "*Dreimännerarbeit*" of Heisenberg, Born, and Jordan which appeared in 1926[27] claimed to resolve a paradox sharply emphasized by Ehrenfest the previous year:[28] while quantization of the proper vibrations in a cavity yielded the Planck formula for spectral energy distribution (as Debye showed in 1911), it produced the *wrong* formula for energy fluctuations in a small volume (as Einstein indicated in 1909 and Lorentz proved in detail a few years later). By introduction of the so-called zero-point energy of the proper vibrations into which the field was analyzed, Jordan—who wrote this section of the paper—claimed to show that the new matrix mechanics, when applied to fields, could produce not only the quadratic (classical wave) term but also the linear (classical particle) term. The latter term results from interference between the zero point and the other oscillations. Einstein, who had been given proofs of the article, was not impressed. In a postcard to Jordan he wrote: "The thing with the fluctuations is rotten [*faul*]. One can indeed calculate the average magnitude of fluctuations with the zero point term $\frac{1}{2}h\nu$, but not the probability of a very large fluctuation" (Einstein to Pascual Jordan, April 6, 1926, Item 13-479).

A discussion of fluctuations had been going on previously between Einstein and Jordan and continued for some time after. In one of his papers on field quantization[29] Jordan tried to answer Einstein's critique, but it is clear that Einstein was never satisfied with the quantum-mechanical explanation of fluctuations or the wave-particle duality.

A couple of quotations illustrate Einstein's attitude:

> As far as 'fluctuations' are concerned, they are completely incompatible with the Maxwell energy tensor, even if the latter does explain a part (for weak fields an insignificant one) of the fluctuations. If one doesn't explain the fluctuations by a swindle

(I mean the renunciation of the description of the individual system), then the radiation must be richer in structure than can be expressed by a wave theory in the conventional sense (Einstein to Schrödinger, June 15, 1950, Item 22-168).

Encouraging a physicist critical of conventional quantum theory, Einstein wrote:

I find it quite remarkable that, by your Gedankenexperiment, you have again drawn people's attention to the fact that the wave-particle duality is a reality, which one should not allow to be deluded away by metaphysical artifices (Einstein to M. Renninger, February 27, 1954, Item 20-036).

Describing to Laue how a study of the fluctuations of a mirror in a cavity filled with black body radiation first convinced him of the inadequacy of Maxwell's theory, he adds:

The quantum theory doesn't help me at all here; it seems to me to be self-deception when one envelops the movement of a plate in a radiation field in the famous probability cloud, and the field as well. Nothing better will come from this (Einstein to von Laue, January 17, 1952, Item 16-167).

To show that Einstein was perhaps not exaggerating in his comment about self-deception which opened this article, let me quote the interview with Peter Debye that I cited earlier.[30] Reminiscing about the Einstein fluctuation result, Debye says:

Einstein had calculated the fluctuations of energy in a space containing radiation. Now, there are two parts to it; if you say that it is all waves, you get a certain fluctuation. But, if you take the entropy, from this, with Planck's formula, you may calculate a second fluctuation which is independent of the fluctuations due to the waves. But that is not a real fluctuation; it is inherent in the quantum formulation.

Bauer: And this caused the trouble. Well, nowadays, we don't feel troubled by this quality.

Debye: Well, of course not. People have become accustomed to it. At that time, one had to try to answer whether an electron was a wave or a particle. Of course, it is both.

Corson: Should we change the subject?

I do not quote this to single out Debye, but just because his comment is so typical of many discussions of the wave-particle duality. Einstein comments on such "explanations": "We know that light has certain characteristics which we designate for short, respectively, as undulatory and corpuscular. It has no meaning to say, it *is* a wave and it *is* a corpuscle. Up to now we just have no reasonable theory which explains all its characteristics. However there is no *contradiction*, any more that it signifies a contradiction that a man feels *and* has weight" (Albert Einstein to Robert Federn, July 17, 1954, Item 59-654).

Even if one fully accepts quantum mechanics, the fluctuation story does not end so simply. A 1931 paper by Heisenberg shows that the calculation in the *Dreimännerabeit* actually *diverges* if one does it right—and not because of a divergent zero-point energy, either. I shall not go into the further history of this interesting problem, the resolution of which from the quantum-mechanical standpoint, I believe, hinges on a specification of the conditions of measurement of the fluctuations.[31]

Recently, the topic of fluctuations has become of central research interest in general relativity, through the study of the Hawking process.[32] Without discussing this fascinating topic, I want to emphasize its significance. For the first time, there appears to be an area in which quantum mechanics and general relativity may be fruitfully applied jointly. Just for this reason I feel there is a need for extreme caution in accepting all the claims that have been made, and for critical analysis of the calculations, and especially of the justification for the application, in this context, of the physical concepts on which these calculations are based.

5. Einstein and the Born Interpretation

Although it is fairly well known that Einstein first introduced transition probabilities into quantum theory, it is not generally realized how large a role he played in the development of the so-called Born or statistical interpretation of the wave function. In his work of 1916–1917, mentioned earlier, Einstein postulated the existence of transition probabilities between the discrete energy levels of an atomic system. Transition probabilities for spontaneous and induced emission and (induced) absorption were assumed to exist by analogy with similar processes in the classical theory of a charged oscillator developed by Planck (by a correspondence principle argument, one would say today) and with the treatment of radioactive decay by Rutherford. Einstein noted this analogy between radioactivity and absorption and emission as early as 1911 in a letter to Besso (*Besso Correspondance*, pp. 26–27). Although he was very pleased with the resulting derivation of the Planck distribution formula, Einstein never regarded his introduction of transition probabilities as anything but a temporary expedient: "The weakness of the theory lies on the one hand in the fact that it does not get us any closer to making the connection with the wave theory [i.e., overcoming the wave-particle duality]; on the other, that it leaves the duration and direction of the elementary processes to 'chance.' Nevertheless I am fully confident that the approach chosen here is a reliable one" (Van der Waerden, *Sources*, p. 76).

In January 1920, he wrote to Born: "That business about causality causes me a lot of trouble, too. Can the quantum absorption and emission of light ever be understood in the sense of the complete causality requirement, or would a statistical residue remain? I must admit that there I lack the courage of my convictions. But I would be very unhappy to renounce *complete* causality. ... (The question whether strict causality exists or not has a definite meaning, even though there can probably never be a definite answer to it.)" (*Born–Einstein Letters*, p. 23).[33]

His doubts about introducing probabilistic assumptions as a matter of principle did not prevent him from discussing a probabilistic interpretation of Maxwell's equations to link the wave and particle concepts. Although these speculations were well known to a number of physicists during the early twenties, Einstein never published anything about his concept of the "ghost-field" ["*Gespensterfeld*"].[34] Lorentz wrote a long letter to Einstein in 1921 discussing this idea and gave a very similar account in his 1922 lectures at the California Institute of Technology, later published as "Problems of Modern Physics."[35] Since the book is readily available, I shall quote only a couple of sentences from it:

> The hypothesis of light-quanta, however, is in contradiction with the phenomena of interference. Can the two views be reconciled? I should like to put forward some considerations about this question, but I must first say that Einstein is to be given credit for whatever in them may be sound. As I know his ideas concerning the points to be discussed only by verbal communication, however, and even by hearsay, I have to take the responsibility for all that remains unsatisfactory (*Problems of Modern Physics*, pp. 156–57).

Now an excerpt from the letter:

> Basic ideas. In emission of light two things are radiated. There is namely:
>
> 1. An interference radiation, which occurs according to the ordinary laws of optics, but still carries no energy. One can, for example, imagine that this radiation consists of ordinary electromagnetic waves, but with vanishingly small amplitudes. As a consequence they cannot themselves be observed; they serve only to prepare the way for the radiation of energy. It is like a dead pattern, that is first brought to life by the energy radiation. [In the book, he says: "On the screen you will have something like an undeveloped photographic image."]
>
> 2. The energy radiation. This consists of indivisible quanta of magnitude $h\nu$. Their path is prescribed by the (vanishingly small) energy flux in the interference radiation, and they can never reach places where this flux is zero (dark interference bands).
>
> In an individual act of radiation the full interference radiation arises, but only a single quantum is radiated, which therefore can only reach one place on a screen placed in the radiation. However, this elementary act is repeated innumerably many times, with as good as identical interference radiation (the same pattern). The different quanta now distribute themselves statistically over the pattern, in the sense that the average number of them at each point of the screen is proportional to the intensity of the interference radiation reaching that point. In this way the observed interference phenomena arise, corresponding to the classical results (Lorentz to Einstein, November 13, 1921, Item 16-544).

Lorentz explicitly attributes the ideas quoted to Einstein.

If these ideas reminded you of the Born interpretation of the wave function, this is not a coincidence. In the second, more detailed, of two papers entitled "The Quantum Mechanics of Collision Processes," received July 21, 1926, Born contrasts Heisenberg's interpretation of matrix mechanics, according to which "an exact representation [*Darstellung*] of processes in space and time is impossible in general" with Schrödinger's interpretation of wave mechanics, according to which a reality similar to that of classical light waves is attributed to de Broglie waves.[36] He continues: "Neither of these two conceptions seems satisfactory to me. I shall attempt to give a third interpretation and test its usefulness for collision processes. Here I connect with an observation of Einstein on the relation of wave field and light quantum; he said more or less, that the waves are only there in order to show the path to the corpuscular quanta, and he spoke in this sense of a ghost-field [*Gespensterfeld*]. This determines the probability for a light quantuum, the carrier of energy and momentum, to take a particular path; however, no energy or momentum belongs to the field itself."

Leaving aside the puzzling question why no such acknowledgment appears in Born's first paper on the subject, written about a month earlier, note that Born made similar acknowledgments on a number of later occasions. The first seems to have been in a letter to Einstein that is not included in the published *Born–Einstein Letters*. On November 30, 1926, Born wrote: "To report about myself, I am quite satisfied as far as physics goes, since my idea to conceive the Schrödinger wave field as a ghost field in your sense, is constantly proving to be better" (Item 8-179).

This might seem to be the whole story: Einstein invented the "ghost-field" interpretation of the Maxwell field, and Born later applied it to the de Broglie–Schrödinger waves, in particular to collision phenomena. So the story appears in Jammer's book, for example, but two footnotes in Heisenberg's first paper on the uncertainty relations suggest there is more to the story. Early in the paper is a footnote which reads: "The present paper arose out of efforts and desires to which others gave clear expression much earlier, before the origin of quantum mechanics. I recall here especially Bohr's works on the fundamental postulate of quantum theory and Einstein's discussion of the connection between wave field and light quantum" (pp. 173–74).[37]

This note makes it clear that Heisenberg knew about Einstein's *Gespensterfeld* idea.[38] Perhaps he had even discussed it with Einstein, although Heisenberg's later reminiscences of his conversations with Einstein do not report such a discussion. The really remarkable footnote, however, comes later: "The statistical interpretation of the de Broglie waves was first formulated by Einstein. (*Sitzungsber. d. preuss. Akad. d. Wiss.* 1925, p. 3)" (p. 176). After citing a paper of Born and Heisenberg, and one by Jordan, in which "this statistical element in quantum mechanics plays an essential role," he continues: "It was mathematically analyzed in a fundamental paper of M. Born (*Zeitschr. f. Physik* 38, 803, 1926) and used for the interpretation of collision phenomena."

Since Heisenberg had previously acknowledged Einstein's work on "the connection between wave field and light quantum" his citation of Einstein as originator of the statistical interpretation of *de Broglie waves* must be taken seriously. The paper he cites is Einstein's famous second paper on "Quantum Theory of the Monatomic Ideal Gas," submitted January 8, 1925, in which he applied what we now call Bose–Einstein statistics to such a gas.[39] Sections 8 and 9 of the paper are of relevance to our discussion. In the first, entitled "The Fluctuation Properties of the Ideal Gas," he uses his ubiquitous fluctuation technique to calculate the average fluctuation in the number of molecules within a small volume of the gas. As in the radiation case, he arrives at two terms: one corresponds to a classical particle fluctuation term, as might be expected; the other corresponds to a fluctuation term arising from interference of classical waves. The wave-particle duality has now reappeared for a corpuscular system.

Einstein states that this result should be taken seriously "because I believe it is a question of more than a mere analogy." He goes on to cite de Broglie's work, known to him from a copy of de Broglie's 1924 thesis sent to him by Paul

Langevin, Einstein's old friend and de Broglie's thesis adviser. This discussion of de Broglie's work is well known; de Broglie acknowledged the role it played in getting a hearing for his ideas:

> In November 1924 I had defended before the Faculty of Science of Paris a Doctoral Thesis in which I had summarized my new ideas on wave mechanics. Paul Langevin had communicated the manuscript of my work to M. Einstein who had immediately perceived its interest. Shortly afterwards, in January 1925 the illustrious scientist presented a note to the Berlin Academy of Sciences in which he stressed the importance of the concepts expounded in my Thesis and deducing numerous consequences from them. This memoir of M. Einstein drew attention to my work, little noticed up to then and for that reason I have always felt that I owed him to a great personal debt for the precious encouragement he thus brought me.[40]

But the following (ninth) section of the paper does not seem to be as well known, perhaps because its title, "Remark on the Viscosity of the Gas at Low Temperatures," does not seem to promise much of general interest. I shall therefore quote from it at length:

> According to the considerations of the previous paragraph, it appears that an undulatory field is connected with every motion [*Bewegungsvorgang*], just as the optical undulatory field is connected with the motion of light quanta. This undulatory field, whose physical nature is for the moment still unclear, must in principle permit its existence to be demonstrated by the corresponding phenomena of its motion [*Bewegungserscheinungen*. This is probably a misprint for *Beugungserscheinungen* diffraction phenomena.]

He goes on to discuss the diffraction of a stream of gas molecules by a slit, presumably following de Broglie, who had given a similar discussion in his thesis. Einstein concludes that, at ordinary speeds, the wavelength is too small to allow observation of such diffraction by a slit. He goes on:

> At low temperatures, λ [the de Broglie wavelength] is of the order of magnitude of σ [the molecular diameter] for the gases hydrogen and helium, and it seems indeed that the influence on the coefficient of viscosity to be expected from the theory makes itself felt.
>
> If the stream of molecules moving with the speed v strikes another molecule, which for convenience we represent as immobile, then this is comparable with the case of a wave train of a certain wavelength λ striking a foil of diameter 2σ. A (*Fraunhofer*) diffraction effect occurs, which is analogous to that yielded by an aperture of the same magnitude. Large diffraction angles occur if λ is of the same order of magnitude as σ or larger. Thus, in addition to deflections due to collision in accord with [classical] mechanics, mechanically inexplicable deflections of the molecules also will occur with frequency comparable to the former, which will diminish the mean free path.

Einstein applies this idea to the interpretation of data on the variation of the viscosity of a gas with its temperature. Note how close Einstein came here to the statistical interpretation of de Broglie waves scattering from a hard sphere. Born was certainly following Einstein's work on gas degeneracy, as his letter to Einstein of July 15, 1925 shows (*Born–Einstein Letters*, p. 83), and also read de Broglie's work "At Einstein's instigation," as he puts it (*Born–Einstein Letters*,

p. 86). Born's reference, in his letter to Einstein of November 30, 1926 (previously quoted), to "my idea, to conceive of the Schrödinger wave field as a ghost field in your sense" thus may be an overstatement of his claim to originality.

This example also illustrates how deeply Einstein was involved in the development of quantum mechanics, both directly and indirectly. I could cite other examples, some well known, like his correspondence with Schrödinger during the development of wave mechanics;[41] others, not so well known, like his correspondence with Heisenberg and Jordan during the development of matrix mechanics.[42] However, I shall only comment on two points that I feel are often misunderstood.

Einstein's attitude towards quantum mechanics is sometimes presented as if Einstein were repelled by matrix mechanics from the start while initially rather attracted to wave mechanics. Actually, his attitude to both was similar: initially intrigued by each, he immersed himself in the details of each approach. He came to feel that—although each contained "a piece of the truth" as he sometimes put it—neither was ultimately satisfactory. This happened *before* the discovery that the two theories were mathematically identical, made independently by Schrödinger and Pauli in mid-1926. Rather than document this process in detail, I shall only quote a remarkable anecdote:

> Hartmut Kallman recalled one of the famed Berlin colloquia at which the relative merits of Heisenberg's and Schrödinger's points of view were being discussed:
>
> "When a Colloquium on this theme was held at the Berlin University, not 20 but 200 physicists were present. People were packed into the room as lectures on Heisenberg's and Schrödinger's theories were given. At the end of these reports Einstein stood up and said: 'Now just listen! Up until now we had no exact quantum theory, and now we suddenly have two. You will agree with me,' he continued, 'that these two exclude each other. Which theory is correct? Perhaps neither is correct.' At that moment—I shall never forget it—[Walter] Gordon stood up and said: 'I have just returned from Zurich. Pauli has proved that both theories are identical'."[43]

6. Observable Quantities

The other comment is on the Heisenberg–Einstein discussion, reported in Heisenberg's book, *Physics and Beyond*,[44] over whether quantum theory should be based exclusively on "quantities observable in principle" [*prinzipiell beobachtbare Grössen*]. I am not certain when this phrase was first used in the physics literature, but Minkowski used a related expression. In his 1908 paper on relativistic electrodynamics[45] ("Die Grundgleichungen für die elektromagnetischen Vorgänge in bewegten Körpern," he speaks of "*lauter beobachtbaren Grössen*" (purely observable quantities). Born, of course, was working with Minkowski at the time, and it is possible that this expression influenced his later thinking about quantum theory. It is more likely, however, that it was Einstein who introduced the phrase "quantities observable in principle" to the relevant circle of physicists. A long tradition associates this concept with the special theory of relativity. The closest he gets to using it, as far as I have been able to determine, is at the beginning of the 1905 paper on the special theory, in which he contrasts "observable phenomena," which only depend on the relative motion of a magnet and a conductor, with the accepted interpretation of Maxwell's theory, which "leads to asymmetries that do not seem to be attached to the phenomena." This clearly suggests a criterion for theories that Einstein made explicit a decade later in a letter to Lorentz, justifying his *general* relativity theory:

> In the description of the relative motion (of an arbitrary type) of two coordinate systems K_1 and K_2, it is immaterial whether I relate K_2 to K_1 or inversely K_1 to K_2. If in spite of this, K_1 is distinguished because, relative to K_1, the general laws of nature are supposed to be simpler than relative to K_2, then this preferential status is a fact without physical cause. Of two things K_1 and K_2, equivalent according to their definition, one is distinguished without physical (in principle accessible to observation) basis. My trust in the consistency of natural processes resists this most forcefully (Einstein to H. A. Lorentz, January 23, 1915, Item 16-436).

Here he uses the phrase "in principle accessible to observation" (*der Beobachtung prinzipiell zugänglich*). A few years earlier he used the expression "quantities observable in principle" (the first use I have found) in comments at the first (1911) Solvay Congress. Discussing his interpretation of Boltzmann's principle, Einstein starts by saying: "If an externally isolated physical system of fixed energy is given, then the system can still assume the most varied states, which are characterized by a number of quantities observable in principle (e.g., volumes, concentrations, the energies of parts of the system, etc.)."[46] After discussing how to define the probability W of each such state Einstein continues: "If one defines W in the indicated way as temporal frequency, then Boltzmann's equation directly contains a physical assertion. It contains a relation between quantities observable in principle."[47]

It was in justification of the principle of general covariance that he again employed this concept. As I have discussed in detail elsewhere,[48] his discovery of the flaw in his "hole" argument against general covariance—an argument that helped to delay formulation of the final gravitational field equations for over two years—led him to emphasize the concept of observability in principle. In a letter to Besso he stated: "Nothing is physically real but the totality of spatio-temporal coincidences. If, for example, physical events were to be built exclusively from the motions of material points, then meetings of the points, i.e., the intersections of their world lines would be the only reality, i.e., observable in principle" (*Besso Correspondance*, p. 64, letter of January 3, 1916).

The first use of the phrase I have located by someone central to the development of the new quantum theory is again in a relativistic context. It occurs in a paper by Pauli on Weyl's unified field theory: "One would like to keep to the introduction into physics of only quantities observable in principle."[49]

By 1923 at the latest, Pauli was applying this criterion to questions of atomic theory, as a letter to Eddington shows.[50] At the end of 1924, he wrote to Bohr: "I believe that the values of the energy and momentum of the stationary states are something much more real than the 'orbits.' The (still unattained) goal must be to deduce these and all other physically real, observable properties of the stationary states from the (whole) quantum numbers and the quantum-theoretical laws" (*Pauli Briefwechsel*, p. 189).

Near the beginning of their paper "On the Quantum Theory of Aperiodic Processes," which just preceded Heisenberg's first paper on what was later called matrix mechanics, Born and Jordan remark: "A fundamental proposition of great significance and fruitfulness asserts that in the true laws of nature only such quantities enter as are observable, determinable in principle."[51]

They add in a footnote: "Thus, the theory of relativity arose from the circumstance that Einstein recognized the impossibility in principle of determining absolute simultaneity of two events taking place at different places."

Even Born's first papers on the statistical interpretation of the wave function do not deviate from this approach. In a 1926 lecture on "Physical Aspects of Quantum Mechanics," he still rejected positions as unobservable:

> Formal quantum mechanics obviously provides no means for the determination of the positions of particles in space and time. ... The quantum theoretical description of the system contains certain assertions about the energy, the momenta, the angular momenta of the system; but does not answer, or at least only answers in the limiting case of classical mechanics, the question of where a certain particle is at a given time. In this respect the quantum theory is in agreement with the experimentalists, for whom microscopic coordinates are also out of reach.[52]

Thus, originally Born did not envisage the interpretation of the amplitude squared of the wave function as proportional to the probability of a position measurement as Linda Wessels and Mara Beller have pointed out.[53]

By the time of Heisenberg's discussions with Einstein, there was thus a considerable tradition citing the authority of Einstein to assert that only quantities observable in principle should be introduced in a physical theory. Heisenberg claimed to base his approach to quantum theory on this principle.[54] In his first paper (July 1925) he states:

> It is well known that the formal rules which are used in quantum theory for calculating observable quantities, such as the energy of the hydrogen atom, may be seriously criticized on the grounds that they contain, as a basic element, relations between quantities that are apparently unobservable in principle, e.g., position and period of revolution of the electron. Thus, these rules lack an evident physical foundation, unless one still wants to retain the hope that the hitherto unobservable quantities may later come within the realm of experimental determination.

After giving reasons why such a hope seems unfounded, he continues:

> In this situation it seems sensible to discard all hope of observing hitherto unobservable quantities, such as the position and period of the electron, and to concede that the partial agreement of the quantum rules with experience is more or less fortuitous. Instead it seems more reasonable to try to establish a theoretical quantum mechanics, analogous to classical mechanics, but in which only relations between observable quantities occur.[55]

Heisenberg was in correspondence with Einstein soon after this. Only Heisenberg's replies are available; it appears from these that, as a result of Einstein's objections, he was anxious for a discussion of his approach. A talk in Berlin gave Heisenberg the occasion for this discussion. In a chapter entitled "Quantum Mechanics and a Talk with Einstein," he gives an admittedly Thucydidean reconstruction of that conversation. Heisenberg reports that he stated:

> We cannot observe electron orbits inside the atom but the radiation which an atom emits during discharges enables us to deduce the frequencies and corresponding amplitudes of its electrons. After all, even in the older physics wave numbers and amplitudes could be considered substitutes for electron orbits. Now, since a good theory must be based on directly observable magnitudes, I thought it more fitting to restrict myself to these, treating them, as it were, as representatives of the electron orbits.
>
> But you don't seriously believe, Einstein protested, that none but observable magnitudes must go into a physical theory?
>
> Isn't that precisely what you have done with relativity? I asked in some surprise. After all, you did stress the fact that it is impermissible to speak of absolute time, simply because absolute time cannot be observed; that only clock readings, be it in the moving reference system or the system at rest, are relevant to the determination of time.
>
> Possibly I did use this kind of reasoning, Einstein admitted, but it is nonsense all the same. Perhaps I could put it more diplomatically by saying that it may be heuristically useful to keep in mind what one has actually observed. But on principle, it is quite wrong to try founding a theory on observable magnitudes alone. In reality the very opposite happens. It is the theory which decides what we can observe (*Physics and Beyond*, p. 63).

This discussion is often cited as the quintessential clash between the positivism of Heisenberg and the realistic approach of Einstein. But that is *not* the reason Heisenberg recounts it. The denouement comes only in the next chapter, in which Heisenberg recalls the discovery of the uncertainty relations in February 1927. Bohr and Heisenberg were then beset by doubts about how to reconcile the quantum-mechanical formalism with the circumstance that:

> The path of the electron through the cloud chamber obviously existed; one could easily observe it. The mathematical framework of quantum mechanics existed as well, and was much too convincing to allow for any changes. Hence it ought to be possible to establish a connection between the two, hard though it appeared to be.
>
> It must have been one evening after midnight when I suddenly remembered my conversation with Einstein and particularly his statement, "It is the theory which decides what we can observe." I was immediately convinced that the key to the gate that had been closed for so long must be sought right here. . . .
>
> We had always said so glibly that the path of the electron in the cloud chamber could be observed. But perhaps what we really observed was something much less. Perhaps we merely saw a series of discrete and ill-defined spots through which the electron had passed. In fact, all we do see in the cloud chamber are individual water droplets which must certainly be much larger than the electron. The right question should therefore be: Can quantum mechanics represent the fact that an electron finds itself approximately in a given place and that it moves approximately with a given velocity, and can we make these approximations so close that they do not cause experimental difficulties? (*Physics and Beyond*, pp. 77–78).

Heisenberg thus says that a *retreat* from his original extreme positivist or operationalist viewpoint, motivated by Einstein's critique, helped him to find the uncertainty relations. I shall not retell Bohr's story of how Einstein originally tried to circumvent the uncertainty relations by a series of ingenious thought experiments.[56] But I must emphasize that he finally came to accept them fully as a limit on possible experimental determination of the properties of a system. Indeed, he often stressed this point in his later discussions of quantum mechanics, as we shall see.

I shall only mention Einstein's contributions to the experimental study of the question of the nature of light. As we have seen, Einstein never believed in what he called the "naive conception" of light quanta as totally independent point particles. After he became convinced, in 1916–1917, that light quanta had the particlelike property of momentum as well as energy, he tried to design a crucial experiment, which would distinguish between any reasonable sort of particle picture of light and the classical undulatory picture. If there ever was a time when Einstein was an "insider" in the physics community, it was during the twenties in Berlin. So his two suggestions were quickly carried out by experimentalists. But in both cases it turned out, upon more careful theoretical analysis of the situation, that the proposed experiment was actually unable to distinguish between the wave and particle pictures of light.[57] The failure of these "crucial" experiments designed by Einstein probably played a role in the evolution of Bohr's views on light. Bohr held firmly to the undulatory picture for a long time. As late as the early 1920s, he regarded Einstein's light quantum concept as an aberration. Bohr was willing to give up conservation of energy for individual interactions between light and matter to preserve an exclusively undulatory model of light (the Bohr–Kramers–Slater theory of 1923).[58] When the results of the Compton–Simon and Bothe–Geiger experiments showed this approach to be untenable, Bohr was forced to take the particle picture more seriously; but he still would not give up the undulatory picture. I believe the demonstration that both pictures led to the same prediction for the outcome of Einstein's experiments was one of the elements that contributed to shaping Bohr's concept of complementarity of wave and particle descriptions.[59] Bohr continued to regard the particle aspect of light, however, as playing a secondary, more formal role compared to the undulatory aspect, the only one to manifest itself in the classical limit. In Bohr's view, the wave aspects of the electron played a similar formal role. Although Einstein never accepted complementarity (saying he had failed to make sense of Bohr's concept of complementarity in spite of repeated attempts to do so), he came to a rather similar conclusion: "I do not believe that the light-quanta have reality in the same immediate sense as the corpuscles of electricity. Likewise I do not believe that the particle-waves have reality in the same sense as the particles themselves. The wave-character of particles and the particle-character of light will—in my opinion—be understood in a more indirect way, not as immediate physical reality." (Einstein to Paul Bonofield, September 18, 1939, Item 6-118.1).

7. Einstein's Critique of Quantum Mechanics

In concluding this essay, I shall discuss the nature of Einstein's critique of quantum mechanics. This question was closely tied up with his hopes for—and doubts about—the future course of physics. As mentioned previously, Einstein was troubled by the inherently probabilistic nature of the quantum-mechanical description of reality. He indicated that he could more easily conceive of a completely chaotic universe than one governed by probabilistic *laws*: "I still do not believe that the Lord God plays dice. If he had wanted to do this, then he would have done it quite thoroughly and not stopped with a plan for his gambling: In for a penny, in for a pound [*Wenn schon, denn schon*]. Then we wouldn't have to search for laws at all" (Einstein to F. Reiche and wife, August 15, 1942, Item 20-190).

He drafted a reply to Born's essay on "Einstein's Statistical Theories": "This article is a moving hymn to a beloved friend, who in his old age (shall we say) unfortunately has succumbed to occultism, one who will not believe in spite of all the evidence, that God plays dice. In one point however, Born does me an injustice, namely, when he thinks that I have been untrue to myself in this respect since earlier I often availed myself of statistical methods. In truth, I never believed that the foundations [*Grundlage*] of physics could consist of laws of a statistical nature" (unpublished draft comment on Born's essay for *Albert Einstein: Philosopher-Scientist*, Item 2-027).

But this issue was not the deepest source of his dissatisfaction with the prevailing interpretation of quantum mechanics: "The sore point [*Der wunde Punkt*] lies less in the renunciation of causality than in the renunciation of the representation of a reality thought of as independent of observation" (Einstein to Georg Jaffe, January 19, 1954, Item 13-405).

It is important to emphasize that he did not see this as a defect of quantum mechanics as such, but as a defect of the prevailing interpretation of the theory as the most complete possible description of an individual system:

> In what relation does the "state" ("*quantum state*") described by a ψ function stand to a definite real situation [*Sachverhalt*] (let us call it "real state")? Does the quantum state characterize a real state (1) completely or (2) only incompletely?
>
> The question cannot be answered at once [*ohne Weiteres*] because indeed every measurement signifies an uncontrollable real intervention [*Eingriff*] in the system (Heisenberg). The real state is thus not immediately accessible to experience and its judgement always remains hypothetical (comparable with the concept of force in classical mechanics, if we do not set up the laws of motion a priori). Assumptions (1) and (2) are therefore both possible in principle. A decision between them can only be reached by investigation and comparison of their consequences.
>
> I reject (1) because this outlook necessitates the assumption that a rigid coupling exists between parts of a system spatially arbitrarily far apart from each other (instantaneous action at a distance, which does not decrease with increasing distance). (*Besso Correspondance*, p. 487, letter of October 8, 1952).

The last sentence is an allusion to the Einstein–Podolsky–Rosen (EPR) argument. Since this argument is well known, I shall only quote a short account by Einstein:

> It is not my opinion that there is a logical inconsistency in the quantum-theory itself and the "paradoxon" does not try to show it. The intention is to show that statistical quantum theory is not compatible with certain principles the convincing power of which is independent of the present quantum theory.
>
> There is the question: Does it make sense to say that two parts A and B of a system do exist independently of each other if they are (in ordinary language) located in different parts of space at a certain time, if there are no considerable interactions between those parts (expressed in terms of potential energy) at the considered time? ... I mean by "independent of each other" that an action on A has no immediate influence on the part B. In this sense I express a principle a)
>
> a) independent existence of the spacially separated. This has to be considered with the other thesis b)
>
> b) the ψ-function is the complete description of an individual physical situation.
>
> My thesis is that a) and b) can not be true together, for if they would hold together the special kind of measurement concerning A could not influence the resulting ψ-function for B, (after measurement of A).
>
> The majority of quantum theorists discard a) tacitly to be able to conserve b). I, however, have strong confidence in a) so that I feel compelled to relinquish b) (Einstein to L. Cooper, October 31, 1949, Item 8-411).

Einstein held that, if one adopted the ensemble or statistical interpretation of the ψ-function, which he identified with the Born interpretation, there was no problem. After a concise exposition of the EPR paper, Einstein continues:

> It became clear to me, meanwhile, how one has to choose the interpretation of the formalism [*des Schemas*] of quantum mechanics, in order that the concept of the real state of an (isolated) system not give rise to any sort of paradox; it is none other than the Born interpretation, about which I do not know for certain if it is represented by Born himself with complete consistency.
>
> It is ... to be expected that behind quantum mechanics there lies a lawfulness and a description which refer to the individual system. That it is not attainable within the bounds of concepts [taken] from classical mechanics is clear; the latter however is in any case outmoded as the foundation [*Fundament*] of physics (Einstein to Gregory Breit, August 2, 1935, Item 6-173).

Elsewhere, he elaborates on the inadequacy of the classical-mechanical starting point of quantum mechanics:

> I do not at all doubt that the contemporary quantum theory (more exactly "quantum mechanics") is the most complete theory compatible with experience, as long as one bases the description on the concepts of material point and potential energy as fundamental concepts. [The difficulties of quantum mechanics] are connected with the fact that one retains the classical concepts of force or potential energy and only replaces the laws of motion by something entirely new. The completeness of the mathematical mechanism of the theory and its significant success turn attention away from the difficulty of the sacrifice that has been made.
>
> To me it seems, however, that one will finally recognize that something must take the place of forces acting or potential energy (or in the Compton effect, of wave fields), something which has an atomistic structure in the same sense as the electron itself. "Weak fields" or forces as active causes will then not occur at all, just as little as mixed states.[60]

In one of his last letters, he again touched on this point, also noting a source of resistance to such objections.

> I believe however that the renunciation of the objective description of "reality" is based upon the fact that one operates with fundamental concepts which are untenable in the long run (like f.i. classical thermodynamics). Quite understandably most physicists resist the idea that we are still very far from a deeper insight into the structure of reality (Einstein to Andre Lamouche, March 20, 1955, Item 15-044).

Did Einstein believe in some sort of hidden variable program to underpin quantum mechanics? Bernard d'Espagnat, after a clear exposition of Einstein's approach to the nonseparability problem in quantum mechanics, maintains that: "Einstein's conclusion was that hidden parameters must exist. These parameters he seems to have conceived as attached to each system, as in the classical case, or as we would say now, as 'local variables'; for in such a way his principle of separability would not be violated."[61]

J. S. Bell expresses similar views in a note, "Einstein and Hidden Variables," defending his position against criticism by Max Jammer: "I had for long thought it quite conventional and uncontroversial to regard Einstein as a proponent of hidden variables and indeed ... as the most profound advocate of hidden variables.[62] A recent review article on the topic by Clauser and Shimony repeats this view.[63]

While there are statements by Einstein that might seem to allow such an interpretation, I believe that this view misses the basic thrust of Einstein's search for a unified field theory:

> In a consistent field theory there is no real definition of the field. ... *A priori* no bridge to the empirical is given. There is not, for example, any "particle" in the strict sense of the word, since it does not fit into the program of representing reality by everywhere continuous, indeed even analytic, functions. ... The upshot is that a comparison with the empirically known can only be expected to come from finding exact solutions of the system of equations, in which empirically "known" structures and their interactions are "reflected." Since this is immensely difficult, the sceptical attitude of contemporary physicists is quite understandable (*Besso Correspondance*, pp. 483–39, letter of April 15, 1950).

> Our situation is this. We stand before a closed box, that we cannot open, and try to discuss what is inside and what is not. The similarity of the theory to Maxwell's is only external, so that we cannot transfer the concept of "force" from this theory to the asymmetric field theory. If this theory is at all useful, then one cannot assume any separation between particles and field of interaction. In addition, there is no concept at all of the *motion* of something more or less rigid. The question here is exclusively: are there singularity-free solutions? Is their energy in particular localized in such a way as demanded by our knowledge of the atomic and quantum character of reality [*Realität*]? The answer to this question is really not attainable with contemporary mathematical methods. Thus I do not see how one can guess whether any sort of action-at-a-distance or any type of object, insofar as we have attained a semi-empirical knowledge of them, can be represented by the theory (Einstein to John Moffat, August 24, 1953, Item 17-394).

If the concept of "hidden variable theory," not to speak of "*local* hidden variable theory," is given a precise meaning—and not used as shorthand for *any* non-quantum-mechanical approach—then what Einstein has in mind is *not* a hidden variable approach. Fuller discussions of this question are found in the book and articles by Max Jammer cited at the beginning of this paper.

Einstein never denied the great explanatory power of quantum mechanics, yet he did not feel this success required acceptance of its conceptual structure as the basis for further progress in theoretical physics. He wrote to Schrödinger, "The wonder [about quantum mechanics] is only that one can represent so much with it, although the most important theoretical source of knowledge, group invariance, finds such incomplete application there. . . . It is the case that a logically coherent theory that is connected appropriately to the real state of affairs usually has great extrapolatory power, even if it is little related to the deeper truth [*der Wahrheit in der Tiefe*]" (Einstein to Schrödinger, July 16, 1946, Item 22-109).

8. General Covariance

The mention of group invariance brings up one of the most profound sources of his skepticism about the ultimate significance of quantum mechanics. Einstein attached primary importance to the principle of general covariance: "You consider the transition to special relativity as the most essential thought of relativity, not the transition to general relativity. I consider the reverse to be correct. I see the most essential thing in the overcoming of the inertial system, a thing that acts upon all processes, but undergoes no reaction. This concept is in principle no better than that of the center of the universe in Aristotelian physics" (Einstein to Georg Jaffe, January 19, 1954, Item 13-405). Even relativistic quantum mechanics is constructed on a nongenerally covariant basis, but:

> Contemporary physicists do not see that it is hopeless to take a theory that is based on an independent rigid space (Lorentz-invariance) and later hope to make it general-relativistic (in some natural way) (Einstein to von Laue, September, 1950, Item 16-147).

> I have not really studied quantum field theory. This is because I cannot believe that special relativity theory suffices as the basis for a theory of matter, and that one can afterwards make a non-generally relativistic theory into a generally relativistic one. But I am well aware of the possibility that this opinion may be erroneous (Einstein to K. Roberts, September 6, 1954, Item 20-049).

Einstein felt it important to explore the possibility that a field theory based upon a continuous manifold, the principle of general covariance and partial differential equations, could provide an explanation of quantum phenomena:

> I am firmly convinced, that the [EPR] dilemma depicted can only be overcome through a quite different outlook on the situation, and indeed by a description that lies much closer to the "classical" than we now hold probable or even conceivable. One must always bear in mind that up to now we know absolutely nothing about the laws of motion of material points from the standpoint of "classical field theory." For the mas-

tery of this problem, however, no special physical hypothesis is needed, but "only" the solution of certain mathematical problems (Einstein to Ernst Cassirer, March 16, 1937, Item 8-394).

To Schrödinger he wrote explaining in more detail why exact solutions were needed:

The quantum facts seem ... to show that arbitrarily weak interactions can produce discrete changes (quantum jumps). Since this seems contradictory for the individual event, one is forced to the statistical interpretation: small forces produce small changes not in the individual things, but in the probability of their states. In reality however it must be something like this, that interactions have just as atomistic a character as the structures on which they act. Then it is all over with the quasi-static interpretation of interactions! Therefore I don't believe that one can advance further with field theory by approximate considerations (weak fields) (Einstein to Schrödinger, July 25, 1951, Item 22-180).

Sometimes, he indicated that a fundamental length might be needed to explain the existence of stable structures in such a field theory:

If one does *not* want to introduce rods and clocks as independent objects into the theory, then one must have a structural theory in which a fundamental length enters, which then leads to the existence of a solution in which this length occurs, so that there no longer exists a continuous sequence of "similar" solutions. This is indeed the case in the present quantum theory, but has nothing to do with its basic characteristics. Any theory which has a universal length in its foundations and, on the basis of this circumstance, qualitatively distinguished solutions of definite extent would offer the same thing with respect to the question envisioned here (Einstein to W. F. G. Swann, January 24, 1942, Item 20-624).

At other times he proposed another idea: the overdetermination of a system of equations to restrict the manifold of possible initial conditions. In "Does Field Theory Offer the Possibility for a Solution of the Quantum Problem," he wrote:

According to our theories up to now the initial state of a system can be freely chosen; the differential equations then give the temporal development. From our knowledge of quantum states ... this trait of the theory does not correspond to reality. The initial state of an electron moving around the nucleus of a hydrogen atom cannot be freely chosen, but the choice must correspond to the quantum conditions. More generally; not only the temporal development but also the initial conditions obey laws.

Can one do justice to this knowledge about natural processes, to which indeed we must ascribe general significance, in a theory based upon partial differential equations? Quite certainly: we need only "overdetermine" the field variables by the equations. That is, the number of differential equations must be greater than the number of field variables determined by them.[64]

Einstein carried on a decades-long search for a classical field theory whose singularity-free solutions could reproduce the atomistic, quantum structure of matter and radiation. In spite of these efforts, he acknowledged the possibility that *any* theory based upon continuum concepts might be inadequate and that totally new mathematical concepts might be needed to explain quantum effects. As early as November 1916, he wrote to a former student:

> But you have correctly grasped the drawback that the continuum brings. If the molecular view of matter is the correct (appropriate) one, i.e., if a part of the universe is to be represented by a finite number of moving points, then the continuum of the present theory contains too great a manifold of possibilities. I also believe that this too great is responsible for the fact that our present means of description miscarry with the quantum theory. The problem seems to me how one can formulate statements about a discontinuum without calling upon a continuum (space-time) as an aid; the latter should be banned from the theory as a supplementary construction not justified by the essence of the problem, which corresponds to nothing "real." But we still lack the mathematical structure unfortunately. How much have I already plagued myself in this way!
>
> Yet I see difficulties of principle here too. The electrons (as points) would be the ultimate entities in such a system (building blocks). *Are there* indeed such building blocks? Why are they all of equal magnitude? Is it satisfactory to say: God in his wisdom made them all equally big, each like every other, because he wanted it that way; if it had pleased him, he could also have created them different. With the continuum viewpoint one is better off in this respect, because one doesn't have to prescribe elementary building blocks from the beginning. Further, the old question of the vacuum! But these considerations must pale beside the overwhelming fact: The continuum is more ample than the things to be described (Einstein to Walter Dällenbach, November 1916, Item 9-072).

He continued to discuss this possibility over the years. In 1935 he wrote to Paul Langevin:

> In spite of all successes of quantum mechanics I do not believe that this method can offer a useable *foundation* [*Fundament*] of physics. I see in it something analogous to classical statistical mechanics, only with the difference that here we have not found the equations corresponding to those of classical mechanics.
>
> In any case one does not have the right today to maintain that the foundation must consist in a *field theory* in the sense of Maxwell. The other possibility, however, leads in my opinion to a renunciation of the time-space continuum and to a purely algebraic physics. Logically this is quite possible (the system is completely described by a number of integers; "time" is only a possible standpoint [*Gesichtspunkt*], from which the other "observables" can be considered—an observable logically coordinated to all the others). Such a theory doesn't have to be based upon the probability concept. For the present, however, instinct rebels against such a theory (Einstein to Langevin, October 3, 1935, Item 15-408).

In his last years, he made a number of rather pessimistic comments about the prospects for a continuum theory:

> In present-day physics there is manifested a kind of battle between the particle-concept and the field-concept for leadership, which will probably not be decided for a long time. It is even doubtful if one of the two rivals finally will be able to maintain itself as a fundamental concept (Einstein to Herbert Kondo, August 11, 1952, Item 13-306).

> I consider it as entirely possible that physics cannot be based upon the field concept, that is on continuous structures. Then *nothing* will remain of my whole castle in the air including the theory of gravitation, but also nothing of the rest of contemporary physics (*Besso Correspondance*, p. 527, letter of August 10, 1954).

> Your objections regarding the existence of singularity-free solutions which could represent the field together with the particles I find most justified. I also share this doubt. If it should finally turn out to be the case, then I doubt in general, the existence of a rational physically useful *field* theory. But what then? Heine's classical line comes to mind "And a fool waits for the answer" (Einstein to André Lichnerowicz, February 25, 1954, Item 16-321).

> I must confess that I was not able to find a way to explain the atomistic character of nature. My opinion is that if the objective description through the field as an elementary concept is not possible, then one has to find a possibility to avoid the continuum (together with space and time) altogether. But I have not the slightest idea what kind of elementary concepts could be used in such a theory (Einstein to Bohm, October 28, 1954, Item 8-050).

At least once, he speculated in somewhat greater detail on what a noncontinuum theory might involve:

> The alternative continuum-discontinuum seems to me to be a real alternative; i.e., there is here no compromise. By discontinuum theory I understand one in which there are no differential quotients. In such a theory space and time cannot occur, but only numbers and number-fields and rules for the formation of such on the basis of algebraic rules with exclusion of limiting processes. Which way will prove itself, only success can teach us.
>
> Physics up to now is naturally in its essence a continuum physics, in spite of the use of the material point, which looks like a discontinuous conceptual element, and has no more right of existence in field description. Its strength lies in the fact that it posits parts which exist quasi-independently, *beside* one another. Upon this rests the fact that there are reasonable laws, that is rules which can be formulated and tested for the invididual parts. Its weakness lies in the fact that it has not been possible up to now to see how that atomistic aspect including quantum relations can result as consequences. On the other hand dimensionality (as four-dimensionality) lies at the foundation of the theory.
>
> An algebraic theory of physics is affected with just the inverted advantages and weaknesses, aside from the fact that no one has been able to propose a possible logical schema for such a theory. It will be especially difficult to derive something like a spatio-temporal quasi-order from such a schema. I cannot imagine how the axiomatic framework of such a physics would appear, and don't like it, when one talks about it in dark apostrophes (*Anredungen*). But I hold it entirely possible that the development will lead there; for it seems that the state of any finite spatially limited system may be fully characterized by a finite number of numbers. This speaks against the continuum with its infinitely many degrees of freedom. This objection is not decisive only because one doesn't know, in the contemporary state of mathematics, in what way the demand for freedom from singularity (in the continuum theory) limits the manifold of solutions (Einstein to H. S. Joachim, August 24, 1954, Item 13-453).

Even near the end of his life, he was on the lookout for new mathematical tools, which could turn such speculations—best kept private—into the basis for a real theory. The well-known mathematician Abraham Fraenkel reports:

> In December 1951 I had the privilege of talking to Professor Einstein and describing the recent controversies between (neo-) intuitionists and their "formalistic" and

> "logicistic" antagonists; I pointed out that the first attitude would mean a kind of atomistic theory of functions, comparable to the atomistic structure of matter and energy. Einstein showed a lively interest in the subject and pointed out that to the physicist such a theory would seem by far preferable to the classical theory of continuity. I objected by stressing the main difficulty, namely the fact that the procedures of mathematical analysis, e.g., of differential equations, are based on the assumption of mathematical continuity, while a modification sufficient to cover an intuitionistic-discrete medium cannot easily be imagined. Einstein did not share this pessimism and urged mathematicians to try to develop suitable new methods not based on continuity.[65]

9. Conclusion

One thing seems certain. The many attempts to reconcile general relativity and quantum theory have not yet brought us anything like a fully successful theory. Einstein expressed himself skeptically about all such attempts: "I do not believe that it will lead to the goal if one sets up a classical theory and then 'quantizes' it. This way was indeed successful in connection with the interpretation of classical mechanics and the interpretation of the quantum facts [*Quantentatsachen*] by modification of that theory on a fundamentally [*prinzipiell*] statistical basis. But I believe that, in attempts to transfer this method to field theories, one will hit upon steadily mounting complications and upon the necessity to multiply the independent assumptions enormously" (Einstein to John Moffat, June 4, 1953, Item 17-390).

The reconcilation of gravitation and quantization is still one of the most intractable problems in contemporary theoretical physics. The most profound comment one can make on it is that nobody has any really sound basis for believing that he or she knows how it will be solved—or bypassed—in the future. We can agree with Einstein: "One thing I have learnt in a long life: it is devilishly hard to get closer to 'Him', if one doesn't want to remain on the surface" (*Besso Correspondance*, p. 439, letter of April 15, 1950).

Addenda

To Section 2, p. 371: A letter from Einstein to Sommerfeld, discovered after this article was written, shows that by 1908 Einstein made clear that he regarded what he later called "principle theories," such as the theory of special relativity, as only provisional steps on the road to what he later called "constructive theories":

> First of all, now, the question of whether I consider the relativistic treatment, e.g., of the mechanics of the electron, as definitive. No, certainly not. To me, it also seems that a physical theory can only be satisfactory if it constructs its structures [Gebilde] from *basic* elements [*elementare* Grundlagen]. The theory of relativity is just as little ultimately satisfying as, e.g., classical thermodynamics before Boltzmann had interpreted entropy as probability. If the Michelson–Morley experiment had not left us in the greatest confusion, no one would have accepted the theory of relativity as a (half-) salvation. Besides, I believe that we are still far from having satisfactory basic elements for electrical and mechanical progress [Vorgänge]. I am led to this pessimistic

> viewpoint primarily as a consequence of endless vain attempts to interpret the second universal constant in Planck's radiation law [i.e., *h*–JS] in an intuitive [anschaulich] way. I even seriously doubt that we shall be able to maintain the general validity of Maxwell's equations for empty space (Einstein to Arnold Sommerfeld, 14 January, 1908, *Collected Papers of Albert Einstein*, vol. 5, *The Swiss Years: Correspondence, 1902–1914* [Princeton University Press, 1993], Doc. 73, pp. 86–89).

To Section 7, pp. 390–393: I have since found a letter, in which Einstein explicitly rejects the "hidden variable" approach to quantum mechanics:

> Dr. Bohm has rediscovered a 30 years old idea of de Broglie and with great acumen enlarged and deepened it. The goal is description of the single system instead of an ensemble to which the individual system belongs. Born's statistical interpretation of the ψ-function is replaced by a rule according to which to a given ψ-function a world line is coordinated, determined by a given initial configuration (point) in the coordinate space. ... I think that it is not possible to get rid of the statistical character of the present quantum theory by merely adding something to the latter without changing the fundamental concepts about the whole structure. Superposition principle and statistical interpretation are inseparably bound together. If one believes that the statistical interpretation should be avoided and replaced it seems one cannot conserve a linear Schrödinger equation which implies by linearity the principle of superposition of "states" (Einstein to Dr. Aron Kupperman, 14 November 1953, Einstein Archive, Control No. 08-036).

Notes

[1] P. Speziali, ed., *Albert Einstein–Michele Besso Correspondance, 1903–1955* (Paris: Hermann, 1972), p. 453 (hereafter referred to as *Besso Correspondance*). I shall give an example later of the sort of thing Einstein might have had in mind when he referred to self-deception.

[2] A copy of the letter is in the Einstein Archive, Hebrew University of Jerusalem. I shall refer to such unpublished items by the Control Index number of the item in the duplicate of this Archive in Mudd Library, Princeton University (in this case, Item 16-167). Laue's was the first such treatise, published in 1911 after Einstein had turned down the publisher's request to write one.

[3] M. Klein, "No Firm Foundation: Einstein and the Early Quantum Theory," in H. Woolf, ed., *Some Strangeness in the Proportion* (Reading: Addison-Wesley, 1980), pp. 161–85.

[4] A. Pais, "Einstein and the Quantum Theory," *Rev. Mod Phys.* **51** (1979); 863. This article is included almost unchanged in A. Pais, *"Subtle is the Lord ..." The Science and the Life of Albert Einstein* (Oxford: Clarendon Press, 1982). See also A. Pais, "Einstein on Particles, Fields and the Quantum Theory," in Woolf, *Some Strangeness in the Proportion*, pp. 197–251.

[5] M. Jammer, *The Conceptual Development of Quantum Mechanics* (New York: McGraw-Hill, 1966); M. Jammer, *Albert Einstein und das Quantenproblem*, in H. Nelkowski et al., eds., *Einstein Symposion Berlin* (Berlin: Springer-Verlag, 1980); G. Holton and Y. Elkana, eds., *Albert Einstein/Historical and Cultural Perspectives* (Princeton: Princeton University Press, 1982), pp. 59–76.

[6] *Besso Correspondance*, p. 238. There is a word following "fame" not transcribed in the printed version. It is illegible in the copy in the Einstein Archive and Professor Speziali kindly informed me that the original is no longer available to him.

[7] The letter is quoted in Carl Seelig, *Albert Einstein und die Schweiz* (Zürich: Europa Verlag, 1952), pp. 77–78. The letter can be dated from its reference to the discovery of the mass-energy equivalence relation. I am indebted to Martin Klein for pointing out this quotation and for encouraging me to publish these speculations.

[8] See A. Kleinert and C. Schoenbeck, "Lenard und Einstein," *Gesnerus* **35** (1978): 318, for the text of both letters cited here.

[9] Of course, we now know that emission of a series of spectral lines is connected with transitions from a number of higher energy states to one lower state or level. I am not claiming that Einstein correctly anticipated the 1913 Bohr model in this letter.

[10] This letter is printed in A. S. Eve, *Rutherford* (New York: Macmillan, 1930), pp. 224–26. The Pickering–Fowler series of spectral lines was first attributed to hydrogen. But the Bohr theory's prediction that it was actually due to singly ionized helium was experimentally confirmed in 1913 by Evans and, later, by Fowler and Paschen.

[11] Letter of September 23, 1913, published in Ulrich Hoyer, ed., *Collected Works of Niels Bohr*, vol. 2, *Work on Atomic Physics, 1912–1917* (Amsterdam: North-Holland, 1982), p. 532. I am indebted to Dr. Erik Rüdinger for a copy of this letter before its publication.

[12] *Über den Aether*, lecture of October 4, 1924, published in A. Einstein, *Verh d. Schweiz. Naturf. Ges.* **105** (1924): 85.

[13] A. Pais, "Reminiscences from Post-war Years," in Rozental, *Niels Bohr* (New York: Interscience, 1964), p. 222.

[14] A. Einstein, "Autobiographical Notes," in P. A. Schilpp, ed., *Albert Einstein: Philosopher-Scientist* (La Salle: Open Court, 1951), pp. 3–94. A new edition with corrected translation appeared in 1979 entitled *Albert Einstein: Autobiographical Notes*. The first edition will hereafter be cited as "Autobiographical Notes." In a draft statement he wrote in the early 1920s, proposing Bohr as a corresponding member of the Prussian Academy of Sciences, Einstein elaborated: "The empirically widely known structure of [atomic] spectra differed so essentially from what was to be expected, according to our old theories, that no one ⟨had the courage⟩ saw a possibility ⟨for saw for a compelling⟩ for a convincing theoretical interpretation of the observed regularities." (The words Einstein crossed out are in angle brackets; the lack of "courage" may well have been a self-reference.)

[15] Cited in Ronald Clark, *Einstein: The Life and Times* (New York: World 1971), p. 169.

[16] T. Kuhn, *Black Body Theory and the Quantum Discontinuity, 1894–1912* (Oxford: Oxford University Press, 1978).

[17] "Peter J. W. Debye: An Interview," *Science* **145** (1964): 554.

[18] A. Einstein, *Ann. Phys.* **23** (1907): 371.

[19] A. Einstein, *Ideas and Opinions* (New York: Crown, 1954), p. 225. It will hereafter be cited as *Ideas and Opinions*.

[20] The German text, as well as an English translation, appear in *Builders of the Universe* (Los Angeles: U. S. Library Association, 1932). I have retranslated the German version, pp. 94–96.

[21] Einstein was clearly in a foul mood at the time this letter was written, shortly after Hitler's triumph at Munich. He added a P.S.: "What do you say to the way the fine democracies [*sauberen Demokratien*] are behaving? I say 'Pfui Teufel!'"

[22] But not classical statistical mechanics proper, which led to disastrous consequences such as the equipartition theorem giving the wrong black body spectrum—the so-called Rayleigh–Jeans law (although Einstein was the first to write it down completely correctly)—and the wrong low temperature behavior of specific heats.

[23] A. Einstein, "Zum gegenwärtigen Stande des Problems der spezifischen Wärme," in W. Nernst, ed., *Die Theorie der Strahlung und der Quanten* (Halle: Krapp Verlag, 1914), p. 353. This edition contains the original German text of Einstein's comments. The proceedings of the 1911 Solvay Congress were first published in French: P. Langevin and M. de Broglie, eds., *La Théorie du Rayonnement et les Quanta* (Paris: Gauther-Villars, 1912). Einstein had already considered and rejected the possibility of a statistical interpretation of the energy conservation principle. In a letter to Jacob Laub of Nov. 4, 1910, Einstein wrote: "Currently I have great hope of solving the radiation problem, and indeed *without light quanta* ... One must renounce the energy principle in its present form."

[24] However, Einstein showed great reluctance to mixing relativistic considerations into his work on quantum theory: witness his over-a-decade-long delay between writing $E = h\nu$ and $p = h\nu/c$. Pais has discussed the reasons for this in detail.

[25] A. Einstein, *Physik. Zeitschr.* **10** (1909): 185.

[26] Einstein published his results three times: A. Einstein, *Verhandl. Deutsch. Phys. Ges.* **18** (1916): 318, *Mitt. Phys. Ges. Zürich* **16** (1916): 47, *Phys. Z.* **18** (1917); 121, the last two being the same paper, translated in B. L. van der Waerden, ed., *Sources of Quantum Mechanics* (Amsterdam: North-Holland, 1967), pp. 63–77. This book will be cited hereafter as van der Waerden, *Sources*.

[27] M. Born, W. Heisenberg, and P. Jordan, *Z. Phys.* **35** (1926): 557, translated in van der Waerdan, *Sources*, pp. 312–85. Fluctuations are discussed in the final section of the paper.

[28] P. Ehrenfest, *Z. Phys.* **34** (1925); 362.

[29] P. Jordan, *Z. Phys.* **45** (1927): 766. See pp. 773–75.

[30] See footnote 17.

[31] See J. J. Gonzalez and H. Wergeland, *Kongel. Norske Videnskab. Selsk. Skr. 1973*, no. 4, "Einstein–Lorentz's Formula for the Fluctuations of Electromagnetic Energy," for a recent review, including reference to Heisenberg's paper. See also N. Bohr and L. Rosenfeld, *Mat.-fys. Medd. Dan. Vid. Selsk.* **12**, no. 8 (1933), for the fundamental paper on the conditions of measurability of field quantities. This paper is translated in R. S. Cohen and J. J. Stachel, eds., *Selected Papers of Leon Rosenfeld* (Dordrecht: Reidel, 1979), pp. 357–400.

[32] For recent suveys of this topic, see D. W. Sciama, "Black Holes and Fluctuations of Quantum Particles: An Einstein Synthesis," in M. Panteleo and F. de Finis, eds., *Relativity, Quanta and Cosmology in the Development of the Scientific Thought of Albert Einstein* (New York: Johnson Reprint Corp., 1979), vol. 2, pp. 681–724; and J. D. Bekenstein, "Gravitation, the Quantum and Statistical Physics," in Y. Ne'eman, ed., *To Fulfill a Vision* (Reading: Addison-Wesley, 1981), pp. 42–59.

[33] M. Born, ed., *The Born–Einstein Letters* (New York: Walker, 1971).

[34] He did use the term at least once in his correspondence. See Einstein to Ehrenfest, January 11, 1922, Item 10-003.

[35] H. A. Lorentz, *Problems of Modern Physics* (Boston: Ginn and Co., 1927; reprint, New York: Dover, 1967).

[36] M. Born, *Zeitschr. f. Physik* **38** (1926): 803.

[37] W. Heisenberg, *Zeitschr. f. Physik* **43** (1927): 172.

[38] A reference to "something which he [Einstein] calls a 'ghost' field of radiation," in H. A. Kramers and H. Holst, *The Atom and the Bohr Theory of Its Structure* (New York: Knopf, 1923) proves that Einstein's idea was well known in Copenhagen in the early twenties.

[39] A. Einstein, *Sitzungsber. preuss. Akad. Wiss., physik.-math. Kl.* (1925): 3.

[40] L. de Broglie, "Meeting with Einstein at the Solvay Council in 1927," in L. de Broglie, *New Perspectives in Physics* (New York: Basic Books, 1962), pp. 180–85.

[41] Part of this correspondence has been published: K. Przibram, ed., *Schrödinger–Planck–Einstein–Lorentz: Briefe zur Wellenmechanik* (Vienna: Springer, 1963), translated as K. Przibram, ed., *Letters on Wave Mechanics: Schrödinger, Planck, Einstein, Lorentz* (New York: Philosophical Library, 1967).

[42] Copies of Heisenberg's and Jordan's letters to Einstein are in the duplicate Einstein Archive. Einstein's letters to both, except for one postcard to Jordan, appear to have been destroyed during the war.

[43] Ulrich Benz, *Arnold Sommerfeld* (Stuttgart: Wissenschafliche Verlagsgesellschaft, 1975), pp. 152–53.

[44] W. Heisenberg, *Der Teil und das Ganze* (Munich: DTV, 1973), translated as *Physics and Beyond* (New York: Harper & Row, 1971). This will be cited hereafter as *Physics and Beyond.*

[45] H. Minkowski, *Gött, Nachr. Math.-Phys. Kl.* (1908): 54.

[46] Einstein, "Zum gegenwärtigen Stande," p. 353. Einstein had used the phrase "variables observable in principle" in a similar context the previous year. See A. Einstein, *Ann. Phys.* **33** (1910): 1275.

[47] Einstein, "Zum gegenwärtigen Stande," p. 355.

[48] J. Stachel, "Einstein's Search for General Covariance, 1912–1915," talk given at the Ninth International Conference on General Relativity and Gravitation, July 17, 1980, to be published. [See this volume, pp. 301–337.]

[49] W. Pauli, *Verhandl. Deutsch. Physik. Ges.* **21** (1919): 742.

[50] A. Hermann, K. V. Meyenn, and V. F. Weisskopf, eds., *Wolfgang Pauli/Wissenschaftliche Briefwechsel*, vol. 1: 1919–1920 (Berlin: Springer-Verlag, 1979). This will be referred to hereafter as *Pauli Briefwechsel.* See pp. 115–19 for the letter to Eddington.

[51] M. Born and P. Jordan, *Zeitschr. f. Physik* **33** (1925): 479.

[52] M. Born, *Nature* **119** (1927): 354, reprinted in M. Born, *Physics in My Generation* (New York: Springer-Verlag, 1969), pp. 6–12.

[53] L. Wessels, "What Was Born's Statistical Interpretation," in P. O. Asquith and R. N. Giesel, eds., *PSA 1980* (East Lansing: P. S. A., 1980), vol. 2; M. Beller, "The Myth of Born's Probabilistic Interpretation" (preprint). I thank Dr. Beller for a copy of this paper.

[54] The historical accuracy of this claim has been challenged: see E. MacKinnon's paper, "Heisenberg, Models, and the Rise of Matrix Mechanics," in R. McCormmach and L. Peyson, eds., *Historical Studies in the Physical Sciences*, vol. 8 (Baltimore: Johns Hopkins University Press, 1977).

[55] W. Heisenberg, *Zeitschr. f. Physik.* **33** (1925): 879. Translation cited from Van der Waerden, *Sources*, pp. 261–62.

[56] For Bohr's account, see N. Bohr, "Discussions with Einstein on Epistemological Problems in Atomic Physics," in Schilpp, *Albert Einstein* (1951), pp. 201–41.

[57] See A. Einstein, *Sitzungsber. preuss. Akad. Wiss. physik.-math. Kl.* (1921): 882, for the first proposed experiment and (1922): 18, for the reanalysis; A. Einstein, *Naturwiss.* **14** (1926); 300, for the second proposed experiment, and A. Einstein, *Sitzungsber. preuss. Akad. Wiss., physik.-math. Kl.* (1926): 234, for the reanalysis.

[58] As mentioned earlier, Einstein had considered and rejected this idea over a decade before.

[59] See Bohr's letter to Einstein of April 13, 1927 (Item 8-084).

[60] A. Einstein, "Introductory Remarks on Basic Concepts," in *Louis de Broglie, Physicien et Penseur* (Paris: Albin Michel, 1953), pp. 4–14.

[61] B. d'Espagnt, "Quantum Non-separability: How the Problem Evolved from Einstein's Time to Ours," Orsay preprint, December, 1976.

[62] In *Frontier Problems in High Energy Physics* (Pisa: Scuola Normale Superiore, n.d.), pp. 41–45.

[63] J. F. Clauser and A. Shimony, *Rep. Prog. Phys.* **41** (1978): 1881.

[64] A. Einstein, *Sitzungsber. preuss. Akad. Wiss., physik.-math. Kl.* (1923): 359.

[65] A. Fraenkel, *Bull. Res. Coun. Israel* **3** (1954): 283.

Einstein and Quantum Mechanics

John Stachel

This paper is divided into five sections. Four of them are largely historical: The first ("A Lost Leader") discusses the charge that Einstein deserted his own earlier approach to the role of probability in physics when he refused to accept the fundamentally indeterministic element in quantum mechanics. The second section discusses Einstein's lifelong search for what he called a "constructive theory" of the quantum, and the third tries to highlight the essential elements in Einstein's critique of quantum mechanics. The fourth section ("The Other Einstein") recounts Einstein's persistent doubts about the entire field program as a foundation for theoretical physics, and his fragmentary suggestions for an alternative program.

The fifth section discusses three lessons that Einstein drew from his successful search for a generally covariant theory of gravitation: the indivisibility of gravitation and inertia; the lack of any prior physical individuation of points in a generally covariant theory before a solution to the field equations is given (no metric —no events); and the radically local character of general relativity in which even the global topology of the manifold depends on the solution to the field equations (no metric —no anything). It attempts to apply these lessons to a few contemporary problems of classical and quantum gravity.

1. "A Lost Leader"[1]

Max Born prefaced an essay on "Einsten's Statistical Theories" (Born 1949) with a few intensely personal words about Einstein:

> He has seen more clearly than anyone before him the statistical background of the laws of physics, and he was a pioneer in the struggle for conquering the wilderness of quantum phenomena. Yet later, when out of his own work a synthesis of statistical and quantum physics emerged which seemed to be acceptable to almost all physicists,

Conceptual Problems of Quantum Gravity
Proceedings of the 1988 Osgood Hill Conference
Abhay Ashtekar and John Stachel, eds.
Birkhäuser, 1991

> he kept himself aloof and sceptical. Many of us regard this as a tragedy —for him, as he gropes his way in loneliness, and for us who miss our leader and standard bearer (Born 1949, pp. 163–164).

In an unpublished reply to Born's essay, Einstein wrote:

> This article is a moving hymn to a beloved friend, who in his old age (shall we say) unfortunately has succumbed to occultism, one who will not believe in spite of all the evidence that God plays dice. In one point, however, Born does me an injustice, namely, when he thinks that I have been untrue to myself in this respect since earlier I often availed myself of statistical methods. In truth, I never believed that the foundation [*Grundlage*] of physics could consist of laws of a statistical nature (cited from Stachel 1986a, p. 374).

Einstein was quite correct in his rejection of Born's charge of apostasy. From his earliest statistical papers on (see Einstein 1989), Einstein's primary definition of the probability of any state of a system was always based on consideration of the temporal evolution of that system. Consider a (bounded) system that passes through a number of states in the course of that evolution. If, during the total time interval T, the system spends the time interval t in some state, then Einstein defines the probability of that state as:

$$\lim_{T \to \infty} t/T.$$

In practice, Einstein then generally assumed something like an ergodic hypothesis so that he could replace time averages with ensemble averages; but this was a secondary step following the primary, time-average definition of probability. Aside from the familiar problems with infinite-time limits, this definition makes the probability observable in principle. Even more important, the definition does not require a knowledge of the form of the dynamical laws that govern the temporal evolution of the system. Early on, Einstein had objected to Planck's definition of probabilities, in particular for states of an oscillator in equilibrium with a radiation field in his derivation of the black-body law, by counting possible complexions of the system. In the absence of a dynamical theory of such a system, Einstein asked, how does one justify the assumption that all complexions are equiprobable?

In order to explain the absorption and emission of light by a quantum system, Einstein later postulated transition probabilities between the states of such a system by analogy with similar probabilities that could be derived for a classical charged oscillator in equilibrium with a radiation field (Einstein 1916). He was well aware of the problems that this step raised for the causality principle, asking in 1920:

> Can the quantum absorption and emission of light ever be understood in the sense of the complete causality requirement, or would a statistical residue remain? I must admit that there the courage of certainty fails me. But I would abandon *complete* causality very, very reluctantly (Einstein to Max Born, 27 January 1920; cited from Born 1971, p. 23, translation modified).

Einstein thus saw no contradiction between employing probabilistic concepts in theoretical physics and maintaining a belief in "complete causality." In some cases, for example in the theory of Brownian motion, it was already possible to see the connection between the probabilistic and the causal viewpoints. In other cases, for example in the quantum theory, the nature of this connection might still be quite obscure. But this did not prevent him from continuing to search for such a connection, a search that continued even after the advent of quantum mechanics (see Section 3).

2. Einstein's Search for a Constructive Theory of the Quantum

In order to appreciate the reasons for Einstein's discontent with quantum mechanics, it is important first to understand the nature of Einstein's concern with the foundations of theoretical physics, to understand what he regarded as a fully satisfactory physical explanation. We can best begin by considering the distinction he enunciated in 1919 between principle-theories and constructive theories:

> We can distinguish between various kinds of theories in physics. Most of them are constructive. They attempt to build up a picture of the more complex phenomena out of the materials of a relatively simple formal scheme from which they start out. Thus the kinetic theory of gases seeks to reduce mechanical, thermal, and diffusional processes to movements of molecules —i.e., to build them up out of the hypothesis of molecular motion. When we say that we have succeeded in *understanding* a group of natural processes, we invariable mean that a constructive theory has been found that covers the processes in question.
>
> Along with this most important class of theories there exists a second, which I will call "principle-theories." These employ the analytic, not the synthetic method. The elements which form their basis and starting points are not hypothetically constructed but empirically discovered ones, general characteristics of natural processes, principles that give rise to mathematically formulated criteria which the separate processes or the theoretical representations of them have to satisfy. Thus, the science of thermodynamics seeks by analytical means to deduce necessary conditions, which separate events have to satisfy, from the universally experienced fact that perpetual motion is impossible.
>
> The advantages of the constructive theory are completeness, adaptability, and clearness, those of the principle theory are logical perfection and security of foundations.
>
> The theory of relativity belongs to the latter class. (Einstein 1919, p. 228).

Since this quotation dates from 1919, one might be inclined to wonder whether it reflects Einstein's outlook a decade and more earlier, during the years in which he was developing his ideas on relativity and the quantum. A recently discovered letter, at any rate, shows that he held quite similar views by the beginning of 1908. He wrote that he considered a physical theory to be satisfactory only

> if it builds its structures out of *elementary* foundations... we are still far from having satisfactory elementary foundations for electrical and mechanical processes... The theory of relativity is just as little ultimately satisfactory as, for example, classical thermodynamics was before Boltzmann had interpreted entropy as probability. If the Michelson–Morley experiment had not put us in the most awkward position, no one would have accepted the theory of relativity as a (partial) salvation (Einstein to Arnold

Sommerfeld, 14 January 1908; German text in Eckert and Pricha 1984, translation cited from Einstein 1989, Introduction).

Einstein's distinction between principle and constructive theories is sometimes interpreted as implying that he thought that principle-theories were *better*. We see from the quotations above that Einstein never held such a view. Principle-theories are more secure, but they are less satisfying. For Einstein, to understand was to have a constructive theory, a *Weltbild* (picture of the world).

It is important to appreciate the emotional significance the achievement of such understanding had for him:

> Man tries to make for himself in the fashion that suits him best a simplified and intelligible picture of the world; he then tries to some extent to substitute this cosmos of his for the world of experience, and thus to overcome it. This is what the painter, the poet, the speculative philosopher and the natural scientist do, each in his own fashion. Each makes a cosmos and its construction the pivot of his emotional life in order to find in this way the peace and security which he cannot find in the narrow whirlpool of personal experience (Einstein 1918, p. 225).

The large emotional charge associated with his views on quantum mechanics, in particular, can perhaps be glimpsed in the following quotation:

> I find the idea that there should not be laws for being [*das Seiende*], but only laws for probabilities, simply monstrous [*scheusslich*] (a nauseatingly indirect description) (Einstein to Tatiana Ehrenfest, 12 October 1938, cited from Stachel 1986a, p. 356).

As you see, this is a very emotional topic for Einstein. To grasp, to comprehend the world is not just an intellectual experience, it is also an emotional experience —the drive which he regards as necessary for really great accomplishments in science. And who are we to quarrel with Einstein's judgement in such matters? You also see that for Einstein, the theory of relativity was in a sense a last resort. It was the failure of his constructive attempts that led him to set up the special theory of relativity, as he explained much later:

> All my attempts . . . to adapt the theoretical foundations of physics to this [new] knowledge failed completely. It was as if the ground had been pulled out from under one, with no firm foundation to be seen anywhere upon which one could have built (Einstein 1949, cited from Einstein 1979, p. 43).

Again, I call your attention to the emotional charge attached to his words. He goes on:

> Reflections of this type made it clear to me as long ago as shortly after 1900 . . . that neither mechanics nor electrodynamics could (except in limiting cases) claim exact validity. Gradually I despaired of the possibility of discovering the true laws by means of constructive efforts based on known facts. The longer and more desperately I tried, the more I came to the conviction that only the discovery of a universal formal principle could lead us to assured results (Einstein 1949, cited from Einstein 1979, p. 49).

It could lead to assured results; but by itself, a theory of principle could not produce an ultimately satisfactory theory —that is, a constructive theory. It was a tool to help progress by providing new guidelines in the further search for constructive theories of matter and light.

I shall not discuss Einstein's work on the theory of relativity any further here. But I maintain that Einstein approached the quantum theory in a very similar spirit. He never regarded the work that he did in 1905 or in subsequent years on what he first called "the quantum hypothesis" (see Einstein 1989, Doc. 14) as an explanation, in his sense, since it did not constitute a constructive theory. Rather, it was a fruitful way of attempting to describe —at least partially —the nature of certain quantum phenomena, and thus to pose the quantum problem in a form amenable to further constructive efforts. In his 1911 report to the first Solvay Congress he says:

> We are all agreed that the so-called quantum theory of today is indeed a useful tool, but no theory in the ordinary meaning of the word, at any rate not a theory that could now be developed in a coherent manner. On the other hand, it has been proven that classical mechanics . . . no longer can be regarded as a usable system for the theoretical representation of all physical phenomena . . . So the question arises: on the validity of which general principles of physics may we hope to rely in the field of concern to us [i.e., quantum phenomena] (cited from Stachel 1986a, p. 358).

He goes on to discuss reliable principles such as the conservation of energy and Boltzmann's principle, but I just want to indicate the flavor of Einstein's approach to these problems. Einstein's strategy was, thus, first to formulate the quantum hypothesis and apply it to as many phenomena as possible. But this step was only a way station in his search for a constructive theory. It is, if you like, a useful formulation of the problem, rather than a solution. In that sense it is analogous to the way he looked upon the relativity principle as a guidepost, a tool to help in the search for constructive theories.

Now I come to his first abortive attempt at a constructive theory in 1909, which took the form of a unified field theory of matter and radiation. This is noteworthy since earlier he had inclined to the idea of entirely replacing fields by particles; or perhaps it is better to say the idea of replacing systems having an infinite number of degrees of freedom by systems with only a finite number of degrees of freedom; for, to Einstein, this was the essential distinction between fields and particles. In fact, the earliest extant account of his views on electromagnetic theory, a recently discovered letter of 1899, reads very much like an adumbration of the later Wheeler–Feynman program: get rid of fields, which are just a way of summing up the direct interactions between particles (see Einstein 1987, Doc. 25). And, you remember, he opens his 1905 quantum paper by pointing out the formal distinction between the way Maxwell's theory treats an electromagnetic system, using infinitely many degrees of freedom, and the way statistical mechanics treats material systems, using a finite number of degrees of freedom. He goes on to suggest that this use of infinitely many degrees of freedom may be responsible for the difficulties with electromagnetic theory that he points out in the paper, in particular what we now call the ultraviolet catastrophe. The first thing he does in the paper is to show that if you take classical statistical mechanics and Maxwell's theory seriously you get a perfectly definite result for the black-body radiation distribution law, a result that was later called the Rayleigh–Jeans law. But this result is obviously incorrect since it results in an infinite total electromagnetic energy in a finite volume of space (see Einstein 1989, Doc. 14).

In more high-falutin' philosophical terms, Einstein's first inclination was to accept an exclusively particle ontology; particles or systems with a finite number of degrees of freedom are the only reality, and one has to reduce the apparent fields in the world to direct interactions between particles. By 1909 Einstein had completely reversed field (to make a bad pun) by adopting a field ontology, and was trying to explain particles (both electrons and light quanta) as some sort of singularities or other cohesive structures in a (nonlinear) field. He did not publish a full theory of this type, but did give a few hints about the sort of theory he was trying to develop. The basic problem he felt was that the quantum of electric charge remained a "stranger" (*Fremdling*) in Maxwell's theory, put in by hand. How could one explain it? As we have seen, he did not accept the idea of simply postulating such structures; one had to construct them in his sense. He opined in 1909 that

> the next phase in the development of theoretical physics will bring us a theory of light that may be regarded as a sort of fusion of the wave and emission theories of light (Einstein 1989, Doc. 60, translation cited from the Introduction).

This is the first hint of what later came to be called the wave-particle duality. But the evidence he brought forth in support of this claim seems to have convinced Einstein by 1909 that what was really needed was some sort of modification of the Maxwell field theory that would allow structures within the field that could be interpreted as explaining the particle-like properties of both radiation —the light quantum, and matter —essentially the electron. Einstein was impressed by the existence of a dimensionless constant formed out of e, h and c —what we now call the fine structure constant. He was a little worried by the fact that its numerical value was not close to one, but much less so than, for example, Lorentz. On the basis of the existence of this constant, he anticipated that

> The same theoretical modification that leads to the elementary quantum [of charge] will lead to the quantum spectrum of radiation as a consequence.

In other words, Einstein expected that the quantum of charge and the quantum of radiation would both emerge from the same theory. Of course, he did not succeed in finding such a theory in 1909, and all attempts have failed to this day. Einstein tried to find (special-) relativistically invariant, nonlinear modifications of Maxwell's equations that would lead to the existence of stable structures in the field: either singularities, or finite, nondissipative concentrations of energy. Although he did not succeed, he wrote encouragingly about the project in 1909:

> I have not yet succeeded . . . in finding a system of equations that I could see was suited to the construction of the elementary quantum of electricity and the light quantum. The manifold of possibilities does not seem to be so large, however, that one should shrink from the task (Einstein 1989, Doc. 56; translation cited from the Introduction).

He does not refer to this project again in his published papers, but in his correspondence during the next few years he oscillates back and forth between hope and despair about its prospects. Phillipp Frank tells about his first visit to Einstein in 1912, while the latter was working in Prague. Einstein's room in the Physics Institute looked out over a park that belonged to a neighboring mental institution. Einstein remarked to Frank, "There you see that portion of the lunatics who are *not* working on the quantum theory" (Frank 1979, pp. 141–143). About this time, he turned his attention more and more toward the search for a relativistic theory of gravitation, which eventually resulted in the general theory of relativity, and published nothing further on quantum problems until 1916.

In later years, after he developed the general theory of relativity, Einstein came to doubt whether synthetic, constructive efforts, starting from contexts closely linked to empirical evidence, could ever lead to the kind of unified theory that he had in mind. He shifted to the search for a formal scheme that, starting from a small number of highly abstract concepts, would lead through a long deductive chain to an explanation of the quantum effects. In a 1953 letter he wrote:

> I am firmly convinced that every attempt to arrive at a rational theory by *synthetic* construction, will have an unsatisfactory result. Only a new basis for all of physics, from which all possible processes can be deduced with logical necessity (as for example is the case with thermodynamics), can bring a convincing solution (Einstein to M. Renninger, 11 June 1953; cited from Stachel 1986a, pp. 359–360).

In other letters from about this time he wrote:

> Now you will understand why I lapsed into my apparently Don Quixotic attempts to generalize the gravitational equations. If one cannot trust Maxwell's equations, and a representation [*Darstellung*] by means of fields and differential equations is indicated, on account of the principle of general relativity, and one has come to despair of arriving at a deeper basis [*Tieferlegung*] of the theory by intuitive [*anschaulich*]-constructive means; then no other sort of effort seems open (Einstein to Max von Laue, 17 January 1952; cited from Stachel 1986a, p. 360).
>
> I do not believe in macro- and micro-laws, but only in (structure) laws of general rigorous validity. And I believe that these laws are *logically simple*, and that reliance on their logical simplicity is our best guide. Thus, it would not be necessary to start with more than a relatively small number of empirical facts. If nature is not arranged in correspondence with this belief then we have altogether very little hope of understanding it more deeply ... This is not an attempt to convince you in any way. I just wanted to show you how I came to my attitude. I was especially strongly impressed with the realization that, by using a semiempirical method, one would never have arrived at the gravitational equations of empty space (Einstein to David Bohm, 24 November 1954; cited from Stachel 1986a, p. 360).

These quotations indicate that the sharp line that Einstein had earlier drawn between theories of principle and constructive theories has become blurred. But the goal of *deducing* quantum phenomena from some unified theory, rather than positing their existence from the outset, remains constant in his approach.

3. Einstein's Critique of Quantum Mechanics

Very often Einstein's critique is presented as if what worried Einstein most about quantum mechanics is its inherent radically-probabilistic element. It is true that he stated he could more easily conceive of a completely chaotic universe than one governed by probabilistic laws.

> I still do not believe that the Lord God plays dice. If he had wanted to do this then he would have done it quite thoroughly, and not stopped with a plan for gambling. In for a penny, in for a pound [*Wenn schon, denn schon*] (Einstein to Mr. and Mrs. Fritz Reiche, 15 August 1942; cited from Stachel 1986a, p. 374.)

In other words if everything is going to have a random element then why not have *everything* random? Why bother with laws of randomness? In that case we would not have to search for laws at all, he adds. I think he remarked somewhere that the only rational occupation for a person who believes in this fundamentally probabilistic element in the universe would be that of croupier.

But in spite of this evidence, I maintain that the probability issue was not the deepest source of his dissatisfaction with the prevailing interpretation of quantum mechanics.

> The sore point [*Der wunde Punkt*] lies less in the renunciation of causality than in the renunciation of a reality thought of as independent of observation (Einstein to Georg Jaffe, 19 January 1954; cited from Stachel 1986a, p. 374).

It is important to emphasize that he did not see this renunciation as a defect of quantum mechanics as such, but as a defect of the prevailing interpretation of that theory which regards it as the most complete possible description of an individual physical system. He accepted the ensemble point of view, which he called the Born interpretation —although I am not sure whether Born saw it that way — as a perfectly reasonable interpretation of quantum mechanics —which does not lead to such problems as that raised in the so-called Einstein–Podolsky–Rosen (EPR) argument. Let me cite a short summary by Einstein of this argument:

> It is not my opinion that there is a logical inconsistency in the quantum-theory itself and the "paradoxon" does not try to show it. The intention is to show that the statistical quantum theory is not compatible with certain principles, the convincing power of which is independent of the present quantum theory.
>
> There is the question: Does it make sense to say that two parts A and B of a system do exist independently of each other if they are (in ordinary language) located in different parts of space at a certain time, if there are no considerable interactions between those parts (expressed in terms of potential energy) at the considered time? ... I mean by "independent of each other" that an action on A has no immediate influence on the part B. In this sense I express a principle (a)
>
> (a) independent existence of the spatially separated. This has to be considered with the other thesis (b)
>
> (b) The ψ-function is the complete description of an individual physical situation.
>
> My thesis is that (a) and (b) cannot be true together, for if they would hold together the special kind of measurement concerning A could not influence the resulting ψ-function for B (after measurement of A).

> The majority of quantum theorists discard (a) tacitly to be able to conserve (b); I, however, have strong confidence in (a) so I feel compelled to relinquish (b) (Einstein to Leon Cooper, 31 October 1949; cited from Stachel 1986a, p. 375).

I think this is a very good summary of the conclusion that Einstein drew from his EPR argument. He was not always well understood, in particular by Born, who badly misunderstood Einstein —so badly that they reached an impasse in their correspondence on the subject. Wolfgang Pauli, who was then at the Institute, intervened in a letter to Born that gives a beautiful example of objective critical evaluation of a position with which one does not agree. Although Pauli disagreed with Einstein and was closer to Born's point of view, he understood that Born was just not tuned in on Einstein's wave length. So Pauli understood the task of explaining to Born what Einstein was talking about.

> ... I was unable to recognize Einstein whenever you talked about him in either your letter or your manuscript. It seems as if you had erected some dummy Einstein for yourself, which you then knocked down with great pomp (Pauli to Born, 31 March 1954; cited from Born 1971, p. 221).

I believe that this occupation or hobby still has many practitioners. It is rather important to understand what Einstein meant before one refutes him. But let me continue with the quotation from Pauli:

> In particular Einstein does not consider the concept of "determinism" to be as fundamental as it is frequently held to be ... he *disputes* that he uses as criterion for the admissibility of a theory the question: "Is it rigorously deterministic?" ... he was *not at all* annoyed with you, but only said that you were a person who will not listen (ibid.).

There are still many people in the world who fit that description —persons who will not listen —but one might try to follow Einstein's example and not get annoyed with them.

In this case the situation did not improve with time in spite of Pauli's letter. John Bell recently characterized Born's summary of his discussion with Einstein, written more than twenty years later when Born edited the letters for publication, in the following words: "Misunderstanding could hardly be more complete" (Bell 1987, p. 144).

To return to the main thread of my discussion, as Pauli understood it, the real stumbling point for Einstein was what has come to be called the nonlocality of quantum mechanics, not its indeterminism. He held that, if one adopted the ensemble or statistical interpretation of the theory, there would be no problem.

> It is ... to be expected that behind quantum mechanics there lies a lawfulness and a description that refer to the individual system. That it is not attainable within the bounds of concepts [taken] from classical mechanics is clear; the latter however is in any case outmoded as the foundation [*Fundament*] of physics (Einstein to Gregory Breit, 2 August 1935; cited from Stachel 1986a, p. 375).

So Einstein felt that quantum mechanics started out with a set of basic concepts that were outmoded. If, then, one could not advance further than quantum mechanics did on the basis of these concepts, this was not really so surprising to him.

On the contrary, as we will see in a moment, what he found surprising is that one could get so far with a set of concepts that are known to be inadequate. He elaborated on this inadequacy of the classical-mechanical starting point of quantum mechanics.

> I do not at all doubt that the contemporary quantum theory (more exactly "quantum mechanics") is the most complete theory compatible with experience, as long as one bases the description on the concepts of material point and potential energy as fundamental concepts (Einstein 1953; p. 6; translation cited from Stachel 1986a, pp. 375–376).

Arthur Fine has noted that while the old Einstein often is considered as a reactionary in physics and the quantum mechanicians are considered to be radicals, from Einstein's point of view, the situation was just the opposite (Fine 1986, Chapter 2). The adherents of quantum mechanics were clinging to the classical-mechanical concepts, and trying to find some way to continue to do physics using only this outmoded conceptual basis, whereas he was searching for a *new* conceptual basis for physics, which he regarded as the radical step. To continue with the quotation from Einstein:

> [The difficulties of quantum mechanics] are connected with the fact that one retains the classical concept of force or potential energy and only replaces the law of motion by something entirely new ... To me it seems, however, that one will finally recognize that something must take the place of forces acting or potential energy ... something that has an atomistic structure in the same sense as the electron itself. "Weak fields" or forces as active causes will not then occur at all, just as little as mixed states (Einstein 1953, pp. 12, 14; translation cited from Stachel 1986a, p. 376).

And in one of his last letters, he again touched on this point:

> I believe, however, that the renunciation of the objective description of "reality" is based upon the fact that one operates with fundamental concepts that are untenable in the long run (like f[or] i[instance] classical thermodynamics). Quite understandably most physicists resist the idea that we are still very far from a deeper insight into the structure of reality (Einstein to André Lamouche, 20 March 1955; cited from Stachel 1986a, p. 376).

Each of us likes to feel that he or she was born into the generation that finally has achieved the great illumination about the nature of reality.

Did Einstein believe in some sort of hidden variable program to underpin quantum mechanics? Bernard D'Espagnat and John Bell, for example, believe that he did, as do John Clauser and my dear colleague Abner Shimony (see Stachel 1986a, p. 376, for citations and references); but I don't think so. While some statements by Einstein seem to allow such an interpretation, in commenting on Bohm's hidden variable theory in 1953, he stated flatly:

> I think that it is not possible to get rid of the statistical character of the present quantum theory by merely adding something to the latter without changing the fundamental concepts about the whole structure (Einstein to Aron Kupperman, 14 November 1953, Item 8-036 in the Einstein archive).

Einstein hoped that such new fundamental concepts would emerge from his search for a unified field theory, which

> owes its origin in part to the conjecture that a rational generally relativistic field theory might offer the key to a more complete quantum theory (Einstein 1953, p. 14).

Let me cite a few more quotations from Einstein that indicate in somewhat more detail what he had in mind. The first is from a letter written in 1950:

> In a consistent field theory there is no real definition of the field . . . *A priori* no bridge to the empirical is given. There is not, for example, any "particle" in the strict sense of the word, since it does not fit into the program of representing reality by everywhere continuous, indeed even analytic functions . . . The upshot is that a comparison with the empirically known can only be expected to come from finding exact solutions of the system, in which empirically "known" structures and their interactions are "reflected." Since this is immensely difficult, the sceptical attitude of contemporary physicists is quite understandable (Einstein to Michele Besso, 15 April 1950, Einstein and Besso, 1972, pp. 438–439; translation cited from Stachel 1986a, p. 376).

Thus, for Einstein, once one had a unified field theory the path to the empirically observable was the great problem: how could one connect the theory with observation? It would certainly not be a matter of simply integrating over certain "hidden variables" in order to recover the concepts of classical mechanics. Here is an excerpt from a letter that explains the problem in the context of his last attempt at a unified theory, the theory of the asymmetrical metric tensor field:

> Our situation is this. We stand before a closed box that we cannot open, and try to discuss what is inside and what is not. The similarity to Maxwell's theory is only external, so that we cannot transfer the concept of "force" from this theory to the asymmetric field theory. If this theory is at all useful, then one cannot assume any separation between particles and field of interaction. In addition, there is no concept at all of the *motion* of something more or less rigid. The question here is exclusively: are there singularity-free solutions? Is their energy in particular localized in such a way as demanded by our knowledge of the atomic and quantum character of reality? The answer to this question is not really attainable with contemporary mathematical methods. Thus I do not see how one can guess whether any sort of action-at-a-distance or any type of object, insofar as we have attained a semiempirical knowledge of them, can be represented by the theory (Einstein to John Moffat, 24 August 1953; cited from Stachel 1986a, p. 377).

Now, if the concept of a hidden-variable theory, not to speak of the concept of a *local* hidden-variable theory, is to be given any precise meaning —and not just used as a description of *anything* that is non-quantum mechanical —then I don't think that we can characterize what Einstein is talking about in these quotations as a hidden-variable theory. Max Jammer has expressed similar views on this question (Jammer 1979, pp. 160–161, Jammer 1982, pp. 72–73).

To summarize and conclude this section, after 1930 Einstein never denied the great explanatory power of quantum mechanics, nor challenged its validity; but he did not agree that this success required acceptance of its underlying conceptual structure as the basis for all further progress in theoretical physics. He wrote to Schrödinger:

> The wonder [about quantum mechanics] is only that one can represent so much with it, although the most important source of knowledge, group invariance, finds such incomplete application there ... It is the case that a logically coherent theory that is connected appropriately to the real state of affairs usually has great extrapolatory power even if it is little related to the deeper truth [*der Wahrheit in der Tiefe*] (Einstein to Schrödinger, 16 July 1946; cited from Stachel 1986a, p. 377).

I will later return to the question of what Einstein meant by this remark about group invariance as the most important source of theoretical knowledge; but first I shall discuss another topic.

4. The Other Einstein

As I mentioned earlier, Einstein initially felt that the use of systems with only a finite number of degrees of freedom might prove the ultimate answer to the problems posed by the existence of the quantum. By 1909 he had shifted to the "unified field theory" point of view, to which, according to the conventional wisdom, he held more or less continuously until the end of his life.

But there is another Einstein who, as early as 1916, once more expressed the conviction that the entire continuum approach ultimately might have to be abandoned, and continued to consider this possibility until the end of his life. You may feel that there is a contradiction here, but as Whitman said:

> Do I contradict myself?
> Very well then I contradict myself.
> (I am large, I contain multitudes.)
> (*Song of Myself*, Whitman 1949, p. 142).

Einstein was certainly large, and "the other Einstein" is one of his aspects, even if one that is not very well known.

Let me cite some quotations that reveal "the other Einstein," and his comments on some of the very problems we have been discussing at this conference. In 1916, he wrote to one of his former students:

> ... you have correctly grasped the drawback that the continuum brings. If the molecular view of matter is the correct (appropriate) one; i.e., if a part of the universe is to be represented by a finite number of moving points, then the continuum of the present theory contains too great a manifold of possibilities. I also believe that this too great is responsible for the fact that our present means of description miscarry with the quantum theory. The problem seems to me how one can formulate statements about a discontinuum without calling upon a continuum (space-time) as an aid; the latter should be banned from the theory as a supplementary construction not justified by the essence of the problem, [a construction] which corresponds to nothing "real." But we still lack the mathematical structure unfortunately. How much have I already plagued myself in this way!
>
> Yet I see difficulties of principle here too. The electrons (as points) would be the ultimate entities in such a system (building blocks). *Are there* indeed such building blocks? Why are they all of equal magnitude? Is it satisfactory to say: God in his wisdom made them all equally big, each like every other, because he wanted it that way;

> if it had pleased him, he could also have created them different? With the continuum viewpoint one is better off in this respect because one doesn't have to prescribe elementary building blocks from the beginning. Further, the old question of the vacuum! But these considerations must pale beside the overwhelming fact: The continuum is more ample than the things to be described (Einstein to Walter Dällenbach, November 1916; cited from Stachel 1986a, p. 379).

He continued to discuss this possibility over the years. Here is an excerpt from a letter of 1935:

> In spite of the successes of quantum mechanics, I do not believe that this method can offer a usable *foundation* [*Fundament*] of physics. I see in it something analogous to classical statistical mechanics, only with the difference that here we have not found the equations corresponding to those of classical mechanics.
>
> In any case one does not have the right today to maintain that the foundation must consist in a *field theory* in the sense of Maxwell. The other possibility, however, leads in my opinion to a renunciation of the time-space continuum and to a purely algebraic physics. Logically this is quite possible (the system is described by a number of integers; "time" is only a possible standpoint [*Gesichtspunkt*], from which the other "observables" can be considered —an observable logically coordinated to all the others. Such a theory doesn't have to be based upon the probability concept. For the present, however, instinct rebels against such a theory (Einstein to Paul Langevin, 3 October 1935; cited from Stachel 1986a, pp. 379–380).

Of course, even more serious than his instinctive rebellion was the fact that he had no idea how to create such a theory. Judging from his letters, he evidently became more and more pessimistic about the continuum point of view. In 1952, for example, he wrote:

> In present-day physics there is manifested a kind of battle between the particle-concept and the field-concept for leadership that will probably not be decided for a long time. It is even doubtful if one of the two rivals finally will be able to maintain itself as a fundamental concept (Einstein to Herbert Kondo, 11 August 1952; cited from Stachel 1986a, p. 380).

He wrote to Besso in 1954:

> I consider it entirely possible that physics cannot be based upon the field concept, that is, on continuous structures. Then *nothing* will remain of my whole castle in the air including the theory of gravitation, but also nothing of the rest of contemporary physics (Einstein to Besso, 10 August 1954, Einstein and Besso, 1972, p. 527; cited from Stachel 1986a, p. 380).

In the same year, he wrote to André Lichnerowicz:

> Your objections regarding the existence of singularity-free solutions which could represent the field together with the particles I find most justified. I also share this doubt. If it should finally turn out to be the case, then I doubt in general the existence of a rational physically useful *field* theory. But what then? Heine's classical line comes to mind: "And a fool waits for the answer" (Einstein to André Lichnerowicz, 25 February 1954; cited from Stachel 1986a, p. 380).

He wrote to Bohm:

> I must confess that I was not able to find a way to explain the atomistic character of nature. My opinion is that if the objective description through the field as an elementary concept is not possible, then one has to find a possibility to avoid the continuum (together with space and time) altogether. But I have not the slightest idea what kind of elementary concepts could be used in such a theory (Einstein to David Bohm, 28 October 1954; cited from Stachel 1986a, p. 380).

The reason Einstein didn't extensively publicize these views is that he felt that, unless one has a real theory, it is useless to publish such speculative remarks. But contemporary physicists may find them fascinating, since they speak to so many of our current concerns.

I have found only one place —but I must admit that I have not looked everywhere —where he speculated in somewhat more detail on what such a non-continuum theory might involve.

> The alternative continuum-discontinuum seems to me to be a real alternative; i.e., there is here no compromise. By discontinuum theory I understand one in which there are no differential quotients. In such a theory space and time cannot occur, but only numbers and number-fields and rules for the formation of such on the basis of algebraic rules with exclusion of limiting processes. Which way will prove itself, only success can teach us.
>
> Physics up to now is naturally in its essence a continuum physics, in spite of the use of the material point, which looks like a discontinuous conceptual element, and has no more right of existence in field description. Its strength lies in the fact that it posits parts which exist quasi-independently, *beside* one another. Upon this rests the fact that there are reasonable laws, that is rules which can be formulated and tested for the individual parts. Its weakness lies in the fact that it has not been possible up to now to see how that atomistic aspect including quantum relations can result as a consequence. On the other hand dimensionality (as four-dimensionality) lies at the foundation of the theory.
>
> An algebraic theory of physics is affected with just the inverted advantages and weaknesses, aside from the fact that no one has been able to propose a possible logical schema for such a theory. It would be especially difficult to derive something like a spatio-temporal quasi-order from such a schema. I cannot imagine how the axiomatic framework of such a physics would appear, and I don't like it when one talks about it in dark apostrophes [*Anredungen*]. But I hold it entirely possible that the development will lead there; for it seems that the state of any finite spatially limited system may be fully characterized by a finite number of numbers. This speaks against the continuum with its infinitely many degrees of freedom. The objection is not decisive only because one doesn't know, in the contemporary state of mathematics, in what way the demand for freedom from singularity (in the continuum theory) limits the manifold of solutions (Einstein to H.S. Joachim, 24 August 1954; cited from Einstein 1986a, pp. 380–381).

I was interested to learn that even near the end of his life Einstein was still on the lookout for new mathematical tools that might turn speculations that he thought it best to keep private into the basis of a real theory. Abraham Fraenkel, the mathematician, reports an interview he had with Einstein in 1951. With the background I have reported, you will see at once why Einstein was so excited by what Fraenkel told him:

> In December 1951 I had the privilege of talking to Professor Einstein and describing the recent controversies between the (neo-) intuitionists and their "formalistic" and "logicistic" antagonists; I pointed out that the first attitude would mean a kind of atomistic theory of functions, comparable to the atomistic structure of matter and energy. Einstein showed a lively interest in the subject and pointed out that to the physicist such a theory would seem by far preferable to the classical theory of continuity. I objected by stressing the main difficulty, namely, the fact that the procedures of mathematical analysis, e.g., of differential equations, are based on the assumption of mathematical continuity, while a modification sufficient to cover an intuitionistic-discrete medium cannot easily be imagined. Einstein did not share this pessimism and urged mathematicians to try to develop suitable new methods not based on continuity (Fraenkel 1954).

Now I have arrived at the point where I can discuss Einstein's remark, cited in the last section, that the most important source of theoretical knowledge in physics, group invariance, finds incomplete application in quantum theory. This brings up one of the most profound sources of Einstein's skepticism about the ultimate status of quantum mechanics. He attached primary significance to the concept of general covariance writing, for example, in 1954:

> You consider the transition to special relativity as the most essential thought of relativity, not the transition to general relativity. I consider the reverse to be correct. I see the most essential thing in the overcoming of the inertial system, a thing that acts upon all processes but undergoes no reaction. This concept is in principle no better than that of the center of the universe in Aristotelian physics (Einstein to Georg Jaffe, January 19, 1954; cited from Stachel 1986a, p. 377).

That is, Einstein puts the a priori postulation of inertial systems on a level with the postulation of a center of the universe in the Aristotelian system. Therefore, he felt a quantum theory constructed on the basis of Galilean or even special-relativistic space-times could not constitute a satisfactory foundation of physics.

> Contemporary physicists do not see that it is hopeless to take a theory that is based on an independent rigid space (Lorentz-invariance) and later hope to make it general-relativistic (in some natural way) (Einstein to Max von Laue, September 1950; cited from Stachel 1986a, p. 378).

He justified himself against criticism that he had neglected the development of relativistic quantum field theory, in similar terms:

> I have not really studied quantum field theory. This is because I cannot believe that special relativity theory suffices as the basis of a theory of matter, and that one can afterwards make a non-generally relativistic theory into a generally relativistic one. But I am well aware of the possibility that this opinion may be erroneous (Einstein to K. Roberts, September 6, 1954; cited from Stachel 1986a, p. 378).

In spite of his statement that he had failed to understand Bohr's concept of complementarity, Einstein did share one element of Bohr's viewpoint: Einstein didn't really believe that the wave aspect of the electron is as fundamental as the particle aspect; and conversely, that the particle aspect of the light quantum is as fundamental as the wave aspect. As you may know, Bohr did not believe in a full equality between the wave and the particle aspects of matter and radiation, respectively. He believed that the classical limit of the quantum theory led to the more important aspect; therefore the particle aspect of the electron was physically more fundamental while the wave aspect was more a mathematical construct. Conversely, for the photon the classical limit was given by the wave theory, and the particle aspect was more a mathematical construct. Einstein once said something very similar:

> I do not believe that the light-quanta have reality in the same immediate sense as the corpuscles of electricity. Likewise I do not believe that the particle-waves have reality in the same sense as the particles themselves. The wave-character of particles and the particle-character of light will —in my opinion —be understood in a more indirect way, not as immediate physical reality (Einstein to Paul Bonofield, 18 September 1939; cited from Stachel 1986a, pp. 373–374).

5. Some Lessons From General Relativity

If one looks at how Einstein developed the general theory of relativity, one sees that he drew a certain number of lessons from the way that he did so, lessons that he subsequently incorporated in his approach to the foundations of theoretical physics. We can consider these lessons, and try to apply them to our problems today. However, those who have not studied the history of the development of the subject are not always aware of these lessons (the converse of this statement is unfortunately not true). I am not claiming that these views on the foundations of physics must be accepted uncritically just because Einstein held them, but only that they are worth considering —that they are not automatically irrelevant either, just because Einstein held them.

The first lesson at which Einstein arrived is the fundamental importance to any theory of gravitation of the equivalence principle, which he very soon expressed by the phrase that inertia and gravitation are *wesensgleich*, the same in their essence or essentially the same. The distinction between gravitation and inertia is not absolute, but it depends on the frame of reference employed. Once the appropriate mathematical tool had been developed, this lesson could be expressed by the statement that the affine connection of space-time cannot be split up uniquely into two parts, one of which would be an inertial connection, the other a gravitational field tensor (remember, the difference between two symmetric connection fields is a tensor field). Until the end of his life, Einstein placed great stress on the significance of the affine connection as the correct mathematical representation of the gravitational-cum-inertial field. Replying to von Laue, who, in preparing the post-World War II revision of his relativity textbook (von Laue 1956), espoused the point of view that only a nonvanishing Riemann tensor represented a real gravitational field, Einstein wrote:

> What characterizes the existence of a gravitational field from the empirical standpoint is the nonvanishing of the Γ^l_{ik} [the components of the affine connection –JS], not the nonvanishing of the R_{ikmn} [the components of the Riemann tensor –JS]. If one does not think in such intuitive [*anschaulich*] ways, one cannot comprehend why some-

> thing like curvature should have anything to do with gravitation at all. In any case, no reasonable person would have hit upon something in that way. The key to the understanding of the equality of inertial and gravitational mass would have been missing (Einstein to von Laue, September 1950; cited from Stachel 1989, p. 89).

A couple of comments are in order. First of all, one may note Einstein's physically intuitive (*anschaulich*) way of thinking. There are several quotations from Einstein about his mode of thought that are quite well-known (see especially Einstein 1945), in which he noted that his mode of thought was dependent on visual and even kinesthetic imagery. So I will quote an account that has not been previously cited for this purpose as far as I know. A student who was taking Einstein's 1917 relativity course at the University of Berlin reported in his diary that in private conversations Einstein said:

> he is a poor calculator, he does not work with abstract ideas ... He visualizes [gravitational] waves with the help of an elastic body that represents an elastically oscillating system. He also applied motions of his fingers for this elastic system to the 3 last types [of real, transverse-transverse gravitational waves, as opposed to the purely coordinate waves], thus in this way one has to make them plausible to oneself ... (excerpts from the diary of Rudolf Jakob Humm, Zentralbibliothek Zürich).[2]

You see that Einstein had adumbrated the Weber bar!

Secondly, I believe that Einstein's emphasis on the fundamental role of the affine connection is much more palatable to physicists today, including many relativists, than it was some years ago. This is because of the current emphasis on gauge theories, in which the fundamental role of the connection is clear. And the theoretical discovery of the thermal ambience of accelerating particle detectors has also emphasized that one cannot make a distinction between quantum and thermal fluctuations in a frame-independent way (see, e.g., Sciama, Candelas, and Deutsch 1981). So I think the viewpoint that the Riemann tensor isn't *everything* in general relativity is more accepted today than it was when I was a student, when a sort of monotheistic belief in the Riemann tensor was accepted gospel among many relativists.

This approach implies that no division of the connection into an inertial and a gravitational part has any absolute validity since such a division is entirely frame-dependent. Obviously, if one *could* make such an absolute division, that is, if one had a tensor of first differential order to represent the gravitational field, one could solve with ease many outstanding problems in general relativity. At the classical level, for example, you would have a locally-defined gravitational energy density. Formal quantization of gravity would become more or less trivial, since you could quantize this tensor field on the background inertial connection.

The first person who seems to have realized how useful such a division would be is Nathan Rosen. In 1939, in order to produce it, he introduced a second, flat metric into the theory, undoing what Einstein had done in 1915. Thus, Einstein's approach can be expressed in the injunction: What God has joined together (inertia and gravitation), let no man put asunder. Einstein gave Rosen a piece of advice about his theory that I think was very good, but which Rosen did not follow: "Throw it in the waste basket."[3]

What I want to emphasize here is that people who attempt to quantize general relativity with the help of a background metric, flat or not, are violating this injunction. They are essentially introducing, whether they realize it or not, an inertial connection —the unique one associated with the background metric —and therewith a gravitational tensor of the first order. As indicated above, if you give me such a gravitational tensor, I too can do wonderful things. But the trouble is that one *doesn't* have such a tensor in general relativity. In addition to the obvious question of what the conformal (i.e., light-cone) structure of the nonquantized background metric has to do with the actual propagation of the quantized gravitational field, such a bimetric approach violates the equivalence principle since it implies the existence of an invariant decomposition of the connection, treats one part as an inertial connection that forms part of the nonquantized background, and treats the other part as a gravitational tensor field that is to be quantized. Those quantum field theorists who approach general relativity as if it were no more than a nonlinear Lorentz-invariant field theory with a particularly nasty gauge group are especially prone to this sin of splitting asunder what belongs together.

Such approaches to quantum gravity can lead to truly comical conclusions. Some time ago I had a discussion with a French physicist who told me that it was simple to quantize general relativity: all one has to do is to take a conformally-flat metric tensor, and quantize the conformal factor. When I asked what made him think that one must employ a conformally flat metric, he replied: "Oh, because if you don't, I have no idea how to quantize general relativity." Such approaches to the quantization problem remind me of Einstein's comment that he didn't think much of a carpenter who, when he had to bore a hole in a piece of wood, looked for the thinnest spot. One should bore the hole where it is needed, no matter how hard the job.

The next lesson involves the meaning of the concept of general covariance. Einstein's interpretation of this concept is closely connected with the question of why almost three years elapsed between the time when he decided that the metric tensor was the correct mathematical representation of gravitation (late 1912) and the time when he finally adopted the generally covariant equations for the gravitational field that we now call the Einstein equations (late 1915). The main obstacle that held him up was his failure to realize —to use modern terminology —that, in a generally covariant theory, although the points of a manifold are *mathematically* individuated by a coordinate system, they are not *physically* individuated before one has a metric tensor field on that manifold. Consider a solution to any set of generally covariant field equations with a given distribution of sources. A second solution to these equations can always be generated from the first by the following prescription: it is identical to the first solution everywhere on the manifold except for a finite region containing none of the sources (a "hole"); in the hole, the second solution is generated from the first by dragging the latter along an arbitrary vector field that vanishes on the boundary of the hole. Originally, Einstein had considered the second solution to be physically distinct from the first, and hence rejected generally covariant field equations because they did not provide a physically unique solution for the gravitational field corresponding to a given source distribution (for more detailed discussion of the "hole" argument see Stachel 1986b, 1987, 1989). Finally, after over two years he realized that two such solutions could be considered as two mathematically distinct exemplifications of the *same* physical solution if one stipulated that the physical properties of the points of the manifold inside the hole in any solution to the field equation depend *entirely* on the nature of the solution; hence, when one creates the second solution by dragging, the physical significance of the points is dragged along with the field. It is this stipulation that constitutes the physical meaning of the concept of general covariance, which, properly speaking, has nothing to do with coordinate transformation. In the case of general relativity, one can express this idea by saying: Until one introduces a metric, a manifold does not consist of physical events but just of mathematical points with *no* physical properties; it is the metric that imparts the physical character of events to the points of a manifold.

One moral of this story is that general relativity is better formulated mathematically using a fiber bundle approach in which the metric is represented by a cross section of the bundle. If you do this, you realize immediately that, in this bundle approach, the points of the base manifold do *not* represent physical events, which instead are represented by mappings from a point on a cross section of the bundle into a point of the base manifold. If you have no cross section (no metric), you have no such mapping and hence no events (see Stachel 1986b for more details of this approach). It follows that *any* point of the base manifold can "represent" any event in an appropriate mapping.

So the problem of a correct mathematical representation of events in a generally covariant theory can be easily solved. But the moral is that if you are trying to formulate a quantum theory of general relativity (or of any generally covariant theory), you must always bear in mind that you do *not* have a manifold of events to start with, before finding a solution to the quantized equations for the metric field. This raises the following question: suppose one *had* a consistent quantum formalism for general relativity (we should only be so lucky!), and had found a solution to the quantized field equations; what could one do with it? If you have a wave function of the universe, for example, how do you interpret it physically? Usually, one interprets the quantum formalism in a background space-time, so that such a question may be answered as follows: You put an apparatus *here* to create a particle *now*, and then you put an apparatus *there* to detect it *then*; one can then use quantum probability amplitudes (such as wave functions) to compute transition probabilities between two such events. But in general relativity you have no a priori *here* and *now* or *there* and *then*. What "here and now" and "there and then" signify are among the things that one must solve the equations in order to determine. I am reminded of Gertrude Stein's last words. She roused herself from a coma to ask worriedly: "What is the answer?" Then she smiled and asked: "What is the question?" and lapsed into her final coma. In a sense, in matters of quantum gravity one must know the answer before one knows the question.

Perhaps one could find a way out of this dilemma if one could specify how to use a part of the system under consideration to measure another part, as Murray Gell-Mann suggested in his talk. But this would have to be done *without* introducing any space-time structure a priori. Gell-Mann clearly did not attempt this; he introduced an underlying space-time structure. If you introduce such a background space-time, there is of course, no problem in doing quantum mechanics on it. But then you haven't solved the problem of what to do *without* such a structure.

This question of the meaning and significance of general covariance is not a trivial problem (as I mentioned above, its solution held Einstein up for over two-and-a-half years in the development of general relativity). I feel that many people who study general relativity today still don't really understand this problem. Some recent Russian papers, for example, have unwittingly reproduced Einstein's hole argument as a criticism of general relativity.

This lack of understanding shows up particularly clearly in perturbative approaches to the theory. When one perturbs a field, one is used to automatically comparing the perturbed value of the field at a point with the value of the unperturbed field at the *same* point in the background space-time. It often requires some effort to realize that, since no a priori physical significance is attached to the points of the manifold in general relativity, no such a priori correspondence exists between points of the perturbed and unperturbed metric tensor fields; i.e., there is no a priori definition of corresponding events. This difficulty is usually masked by the use of a coordinate system: equal coordinate values are tacitly assumed to define corresponding points. But this amounts to the unacknowledged introduction of an additional "rigid" geometrical structure (see Geroch 1968), which raises the problem of whether any physical significance can be attached to the introduction of such a structure.

One might think that this problem is avoided by starting from a flat space-time solution to the gravitational field equations. Surely, it seems, one knows what one means by a definite point in Minkowski space-time! Indeed one does (leaving aside the problem of possible global topological complications), if, a priori, one is dealing with the space-time structure of the special theory of relativity. Here, points of Minkowski space-time may be physically identified by means of some additional physical elements introduced into the theory, such as rigid rods and clocks, trajectories of free particles, light rays, and so forth. These additional elements may be introduced without any problem in special relativity precisely *because* the space-time structure is given a priori, and so by definition is unaffected by the presence of additional particles or fields. Flat space as a solution to the field equations of general relativity, however, cannot be assimilated without further ado to special-relativistic Minkowski space-time; it is just as subject to the problem of physical individuation of its points as any other solution to the gravitational field equations. (In fact it is *more* subject than most; since the Riemann tensor vanishes, no invariants of the Riemann tensor can be used for this purpose). In particular, how the points of a perturbed metric are to be identified with the points of an unperturbed flat metric is an open question —unless the whole theory is treated as if it "really' were a special-relativistic theory and not a generally covariant one. In that case, however, it is not clear why one should have singled out the Einstein theory in the first place from the myriad of possible nonlinear Poincaré-invariant theories.[4]

Finally, there is a third issue which I call the "no metric, no nothing" point of view. To quote again from Einstein:

> On the basis of general relativity ... space as opposed to 'what fills space' ... has no separate existence ... If we imagine the gravitational field, i.e., the functions g_{ik} to be removed, there does not remain a space of the type [of Minkowski space in special relativity –JS] but absolutely *nothing*, not even a 'topological space' [i.e., a manifold –JS]. For the functions g_{ik} describe not only the field, but at the same time also the topological and metrical structural properties of the manifold ... There is no such thing as empty space, i.e., a space without a field. Space-time does not claim existence on its own, but only as a structural quality of the field (Einstein 1952).

Einstein was well aware that his views on this question were far from universally shared:

> It required a severe struggle to arrive at the concept of independent and absolute space, indispensable for the development of theory. It has required no less strenuous exertions subsequently to overcome this concept —a process which is by no means as yet completed (Einstein 1954b).

Even relativists have not yet fully adopted the point of view, "no metric, no nothing." If you look at the way the general theory of relativity is formulated mathematically in even the most careful treatises, for example, you see this clearly. They start out by introducing a global manifold (the points of which are usually identified forthwith with events —I have already discussed that problem), and then put such structures on this manifold as the metric tensor field. Is that the way that any one of us actually goes about solving the field equations of general relativity? Of course not. One first solves them on a generic patch, and *then* one tries to maximally extend the local solution (using some criteria for acceptable extensions) from that patch to a global manifold, which is not known ahead of time. Before solving the field equations, one generally doesn't know the global manifold on which the solution will turn out to be maximally extended. So we are pulling a swindle when we tell students, as our definitions imply, that you first pick the manifold and then solve the field equations on it.

The fiber bundle approach mentioned above doesn't solve this problem because it also assumes a base manifold with given global topology to start with. So I believe that it is an important problem to formulate general relativity mathematically in such a way that it is clear from the outset that finding the maximally extended global manifold (whatever criteria are used to define such a maximal extension) is part of the problem of finding a solution.

I am *not* saying, of course, that each topology leads to a unique metric, but rather, that no metric implies no topology. You don't start out with a topology and then look for a metric; you look for a local solution to the field equations on a patch, and then investigate how far that solution patch can be extended. One will not necessarily get a unique answer for the global topology, as the example of flat space shows. But one does not start out with a global topology and then look for *all* the metrics solving the field equations that are compatible with it. The important point is that if one wants to consider *all* solutions to the field equations —or even a subclass wide enough to include solutions on topologically inequivalent manifolds —one must have a mathematical structure that allows this to be done.

Although the two ideas are not exactly the same, the moral I want to draw from this final point may be related to something that Chris Islam talked about. Suppose you take really seriously the point of view that there is something fundamentally local about the way general relativity approaches a problem. Then there is another fundamental tension between the basic approaches of general relativity and of quantum mechanics since quantum mechanics, in a deep sense, is fundamentally global in its approach to problems. It doesn't make much sense to talk about the wave function on one patch of space-time, or the sum over all paths on one patch of space-time. In solving a quantum-mechanical problem, you have to consider the *whole* manifold from the beginning. The conventional mathematical approach to general relativity, which starts with a manifold, masks this tension. If we develop a mathematical formulation of general relativity that emphasizes the element of locality from the beginning, it would emphasize this contrast more sharply. Such an emphasis on the tension may be a necessary stage in finding its ultimate resolution.

Indeed, this way of thinking also suggests new possibilities. Perhaps the existence of a global manifold is just a special case. Maybe in the ultimate theory all the patches won't always fit together to form a manifold, except in the classical limit. Here is where my point of view may connect with some of Isham's ideas. If you don't build in a global manifold at the beginning, perhaps you are better off, because you may have to get rid of it in the end anyway.

NOTES

[1] See Shelley 1901, p. 821.

[2] The excerpts from this diary, as printed in Seelig 1960, pp. 258–259, differ markedly from the text of the diary itself.

[3] On 21 February 1939, Einstein wrote to Rosen: "If you had come to me with such a proposal in the days when we were still engaged in happy daily collaboration, I would have stretched out my splendid tongue at you. Why? . . . —this is the worst thing —one would be longer have any reasonable possibility of assimilating inertia to gravitation as the same thing. . . . So throw it peacefully in the waste basket" ["Wenn Sir mir mit einem solchen Vorschlag gekommen wären, als wir noch täglich zu froher Arbeit zusammenkamen, hätte ich Ihnen meine prächtige Zunge lang herausgestreckt. Warum? . . . —und dies ist das Schlimmste —hat man keine vernünftige Möglichkeit mehr, Trägheit und Schwere als das Gleiche aufzupassen. . . . Also werfen Sie es ruhig in den Papierkorb"] (Einstein Archive, Control Index No. 20-233).

[4] Einstein made this point in a follow-up letter to Nathan Rosen on the latter's bimetric theory: "If one assumes that the second (Euclidean) metric does not express any physical reality, then the entire formalism is empty and misleading. If it does express such a reality, then one is led back to the study of special relativity with its entire manifold of theoretical possibilities. In that case, it is entirely arbitrary to describe gravitation by a *tensor*, and

even more so to give any sort of preference to the *covariant* derivatives formed with this tensor over other special-relativistic modes of tensor-formation. There are then innumerable logically equally justified possibilities, from which you have selected one quite arbitrarily." [Nimmt man an, dass die zweite (euklidische) Metrik keine physikalische Realität ausdrückt, so ist der ganze Formalismus leer und irreführend. Drückt sie aber eine Realität aus, so wird man auf das Studium der speziellen Relativität zurückgeführt mit seiner ganzen Mannigfaltigkeit theoretischer Bildungen. Dann ist es ganz willkürlich, die Gravitation durch einen *Tensor* zu beschreiben und erst recht, den mit diesem Tensor gebildeten *kovarianten* Ableitungen irgend welchen Vorzug vor anderen spezial-relativistichen Tensorbildungen den Vorzug zu geben. Es gibt dann unzählige logisch gleichberechtigte Möglichkeiten, von denen Sie ganz wilkúrlich eine herausgegriffen haben"] (Einstein to Nathan Rosen, 6 March 1939; Einstein Archive, Control Index No. 20-235).

References

Bell, John (1987). *Speakable and Unspeakable in Quantum Mechanics —Collected papers on quantum philosophy.* Cambridge: Cambridge University Press.

Born, Max (1949). "Einstein's Statistical Theories." In Schilpp 1949, 163–177.

—— (1971). *The Born–Einstein Letters —Correspondence between Albert Einstein and Max and Hedwig Born from 1916 to 1955 with commentaries by Max Born.* New York: Walker and Company.

Eckert, Michel and Pricha, Willibald (1984). "Die erste Briefe Albert Einsteins an Arnold Sommerfeld." *Physikalische Blätter* 40: 29–34.

Einstein, Albert (1916). "Strahlungs–Emission und -Absorption nach der Quantentheorie." *Deutsche Physikalische Gesellschaft, Verhandlungen* 18: 318–323.

—— (1918). "Motiv des Forschens." Cited from translation in Einstein 1954a, pp. 224–227.

—— (1919). "What is the Theory of Relativity?" Cited from Einstein 1954a, pp. 227–232.

—— (1945). "A Testimonial from Professor Einstein." In Jacques Hadamard, *An Essay on the Psychology of Invention in the Mathematical Field.* Princeton: Princeton University Press, Appendix II, pp. 142–143. Reprinted as "A Mathematician's Mind" in Einstein 1954a, pp. 25–26.

—— (1949). "Autobiographical Notes." In Schilpp 1949, pp. 2–94. Cited from the corrected reprint, Einstein 1979.

—— (1952). "Relativity and the Problem of Space." In ibid., *Relativity: the Special and the General Theory.* New York: Crown, pp. 135–157.

—— (1953). "Introductory Remarks on Basic Concepts." In *Louis de Broglie — Physicien et Penseur.* André George, ed. Paris: Editions Albin Michel, pp. 4–15.

—— (1954a). *Ideas and Opinions.* Carl Seelig, ed. New York: Crown.

—— (1954b). "Foreword." In Jammer, M., *Concepts of Space.* Cambridge, Mass: Harvard University Press, pp. xiii–xvi.

—— (1979). *Autobiographical Notes —A Centennial Edition.* Paul Arthur Schilpp, transl. and ed. LaSalle and Chicago: Open Court Pub. Co.

—— (1987). *Collected Papers of Albert Einstein.* vol. 1, *The Early Years (1879–1902).* John Stachel et al., eds. Princeton: Princeton University Press.

—— (1987). *Collected Papers of Albert Einstein.* vol. 2, *Writings (1901–1909).* John Stachel et al., eds. Princeton: Princeton University Press.

Einstein, Albert and Besso, Michele (1972). *Correspondance 1903–1955.* Pierre Speziali, trans. and ed. Paris: Hermann.

Fine, Arthur (1986). *The Shaky Game —Einstein, Realism, and the Quantum Theory.* Chicago and London: The University of Chicago Press.

Fraenkel, Abraham (1954). *Bulletin of the Research Council of Israel* 3: 283–289.

Frank, Philipp (1979). *Albert Einstein —Sein Leben und Seine Zeit.* Braunschweig-Wiesbaden: Friedr. Vieweg & Sohn.

Geroch, Robert (1969). "Limits of Space-times." *Communications in Mathematical Physics* 13: 180–193.

Jammer, Max (1979). "Albert Einstein und das Quantenproblem." in *Einstein Symposion Berlin.* H. Nelkowski et al., eds. Berlin–Heidelberg–New York: Springer-Verlag, pp. 146–167.

—— (1982). "Einstein and Quantum Physics." In *Albert Einstein —Historical and Cultural Perspectives.* Gerald Holton and Yehuda Elkana, eds. Princeton: Princeton University Press,pp. 59–76.

Schilpp, Paul Arthur, ed. (1949). *Albert Einstein: Philosopher-Scientist.* Evanston, Illinois: The Library of Living Philosophers.

Sciama, Dennis W.; Candelas, P.; and Deutsch, D. (1981). "Quantum Field Theory, Horizons and Thermodynamics." *Advances in Physics* 30: 327–366.

Seelig, Carl (1960). *Albert Einstein —Leben und Werk eines Genies unserer Zeit.* Zürich: Europa Verlag.

Shelley, Percy Bysse (1901). *The Complete Poetical Works of Shelley —Cambridge Edition.* Boston: Houghton Mifflin.

Stachel, John (1986a). "Einstein and the Quantum: Fifty Years of Struggle." In *From Quarks to Quasars —Philosophical Problems of Modern Physics.* Robert G. Colodny, ed. Pittsburgh: University of Pittsburgh Press, pp. 349–385. [See this volume, pp. 367–402].

—— (1986b). "What a Physicist Can Learn from the History of Einstein's Discovery of General Relativity." In *Proceedings of the Fourth Marcel Grossmann Meeting on General Relativity.* Remo Ruffini, ed. Amsterdam: Elsevier, 1857–1862.

—— (1987). "How Einstein discovered general relativity: a historical tale with some contemporary morals." In *General Relativity and Gravitation —Proceedings of the 11th International Conference on General Relativity and Gravitation.* M.A.H. MacCallum, ed. Cambridge: Cambridge University Press, pp. 200–208. [See this volume, pp. 293–300].

—— (1989). "Einstein's Search for General Covariance, 1912–1915." In *Einstein Studies, vol. 1: Einstein and the History of General Relativity.* Don Howard and John Stachel, eds. Boston–Basel–Berlin: Birkhäuser, pp. 63–100. [See this volume, pp. 301–338].

von Laue, Max (1956). *Die Relativitätstheorie.* vol. 2. *Die allgemeine Relativitätstheorie.* 4th edn. Braunschweig: Vieweg.

Whitman, Walt (1949). *The Inner Sanctum Edition of the Poetry and Prose of Walt Whitman.* Louis Untermeyer, ed. New York: Simon and Schuster.

Einstein's Light-Quantum Hypothesis, or Why Didn't Einstein Propose a Quantum Gas a Decade-and-a-Half Earlier?

John Stachel

1. Introduction

In his "On the Quantum Theory of the Monatomic Ideal Gas, Second Paper," Einstein proposes a novel quantum theory of an ideal gas,

> based on the hypothesis of a far-reaching formal relation between radiation and gas. According to this theory, the degenerate gas deviates from the gas of [classical] statistical mechanics in a way that is analogous to that in which radiation obeying Planck's law deviates from radiation obeying Wien's law. If Bose's derivation of Planck's radiation formula [given in Bose 1924] is taken seriously, then this theory of the ideal gas is inevitable; for if it is justified to regard radiation as a quantum gas, then the analogy between quantum gas and gas of molecules must be complete (Einstein 1924b, p. 3).

His third paper on the subject reiterates this point:

> This theory seems justified if one starts with the conviction that (apart from its property of being polarized) a light quantum differs essentially from a monatomic molecule only in that its rest mass is vanishingly small (Einstein 1925, p. 18).

Einstein had discussed a formal analogy between matter and radiation as early as 1901. In Einstein 1905a, he showed that "radiation obeying Wien's law" behaves thermodynamically as if it were composed of classical particles and proposed his light quantum hypothesis. He had been pondering the significance of Planck's law since 1901, and Einstein 1906 gives a derivation of it showing that "Planck's theory implicitly makes use of the ... light quantum hypothesis" (p. 199). Why then did he not hit upon the idea of a quantum gas and apply it to both

Einstein Studies, vol. 8
Einstein: The Formative Years, pp. 231–251

massless light quanta and massive particles well before receiving Bose's work in 1924?[1]

I regard such counterfactual questions as useful: consideration of alternate historical scenarios can teach us a lot about the actual course of history (Stachel 1994a). In the present case, as in *The Hound of the Baskervilles*, the important question may be why the dog did *not* bark. So, in an attempt to shed more light on the precise meaning of his 1905 light quantum hypothesis, I shall offer some speculations on why Einstein did *not* develop the theory of an ideal quantum gas by 1909.

Those familiar with the work of Klein (see especially 1963, 1964, 1977), Kuhn (1978, chapter 7), and Darrigol (1988, 1991) on Einstein and blackbody radiation will realize my debt to them for many fundamental insights into Einstein's early work on the quantum that are taken for granted here. However, none of them raised this question. Neither did Mehra and Rechenberg but, as we shall see, their discussion suggests a significant new insight into the influence of Wien's work on Einstein that bears directly on this question.[2]

Einstein 1925 mentions two possible objections to his proposed theory of a quantum gas. As applied to blackbody radiation, Planck's formula for the density of blackbody radiation,

$$\varrho(\nu, T) = \frac{8\pi h\nu^3}{c^3[\exp(h\nu/kT) - 1]},$$

may be split into two factors, each of which is related to one of the two objections:

(1) The first objection is related to the already-mentioned matter-radiation analogy, which Einstein notes is not universally accepted. (He may well have had Bohr in mind, who refused to countenance the light quantum hypothesis—see Stachel 1999). It is related to the factor $8\pi\nu^2/c^3$, which Bose showed can be interpreted as the number of cells in the phase space of a gas of light quanta, treated as particles.[3]

(2) The second objection is to the method of counting equiprobable cases for a quantum gas, massless or massive:

> The statistical method applied by Bose and myself is by no means indubitable, but on the contrary only appears to be justified *a posteriori* by its success in the case of radiation (Einstein 1925, p. 18).

It is related to the factor $h\nu/[\exp(h\nu/kT) - 1]$, which Bose interprets as the average energy per cell when a light quantum gas is in thermal equilibrium. Its value depends on the method of counting equiprobable distributions of light quanta among the cells. From this count one computes the probability W and hence (using Boltzmann's principle) the entropy S of the state of a gas, which is then maximized to determine the thermal equilibrium state.

By examining the role of these two issues in Einstein's work between 1901 and 1909, we may hope to understand why he did not develop the idea of a quantum gas; his ready acceptance and quick application of the idea to massive particles after Bose developed it for light quanta; and why he felt much still remained obscure about the reasons for the idea's success.

2. The Matter-Radiation Analogy

Einstein's speculations on a possible analogy between matter and radiation can be traced back to 1901. Just after he graduated from the Zurich Polytechnic, he wrote his fiancée, Mileva Marić:

> Recently the idea came to me that in the production of light perhaps a direct transformation of motional energy into light takes place due to the parallelism kinetic energy of molecules—absolute temperature—spectrum (spatial radiant energy in the state of equilibrium) (Einstein to Marić, 30 April 1901, Einstein 1987, Doc. 101, p. 295).

Since he was already convinced of the correctness of atomism in the case of matter, such an analogy would suggest the desirability of a similar model of radiation.

He developed such a model in Einstein 1905a, his first light quantum paper, which starts by stressing

> the deep-going formal difference . . . between the theoretical representations physicists have formed of gases and other ponderable matter, and that of Maxwell's theory of electromagnetic processes in so-called empty space.

While the representation of a state of matter only involves specification of

> the positions and velocities of a large but still finite number of atoms and electrons, we require continuous spatial functions for the determination of the electromagnetic state of a space, so that a finite number of quantities is not to be regarded as sufficient for determining the electromagnetic state of a space (Einstein 1905a, p. 151).

As a consequence of this formal distinction, the energy of matter is supposed to be the sum of a finite number of terms, namely a sum over energies associated with the particles of matter, while the energy of the electromagnetic field is supposed to be continuously distributed throughout space. But:

> It seems to me indeed that observations concerning "blackbody radiation," photoluminescence, the generation of cathode rays by ultraviolet light [photoelectric effect] and other groups of phenomena concerned with the generation and transformation of light appear to be better understandable on the assumption that the energy of light is discontinuously distributed in space. According to the assumption contemplated here, the light emitted as radiation from a point is not continuously distributed over greater and greater spaces, but consists of a finite number of energy quanta localized at points of space, which move without dividing and can only be absorbed and emitted as a whole (p. 151).

Einstein was undoubtedly impelled to make this assumption by the phenomena connected with the emission and absorption of radiation that he mentions. But he was also impelled by his conviction that the ether concept, which underlay the traditional concept of the electromagnetic field, must be abandoned in order to develop a consistent electrodynamics of moving bodies, a conviction he soon published in his first paper on relativity (Einstein 1905b). As he put it in 1910:

"Without the ether, energy continuously distributed throughout space seems to me an absurdity (*Unding*)" (Einstein 1910a, p. 177). He raised this point as a strong argument in favor of his conviction that the quantum hypothesis could not be restricted to the material entities interacting with the radiation field, as Planck and others still hoped, but must apply to the field itself, as suggested in Einstein 1905a. The main argument "concerning 'blackbody radiation' " in this paper consists of a demonstration that:

> Monochromatic radiation of low density (within the range of validity of Wien's radiation formula [i.e., when ν/T is large and hence, in modern notation, $\rho(\nu, T) = (8\pi h\nu/c^3)\exp(-h\nu/kT)$], behaves thermodynamically as if it consisted of quanta of energy that are independent of each other and of the magnitude [$h\nu$ in modern notation]. (Einstein 1905a, p. 161).[4]

Einstein shows this by an ingenious inversion of Boltzmann's principle (Einstein gave $S = k \ln W$ this name), a trick he often used thereafter. Instead of reading off the entropy S from a calculation of the probability W, he read off the probability W from the known expression for the entropy S of radiation obeying Wien's law. This leads to an expression for the probability W that, at any moment, the total energy E of the radiation is concentrated in a subvolume υ of the total volume V:

$$W = (\upsilon/V)^{E/h\nu}.$$

Now he introduces the analogy between radiation and matter, by considering a gas consisting of n particles in thermal equilibrium in a similar volume V. If each particle moves according to any law of motion that does not pick out a preferred point or direction in V, then the probability w that, at a given moment, that particle will be in the subvolume υ is clearly: $w = (\upsilon/V)$. If the particles are statistically independent of each other (as are classical particles in an ideal gas), then the probability that, at some moment, all n particles are in υ is clearly the product of the probabilities for each of the n particles:

$$W = w^n = (\upsilon/V)^n.$$

If one takes the analogy between radiation and matter seriously, then comparison of the two formulae for W yields Einstein's conclusion: Radiation in the Wien region behaves like a gas of independent particles, each with energy $h\nu$.[5]

Einstein's argument is easily inverted (and indeed was often interpreted by his contemporaries in this sense): a gas of independent light quanta in thermal equilibrium (its state of maximum entropy) obeys Wien's radiation law.[6] Then one might wonder why didn't Einstein attempt to develop a similar analogy based on Planck's radiation formula?[7] Why did he not look for a way of modifying the statistical treatment of a gas of light quanta that leads to the Planck formula when one maximizes the entropy—in other words, why did Einstein himself not develop Bose's statistics? Neither Kuhn nor Darrigol tries to explain why Einstein initially confined himself to Wien's law, and Klein just hazards a guess: "Einstein based his calculations on this Wien distribution, perhaps because of its greater simplicity" (Klein 1977, p. 24, repeating Klein 1963, p. 68).

3. Why Wien?

Mehra and Rechenberg 1982 offers a detailed discussion of this question, drawing attention to Wien 1900, a previously-neglected source of Einstein's approach in the 1905 paper; but I do not entirely agree with their interpretation of Wien's paper. It was written in response to the measurements of Lummer and Pringsheim, which made it clear that Wien's radiation formula fails at short wave lengths. In his response, Wien stresses that he always expected a difference between the behavior of blackbody radiation at short and long wavelengths.[8] Mehra and Rechenberg claim that Wien expected the departure from classical results at short wavelengths to be due to the failure of classical electrodynamics at these wavelengths.[9] But, on my reading, Wien attributed this departure to the atomistic structure of matter. Since Wien 1900 is not well known, and Mehra and Rechenberg do not cite it at any length, I shall do so:

> I must first of all emphasize that, in contrast to Mr. Planck, I still hold to my earlier-expressed viewpoint, that long and short wavelength electromagnetic waves do not represent just a quantitative difference in their relation to thermal radiation. In the case of absorption, it is indeed generally accepted that, for long wave-lengths, it can be represented by a single vector, or what amounts to the same thing, by assuming the continuity of matter; but in the case of short wave-lengths, on the contrary, the molecular constitution of matter is important. The very same thing must also hold for emission. Therefore, I consider it from the outset as improbable that a radiation law based on molecular assumptions should also be valid for very long wavelengths. The agreement with experiment for short wave-lengths obviously speaks in favor of the approximate validity of the assumptions made [in the derivation of Wien's law] for not-too-long wavelengths.
>
> Bearing this in mind, I consider it not very promising to base a generally-valid radiation formula on molecular hypotheses, as long as a purely thermodynamic derivation is impossible. Rather, the viewpoint of Mr. Paschen, communicated by letter,[10] seems to me more promising, to investigate the representation of the radiation formula for long wavelengths independently of that valid for short wavelengths, just as one uses different formulas for slow electric oscillations than for very rapid ones (Wien 1900, pp. 537–538).

Mehra and Rechenberg conclude:

> Einstein, in his analysis of the foundations of Planck's radiation formula in 1905, picked up Wien's idea of 1900. That is, he assumed that the heat radiation consisted of two parts: the long-wavelength radiation, described by the known laws of electrodynamics, and the short-wavelength radiation, determined by laws still unknown. He wanted to discover the nature of these unknown laws (Mehra and Rechenberg 1982, p. 75).

I agree with these comments, except for two points. The first and more important is the claim, already mentioned, that Einstein's idea of using the Wien formula to "discover . . . these unknown laws" of "short-wavelength radiation" just picked up "Wien's idea." As further (in addition to the quotation above), and hopefully conclusive, evidence that "Wien's idea" was to discover the nature of molecular structure rather than that of radiation, I cite his statement: "I have expressly said that I expect an insight into molecular theory from the testing of the [Wien] law by experiment" (Wien 1900, p. 537).

The other objection is that one cannot describe Einstein's 1905 paper as "an analysis of the foundations of Planck's radiation formula." To the contrary, it is noteworthy how extremely reticent Einstein is in that paper about Planck's radiation formula and the theory behind it.[11] He mentions "the Planck formula" only once (it "satisfies all experiments up to now") in the course of a demonstration that Planck's derivation of Avogadro's number (which Einstein calls the elementary quantum) is "to a certain extent independent of Planck's theory of 'blackbody radiation' "—his only mention of that theory! He is at pains to show that Planck's result depends only on the classical limit of Planck's formula (now called the Rayleigh–Jeans formula), which Einstein had already derived in this paper (see below).

Einstein's reticence concerning Planck's work may be explained by a curious remark near the beginning of his second quantum paper: "On the Theory of Light Production and Absorption" (Einstein 1906). After briefly summarizing his light quantum hypothesis in the first paragraph, he goes on: "At the time [of his 1905 paper] it appeared to me as if Planck's theory of radiation in certain respects was counterposed to my work" (p. 350). In other words, in 1905 he believed that Planck's law was *not* compatible with light quanta, and only in 1906 was he able to give an argument showing "that Planck's theory makes implicit use of the above-mentioned light quantum hypothesis" (p. 350).

Einstein 1905 actually contains an implicit critique of Planck.[12] It shows (p. 44), that the accepted foundations of classical statistical mechanics and Maxwell's electrodynamics lead inevitably to a classical distribution law that is valid (as one would expect on the basis of Wien's argument) for long wave lengths, but must fail for short ones since it leads to a divergence of the total radiant energy. Since Planck claimed to base his work on these same foundations, the clear implication is that he should have arrived at the same law of limited validity.

When Einstein uses a radiation law for short wavelengths to argue from it to the light quantum hypothesis, it is (as noted above) Wien's law: "We take this formula as the basis of our calculations, keeping in mind however that our results are only valid within certain [short-wavelength] limits" (p. 157).

Thus, his theoretical analysis indeed realizes Wien's program of analyzing separately the long- and the short-wavelength behavior of the radiation. But, it was Einstein who made the decisive shift: from Wien's concern with the atomic structure of matter interacting with radiation to an argument for the discontinuous structure of the radiation itself.

What might have suggested this shift from a focus on the matter-radiation interaction to the radiation itself? I suggest it may have been the displacement law. Wien himself notes that, as a result of his displacement law,[13] the wavelength at which his radiation law breaks down "will shift to ever smaller wavelengths with rising temperature" (Wien 1900, p. 537). The same will hold true for the wavelength at which the classical (Rayleigh–Jeans) law breaks down. But if the atomic structure of matter were responsible for this breakdown, one would expect the breakdown to occur at some fixed wavelength related to this atomic structure: the atomic diameter, interatomic distance, or some such parameter.

What has all this to do with Einstein's reasoning in 1905? What evidence is there that he was familiar with Wien's work? One answer is that his first published discussion of blackbody radiation, in Einstein 1904, invokes the displacement law. He investigates energy fluctuations in blackbody radiation, assuming that statistical-mechanical concepts are applicable to this radiation. Using a formula he had developed earlier in the paper for energy fluctuations, he shows that the volume, for which such fluctuations are of the same order of magnitude as the energy average, has a linear dimension that varies inversely with T, with roughly the same numerical coefficient as occurs in Wien's displacement law. Now, if its fluctuations are as big as the energy itself, the concept of a continuous energy distribution breaks down; and Einstein's result, when combined with Wien's displacement law, suggests that this breakdown starts to occur at a certain wavelength associated with the radiation. To someone familiar with Wien's 1900 paper, this might well suggest that the failure of the classical (Rayleigh–Jeans) formula at short wavelengths is associated with a discontinuity in the structure of the radiation, rather than in that of the matter with which it is in equilibrium. But Wien's ideas about the non-classical nature of matter-radiation interactions at short wavelengths may still have played a significant role in shaping Einstein's work: The "heuristic point of view" of its title, as well as all the tests of the light quantum hypothesis suggested in Einstein 1905a, are "concerning the production and transformation [i.e., emission and absorption] of light" by interactions with matter.

What is the likelihood that Einstein read Wien 1900? We know that in 1899 he was so impressed by another work by Wien that he wrote to him (see Einstein 1987, Doc. 57, p. 234). Around 1900, he appears to have read the *Annalen der Physik* regularly; and by 1904–1905 he was evidently familiar with both Wien's formula and displacement law. Taken together with the striking similarity between the research program suggested by Wien 1900 and the one Einstein carried out in 1905—as modified in light of his 1904 result—there is good reason to conjecture that Einstein was familiar with Wien 1900.

4. Elucidating Einstein

This is all I have to add to the well-known discussions of the light quantum hypothesis. Returning to the question raised in the introduction, I can suggest two reasons why Einstein did not immediately develop the idea of a "light quantum gas," that is, a particle model of blackbody radiation:

(1) The analogy between matter and radiation does not demand that one adopt a particle model for radiation;

(2) As indicated above, the particle model of a light quantum gas, taken together with classical statistics (the only kind then known), leads inexorably to the Wien formula.

I shall elaborate on each of these reasons. Einstein's 1905 arguments for light quanta offer no warrant for attributing particle-like properties (other than energy) to them. The argument by analogy with ponderable matter implies that electromagnetic radiation should be modeled as a system whose energy is concentrated at a number of points rather than spread throughout space; but that, of course, allows other possibilities besides a particle model (see below).

His first relativity paper (Einstein 1905b) does contain elements that suggest the possibility of a particle model of radiation. It is surely not a casual remark when he states: "It is noteworthy that the energy and frequency of a light complex change according to the same law with a change in the state of motion of an observer" (Einstein 1905b, p. 299).

But at this stage he was extremely circumspect in mixing quantum and relativistic considerations, and does not remark on the obvious significance of their similar behavior under Lorentz transformations for the possibility of identifying his hypothetical light quantum, for which $E = h\nu$, with such a relativistic "light complex." Nor does he remark on the implication that a light quantum might have momentum, as does a "light complex" according to Maxwell's theory.[14]

Similarly, 1905b notes that, according to relativistic kinematics, "the velocity of light V cannot be altered by composition with any 'subluminal velocity' "; but Einstein does not note something that must have been equally obvious to him: This result allows for a relativistic emission theory of light. Taken together with classical kinematics, such an emission theory makes the velocity of light different—and even direction-dependent—in different inertial frames of reference; but Einstein's result shows that relativistic kinematics eliminates this problem. Since emission theories of light were traditionally associated with particle models, this meant that a particle model of light is compatible with special-relativistic kinematics. But again, an emission theory does not require a particle model: other types of emission theory are possible as Ritz's contemporary work showed. Einstein was quite aware that, whatever they were, light quanta could *not* be statistically independent of each other as are "normal" gas particles. For (as mentioned earlier) it follows from reversing the line of argument in the 1905 quantum paper that, if the light quanta *were* independent, the usual probabilistic calculation of the entropy would lead to *Wien's* law for a light quantum gas.

In 1909, Einstein explicitly stated his objections to a particle model of light:

> Indeed, I am not at all of the opinion that one should think of light as composed of quanta localized in relatively small spaces and independent of each other. This would indeed be the most convenient explanation of the Wien end of the radiation formula. But the division of a light ray at the surface of a refracting medium by itself completely forbids this conception. A light ray can divide, but a light quantum cannot divide without change of frequency (Einstein to Lorentz, 23 May 1909, Einstein 1993b, Doc. 163, p. 193).

5. Einstein's Critique of Complexions

Einstein had strongly-held views about physical probabilities: For him, the time-ensemble definition of the probability of the state of a physical system was primary, as he made clear in 1903:

> Consider a physical system, which can be represented by equations [determining the time rate of change of its state variables as a function of their instantaneous values] and having energy E, during the time interval T starting from an arbitrary initial time. Imagine an arbitrary region Γ of the state variables $p_1 \ldots p_n$ chosen, so that at a definite moment during the time interval T the variables $p_1 \ldots p_n$ either lie inside this region or outside it; they will hence lie in the chosen region Γ, during a definite part of the interval T, which we shall call τ. Our condition then takes the form: If $p_1 \ldots p_n$ are the state variables of a physical system, that is of a system that assumes a stationary state, then the quantity τ/T assumes a definite limiting value for $T = \infty$ for every region Γ (Einstein 1903, pp. 171–172).

Any *combinatorial* argument, based on counting the number of complexions of the elements composing the system that correspond to a given physical state, must be supplemented by a *dynamical* argument proving that these complexions are equally probable before such a count could be used to define the W needed in Boltzmann's principle. Einstein did not name names in 1905, but made the criticism clear:

> In calculations of entropy based on molecular-kinetic theory, the word "probability" is often used with a meaning that does not agree with the definition of probability given in probability theory. In particular, "cases of equal probability" are often hypothetically postulated in cases where the theoretical model being applied suffices for their deduction, rather than hypothetical postulation. In a separate article, I will show that the so-called "statistical probability" is all that is needed for the treatment of thermal processes, and hope thereby to overcome a logical difficulty that still stands in the way of the application of Boltzmann's principle (Einstein 1905a, p. 140).[15]

Indeed, his next paper on radiation theory, Einstein 1906, showed that, if one treated Planck's oscillators as systems capable of existing only in states whose energies are integral multiples of $h\nu$, where ν is the frequency of the oscillator, and whose energy changes are discontinuous during absorption and emission of radiation, then the statistical interpretation of Planck's probability W could be maintained.

By 1909, Einstein evidently felt able to criticize Planck openly. After summarizing his time-ensemble definition of the probability of a state, he continues:

> Starting from this definition, one can show that the following equation for the entropy S must hold
>
> $$S = R/N \ln W + \text{const},$$
>
> where the constant is the same for all states of equal energy. Neither Boltzmann nor Planck have given a definition of W. Proceeding quite formally, they set $W =$ the number of complexions of the state in question.
>
> If one now requires that these complexions be equally probable, where the probability of a complexion is defined analogously to how we defined the probability of a state [above], then one arrives at exactly the definition of the probability of a state

given [above]; except that the logically unnecessary element complexion has been used in the definition.

Although the relation given between S and W only holds when the probability of a complexion is defined in the way indicated or an equivalent one, neither Boltzmann nor Planck has defined the probability of a complexion. ... In the oscillator theory of radiation Planck was not free to choose his complexions. He could only postulate the pair of equations

$$S = R/N \ln W$$

and

$$W = \text{number of complexions}$$

if he added the condition that the complexions must be so chosen that, in his theoretical model, they are found to be equally probable on the basis of statistical considerations. In this way, he would have been led to the [Rayleigh–] Jeans formula (Einstein 1909, pp. 544–545).

Leaving aside the question of whether Einstein's inclusion of Boltzmann in his criticism is justified,[16] Planck's original derivation of the blackbody spectrum applied such combinatorial considerations to the energy of an ensemble of material oscillators in thermal equilibrium with the radiation field, and just assumed all complexions to be equiprobable.

A decade later, Debye applied similar considerations to the radiation field itself, considered as an ensemble of its normal modes confined to a cavity. In both cases, the problem was to divide the fixed total energy into quanta of magnitude $h\nu$ and distribute them among the oscillators—the material oscillators (Planck), or the normal modes of the radiation field (Debye). A complexion here is just the name for one such way of distributing the total energy. The counting of complexions is based on the assumption that, while material oscillators or normal modes of the radiation field are *distinguishable*, quanta of energy of a given frequency, being just aliquot portions of the total energy at that frequency, are *indistinguishable*. If there are seven quanta of energy associated with one oscillator or normal mode and five with another, it makes no sense to ask *which* seven or *which* five quanta of energy are involved.

Debye, like Planck, just *assumed* that the number of such complexions is a measure of the probability of a state; that is, that each complexion is equally probable. And here Einstein's criticism applies. They offered no dynamical argument why such complexions should be equiprobable in *his* sense (time ensemble) of probability. Indeed, as Einstein pointed out in 1906, Planck had to arbitrarily (from the point of view of classical theory) restrict himself to those complexions that correspond to energies that are integral multiples of $h\nu$ for each frequency ν; and then assume the equiprobability of all such arbitrarily-restricted complexions. As Planck had long recognized, his procedure could only be justified *a posteriori* by its success in producing Planck's formula.[17]

An attempt to turn Debye's argument, based on quanta of energy of the wave field, into an argument based on the model of a gas of particles (light quanta), faces two problems:

(1) How to justify the count of the number of possible states of the gas corresponding to a given energy (or frequency) of a gas particle—this is the problem of the first factor, mentioned in the Introduction.

(2) How to justify the counting of complexions corresponding to a given state of the gas as a counting of equal probabilities—this is the problem of the second factor.

If the first problem is solved, the second only becomes more acute. For, in addition to Einstein's *general* objection to the counting of complexions, a *new* objection arises when such a count of complexions in the Planck–Debye way is reinterpreted as a count of complexions for *particles*. While the energy elements that comprise the total energy are indistinguishable, so that it makes no sense to ask *which* energy quanta are in a given state, how can one avoid this question for particles? Classically, one *cannot*: it seemed part of the nature of the particle concept that such a question makes sense. As was clearly understood at the time (see Ehrenfest 1911), to give it up means giving up the statistical independence of non-interacting particles.

No wonder that even Einstein paused before entering such a statistical morass. Rather than an independent-particle model for light quanta, in 1909 he was searching for a model in which a light quantum represents some sort of singularity in a (nonlinear) electromagnetic field, the motion of which would be "guided" by the equation of motion of the field. Remember that by 1905 he realized that the low-frequency (Rayleigh–Jeans) behavior of the radiation field is correctly described by the application of the equipartition theorem to (low-frequency) classical oscillators in equilibrium with the field; so there is no need to modify the classical wave picture of emission and absorption of radiation for these frequencies. In 1906, Laue had shown that, for coherent bundles of radiation, the addition theorem for entropy must fail if Planck's theory of radiation is to be consistent with the second law of thermodynamics. But if the bundles consisted of independent quanta, the addition theorem would have to hold. Laue considered this a decisive argument against the assumption that the radiation field is composed of light quanta, which he tacitly identified with particles;[18] but Einstein must have taken it as another argument (if one were needed) against the picture of the radiation field as composed of *independent* quanta.

I hope it is now clear that Einstein understood the problems facing an attempt to formulate a quantum gas theory long before he received Bose's paper in 1924. I have already discussed that period and its aftermath in Stachel 1994b, so I shall be brief. Bose rushed in where angels feared to tread, solving the first problem in setting up a theory of a light quantum gas (the counting of states in phase space) very neatly. He solved the second problem (the counting of complexions) without realizing that he was doing something new and revolutionary: He just repeated Planck and Debye's counting method, without realizing that he was abandoning classical (Boltzmann) statistics. Einstein, however, was well aware of what Bose had done, and that this method of counting implied a new, non-classical statistics for light quanta. Since it (like Planck's original method) was justified *a posteriori* by its success in producing Planck's formula, Einstein immediately proceeded to apply the new counting method (i.e., the new statistics that we now call Bose–Einstein) to a quantum gas of massive particles (with the small modification needed in this case, because the particle number is conserved for massive particles). He showed that, as the temperature is lowered, such a gas must undergo a phase change, which is now called Bose–Einstein condensation. Perhaps even more important, he showed that the fluctuations of his quantum gas, like those of the blackbody gas, consisted of two terms, a wave term and a particle term; this provided additional evidence to support extension of the ideas of wave-particle duality to massive particles just at the time when DeBroglie and Schrödinger were occupying themselves with such problems.

Einstein was also well aware that the new statistics implies a peculiar statistical entanglement between the particles of his quantum gas. He still hoped to interpret the new statistics as evidence of some novel type of physical interaction between the particles (see Einstein 1924c)—but that is another story, which Don Howard has told elsewhere (see Howard 1990 and Stachel 1997).

6. Appendix: Natanson Neglected?

Kastler 1983 claims that Einstein could have developed the theory of a quantum gas on the basis of Natanson's work, which he overlooked at the time (1911):

> Probably at this time (1911) Einstein's mind was fully occupied by the puzzle of gravitation. Only when he had solved this problem (1916) did he turn his attention again to quantum theory. ... And so it happened that it took another 13 years [after 1911] before Einstein, inspired by Bose's paper, worked out the statistics of indistinguishable particles and transposed it from photons to atoms and molecules (Kastler 1983, p. 621).

First of all, it is not correct to say Einstein did nothing in quantum theory between 1911 and 1916 (see the *Collected Papers* vols. 3, 4, and 6, Einstein 1993a, 1995, 1996). And even if Kastler were correct about the period before 1916, Natanson was in Berlin in 1915–1916 and Einstein was in contact with him. So it is difficult to accept Kastler's explanation.

Let us look, then, at what Natanson actually did. He was concerned with the problem of counting the number of equiprobable ways of distributing "discrete elements or units of energy" over a number of "receptacles of energy," as he calls them "for brevity." He is quite definite in asserting that these "receptacles" are the "ultimate particles of which matter consists ... capable only of absorbing, containing and emitting amounts of energy which are multiples of these finite and determinate units" (Natanson 1911a, p. 134). He goes on to point out that the result of such counting depends on whether or not one assumes that "we can identify either receptacles or energy-units" (p. 135). His argument is directed at Planck's earlier derivations of the average energy of a system of such "receptacles of energy," which Planck took to be charged oscillators in equilibrium with the radiation field at some definite temperature. And indeed, when Planck 1911 mentions Natanson 1911b, it is to claim that his method of calculating the average energy of a system of N such oscillators is now "completely unambivalent and in particular no longer contains the indefiniteness about which L. Natanson has recently spoken with justification" (Planck 1911, p. 277). So neither Planck nor Natanson was concerned with the radiation field itself, or the distribution of particles (of radiation or of any other kind) over cells in phase space, as Kastler claims:

> He [Natanson] defines the different ways for distributing discrete particles over discrete cells and demonstrates the possibility of three modes of distribution, depending on whether the particles, on the one hand, and the cells, on the other, are indistinguishable or distinguishable entities (Kastler 1983, p. 620).

Kastler has misidentified Natanson's "energy units" as particles, and Natanson's "ultimate particles" or "receptacles of energy" as cells in phase space. Once we eliminate this confusion, there is no reason to question Einstein's lack of public reaction to Natanson's paper. Even assuming he was familiar with it, it did not remove his objection to combinatorial techniques, old or new, as a way of defining probabilities.

> For his foundation Planck assumes a perfectly definite rule for evaluating the chances; and by a distinct appeal to experience, we find *a posteriori* that the proposition is true. But no one apparently has ever attempted to justify its correctness on general principles and so far as I can see it cannot be done (Natanson 1911a, p. 139).

As we saw in section 5, as early as 1906 Einstein had found a way of justifying Planck's results by a statistical argument applied to oscillators with discrete energies. And there is no reason why Natanson's work should have suggested that Einstein reconsider his conclusion that a particulate model of radiation must lead to Wien's law. As emphasized above, Natanson was concerned with distribution of energy units over material entities, not light quanta over cells in phase space. Ehrenfest, who did consider the latter problem (see Ehrenfest 1911, Ehrenfest and Onnes 1914), had just demonstrated that Planck's combinatorics is incompatible with the concept of statistically-independent particles—the only kind that Einstein had been considering before Bose came along.

ACKNOWLEDGMENTS. I thank Olivier Darrigol for a critical reading of the text.

NOTES

[1] This question first occurred to me in the course of writing Stachel 1994b, which discusses Einstein's reaction to Bose's work.

[2] For their discussion of Einstein's early work, see Mehra and Rechenberg 1982, pp. 59–83.

[3] "$8\pi\nu^2/c^3 V\, d\nu$ can be interpreted as the number of elementary cells of the six-dimensional phase space for the [light] quanta" (Bose 1924, p. 384). Here V is the volume of the cavity.

[4] Einstein 1905a does not use either Planck's constant h or Boltzmann's constant k; but rather two equivalent constants. This is of some significance to an understanding of Einstein's approach; but this has been elucidated in Klein 1963, so I shall not repeat his

discussion. Einstein 1910b, which sketches his 1905 argument again (in some respects more carefully—see the next footnote, does use h, but still uses R/N instead of k.

[5] In 1910, he repeats the argument with a more cautious conclusion, not using the word "particle." After showing that "for sufficiently dilute radiation," $W = \left(\frac{V}{V_0}\right)^{\frac{\varepsilon_0}{h\nu}}$ is "the probability that at a given instant all the energy E_0 is included in a volume V," he concludes: "one can easily show that this expression is not compatible with the superposition principle. The radiation behaves ... as if its energy were localized in $E_0/h\nu$ points moving independently of each other" (Einstein 1910b, p. 251).

[6] Renn 1993, p. 334, has emphasized the invertibility of Einstein's 1905 argument.

[7] As is well known, when he did discuss energy fluctuations in the Planck distribution in Einstein 1909, he showed that they are the sum of the two terms: one could be interpreted as a particle term and the other as a wave term. While it is not impossible that he first made this calculation much earlier, as is suggested by some comments in his "Autobiographical Notes" (see Einstein 1979, p. 48), in the next section we shall argue that initially he was suspicious of Planck's formula and had good reason to start with Wien's formula. This argument replaces my earlier speculation in Stachel 1994b, that Einstein did attempt such a derivation.

[8] In support of this claim he cites Wien 1893b which argues on the basis of the molecular structure of matter that there should be an upper limit to the wave lengths that occur in the blackbody radiation from rigid bodies. It is possible that Einstein followed up this reference, but this point is not important for my argument.

[9] "Wien had stated ... that the heat radiation satisfying his law could not be explained on the basis of the electrodynamic theory; rather it exhibited a structure which was different in quality from that of the long-wavelength radiation, the radiation which obeyed the laws of classical electrodynamics" (Mehra and Rechenberg 1982, p. 75).

[10] It would be interesting to search for this letter, or other evidence of Paschen's views. Lacking this, I shall here refer to Wien's views, leaving open the question of how far they reflect Paschen's.

[11] Klein 1977, p. 29 noted this, as do the Editors of the Einstein Papers: "Among Einstein's papers on the quantum hypothesis, the 1905 paper is unique in arguing for the notion of light quanta without using either the formal apparatus of his statistical papers or Planck's law" (see the headnote, "Einstein's Early Work on the Quantum Hypothesis," Einstein 1989, p. 139).

[12] I agree with Darrigol 1988, which interprets a later remark of Besso as a claim that in 1904–1905 he persuaded Einstein not to criticize Planck explicitly (p. 57). Below we shall quote another passage from Einstein 1905a, cited by Darrigol as an implicit criticism of Planck.

[13] Wien 1894 shows by a thermodynamic argument that the energy spectrum must be of the form $\lambda^{-5} f(\lambda T)$ where λ is the wavelength and T the absolute temperature of the radiation. Thus, any particular wavelength characterizing the spectrum, such as λ_m, the wavelength at the energy maximum, will vary as $1/T$ times a universal coefficient, a result that became known as Wien's displacement law.

[14] It was only in 1909 that Einstein gave the first of a series of arguments that culminated in 1916 with the definite attribution of momentum to light quanta.

[15] I agreed with Darrigol 1988, which sees the omission of Planck's name as a strategic move by Einstein. See note 12.

[16] See the paper by Jürgen Renn in this volume for a discussion of Einstein's knowledge of Boltzmann's work.

[17] In 1897, he wrote: "The probability of each [macro-] state is proportional to the number of complexions corresponding to it, or, in other words, any particular complexion has the same probability as any other complexion. . . . This statement is the central point of the whole theory under discussion; it can be proved in the final analysis only by experience" (cited from Kastler 1983, p. 616).

[18] See Max Laue to Albert Einstein, 2 February 1906, Einstein 1993, Doc. 37, pp. 41–42.

References

Bose, Satyendra Nath (1924). "Wärmegleichgewicht im Strahlungsfeld bei Anwesenheit der Materie." *Zeitschrift für Physik* 27: 384–392.

Darrigol, Oliver (1988). "Statistics and Combinatorics in Early Quantum Theory." *Historical Studies in the Physical and Biological Sciences* 19: 1–80.

—— (1991). "Statistics and Combinatorics in Early Quantum Theory, II: Early Symptoms of Indistinguishability and Holism." *Historical Studies in the Physical and Biological Sciences* 21: 237–298.

Debye, Peter (1910). "Der Wahrscheinlichkeitsbegriff in der Theorie der Strahlung." *Annalen der Physik* 33: 1427–1434.

Ehrenfest, Paul (1911). "Welche Züge der Lichtquantenhypothese spielen in der Theorie der Wärmestrahlung eine wesentliche Rolle." *Annalen der Physik* 36: 91–118.

Ehrenfest, Paul and Kamerlingh Onnes, Heike (1914). "Vereinfachte Ableitung der kombinatorischen Formel, welche der Planckschen Strahlungstheorie zugrunde liegt." *Annalen der Physik* 46: 1021–1024.

Einstein, Albert (1903). "Eine Theorie der Grundlagen der Thermodynamik." *Annalen der Physik* 11: 170–187. Reprinted in Einstein 1989, Doc. 4, pp. 77–94.

—— (1904). "Zur allgemeinen molekularen Theorie der Wärme." *Annalen der Physik* 14: 354–362. Reprinted in Einstein 1989, Doc. 4, pp. 99–107.

—— (1905a). "Über einen die Erzeugung und Verwandlung des Lichtes betreffenden heuristischen Gesichtspunkt." *Annalen der Physik* 17: 132–148. Reprinted in Einstein 1989, Doc. 14, pp. 150–166.

—— (1905b). "Zur Elektrodynamik bewegter Körper." *Annalen der Physik* 17: 891–921. Reprinted in Einstein 1989, Doc. 23, pp. 276–306.

—— (1906). "Zur Theorie der Lichterzeugung und Lichtabsorption." *Annalen der Physik* 20: 199–206. Reprinted in Einstein 1989, Doc. 34, pp. 350–357.

—— (1909). "Zum gegenwärtigen Stand des Strahlungsproblems." *Physikalische Zeitschrift* 10: 185–193. Reprinted in Einstein 1989, Doc. 56, pp. 542–550.

—— (1910a). "Response to Manuscript of Planck." Written before 18 January 1910. In Einstein 1993a, Doc. 3, pp. 177–178.

——— (1910b). "Sur la théorie des quantités lumineuses et la question de la localisation de l'énergie électromagnétique." *Archives des sciences physiques et naturelles* 29: 525–528. Reprinted in Einstein 1993a, Doc. 5, pp. 249–252.

——— (1924a). "Quantentheorie des einatomigen idealen Gases." *Preussische Akademie der Wissenschaften* (Berlin). *Physikalisch-mathematische Klasse. Sitzungsberichte*: 261–267.

——— (1924b). "Quantentheorie des einatomigen idealen Gases. Zweite Abhandlung." *Preussische Akademie der Wissenschaften* (Berlin). *Physikalisch-mathematische Klasse. Sitzungsberichte* (1925): 3–14.

——— (1924c). Über den Aether." *Schweizerische naturforschende Gesellschaft. Verhandlungen* 105: 85–93.

——— (1925). "Quantentheorie des idealen Gases." *Preussische Akademie der Wissenschaften* (Berlin). *Physikalisch-mathematische Klasse. Sitzungsberichte*: 18–25.

——— (1979). *Autobiographical Notes*. Paul Arthur Schilpp, transl. and ed. (La Salle, IL and Chicago: Open Court).

——— (1987). *The Collected Papers of Albert Einstein*. Vol. 1, *The Early Years, 1879–1902*. John Stachel, et al., eds. Princeton: Princeton University Press.

——— (1989). *The Collected Papers of Albert Einstein*. Vol. 2, *The Swiss Years: Writings, 1900–1909*. John Stachel, et al., eds. Princeton: Princeton University Press.

——— (1993a). *The Collected Papers of Albert Einstein*. Vol. 3, *The Swiss Years: Writings, 1909–1911*. Martin J. Klein, et al., eds. Princeton: Princeton University Press.

——— (1993b). *The Collected Papers of Albert Einstein*. Vol. 5, *The Swiss Years: Correspondence, 1902–1914*. Martin J. Klein, et al., eds. Princeton: Princeton University Press.

——— (1995). *The Collected Papers of Albert Einstein*, Vol. 4, *The Swiss Years: Writings, 1912–1914*. Martin J. Klein, et al., eds. Princeton: Princeton University Press.

——— (1996). *The Collected Papers of Albert Einstein*, Vol. 6, *The Berlin Years: Writings 1914–1917*. Martin J. Klein, et al., eds. Princeton: Princeton University Press.

Howard, Don (1990). "'Nicht sein kann was nicht sein darf,' or the Prehistory of EPR, 1909–1935: Einstein's Early Worries About the Quantum Mechanics of Composite Systems." In *Sixty-two Years of Uncertainty: Historical, Philosophical, and Physical Inquiries into the Foundations of Quantum Physics*. Arthur Miller, ed. New York: Plenum, pp. 61–111.

Kastler, Alfred (1983). "On the Historical Development of the Indistinguishability Concept for Microparticles." In *Old and New Questions in Physics, Cosmology, Philosophy, and Theoretical Biology*. Alwyn van der Merwe, ed. New York and London: Plenum, pp. 607–623.

Klein, Martin (1963). "Einstein's First Paper on Quanta." *The Natural Philosopher*. Vol. 1. Daniel E. Gershenson and Daniel A. Greenberg, eds. New York: Blaisdel, pp. 59–85.

——— (1964). "Einstein and the Wave-Particle Duality." *The Natural Philosopher*. Vol. 3. Daniel E. Gershenson and Daniel A. Greenberg, eds. New York: Blaisdell, pp. 3–49.

—— (1970). *Paul Ehrenfest*. Vol. 1, *The Making of a Theoretical Physicist*. Amsterdam: North-Holland.

—— (1977). "The Beginnings of the Quantum Theory." In *Storia della Fisica del XX Secolo. Rendiconti della Scuola Internazionale di Fisica "Enrico Fermi."* LVII Corso. Charles Wiener, ed. New York and London: Academic, pp. 1–39.

Kuhn, Thomas S. (1978). *Black-Body Theory and The Quantum Discontinuity, 1894–1912*. Oxford: Clarendon Press; New York: Oxford University Press.

Laue, Max (1906). "Zur Thermodynamik der Interferenzerscheinungen." *Annalen der Physik* 20: 365–378.

Mehra, Jagdish and Rechenberg, Helmut (1982). *The Historical Development of Quantum Theory*. Vol 1, part 1, *The Quantum Theory of Planck, Einstein, Bohr and Sommerfeld: Its Foundation and the Rise of its Difficulties 1900–1925*. New York, Heidelberg, and Berlin: Springer-Verlag.

Natanson, Ladisłas (1911a). "On the Statistical Theory of Radiation." *Bulletin de l'Academie des Sciences de Cracovie (A)*: 134–148.

—— (1911b). "Über die statistische Theorie der Strahlung." *Physikalische Zeitschrift* 12: 659–665 (German translation of Natanson 1911a).

Pais, Abraham (1982). *'Subtle is the Lord ...' The Science and the Life of Albert Einstein*. Oxford: Clarendon Press; New York: Oxford University Press.

Planck, Max (1900). "Zur Theorie des Gesetzes der Energieverteilung im Normalspectrum." *Deutsche physikalische Gesellschaft* Berlin. *Verhandlungen* 2: 237–245.

—— (1911). "Die Gesetze der Wärmestrahlung und die Hypothese der elementaren Wirkungsquanten." Report to the 1911 Solvay Congress. Cited from: Max Planck. *Physikalische Abhandlungen und Vorträge*, vol. 2. Braunschweig: Friedrich Vieweg & Sohn, 1958, pp. 269–286.

Renn, Jürgen (1993). "Einstein as a Disciple of Galileo: A Comparative Study of Concept Development in Physics." *Science in Context* 6: 311–341.

Stachel, John (1986). "Einstein and the Quantum: Fifty Years of Struggle." In *From Quarks to Quasars: Philosophical Problems of Modern Physics*. Robert Colodny, ed. Pittsburgh: University of Pittsburgh Press, pp. 349–385. [See this volume, pp. 367–402].

—— (1994a). "Scientific Discoveries as Historical Artifacts." In *Trends in the Historiography of Science*. Kostas Gavroglu, ed. Dordrecht: Kluwer, pp. 139–148.

—— (1994b). "Einstein and Bose." [See this volume, pp. 519–538].

—— (1997). "Feynman Paths and Quantum Entanglement: Is There Any More to the Mystery?" In *Potentiality, Entanglement, and Passion-at-a-Distance*. R. S. Cohen et al., eds. Dordrecht: Kluwer, pp. 245–255.

—— (1999). "Bohr and the Photon." In *Going Critical: Selected Papers*. Dordrecht: Kluwer (forthcoming).

Wien, Wilhelm (1893a). "Eine neue Beziehung der Strahlung schwarzer Körper zum zweiten Hauptsatz der Wärmetheorie." *Königlich Preussische Akademie der Wissenschaften* (Berlin). *Sitzungsberichte*: 55–62.

—— (1893b). "Die obere Grenze der Wellenlängen, welche in der Wärmestrahlung fester Körper vorkommen können; Folgerungen aus dem zweiten Hauptsatz der Wärmetheorie." *Annalen der Physik* 49: 633–641.

—— (1894). "Temperatur und Entropie der Strahlung." *Annalen der Physik* 52: 132–165.

—— (1896). "Über die Energievertheilung im Emissionsspectrum eines schwarzen Körpers." *Annalen der Physik* 58: 662–669.

—— (1900). "Zur Theorie der Strahlung schwarzer Körper. Kritisches." *Annalen der Physik* 3: 530–539.

Part VII

Einstein and Others

Einstein and Newton

John Stachel

To anyone familiar with the history of modern science, the phrase "miraculous year" in the title immediately calls to mind its Latin counterpart "annus mirabilis," long used to describe the year 1666, during which Isaac Newton laid the foundations for much of the physics and mathematics that revolutionized seventeenth-century science. It seems entirely fitting to apply the same phrase to the year 1905, during which Albert Einstein not only brought to fruition parts of that Newtonian legacy, but laid the foundations for the break with it that has revolutionized twentieth-century science.

But the phrase was coined without reference to Newton. In a long poem entitled *Annus Mirabilis: The Year of Wonders, 1666*, John Dryden, the famed Restoration poet, celebrated the victory of the English fleet over the Dutch as well as the city of London's survival of the Great Fire. The term was then used to celebrate Newton's scientific activities during the same year—a year in which he laid the foundations of his version of the calculus, his theory of colors, and his theory of gravitation.[1] Here is Newton's own (much later) summary of his accomplishments during this period:

> In the beginning of the year 1665 I found the Method of approximating series & the Rule for reducing any dignity [power] of any Binomial into such a series [i.e., the binomial theorem]. The same year in May I found the method of Tangents . . . , & in November had the direct method of fluxions [i.e., the differential calculus] & the next year in January had the Theory of colours & in May following I had entrance into [th]e inverse method of fluxions [i.e., the integral calculus]. And the same year I began to think of gravity extending to [th]e orb of the Moon & (having found out how to estimate the force with w[hi]ch [a] globe revolving within a sphere presses the surface of a sphere [i.e., the centrifugal force]): from Kepler's rule of the periodical times of the Planets being in sesquialterate proportion of their distances from the centers of their Orbs [i.e., Kepler's third law], I deduced that the forces w[hi]ch keep the Planets

Excerpt from the Introduction
to Einstein's Miraculous Years

> in their Orbs must [be] reciprocally as the squares of their distances from the centers about w[hi]ch they revolve: & thereby compared the force requisite to keep the Moon in her Orb with the force of gravity at the surface of the earth, & found them answer pretty nearly. All this was in the two plague years of 1665 & 1666. For in those days I was in the prime of my age for invention & minded Mathematicks & Philosophy more then [*sic*] at any time since.[2]

More recently, the term *annus mirabilis* has been applied to the work of Albert Einstein during 1905 in an effort to establish a parallel between a crucial year in the life of the founding father of classical physics and of his twentieth-century successor.[3] What did Einstein accomplish during his miraculous year? We are fortunate in having his own contemporary summaries of his 1905 papers. Of the first four he wrote to a close friend:

> I promise you four papers ..., the first of which I could send you soon, since I will soon receive the free reprints. The paper deals with radiation and the energetic properties of light and is very revolutionary, as you will see. ... The second paper is a determiniation of the true sizes of atoms from the diffusion and viscosity of dilute solutions of neutral substances. The third proves that, on the assumption of the molecular [kinetic] theory of heat, bodies of the order of magnitude of 1/1000 mm, suspended in liquids, must already perform an observable random movement that is produced by thermal motion; in fact, physiologists have observed motions of suspended small, inanimate, bodies, which they call "Brownian molecular motion." The fourth paper is only a rough draft at this point, and is an electrodynamics of moving bodies, which employs a modification of the theory of space and time; the purely kinematical part of this paper will surely interest you.[4]

Einstein characterized the fifth paper in these words:

> One more consequence of the paper on electrodynamics has also occurred to me. The principle of relativity, in conjunction with Maxwell's equations, requires that mass be a direct measure of the energy contained in a body; light carries mass with it. A noticeable decrease of mass should occur in the case of radium. The argument is amusing and seductive; but for all I know, the Lord might be laughing over it and leading me around by the nose.[5]

The parallels are clear: each man was in his mid-twenties; each had given little previous sign of the incipient flowering of his genius; and, during a brief time span, each struck out on new paths that would ultimately revolutionize the science of his times. If Newton was only twenty-four in 1666 while Einstein was twenty-six in 1905, no one expects such parallels to be perfect.

While these parallels cannot be denied, upon closer inspection we can also see differences—much more significant than the slight disparity in age—between the activities of the two men during their *anni mirabiles* and in the immediate consequences of their work. The first striking difference is the one between their life situations: rejected by the academic community after graduation from the Swiss Polytechnical School in 1900, by 1905 Einstein was already a married man and an active father of a one-year-old son, obliged to fulfill the demanding responsibilities of a full-time job at the Swiss Patent Office. Newton never married (there is speculation that he died a virgin), and he had just taken his bachelor's degree but was still what we would call a graduate student in 1666. Indeed, he had been temporarily freed of even his academic responsibilities by the closure of Cambridge University after outbreaks of the plague.

Next we may note the difference in their scientific standing. Newton had published nothing by 1666, while Einstein already had published five respectable if not extraordinary papers in the prestigious *Annalen der Physik*. Thus, if 1666 marks the year when Newton's genius caught fire and he embarked on independent research, 1905 marks the year when Einstein's already matured talents manifested themselves to the world in a burst of creativity, a series of epoch-making works, all of which were published by the *Annalen* either in that year or the next. None of Newton's activities in 1666 found their way into print until much later: "The first blossoms of his genius flowered in private, observed silently by his own eyes alone in the years 1664 to 1666, his *anni mirabiles*."[6] The reasons for Newton's evident lack of a need for recognition—indeed, his pronounced reluctance to share his ideas with others, as his major works had to be pried from his hands by others—have long been the topic of psychological, even psychopathological, speculation.

It took a few years—an agonizingly long time for a young man eager for recognition—for Einstein's achievements to be fully acknowledged by the physics community. But the process started almost immediately in 1905; by 1909 Einstein had been called to a chair of theoretical physics created for him at the University of Zurich, and he was invited to lecture at the annual meeting of the assembled German-speaking scientific community.

Thus, if 1905 marks the beginning of the emergence of Einstein as a leading figure in the physics community, Newton remained in self-imposed obscurity well after 1666. Only in 1669, when at the urging of friends he allowed the limited circulation of a mathematical manuscript divulging some parts of the calculus he had developed, did "Newton's anonymity begin to dissolve."[7]

Another striking difference between the two is in their mathematical talents. Newton manifested his mathematical creativity from the outset. "In roughly a year [1664], without the benefits of instruction, he mastered the entire achievement of seventeenth-century analysis and began to break new ground. ... The fact that he was unknown does not alter the fact that the young man not yet twenty-four, without benefit of formal instruction, had become the leading mathematician in Europe."[8]

Newton was thus able to create the mathematics necessary to develop his ideas about mechanics and gravitation. Einstein, while an able pupil and practitioner, was never really creative in mathematics. Writing about his student years, Einstein said:

> The fact that I neglected mathematics to a certain extent had its cause not merely in my stronger interest in the natural sciences than in mathematics but also in the following peculiar experience. I saw that mathematics was split up into numerous specialties, each of which could easily absorb the short lifetime granted to us. Consequently, I saw myself in the position of Buridan's ass, which was unable to decide upon any particular bundle of hay. Presumably this was because my intuition was not strong enough in the

> field of mathematics to differentiate clearly the fundamentally important, that which is really basic, from the rest of the more or less dispensable erudition. Also, my interest in the study of nature was no doubt stronger; and it was not clear to me as a young student that access to a more profound knowledge of the more basic principles of physics depends on the most intricate mathematical methods. This dawned upon me only gradually after years of independent scientific work.[9]

Fortunately, for his works of 1905 he needed no more mathematics than he had been taught at school. Even so, it was left to Henri Poincaré, Hermann Minkowski, and Arnold Sommerfeld to give the special theory of relativity its most appropriate mathematical foundation.

When a really crucial need for new mathematics manifested itself in the course of his work on the general theory of relativity, Einstein had to make do with the tensor calculus as developed by Gregorio Ricci-Curbastro and Tullio Levi-Civita and presented to Einstein by his friend and colleague, Marcel Grossmann. This was based on Riemannian geometry, which lacked the concepts of parallel displacement and affine connection that would have so facilitated Einstein's work. But he was incapable of filling this mathematical lacuna, a task that was accomplished by Levi-Civita and Hermann Weyl only after the completion of the general theory.

Returning to Newton: in some respects he was right to hesitate about publication in 1666. "When 1666 closed, Newton was not in command of the results that have made his reputation deathless, not in mathematics, not in mechanics, not in optics. What he had done in all three was to lay foundations, some more extensive than others, on which he could build with assurance, but nothing was complete at the end of 1666, and most were not even close to complete."[10]

His work on the method of fluxions (as he called the calculus), even if incomplete, was worthy of publication and would have been of great service to contemporary mathematicians had it been available to them. His work in physics was far less advanced. His experiments on the theory of colors were interrupted by the closing of the university, and after his return to Cambridge in 1667 he spent a decade pursuing his optical investigations. Nevertheless, a more outgoing man might have published a preliminary account of his theory of colors in 1666. But in the case of gravitation, after carefully reviewing the evidence bearing on Newton's work on this subject through 1666, the physicist Leon Rosenfeld concluded that "it will be clear to every scientist that Newton at this stage had opened up for himself an exciting prospect, but had nothing fit to be published."[11] It is also clear that, in thinking about mechanics, he had not yet arrived at a clear concept of force—an essential prerequisite for the development of what we now call Newtonian mechanics. He had given "a new definition of force in which a body was treated as the passive subject of external forces impressed upon it instead of the active vehicle of forces impinging on others." But: "More than twenty years of patient if intermittent thought would in the end elicit his whole dynamics from this initial insight."[12]

To sum up, in the case of Newton, in 1666 we have a student, working at his leisure, a mature genius in mathematics, but whose work in physics, however genial, was still in its formative stages. In the case of Einstein, in 1905 we have a man raising a family and pursuing a practical career, forced to fit physics into the interstices of an already-full life, yet already a master of theoretical physics ready to demonstrate that mastery to the world.

NOTES

[1] The phrase *anni mirabiles* (years of wonders) has been applied with more accuracy to the years 1664–1666 by Newton's biographer Richard Westfall in *Never at Rest/A Biography of Isaac Newton* (Cambridge, U.K.: Cambridge University Press, 1980; paperback edition, 1983), p. 140. This book may be consulted for generally reliable biographical information about Newton's life.

[2] I. Bernard Cohen, *Introduction to Newton's 'Principia'* (Cambridge, Mass.: Harvard University Press, 1971), p. 291.

[3] See, for example, Albrecht Fölsing, *Albert Einstein/A Biography*, tr. by Ewald Osers (New York: Viking, 1997), p. 121: "Never before and never since has a single person enriched science by so much in such a short time as Einstein did in his *annus mirabilis*." This book may be consulted for generally reliable biographical information about Einstein, but its scientific explanations should be treated with caution. For an account of Einstein's scientific work organized biographically, see Abraham Pais, *Subtle is the Lord ...: The Science and the Life of Albert Einstein* (Oxford: Clarendon Press; New York: Oxford University Press, 1982).

[4] Einstein to Conrad Habicht, 18 or 25 May 1905, *The Collected Papers of Albert Einstein* (Princeton, N.J.: Princeton University Press, 1987–), cited hereafter as *Collected Papers*, vol. 5 (1993), doc. 27, p. 31. Translation from Anna Beck, tr., *The Collected Papers of Albert Einstein: English Translation* (Princeton University Press, 1987–), cited hereafter as *English Translation*, vol. 5 (1995), p. 20; translation modified.

[5] Einstein to Conrad Habicht, 30 June–22 September 1905, *Collected Papers*, vol. 5, doc. 28, p. 33; *English Translation*, p. 21; translation modified. Forty years later, when the explosion of the first atomic bombs brought the equivalence between mass and energy forcefully to the world's attention, Einstein might have wondered just what sort of trick the Lord had played on him.

[6] Westfall, *Never at Rest*, p. 140.

[7] Ibid., p. 205.

[8] Ibid., pp. 100, 137.

[9] Albert Einstein, *Autobiographical Notes*, Paul Arthur Schilpp, ed. and trans. (LaSalle, Ill.: Open Court, 1979), p. 15.

[10] Westfall, *Never at Rest*, p. 174.

[11] "Newton and the Law of Gravitation," *Arch. Hist. Exact Sci.* 2 (1965): 365–386, reprinted in Robert S. Cohen and John J. Stachel, eds., *Selected Papers of Leon Rosenfeld* (Dordrecht/Boston: Reidel, 1979), p. 146.

[12] Citation from Westfall, *Never at Rest*, p. 146.

Eddington and Einstein

John Stachel

In this paper I shall discuss Eddington's work in the field of relativity and Einstein's reaction to it. But it must be stressed that his work in this area, important as it was, is the lesser part of Eddington's life's work. His greatest and most enduring contributions were in the fields of astonomy and astrophysics. His biographer, A. V. Douglas,[1] says: "After 1906 astronomy was the dominant interest in Eddington's life and, to a most unusual extent, his work was his life and his life was his work."

After Eddington's death, Harlow Shapley wrote: "It is probably correct to say that during the past 25 years Eddington's influence on fellow astronomers has been greater than that of any other man" (Douglas, p. 185).

H. N. Russell wrote in *Nature*: "A great master is gone, a master in many fields, but most of all in astrophysics" (Douglas, p. 185).

In one key episode of his life, Eddington's interest in astronomy and in relativity came together. I am referring, of course, to his participation in the famous 1919 eclipse expedition, which so decisively altered the public careers of both Eddington and Einstein. I shall begin by recalling some events in Eddington's life which led up to this fateful role, and some lesser known subsequent events. I shall not discuss the expedition proper nor the official announcement of its results at the joint Royal Society–Royal Astronomical Society Meeting. Excellent accounts of these events are available.[2]

Then I shall discuss Eddington's approach to relativity and his unified theory of gravitation and electromagnetism which played a significant role in shaping Einstein's subsequent work on unified field theories.

Having alluded to cosmology in the course of earlier comments, I will next discuss Eddington's role in the great turn in cosmology which took place around

E. Ullmann-Margalit, ed.
The Prism of Science, The Israel Colloquium:
Studies in History, Philosophy and Sociology of Science 2

1930,[3] leading to acceptance of expanding models of the universe as the new orthodoxy.

In the course of the paper I shall quote from the Einstein–Eddington correspondence as well as some of Einstein's evaluations of Eddington's work.

I shall not discuss Eddington's later theoretical speculations, which he included under the heading of "Relativity." Nor shall I do more than allude to Eddington's philosophical view. This is no loss, given my ignorance of the subject and the topics of the other papers in this symposium.

As a schoolboy of fifteen, Eddington wrote an essay on solar eclipses based on his talk to a meeting of the Brymelyn Literary Society a few months earlier.[4] Five years later, during his second year at Cambridge (1903), he read a paper on "the velocity of gravitation" to the Trinity Mathematical Society, and in 1906 he read a paper on gravitation to the $\nabla^2 V$ Club (Douglas, p. 9).

Although he studied physics at Cambridge, in 1906 he accepted a position as Chief Assistant at the Royal Observatory in Greenwich. There he served his apprenticeship in the astronomical profession between 1906 and 1913. It is significant that in 1909 he rejected a feeler from Arthur Schuster, sounding him out about the Chair of Theoretical Physics at Manchester.

It was in 1912 that he obtained his first chance to participate in a solar eclipse expedition, fourteen years after he had written that such expeditions were "frequently accompanied by some of the greatest astronomers in the world in order to make observations." It is not widely known that an expedition to Brazil was the first to include in its research program a test of Einstein's 1911 prediction of a deflection in the apparent position of a star in the gravitational field of the sun.[5] I hasten to add that it was not Eddington who wanted to test the Einstein prediction, but C. D. Perrine, an American astronomer who had become Director of the Argentine National Observatory in 1909. Stimulated by his contacts with Erwin Freundlich, Perrine had taken the Lick Observatory's 40-foot camera to Brazil, hoping to obtain photographs suitable for observation of the apparent deflection. However, the rain god did not cooperate. Both the English group at Passa Quatro and Perrine at Christina "suffered a total eclipse instead of observing one" as Perrine put it.[6] Eddington gives a little more detail in one of his accounts of the expedition.[7]

> The first impression of the climate of Passa Quatro was extremely favourable. Although situated within the tropics, its elevation of 3000 feet made it pleasantly cool or even frosty at nights. The days were generally fresh and the Sun seldom oppressive. At times the sky had that perfect transparency which the observer longs for and so seldom finds. More often the conditions, though not ideal, were such as would have permitted entirely successful observations of the eclipse. Occasional wet or cloudy days intervened, but, roughly speaking, the chances of success were estimated at 3 to 1 in favour. Unhappily, on the day before the eclipse an entirely different type of weather set in and the week which followed was one of almost unbroken cloud and rain.
>
> The hopelessness of the weather on the eclipse-morning was soon apparent, at least to all except the observers. A heavy cloud had settled over the whole district and

rain was falling steadily through the saturated air. Everything was made ready in case at the last moment a break should appear in the clouds; but as the darkness came on, the rainfall increased if anything, and the coelostats were never uncovered. The rapid increase of the darkness a few seconds before totality was very striking, as was also the almost instantaneous brightening when it was over.

It seems likely that Eddington first learned of Einstein's 1911 work from Perrine during this Brazilian fiasco. At any rate, he first referred to it in print, after he had returned to Cambridge in 1913 as Plumian Professor of Astronomy and Experimental Philosophy, in an article entitled "Some Problems of Astronomy. XIX. Gravitation," published in February 1915.[8] This was well before Einstein worked out the final version of his field equations in November of the same year, and of course even longer before the contacts with de Sitter, often credited with first bringing Einstein's work to Eddington's attention. The last paragraphs of this 1915 article show familiarity with the special theory of relativity, as well as Einstein's 1911 paper:

> ... does gravitation conform to the Principle of Relativity? Without attempting to express this principle in a rigorous or philosophical form, we may say that it expresses that a terrestrial observer can detect no difference in the sequence of phenomena whether the solar system is at rest in the aether or rushing through it at many thousand miles per second. So far as optical and electrical phenomena are concerned, this is mainly a mathematical deduction from the ordinary equations of the subject, aided by some experimental results as to the deformation of moving electrons. To bring in gravitation is a pure assumption as yet; but unless gravitation is affected by motion in the way required by relativity, the compensation for astronomical phenomena in a moving system cannot be perfect. This question has an important bearing on the considerations of the last paragraph, for relativity, if we accept the principle, points out a way of modifying, or rather completing, the law of gravitation for bodies in motion. Unfortunately, however, the terms depending on the velocities are not determined uniquely, and there is more than one way of satisfying the conditions.
>
> There are two ways in which we might hope to test whether gravitation conforms to the principle of relativity: (1) we could test directly whether the compensation of astronomical phenomena is so perfect that the absolute translation of the solar system through the aether has no effect, or (2) we could test whether the relative motions of bodies within the solar system conform to the modified law of gravitation required by the principle. The eclipses of Jupiter's satellites have been suggested as material for the first test; it could not in any case be very decisive, because we do not know how large a velocity of translation we ought to look for. The motions of the perihelia and nodes of the inner planets would be the best material for the second test, but they are already affected by anomalies which certainly are mainly of a different nature. Both tests appear to be beyond practical possibility at present.
>
> This does not quite exhaust the tests. There is another result of the principle of relativity of remarkable interest. Einstein has shown that on this theory a wave of light will travel more slowly when it enters an intense gravitational field of force. This must lead to a refraction of waves of light passing near a massive body. For instance, a star seen close to the limb of the Sun would be apparently displaced $0''.83$ away from the Sun's limb. The result is a natural (though, I understand, not actually an inevitable) deduction from the principle of relativity. It would be extremely difficult to detect

> this deflection even during a total eclipse; but attempts are being made to find it. A decisive result, whether positive or negative, would be of remarkable importance—the first definite advance in our knowledge of gravitation for over two hundred years. A positive result would mean that gravitation has been pulled down from its pedestal, and ceases to stand aloof from the other interrelated forces of nature.

It is apparent that Eddington's interest in Einstein and the eclipses had been aroused well before de Sitter appeared on his horizon in 1916. If his imagination then caught fire, it was because plenty of well-prepared kindling was at hand. In a letter of June 11, 1916 to de Sitter, he wrote:

> I am immensely interested in what you tell me about Einstein's theory... Hitherto I had only heard vague rumors of Einstein's new work. I do not think anyone in England knows the details of the paper.[9]

In a letter of July 4, 1916 Eddington informed de Sitter:

> We are having a discussion at the British Association [for the Advancement of Science—the BAAS] on Gravitation—at Newcastle, Dec. 5–8. I wish we could have invited you to come over to take part; but we are not inviting any foreign guests this year because Newcastle is a "restricted area" and aliens are not allowed in it. ... I feel sure you will allow me to make use of the papers you send, in making my contribution to the discussion. So far as I can make out, no one in England has yet been able to see Einstein's paper and many are very curious to know the new theory. So I propose to give an account of it at the Meeting.[10]

Eddington is often portrayed as the first person in England to try to test the light deflection prediction. Yet a published paper, as well as correspondence with Einstein, makes it clear that the Lindemanns, father and son, had independently undertaken such a test in 1916, based on the final (1915) version of Einstein's theory.

In a letter of November 23, 1919 to Einstein, the elder Lindemann (A. F.) wrote:

> No doubt you still remember our discussion in Brussels in 1913 about the possibility of observing a star near enough to the sun and measuring its relative position in order to test your gravitational postulate. We took up the thing here and invented our daylight star photography and finally photographed stars up to 3–2 magnitude at $10^{\mathrm{h}}40^{\mathrm{m}}$ on the morning of October 26, 1916.

After discussing why this and other attempts failed, he continues:

> In our article we called special attention to solar eclipses and immediately after our paper the particular investigation of the question was undertaken by Sir F. W. Dyson, Astronomer Royal at Greenwich and carried through with governmental funding. Professor Eddington has most particularly occupied himself with your theory and is a most amiable person without the modern prejudices which are unfortunately so strongly developed among many.

The article referred to in the letter is "Daylight Photography of Stars as a Means of Testing the Equivalence Postulate in the Theory of Relativity." It is dated December 4, 1916 and appeared in the December 1916 issue of the *Monthly Notices*.[11] A footnote on page 141 refers specifically to the value for the deflection given in Einstein's "Grundlage der allgemeinen Relativitätstheorie," noting that "In his first paper on the subject Professor Einstein gave half these values."

In a subsequent letter of August 18, 1920, Lindemann said:

> ... we do not at all claim to have been the cause of the solar eclipse expedition of May 29, 1919—I only wanted to say that our report on the possibility of using stellar photography in daylight with special reference to your theory *before* the total eclipse of May 29, 1919 perhaps supported the thing and in any case drew many people's attention to it.

The first Lindemann letter and other contemporary documents suggest that Sir Frank Watson Dyson was a moving spirit behind the 1919 eclipse expedition. It has been stated that "... [Dyson's] acquaintance with the theory and interest in it doubtless came from Eddington,"[12] but this is not proved by any evidence known to me. Sir James Jeans, in presenting the RAS Gold Medal for 1925 to Dyson (Eddington had received it the previous year), said: "... if either [Eddington or Dyson] had failed to play his part, either from want of vision, of enthusiasm, or of capacity for seizing the right moment, I doubt if the expeditions would have gone at all. ..."[13]

Eddington's first published paper devoted to the general theory of relativity, "Gravitation and the Principle of Relativity," was published in *Nature*. It is presumably based on his talk on the same subject to the BAAS.[14] It starts on a characteristically Eddingtonian note:

> According to the principle of relativity in its most extended sense, the space and time of physics are merely a mental scaffolding in which for our own convenience we locate the observable phenomena of Nature. Phenomena are conditioned by other phenomena according to certain laws, but not by the space-time scaffolding, which does not exist outside our brains.

Given this approach to the theory, it is not surprising that Eddington later states:

> No attempt is made to explain the cause of gravitation—as a kink in space or anything of that nature. But the extended law of gravitation is determined, to which Newton's law is an approximation under ordinary conditions. It has long been suspected that there must be some modification of the law when the bodies concerned are in rapid relative motion; moreover, the "mass" of a moving body no longer has a unique meaning, so that a further definition, if not extension, of Newton's law is clearly needed. Now, although we do not seek a *cause* of gravitation in the properties of space, it may well happen that the *law* of gravitation is determined by these properties. The inverse-square law represents the natural weakening of an effect through spreading out in three dimensions; we may say that it is determined by the properties of Euclidean space. There is, therefore, nothing unreasonable in proceeding, as Einstein does, to examine whether a more extended law is suggested by the properties of generalised space—that is, by geometry.

Eddington had already adopted a conventionalist view of the laws of geometry, apparently based on his reading of Poincaré. In an obituary of Poincaré, he had written:[15]

> In no subject has the influence of Poincaré been more widely felt than in his philosophy of science and mathematics. Much of his more technical work is hardly yet appreciated, and it will be many years before the new methods he introduced will reap their full harvest; but his philosophical books, *La Science et l'Hypothèse*, *La Valeur de la Science*, *Science et Méthode*, written in his delightfully clear style, have met immediately with a multitude of readers. They have come to a world which was eager for them. Many scientists are repelled from the ordinary philosophic works, feeling that a hopeless difference of standpoint divides them from the author from the outset. To such the writings of Poincaré come as a complete revelation. His way of thought seems to fill exactly the need of which they are conscious. There must be many of the younger generation of physicists and mathematicians whose philosophical ideas are based almost wholly on his teaching. The view which is particularly associated with the name of Poincaré is his attitude towards space and time and geometry; to him space is neither Euclidean nor non-Euclidean; the postulates of geometry are not to be derived from experiment. No system of geometry is forced on us; we may adopt whichever we will. But a certain geometry may be more convenient than any other, because it enables the facts of nature to be expressed with greater simplicity. Euclidean geometry is convenient because, when it is adopted, we find that light travels in straight lines, and that ordinary bodies are more or less rigid. If, as the principle of relativity tells us, it is found that bodies are not rigid, but systematically change their shape when they are differently oriented, we need not change our geometry though it may be more convenient to do so. "Geometry is a convention, a sort of cloak badly fitted between our love of simplicity and our desire not to depart too far from what our instruments teach us." Poincaré applied this principle to other physical ideas. "What shall be our position, faced with these new conceptions? Are we going to be forced to modify our former conclusions? Not at all: we adopted a convention because it seemed convenient, and we said that nothing could compel us to abandon it. Today certain physicists want to adopt a new convention. They are not forced to do so; they consider the new convention more useful, that is all; and those who are not of the same opinion can rightly retain the old if they do not care to disturb their old habits of thought. Between ourselves, I think they will do so for a long time to come."

Eddington's first full account of general relativity is given in his *Report on the Relativity Theory of Gravitation* in 1918, reprinted in a second edition (1920) after the eclipse expedition announcement made relativity theory all the rage.[16] In the closing paragraph (53) of this *Report*, Eddington reiterates his argument that Einstein's law of gravitation does not constitute an explanation. He follows this with a remarkable argument for what would now be called a quantum theory of gravity—the first such argument of which I know (however, Planck had already derived the fundamental length now known as Planck's length in 1900):

> In this discussion of the law of gravitation, we have not sought, and we have not reached, any ultimate explanation of its cause. A certain connection between the gravitational field and the measurement of space has been postulated, but this throws light rather on the nature of our measurements than on gravitation itself. The relativity theory is indifferent to hypotheses as to the nature of gravitation, just as it is indifferent to hypotheses as to matter and light. We do not in these days seek to explain the behaviour of natural forces in terms of a mechanical model having the familiar characteristics of matter in bulk; we have to accept some mathematical expression as

an axiomatic property which cannot be further analysed. But I do not think we have reached this stage in the case of gravitation.

There are three fundamental constants of nature which stand out pre-eminently—The velocity of light, $3.00 \cdot 10^{10}$ c.g.s. units; dimensions LT^{-1}. The quantum, $6.55 \cdot 10^{-27}$ c.g.s. units; dimensions, ML^2T^{-1}. The constant of gravitation, $6.66 \cdot 10^{-8}$ c.g.s. units; dimensions $M^{-1}L^3T^{-3}$.

From these we can construct a fundamental unit of length whose value is

$$4 \cdot 10^{-33}\text{cms.}$$

There are other natural units of length—the radii of the positive and negative unit electric charges—but these are of an altogether higher order of magnitude.

With the possible exception of Osborne Reynold's theory of matter, no theory has attempted to reach such fine-grainedness. But it is evident that this length must be the key to some essential structure. It may not be an unattainable hope that some day a clearer knowledge of the processes of gravitation may be reached; and the extreme generality and detachment of the relativity theory may be illuminated by the particular study of a precise mechanism.

Considering Eddington's later passionate attachment to the cosmological constant, it is amusing to note his cool attitude toward it in the first edition of the *Report* (pp. 90–91). Apparently this was due in some measure to Einstein's original association of the cosmological constant with the Machian argument:

The followers of Mach maintain that if there were no matter there could be no inertial frame, and it appears that this is Einstein's reason for preferring his own suggestion. In his theory if all matter were abolished, R would become zero and the world would vanish to a point. There is something rather fascinating in a theory of space by which, the more matter there is, the more room is provided. It is satisfactory, too, from Einstein's standpoint, because he is unwilling to admit that a thinkable space without matter could exist. For our part, we feel equally unwilling to assent to the introduction of vast quantities of world-matter, which (to quote de Sitter) "fulfils no other purpose than to enable us to suppose it not to exist."

Eddington was slightly more sympathetic to de Sitter's cosmology:

De Sitter's hypothesis thus appears to present the greatest advantages; but it will not satisfy the followers of Mach's philosophy. He derives his inertial-frame from the spherical property of space-time which in turn is derived from the slightly modified law of gravitation; it is not determined by anything material (*Report*, p. 90).

It thus did not require the "assumption of the existence of vast quantities of matter not yet recognized" (p. 88).

By the time of the second edition of the *Report*, dated December 1919, Eddington had come under the spell of Weyl's unified theory of gravitation and electromagnetism, noting in the Preface: "... it appears to favor Einstein's view of the curvature of space, which has been treated, perhaps too unsympathetically, in Chapter VIII" (*Report*, p. xi). He wrote in a similar vein to Einstein—but I am getting ahead of my story.

Matter in general received short shrift in Eddington's conception of the general theory. As he summarized his view in the Preface to the second edition (*Report*, p. xi):

The writer holds the view that the fundamental equations of gravitation (35·8), which on this theory are the sole basis of mechanics, should be regarded as a definition of matter rather than as a law of nature. We need not suppose that the gravitational field has *in vacuo* some innate tendency to arrange itself according to the law $G_{\mu\nu} = 0$; we should rather say that in regions of the world where this state happens to exist we perceive emptiness; and where the equations fail, the failure of the equations is itself the cause of our perception of matter. Matter does not cause the curvature (G) of space-time; it *is* the curvature. Just as light does not cause electromagnetic oscillations; it is the oscillations.

Another noteworthy feature of the *Report* (p. 66) is its clear statement that the laws of motion of a particle, treated as a singularity of the gravitational field, must follow from the gravitational field equations:

It is startling to find that the whole of the dynamics of material systems is contained in the law of gravitation; at first sight gravitation seems scarcely relevant in much of our dynamics. But there is a natural explanation. A particle of matter is a singularity in the gravitational field, and its mass is the pole-strength of the singularity; consequently the laws of motion of the singularities must be contained in the field-equations, just as those of electromagnetic singularities (electrons) are contained in the electromagnetic field-equations.[17]

This insight is often still attributed to Einstein and Grommer in 1927. It is worth noting that, although Eddington appears to be the first to have stated it, he was not the only one to do so before 1927.

I shall not retell the oft-told tale of the eclipse expeditions of 1919, culminating in the November 6 joint meeting of the Royal Society and the Royal Astronomical Society, at which the results were officially—and dramatically—announced.[18]

Einstein learned about the favorable preliminary results of the expedition in a cable from Lorentz in September 1919 (Item 16-478).[19] Ehrenfest had hoped to bring Einstein and Eddington together during one of Einstein's then rather frequent trips to Holland, but it proved impossible to synchronize their visits (Ehrenfest to Einstein, October 5, 1919, Item 9-439). Einstein stated that he "would naturally have been very happy to see Eddington," but decided not to postpone his trip (Einstein to Ehrenfest, October 15, 1919, Item 9-444), which took place in late October. While there, Einstein "learned ... that the confirmation found by Eddington is also a complete one quantitatively,"[20] and wrote to Eddington. This letter has not been found, but Eddington's reply (Item 9-260) is in the Einstein Archive.[21]

One of the less happy by-products of the furor resulting from the eclipse expedition was the failure of the Royal Astronomical Society to award its Gold Medal in 1919. We can follow the story through Eddington's letters to Einstein. In his first letter of December 1, 1919, he wrote:

Our results were announced on November 6; and you probably know that since then all England has been talking about your theory. It has made a tremendous sensation; and although the popular interest will die down, there is no mistaking the genuine enthusiasm in scientific circles and perhaps more particularly in this University.

> It is the best possible thing that could have happened for scientific relations between England and Germany. I do not anticipate rapid progress towards official reunion, but there is a big advance towards a more reasonable frame of mind among scientific men, and that is even more important than the renewal of formal associations (Item 9-260).

The Council of the Royal Astronomical Society met on December 12, 1919 and January 9, 1920 to consider the award of its Gold Medal; Eddington describes what happened in his letter of January 21, 1920:

> I am sorry to say an unexpected thing has happened and at the meeting on January 9 the council of the R.A.S. rejected the award, which had been carried by quite a large majority at the previous meeting.
>
> The facts (which are confidential) are that three names were proposed for the Medal. You were selected by an overwhelming majority in December. Meanwhile the "irreconcilables" took alarm, mustered up their full forces in January, and managed to defeat the confirmation of the award in January. So for the first time for about 30 years no Gold Medal will be awarded this year! I confess I was very much surprised when the motion was proposed and carried originally (it was proposed by two men who during the war have been violently "patriotic"); but until a day or two before the January meeeting we all regarded the confirmation as a matter of course.
>
> I did not write before as I was doubtful if Ludlam would see you. I am sure that your disappointment will not be in any way personal; and that you will share with me the regret that this promising opening of a better international spirit has had a rebuff from reaction. Nevertheless I am sure the better spirit is making progress.
>
> Mr. Ludlam mentioned the possibility that you might come over about May. We should be delighted to have you and your wife to stay with us at the Observatory; and you would get a most cordial welcome in Cambridge, and especially at my own college, Trinity (the college of Newton). You will see that there would be some awkwardness in visiting the Royal Astr. Society after what has happened (Item 9-264).[22]

Meanwhile, E. B. Ludlam of Cambridge, who visited Berlin in January, had informed Einstein of the impending award. Einstein wrote to Max Born on January 27:[23] "By the way, I am going to England in the spring, to have a medal pressed into my hand and to have a closer look at the other side of the tomfoolery." After receiving Eddington's letter of January 21 (a similar letter of apology came from Ludlam), he wrote back on February 2:

> Today I received your letter which pleased me uncommonly, in spite of the tragi-comic ending of the medal affair. How little this signifies compared to the dedicated and successful work that you and your friends have dedicated to the theory of relativity and its testing. Now it will give me even more pleasure to accept your friendly invitation, since my trip will have a purely private character (Item 9-272).

But Einstein did not visit England in 1920. On June 11, he wrote to Eddington, apologizing for not coming even though he had promised to do so. He gave two reasons: lack of time and "the feeling that a trip to England at this time would appear to English and German colleagues as a distasteful attempt of a *captatio benevolentiae*; I wanted to avoid that."

Einstein first visited England in 1921, on his return from a trip to the United States, which was anything but private. I refer you to Clark[24] and Douglas for details of this and subsequent contacts with Eddington during Einstein's visits to England. It was not until June 1930 that Einstein finally spent a week with Eddington at his Cambridge home.

The Gold Medal affair has a reasonably happy ending. On January 21, 1926, Eddington was able to write:

> First let me say that it was a great pleasure to me that the Royal Astronomical Society decided to confer its Gold Medal on you. You have no doubt receive a letter from the President, Jeans. I had not much to do with this decision because the real decision was taken at a Council meeting during my absence in Leiden during December, but I was hoping that they would do this (Item 9-287).

After this extended digression on the personal contacts between Eddington and Einstein, I now return to the discussion of Eddington's work on relativity. In *Space Time and Gravitation*,[25] written in response to the tremendous demand in England for a semi-popular account of Einstein's theories, Eddington developed his skeptical approach to matter into a fairly full anticipation of what Wheeler was later to call geometrodynamics. Rather than explaining gravitation as an effect of matter, he sought to explain those properties ordinarily interpreted as material in terms of the metrical structure of space-time:

> When we perceive that a region contains matter we are recognising the intrinsic curvature of the world; and when we believe we are measuring the mass and momentum of the matter (relative to some axes of reference) we are measuring certain components of world-curvature (referred to those axes). The statistical averages of something unknown, which have been used to describe the state of the world, vary from point to point; and it is out of these that the mind has constructed the familiar notions of matter and emptiness.
>
> The law of gravitation is not a law in the sense that it restricts the possible behaviour of the substratum of the world; it is merely the definition of a vacuum. We need not regard matter as a foreign entity causing a disturbance in the gravitational field; the disturbance is matter (*Space Time and Gravitation*, pp. 198–199).

Eddington concludes tentatively that:

> Mind filters out matter from the meaningless jumble of qualities, as the prism filters out the colours of the rainbow from the chaotic pulsations of white light. Mind exalts the permanent and ignores the transitory; and it appears from the mathematical study of relations that the only way in which mind can achieve her object is by picking out one particular quality as the permanent substance of the perceptual world, partitioning a perceptual time and space for it to be permanent in, and, as a necessary consequence of this Hobson's choice, the laws of gravitation and mechanics and geometry have to be obeyed. Is it too much to say that mind's search for permanence has created the world of physics? So that the world we perceive around us could scarcely have been other than it is? ...
>
> Are there then no genuine laws in the external world? Laws inherent in the substratum of events, which break through into the phenomena otherwise regulated by the despotism of the mind? We cannot foretell what the final answer will be; but, at

present, we have to admit that there are laws which appear to have their seat in external nature. The most important of these, if not the only law, is a law of atomicity. Why does that quality of the world which distinguishes matter from emptiness exist only in certain lumps called atoms or electrons, all of comparable mass? Whence arises this discontinuity? At present, there seems no ground for believing that discontinuity is a law due to the mind; indeed the mind seems rather to take pains to smooth the discontinuities of nature into continuous perception. We can only suppose that there is something in the nature of things that causes this aggregation into atoms. Probably our analysis into point-events is not final; and if it could be pushed further to reach something still more fundamental, then atomicity and the remaining laws of physics would be seen as identities. This indeed is the only kind of explanation that a physicist could accept as ultimate. But this more ultimate analysis stands on a different plane from that by which the point-events were reached. The world *may* be so constituted that the laws of atomicity must necessarily hold; but, so far as the mind is concerned, there seems no reason why it should have been constituted in that way. We can conceive a world constituted otherwise. But our argument hitherto has been that, however the world is constituted, the necessary combinations of things can be found which obey the laws of mechanics, gravitation and electrodynamics, and these combinations are ready to play the part of the world of perception for any mind that is tuned to appreciate them; and further, any world of perception of a different character would be rejected by the mind as unsubstantial (*Space Time and Gravitation*, pp. 198–199).

He was even willing at that time (1920) to attribute the failure to master the quantum to this contrast:

It is one thing for the human mind to extract from the phenomena of nature the laws which is has itself put into them; it may be a far harder thing to extract laws over which it has had no control. It is even possible that laws which have not their origin in the mind may be irrational, and we can never succeed in formulating them. This is, however, only a remote possibility; probably if they were really irrational it would not have been possible to make the limited progress that has been achieved. But if the laws of quanta do indeed differentiate the actual world from other worlds possible to the mind, we may expect the task of formulating them to be far harder than anything yet accomplished by physics (*Space Time and Gravitation*, p. 200).

In spite of these doubts and hesitations, he ends the book with the famous passage:

We have found a strange foot-print on the shores of the unknown. We have devised profound theories, one after another, to account for its origin. At last, we have succeeded in reconstructing the creature that made the foot-print. And Lo! it is our own (*Space Time and Gravitation*, p. 201).

When asked, Einstein recommended *Space Time and Gravitation* for translation, but he was basically out of sympathy with Eddington's outlook on relativity theory.

Eddington's book is among the most original and intelligent [most interesting is crossed out] which have so far appeared in this field. I am quite ready to write a short foreword to your planned edition (Item 9-276).

When Sommerfeld wrote that "I know of no book that is as well written," Einstein replied:[26]

> Eddington's book is really extraordinarily ingenious [*geistvoll*]. But the tendency to conceive of the laws of nature merely as ordering schemata, serving to separate those cases which fall under them from those which do not, I really cannot approve. It is also objectionable that he treats relativity theory too much as *logically* necessary. God might also have decided, instead of the relativistic aether to create one absolutely at rest. This holds in particular if he should have, à la de Sitter, set up the aether in (essential) independence of matter, to which belief Eddington rather leans; in that case, however, the aether receives an "absolute" function. It is strange that most heads lack an organ for judging matters like this.

The most important influence Eddington's work had on Einstein was in connection with the question of a unified field theory. Einstein was searching for a physical theory that would not only provide an organic unification of the gravitational and electromagnetic fields, but hopefully would also provide a theory of matter, in particular an explanation of the appearance of quantum effects. Eddington's motivation was rather more mathematical and epistemological. In particular, he interpreted his unified theory as providing an *explanation* of gravitation, as opposed to the mere *description* provided by Einstein's 1915 theory. Before going into this, I must say a few words about the mathematical side of the theory. After 1915, it had been realized by several mathematicians that metrical and affine structures on a manifold were logically independent, even though there exists a unique symmetric affine connection associated with each metric. Weyl was the first to take advantage of the possibilities arising from this distinction to set up a physical theory, in which an additional element—a vector field—is needed, besides the metric field, in order to determine the affine structure of space-time. He identified this vector field physically with the electromagnetic potentials.

Noting the very special nature of Weyl's generalization, Eddington started by assuming that there was no *a priori* connection between the metric and an initially arbitrary symmetric connection. He observed that the curvature or Riemann tensor and its contraction, usually referred to today as the Ricci tensor, may be formed from the affine connection without any metric. This Ricci tensor, however, will in general not be symmetric, even though the connection is. What has this to do with physics? Eddington noted that an affine connection enables us to compare tensors at neighboring points (in particular, to say when two neighboring vectors are parallel). He regarded the possibility of such a comparison between quantities at neighboring points in space-time as the minimum element necessary to do any physics: "For if there were no comparability of relations, even the most closely adjacent, the continuum would be divested of even the rudiments of structure and nothing in nature could resemble anything else."[27]

The purely affine theory discussed so far involves no metric, and thus provides no basis for a comparison of lengths at different points, even neighboring ones. Such a comparison is provided by what Eddington, following Weyl's terminology, calls a gauge. Thus, the pure affine theory is a gauge-invariant theory. Yet, he notes:

> We *do* compare lengths on the sun and on the earth. ... This natural gauge system [i.e. metric] seems at first like an excrescence on the purely geometric [i.e. affine] theory, diverting it from its natural development, but necessary in order to bring it in touch with physics (p. 105).

Where are we to find such a "natural gauge [metric] of the world which is determined by measures of space and time made with material optical appliances"? Eddington answers:

> Any apparatus used to measure the world is itself part of the world, so that a natural gauge represents the world as self-gauging. *This can only mean that the tensor* $g_{\mu\nu}$ [the metric] ... *is not extraneous and arbitrary, but is a tensor already contained in the world geometry* (p. 111).

"There is only one such tensor available": the symmetric part of the Ricci tensor. Thus, up to a "universal constant [proportionality factor] λ," introduced "for convenience in order to remain free to use the centimeter instead of the natural unit of length" defined by the Ricci tensor, the symmetric part of the Ricci tensor and the metric tensor are the same thing. The field equation

$$R_{\mu\nu} = \lambda g_{\mu\nu}$$

is just a consequence of the way we measure the world!

In a 1922 popular lecture, Eddington tried to explain nontechnically why this constituted an explanation of gravitation:[28]

> Einstein's law of gravitation is usually expressed as a set of ten very lengthy differential equations: these equations are exactly equivalent to the geometrical statement that "the radius of spherical curvature of any 3-dimensional section of the 4-dimensional world is a universal constant length, the same for all points of the world and for all directions of the section". The law therefore implies that the world has a certain type of homogeneity and isotropy (not, however, the *complete* homogeneity and isotropy of a sphere). To explain the law of gravitation and the phenomena governed by it, we have to explain how this isotropy and homogeneity is secured. Our explanation is that the homogeneity and isotropy is not initially in the external world, but *in the measurements which we make of it*. It is introduced in all our operations of measurement, because the appliances which we use for measurement are themselves part of the world. In the earlier part of this lecture we saw that the contraction of the arm turned from horizontal to vertical is not detected by measurements made with a yard-measure which shares the contraction; in the same way any anisotropy of the world does not appear in measurements of it by appliances which, being part of the world, share the same anisotropy. The law of gravitation therefore arises from the fact that a certain type of non-homogeneity and non-isotropy of the world cannot come into observational experience, because it is necessarily eliminated in all observations and measurements made with material appliances. The orderly phenomena of gravitation are due to the *absence* of certain conceivable effects. We have been trying to find a key to the mystery; but the secret of the lock lies not in the key but in the wards.

By this explanation Eddington wedded himself to the gravitational field equations with a cosmological constant, as λ is called. It was originally introduced by Einstein in 1917, to permit a cosmological model with two major features: (1) it is static; (2) matter completely determines the metric (Mach's principle). Einstein later abandoned λ gladly when the first demand (staticity) proved observationally incorrect (interpretation of the cosmological red shift as an expansion of space), and his growing faith in a unified field approach led him to abandon the Machian demand: matter was to be interpreted as a particular *configuration* of the field rather than as the *cause* of the field. But for Eddington, dropping the cosmological constant would undercut his whole understanding of gravitation. After 1922 he referred to the gravitational equations with cosmological constant as if there were no question of their correctness. I shall return to this cosmological side of the story later.

What was Einstein's reaction to Eddington's unified theory? He had reacted negatively to Weyl's 1918 work when he realized that it implied that atoms, for example, could not be natural clocks measuring proper time intervals along their world-lines. Thereafter, he regarded Weyl's theory as a mathematically beautiful but physically irrelevant structure. His correspondence with Weyl included some quite sharp exchanges on this question, though always in a personally amicable spirit. Eddington evidently gave Einstein a copy of his 1921 paper during the latter's visit to England in that year.[29] Einstein's initial response was similar to his view of Weyl's theory. In a letter to Weyl, he called it "beautiful but physically meaningless."[30]

He elaborated in another letter:[31] "With Eddington's work [*Ausführungen*] I feel the same way as with Mie's theory; it is a beautiful framework, about which one absolutely does not see how it has to be filled in."

By January 1923, on his return trip from Japan, he had fallen under the spell of Eddington's idea. He wrote to Bohr from shipboard:[32]

> I believe that I have finally understood the connection between electricity and gravitation. Eddington has come closer to the truth than Weyl.

In the same month, still on board ship, he completed a paper on a new approach to Eddington's theory. Weyl wrote a detailed letter to Einstein, criticizing Einstein's paper, but ending on a triumphant note:[33]

> On the whole [*Im ganzen*] I am delighted naturally, that I am meeting you on the same purely speculative paths, against which you always protested before.

Einstein replied almost apologetically:[34]

> I absolutely must publish, because Eddington's train of thought must necessarily be thought through to the end. I believe now also that all these attempts on a purely formal basis will not advance physical knowledge [*Erkenntnis*]. Perhaps field theory has already given everything which lies within its possibilities.

Three days later he added:[35]

> On the whole a rather resigned mood regarding the entire problem has come over me. Mathematics is well and beautiful, but nature leads us around by the nose. In addition there is something comical here. I wanted to distance myself from you and have—apart from the sign of the cosmological term—arrived at the same equations as you with your special variational principle. ... The whole train of thought must be thought

through and is of noteworthy beauty; but above it stands the marble smile of pitiless nature that has bestowed on us more longing than wit.

To see what Einstein found so attractive in Eddington's work, let us turn to his papers on it. In addition to three in the *Sitzungsberichte* of the Prussian Academy,[36] he published a paper in *Nature*.[37] I shall quote from the latter. First of all, Einstein makes it clear that he does not accept Eddington's assertion that the 1915 theory merely describes gravitation:

> These magnitudes $g_{\mu\nu}$ determine the behaviour of measuring-rods and clocks with reference to the coordinate system, as well as the gravitational field. Thus far we are able to say that, from its foundation, the general theory of relativity explains the gravitational field.

The problem of concern to him is:

> Is it possible to generalise the mathematical foundations of the theory in such a way that we can derive from them not only the properties of the gravitational field, but also those of the electromagnetic field?

The mathematical work of Levi-Civita and its applications by Weyl and Eddington suggest that:

> Instead of starting off from the metrical relations . . . and deriving from this the coefficients of the affine relation . . . we proceed from a general affine relation . . . without postulating [the metrical relation]. The search for the mathematical laws which shall correspond to the laws of Nature then resolves itself into the question: What are the formally most natural conditions that can be imposed on an affine relation?. . .
>
> The first step in this direction was taken by H. Weyl . . . Eddington attacked the problem in a more radical manner. He proceeded from an affine relation . . . and sought to characterise this without introducing into the basis of the theory anything derived from . . . the metric. The metric was to appear as a deduction from the theory.

Einstein proceeds to show how Eddington's results could be obtained from a variational principle by choosing what he calls "The simplest possibility" as a Lagrangian density. He concludes: "The theory supplies us, in a natural manner, with the hitherto known laws of the gravitational field and of the electromagnetic field, as well as with a connection as regards their nature of the two kinds of fields; but it brings us no enlightenment on the structure of electrons."

In 1922 Eddington wrote the first treatise on the general theory in English, his justly-celebrated *The Mathematical Theory of Relativity*.[38] This book played a major role in introducing that theory to the English-speaking scientific public, just as Eddington's earlier lectures and book (*Space Time and Gravitation*) had for a broader public. Dirac's memoir on the "Early Years of Relativity" gives us a charming picture of how the book was received by English science students of his generation:[39]

> We really had no chance to understand relativity properly until 1923, when Eddington published his book, *The Mathematical Theory of Relativity*, which contained all the information needed for a proper understanding of the basis of the theory. This mathematical information was interspersed with a lot of philosophy. Eddington had his own

> philosophical views, which, I believe, were somewhat different from Einstein's, but developed from them.

Dirac is certainly right about the difference, although I should not say that Eddington's views developed from Einstein's.

There is no time to go into the many elegant and novel features of Eddington's treatment of the subject. Suffice it to say that the book is still worth reading by a contemporary relativist.[40]

When the second edition of this book was translated into German, Einstein wrote an "Appendix," which recapitulated his method of arriving at Eddington's theory, but ended with an even more pessimistic estimate of its physical significance than in his earlier papers. As he put it in a letter to Besso:[41] "I am firmly convinced that the whole train of thought Weyl–Eddington–Schouten leads to nothing physically useful. . . ."

Einstein soon launched several other attempts to set up unified field theories. But one thing stayed with him as a result of his work on Eddington's theory, and even grew in its influence on his thinking in later years: the role of the affine connection as a starting point in the attempt to construct a mathematically simple unified field theory. As Tonnelat notes in her survey of unified field theories:[42]

> This attempt of Einstein in 1923 heralded certain aspects of his last unified theory. But it had not freed itself sufficiently from the assumptions on which the theories of H. Weyl and A. S. Eddington rested. It assumed *a priori* the existence of a symmetric affine connection, which implies the absence of torsion. This simplification allows the easy calculation of the affine connection, but precludes notable possibilities for the theory.

In 1925 Einstein took the next logical step by dropping the assumption that both affinity and metric were symmetric.[43]

By early 1926 Einstein had fallen away from the affine faith. In a paper submitted on January 9, he wrote: ". . . through manifold failures I have come to the conviction that one does not come any nearer to the truth on this path."[44] He wrote to Eddington about this time: "I am now firmly convinced of the impossibility of deriving the laws of electricity by means of a generalization of the Riemannian conceptual schema, unfortunately. It was all a vain effort."[45] He continued to work on various other possibilities for a unified field theory. An attempt at a theory based on the concept of distant parallelism evoked the following rebuke from Eddington:[46]

> For my own part I cannot readily give up the affine picture, where gravitational and electrical quantities supplement one another as belonging respectively to the symmetrical and antisymmetrical features of world measurement; it is difficult to imagine a neater kind of dovetailing. Perhaps one who believes that Weyl's theory and its affine generalisation afford considerable enlightenment, may be excused for doubting whether the new theory offers sufficient inducement to make an exchange.

Within a few years, Einstein had also dropped distant parallelism, wandering for over a decade in the desert of other unified field theories (such as Kaluza–Klein), before finally returning to the affine oasis in 1945, which he never forsook up to his death.

When Eddington's biographer interviewed Einstein in January 1954, he paid the following tribute to Eddington—amidst a number of critical remarks as we shall see later:[47] "He was one of the first to recognize that the displacement field (i.e., the affine connection) was the most fundamental concept of general relativity theory, for this concept allows us to do without the inertial system."

As noted above, in 1917 Einstein added the cosmological or λ-term to his gravitational field equations to get a cosmological solution that was both static and such that the material content determined the metric (Mach's principle). De Sitter soon showed that a solution to these equations existed without any matter, but Einstein adhered to his static model. In 1922, the Russian mathematician A. A. Friedmann showed that non-static solutions to the field equations (with and *without* the λ-term) existed. His solutions included both Einstein's static and de Sitter's universe, but in general represented three-dimensional spaces of constant but *time-dependent* radius of curvature.[48] These could represent either expanding or contracting universes, and (in some models) even cyclic expansions and contractions. Friedmann's paper came to Einstein's attention. He thought he had found a mathematical flaw in Friedman's argument, and said so in print. When he became convinced that the error was his not Friedmann's, he retracted his mathematical objection, but stuck to his static cosmological model. How little impressed he was by Friedmann's models can be seen from the final clause of his draft retraction, which (fortunately for him) Einstein deleted before it was printed:[49]

> It follows that the field equations, besides the static solution, permit dynamic (that is, varying with the time coordinate) spherically symmetric solutions for the spatial structure, [to which a physical significance can hardly be ascribed.],

the bracketed portion being crossed out in the manuscript.

Friedmann's papers are not mentioned in Eddington's book *The Mathematical Theory of Relativity* nor in any other treatise on relativity from the twenties that I have examined. A special case of Friedmann's metrics was rederived independently by Lemaitre in the mid-twenties, and they were rederived by H. P. Robertson toward the end of the decade. De Sitter's first paper on his model had noted the possible connection between the cosmological red shift predicted by this model and the then extremely scanty observations of galactic red shifts. Yet no one had forcefully put the case to the astronomical public for expanding models of the universe as an explanation of the galactic red shifts before Lemaitre brought his work to Eddington's attention in 1930. Here is the story in Eddington's words:[50]

> Lemaitre's paper came to light again in 1930 in connection with an informal communication given at a meeting of the Society by de Sitter. Making (as I think) a rather excessive estimate of the masses of the nebulae, de Sitter propounded the dilemma that the actual universe apparently contained enough matter to make it an Einstein world and enough motion to make it a de Sitter world. This naturally called attention to the need of intermediate solutions for handling such a question. Some of my remarks in the discussion were reported, and Lemaitre (who had once been a student of mine) wrote to me on the subject, drawing attention to his paper; I was at that time

> working with G. C. McVittie on the question whether Einstein's world is stable, using Robertson's paper as a basis. The elegance and completeness of Lemaitre's work was recognised by de Sitter and myself, and it has been the subject of papers by both of us. I pointed out that it was immediately deducible from his formulae that Einstein's world is unstable, so that an expanding or a contracting universe is an inevitable result of Einstein's law of gravitation.

Eddington was, of course, still wedded to the cosmological constant for the epistemological reason discussed above. He therefore only considered expanding models that satisfied the field equations with λ-term. He wanted λ_E, the Einstein model's value of the cosmological constant. So he favored a model which started from the (unstable) static Einstein model and expanded—once perturbed—until it asymptotically approached the de Sitter model. He did not like the idea of expansion from a singular initial state, which Lemaitre soon came to favor. But Eddington's most important service to the expanding universe was propagandistic. Once he became convinced of the need for an expanding model, he used his impressive powers of influence and persuasion in a series of technical, semi-popular and popular articles, a book and even a radio broadcast, preaching this gospel to both the scientifically literate and illiterate.[51]

From the outset, Lemaitre correctly interpreted his expanding model as implying a cosmological galactic red shift; Robertson did not. Along with just about everyone else, he held fast to the dogma that the universe must be static. Yet, within a year or so, Eddington helped to make the expanding universe the new orthodoxy. Any attempt to interpret all or part of the observed red shifts as due to any other cause or causes had—and still has—to face the formidable opposition of the astronomical establishment. Of course, the announcement of Hubble's law in 1929 helped enormously in this campaign.[52]

Einstein probably became aware of this revolution in cosmology during his visit to Cal Tech (December 31, 1930 to March 4, 1931). His travel diary shows that he started to discuss cosmology with Tolman almost immediately upon arriving in Pasadena.[53] On March 1 he wrote to Besso:[54]

> The Mount Wilson Observatory people are excellent. They have recently found that the spiral nebulae are spatially approximately uniformly distributed and show a strong Doppler effect proportional to their distance, which follows without constraint from the theory of relativity (without cosmological constant).

Almost immediately upon his return to Berlin, Einstein presented a paper to the Prussian Academy in which he publicly withdrew his support for the cosmological constant which he called "anyway theoretically unsatisfactory":[55]

> It is above all noteworthy that general relativity seems able to take account of Hubble's new facts more naturally [*ungezwungener*] (that is without λ-term) than the now empirically rejected [*in die Ferne gerückten*] postulate of the quasi-static nature of space.

In January of the following year, while both were visiting Cal Tech, he and de Sitter wrote a paper, in which they agreed on dropping the λ-term as unnecessary, and pointed out that a spatially flat expanding model was the simplest one compatible with the existence of a non-zero density of matter.[56]

Eddington was horrified by Einstein's apostasy. In his Presidential Address to the Physical Society, delivered on November 6, 1931, he rehearsed his epistemological argument for the λ-term:[57]

> The ratio of the metre to the radius of curvature is determined by λ. If λ is zero the ratio is zero and the connection breaks down. We are left with a space which does not fulfil the first conditions of a medium of measurement; and the relativity theory is laid open to criticisms such as have been brought forward by Prof. Whitehead (mistakenly, I think, as regards the existing theory) ıs failing to provide a "basis of uniformity" for spatial measurement. For this reason the cosmical term $\lambda g_{\mu\nu}$ is essential in relativity theory. When it was first introduced it might have been regarded as a fancy addition, but it is now seen to be indispensable. I see that Einstein has recently proposed to take λ equal to 0; that seems to me an incredibly retrograde proposal.

A remarkable interchange of roles had thus taken place. Einstein had introduced the cosmological constant in order to satisfy what he then thought to be necessary criteria for a cosmological model. De Sitter soon proved that another solution with the λ-term existed. These two solutions were regarded as the main cosmological rivals for over a decade. Eddington was initially skeptical of the whole cosmological enterprise; but after he developed his epistemological argument for λ, it became the cornerstone of his whole approach to relativity. When expanding cosmological models came into favor, largely through his efforts, he naturally adopted one which allowed him his λ-term, with the Einstein value, which by then he was trying to connect with his interpretation of quantum theory.[58] Einstein, on the other hand, once convinced that one could not have a static model anyway because of the need to explain the galactic red shifts, saw no need for λ at all. Now Eddington was trying to connect his cosmological model with a theory of matter, of course on a quite different basis from what Einstein had been trying in his Machian phase; while Einstein had given up the attempt to found a cosmological model on a Machian view of inertia and was trying to explain matter as an aspect of a unified field, rather like Eddington's original "geometrodynamical" interpretation of general relativity. Each regarded the other as dogmatic and inconsistent, and maintained this position until the end of his life.

When asked about Eddington's *Fundamental Theory*, Einstein replied:[59]

> He has always seemed to me an uncommonly ingenious [*geistreich*] but uncritical man. (Shape up or else . . .) [*Reim' dich oder ich fress dich . . .*] . . . With his philosophy he reminds me of a prima ballerina, who doesn't herself believe in the justification [*Berechtigung*] for her elegant leaps.

Notes

1 A. V. Douglas, *The Life of Arthur Stanley Eddington* (London: Thomas Nelson and Sons, 1956). It will be cited hereafter as Douglas.

2 See, in particular, J. Earman and C. Glymour, "Relativity and Eclipses: the British Eclipse Expeditions of 1919 and Their Predecessors," *Historical Studies in the Physical Sciences* 11 (1980):49, and D. Moyer, "Revolution in Science: The 1919 Eclipse Test of

General Relativity," in: *On the Path of Albert Einstein*, eds. A. Perlmutter and L. F. Scott (New York: Plenum Press, 1979), pp. 55–101.

[3] See the summary of his work he prepared in the thirties, printed in Douglas, pp. 189–192.

[4] The manuscript of "A Total Eclipse of the Sun," dated June 30, 1898, is among the Eddington Papers held in Trinity College Library, Cambridge. See Loren Graham's article "Eddington and the English-Speaking World," *Between Science and Values* (New York: Columbia University Press, 1981). The first footnote to this article contains a list of some archives holding Eddington papers.

[5] This was Einstein's 1911 prediction of an apparent deflection of .83″ for grazing incidence, one-half of his later prediction based on the final (1915) field equations. It is amusing to speculate what might have happened if this expedition—or the subsequent 1914 Crimean expedition—had been able to test the original (1911) prediction.

[6] C. Perrine, *Astron. Nachr.* 219 (1923):281. Perrine belatedly wrote this account as a result of the great interest generated by the later confirmations of Einstein's prediction.

[7] A. S. Eddington, *The Observatory* 36 (1913):63. See also *ibid.* 35 (1912):328; and A. S. Eddington and C. Davidson, *Monthly Notices of the Royal Astronomical Society* (*MNRAS*) 73 (1913):386.

[8] A. S. Eddington, *MNRAS* 38 (1915):93.

[9] I am indebted to Dr. A. J. F. Gogelein, Director of the Boerhave Museum, Leiden, for making copies of Eddington's letters to de Sitter in that museum available to me. The reference to Einstein's paper is almost surely to "Die Grundlagen der allgemeinen Relativitätstheorie," *Annalen der Physik* 49 (1916):769.

[10] Eddington is here referring to the three-part paper de Sitter wrote on Einstein's theory, which was published in the *MNRAS* 76 (1916):699; 77 (1916): 155; 78 (1917):3.

[11] A. F. Lindemann and F. A. Lindemann, *MNRAS* 77 (1916):140.

[12] Earman and Glymour, *op. cit.*, p. 71.

[13] Quoted from W. H. McCrea, "Einstein: Relations with the Royal Astronomical Society," *Quart. Journ. RAS* 20 (1979):255.

[14] A. S. Eddington, *Nature* 98 (1916):328. This was the issue of December 28, 1916. Also see the account of the BAAS Meeting in *Nature* 98 (1916):120.

[15] A. S. Eddington, *MNRAS* 73 (1913):223.

[16] A. S. Eddington, *Report on the Relativity Theory of Gravitation* (London: Fleetway Press, 1920), hereafter referred to as *Report*.

[17] Eddington is incorrect in stating that the motion of a singularity in the electromagnetic field follows from the electromagnetic field equations. The law of conservation of energy-momentum for the combined matter-field system must also be assumed to derive the equations of motion.

[18] In addition to the reference in footnote 2, see A. N. Whitehead, *Science and the Modern World* (London: MacMillan, 1925), p. 13, for an eyewitness account of this dramatic meeting.

[19] Documents cited from originals or copies in the Einstein Archive, Hebrew University, Jerusalem will hereafter be cited by their Control Index numbers in the duplicate of that Archive in the Mudd Library, Princeton University.

[20] Einstein to Carl Stumpf, November 3, 1919; quoted in A. Pais, "*Subtle is the Lord ...*" (Oxford: Clarendon Press, 1982), p. 306.

[21] The two remained in fairly regular correspondence until the end of 1921. Seven letters between December 1, 1919 and November 12, 1921 (four from Eddington and three from Einstein), the most intensive period of their correspondence, are preserved in the Einstein Archive. Eight additional letters from Eddington to Einstein, dated between June 22, 1923 and August 26, 1930, but no additional Einstein letters, are in the Archive. Three letters from Einstein to Eddington are in the Fitzwilliam Museum, Cambridge. The first is dated June 11, 1920; the other two are undated, but appear to date from late 1925 and early 1926. There is also a postcard dated October 20, 1925.

[22] W. H. McCrea has denied the accuracy of Eddington's account. In the article cited in footnote 13, he states: "Eddington was a great scientist, but he failed to understand the motives and judgements of his colleagues, on this occasion at any rate."

[23] Max Born (ed.), *The Born–Einstein Letters* (New York: Walker & Co., 1971), p. 22.

[24] Ronald Clark, *Einstein The Life and Times* (New York and Cleveland: World Publishing Co., 1971).

[25] A. S. Eddington, *Space Time and Gravitation: An Outline of the General Relativity Theory* (Cambridge: Cambridge University Press, 1920), hereafter referred to as *Space Time and Gravitation.*

[26] See A. Hermann (ed.), *Albert Einstein/Arnold Sommerfeld Briefwechsel* (Basel-Stuttgart: Schwabe & Co., 1968), pp. 109–110 for Sommerfeld's comment and Einstein's reply.

[27] A. S. Eddington, *Proc. Roy. Soc.* A99 (1921):104. Quotation from p. 106.

[28] A. S. Eddington, *The Theory of Relativity and Its Influence on Scientific Thought* (Oxford: Clarendon Press, 1922). He also wrote a semi-popular article, "Can Gravity be Explained," expounding this view: A. S. Eddington, *Scientia* 33 (1923):313.

[29] Einstein to Weyl, June 12, 1921 (Item 9-277).

[30] Einstein to Weyl, September 5, 1921, Einstein–Sammlung ETH, Zurich, no. 551.

[31] Einstein to Weyl, June 6, 1922 (Item 24-071).

[32] Einstein to Bohr, January 11, 1923. The letter is printed in translation in: *Einstein: A Centenary Volume*, ed. A. P. French (Cambridge: Harvard University Press, 1979), p. 274.

[33] Weyl to Einstein, May 18, 1923 (Item 24-074).

[34] Einstein to Weyl, postmarked May 23, 1923 (Item 24-080).

[35] Einstein to Weyl, May 26, 1923 (Item 24-083).

[36] A. Einstein, *Sitzungber. preuss. Akad. Wiss., physik-math. Kl.* 1923:32, 76, 137.

[37] A. Einstein, *Nature* 112 (1923):448.

[38] A. S. Eddington, *The Mathematical Theory of Relativity* (Cambridge: Cambridge University Press, 1923). Its Preface is dated August 10, 1922.

[39] P. A. M. Dirac, in: *Albert Einstein: Historical Perspectives*, eds. G. Holton and Y. Elkana (Princeton: Princeton University Press, 1982), p. 79. Quotation from p. 82. Dirac's paper also notes Eddington's role as a popularizer of relativity theory.

[40] I must note a misunderstanding of one of Eddington's remarks in Pais' recent book on Einstein, A. Pais, *Subtle is the Lord* ... (Oxford: Clarendon Press, 1982), which I fear may mislead those who have not read Eddington. Pais states on p. 284 that "Eddington

... believed that [gravitational] waves were spurious and 'propagate ... with the speed of thought."' What Eddington actually says on p. 130 is that "We can 'propagate' coordinate changes with the *speed of thought* ..." What he is concerned about here is an invariant separation of coordinate effects from physical ones. He had a justified doubt about Einstein's treatment of gravitational waves, which was based upon the use of certain coordinate conditions. In line with his emphasis on the fundamental importance of the Riemann tensor, Eddington wrote a paper later in 1922 in which he used propagation of the Riemann tensor as a coordinate-invariant characterization of gravitational waves. This is quite a modern point of view, which he was the first to adopt. See A. S. Eddington, *Proc. Roy. Soc.* A 102 (1922):268.

[41] Einstein to Besso, June 6, 1925; in: *Albert Einstein–Michele Besso Correspondance 1903–1955*, ed. P. Speziali (Paris: Hermann, 1972), p. 204; see also the letter of December 25, 1925 on p. 215.

[42] M. A. Tonnelat, *Les théories unitaires de l'electromagnétisme et gravitation* (Paris: Gauthier-Villars, 1965), p. 273.

[43] A. Einstein, *Sitzungsber. preuss. Akad. Wiss. physik.-math. Kl.* 1925, 414.

[44] A. Einstein, *Math. Annalen* 97 (1926):99. The quotation is from p. 100.

[45] This undated letter is probably in reply to Eddington's letter of January 22, quoted earlier, informing Einstein that he had won the Royal Astronomical Society's Gold Medal. Eddington again invited Einstein to visit him and his sister in Cambridge. Einstein declined on the grounds that verbal communication would be difficult, adding: "I have such a desire to talk with you that it would be worthwhile for me for that reason alone to learn to speak English." The letter is in the Fitzwilliam Museum, Cambridge.

[46] A. S. Eddington, *Nature* 123 (1929): 280.

[47] Douglas, p. 45. For an elaboration of this point, see Einstein's letter to Besso of August 10, 1954, translated in A. P. French (ed.), *Einstein/A Centenary Volume* (Cambridge: Harvard University Press, 1979), pp. 267–269.

[48] A. A. Friedmann, *Zeitschr. f. Physik* 11 (1922):326. The original paper only considered three-spaces of constant positive curvature, but Friedmann considered those of negative curvature in a second paper.

[49] See document 1-026 in the Einstein Archive for the manuscript, A. Einstein, *Zeitschr. f. Physik* 16 (1923):228 for the printed text.

[50] A. S. Eddington, *MNRAS* 91 (1931):412; quotation from p. 414.

[51] The first and most fundamental of his papers is A. S. Eddington, *MNRAS* 90 (1930): 668.

[52] I am indebted to J. D. North, *The Measure of the Universe* (Oxford: Clarendon Press, 1965), J. Merleau-Ponty, *Cosmologie du XX siècle* (Paris: Gallimard, 1965), and George Ellis, "The Expansion of the Universe and the de Sitter Effect" (unpublished) for a number of features of my account of this sudden switch of paradigms.

[53] Entry of January 3, 1931: "Doubts about correctness of Tolman's work on cosmological problem. Tolman, however, was in the right." (Item 5-256).

[54] Einstein to Besso, March 1, 1930. See *Einstein–Besso Correspondance* (footnote 41 above).

[55] The paper was presented to the session of April 16, 1931, and published in A. Einstein, *Sitzungber. d. preuss. Akad. Wiss., physik.-math. Kl.* 1931, 235.

[56] A. Einstein and W. de Sitter, *Proc. Nat. Acad. Sci.* 18 (1932):213.

[57] A. S. Eddington, *Proc. Phys. Soc.* 44 (1932):1.

[58] See A. S. Eddington, "On the Value of the Cosmological Constant," *Proc. Roy. Soc.* A133 (1931):605.

[59] Einstein to Besso, July 29, 1953. See *Einstein–Besso Correspondance*, p. 500 (footnote 41 above).

Einstein and Infeld: Seen Through Their Correspondence

John Stachel

"Einen zweiten Infeld — gibt es nicht"
(Albert Einstein, 13 August 1949)

This talk is based exclusively on an examination of the Infeld correspondence files in the Einstein Archive. Some secondary sources, notably Infeld's autobiographical writings, were consulted for help in elucidating the correspondence.

The first item in the Einstein–Infeld files is not from either. It is a letter from Infeld's friend, the philosopher of science Josef Winternitz, to Einstein (13 December 1927), requesting him to read and respond to a paper "Zur Feldtheorie der Elektrizität und Gravitation" (published version, "Zur Feldtheorie von Elektrizität und Gravitation," Eryk Infeld Bibliography, item 4, hereafter cited as EI Bibl. 4; this Bibliography appears after this article in *Acta Physica Polonica* B30 (1999): 2904–2908) that Infeld had previously sent to Einstein.

> I know Mr. Infeld as a very gifted theoretical physicist, because in 1920/21 we, together with Messrs. Leo Szilard and Grolmann formed a circle for the purpose of working through the theory of relativity.

This letter serves to remind us that Infeld's first contact with Einstein was during a visit to Berlin in 1920–1921. Turning to *Quest* (I cite the second edition, New York: Chelsea, 1980, as Q), we find an account of that visit:

> I became weary of the political atmosphere in Poland, of the growing anti-Semitism and the tensions which were gradually diffusing into the university. ... To get away one had to have money. I talked to my father. He agreed immediately to keep me in Germany for a year. ... So I went to Germany, intending to study at the University of Berlin, at that time one of the best universities of Europe... But it was not so simple. If there was a nation hated more than any other at this time in Germany, it was Poland. ... In Poland I was a Jew and not a Pole. But here in Germany I was a Pole, a member of a hostile nation. I learned that it was impossible for a Pole to be admitted to the University without powerful outside influence. (Q, p. 90).

Talk at the Infeld Centennial Conference,
Warsaw, June 1998,
Acta Physica Polonica, B30 (1999): 2879–2904.

He tried to secure the needed influence through letters of introduction from "influential people of the Jewish community in Cracow" to similarly-placed members of the Berlin Jewish community. Here again he met with prejudice:

> I learned about the superior attitude of German Jews to any other Jews in the world and especially to Polish Jews. Among Polish Jews, in turn, those from the Austrian part of Poland, from Galicia, were regarded as most inferior (Q, p. 90).

Finally, someone suggested:

> If you are a physicist, why don't you go to see Einstein? Maybe he will help you. (Q, p. 91).

A telephone call to Einstein's house produced an invitation to "come right now." He did so, and "was shown into a waiting room full of heavy furniture" by Mrs. Einstein—Einstein's second wife Elsa, the source of the Biedermeyer furniture—who explained that Einstein was occupied with "a Chinese minister of education" (remember this is shortly after world notoriety descended upon Einstein in December 1919), and invited Infeld to wait.

> I waited, my cheeks burning with excitement. A few minutes later a young man with a thin vivid face and smiling eyes entered the room and sat down opposite me. (Q, p. 91).

The young man was carrying a copy of Weyl's *Raum-Zeit-Materie*, and he and Infeld started to chat. This is how Infeld met Winternitz, identified in *Quest* only as "Joseph." Winternitz knew Einstein from his days in Prague (1911–12), where Winternitz's father was also a Professor. Einstein soon

> opened the door of his study to let the Chinese gentleman out and me in. Einstein was dressed in a morning coat and striped trousers with one important button missing (Q, p. 92).

Infeld explained his predicament to Einstein, who listened carefully and replied:

> 'I should be very glad to give you a recommendation to the Ministry of Education. But my signature does not mean anything.'
>
> 'Why?'
>
> 'Because I have given very many recommendations, and'—here he lowered his voice to a confidential tone—'they are anti-Semites' (Q, p. 92).

Einstein finally decided to write to Planck, asking him to write a recommendation for Infeld. But in spite of this, Infeld's application for permission to study at the University of Berlin was rejected. After considerable further efforts by Jewish dignitaries, who now knew Infeld had Einstein on his side, Infeld was allowed to register as a "special student," entitled to take courses even though they did not count towards a degree.

At the University, he renewed his contacts with Winternitz, and together they:

> organized a small group in which we discussed carefully the more difficult and subtle points of relativity theory. One of the members of our circle was [Jacob] Grommer, who had published a few papers with Einstein (Q, p. 96).

In the Epilogue to *Quest* (Q, p. 351), Infeld confirms that a third member was Leo Szilard. This is evidently the discussion group referred to in Winternitz's letter; it seems that the "Grolmann" mentioned is an incorrect reference to Grommer. Winternitz gave special thanks to Infeld in the preface to his book, based in large part on these discussions (as well as his conversations with Einstein).

> I would not have penetrated so deeply into the understanding of the theory [of relativity] without the collaboration of my dear friends, the young physicists L. and F. Infeld, which I here gratefully acknowledge (*Relativitätstheorie und Erkenntnislehre*, Leipzig: B.G. Teubner, 1923, p. v).

The reference to L. and F. Infeld reminds us that Leopold and Fanny Infeld (I am grateful to Eryk Infeld for providing Mrs. Infeld's first name) were married when he was twenty, and she evidently accompanied him to Berlin. The marriage lasted for some ten years, ending while Infeld was working in Warsaw (see "Bronia," p. 126 in *Why I Left Canada*, tr. Helen Infeld, ed. Lewis Pyenson, Montreal, London: Queen's University Press, 1978, hereafter cited as *Why*. . .).

Returning to 1927, Infeld was now in Warsaw. After returning to Krakow, he obtained his doctorate in 1921 under Prof. Wladyslaw Natanson. Infeld recalls that it was the first in theoretical physics in Poland (Q, p. 102), and states that he was Natanson's only doctoral student (Q, p. 348). He worked for the next eight years in gymnasia (secondary schools), first as a teacher of physics in Bedzin (Q, p. 348), and then for a few years as headmaster of the Jewish gymnasium in Konin ("Konin," in *Why*. . ., pp. 130–135). Then he managed to secure a post in Warsaw at a Jewish girls' gymnasium.

There, he started to work again in earnest on physics, in particular on a unified theory of gravitation and electromagnetism, and sent Einstein a copy of his work. Receiving no reply, he must have appealed to Winternitz, whose letter to Einstein produced the desired effect. Einstein wrote Infeld on 17 December, apologizing for not having been able to read his work in more detail due to overwork.

> But now I may tell you that after innumerable efforts I have arrived at the conviction that one cannot attain a satisfactory theory of electromagnetic phenomena starting with the affine connection. In particular, concerning your assumption of a non-symmetric metric tensor, this road also does not seem practicable to me.

Introduction of the antisymmetric part of the metric

> *alongside the symmetrical part represents a quite independent postulation, so that a unification appears as quite external. For the moment, it looks as if Kaluza's method of introducing a formal fifth dimension still has the most to offer.*

As I am sure many of you know, Einstein's opinions on this question vacillated a number of times over the years; his last unified field theory was based on a non-symmetric metric and connection.

Infeld replied on 6 January 1928:

> I thank you, Hochverehrter Herr Professor, for your letter and for the paper, receipt of which gave me real pleasure. ... I had already become convinced that the non-symmetry of the metric tensor in the form that I gave should not be assumed.

> Yet I believe that an (affine) unified theory of gravitation and electricity can be constructed . . .

He went on to give further details of a new version of his theory, concluding:

> I would be much indebted to you if you were inclined to give me your opinion of the assumptions I have indicated.

A brief note in Einstein's hand on Infeld's letter indicates the nature of his "critical reply," which is not in the Archive.

Infeld replied on June 2, 1928:

> I thank you very much . . . for the letter that I received from you in January of this year. I permit myself to report briefly to you the results concerning the unified field theory that follow from my assumptions.

After a discussion of these, he concludes:

> Allow me at the same time to thank you most heartily for the good will that you have shown towards me. I am very happy that in July I shall most probably be able to attend the summer course in physics, which will enable me to deepen my understanding of your latest works through your lectures.

It is not clear to which summer course Infeld is referring. At any rate, Einstein spent July 1928 on the Baltic sea coast, still recuperating from an episode of heart trouble earlier in the year, and gave no lectures during that month.

On 4 October 1928, Infeld again wrote Einstein, commenting on the latter's paper on Riemannian geometry with distant parallism. He pointed out that:

> The geometry developed there by you is a special case of non-Riemmanian geometry, and as such is treated in L. P. Eisenhart's book, *Non-Riemannian Geometry*, pp. 47–50.

He detailed the nature of the special case and pointed out some similarities with his own theory.

In 1929, Infeld attempted to get a fellowship from the International Education Board (Rockefeller Foundation). On 14 January he wrote Einstein for help:

> I take the liberty of turning to you in the following personal matter:
>
> The conditions in which I work here make any scientific work extraordinarily difficult. For reasons that I do not want to go into here in any detail, and that are certainly well known to you, very esteemed Herr Professor, the possibilities for me here scientically are very limited.
>
> My Promotor, Prof. Natanson, in order to allow me to study abroad has written a letter of recommendation to the International Education Board, a copy of which I enclose. I very much want to spend at least a part of my study period in Berlin, in order to work my way into the field of quantum mechanics. Prof. Schrödinger has said that he is ready to accept me as a stipend holder. The probability of obtaining a stipend, upon which so much depends for me, would certainly be considerably greater, if you, very honored Herr Professor, would have the goodness to send a few lines of recommendation to the International Education Board. I would have approached you through Professor Natanson, very honored Herr Professor, if Prof. Natanson was not so very sick at the moment.
>
> Finally, I would like to once more justify turning to you with this request by the importance of this matter to me.

On 18 January, Einstein wrote the requested "few lines":

> I permit myself to most warmly support the application of Dr. Infeld. In particular, through repeated scientific correspondence with Dr. Infeld, I know that he has had careful training in the field of theoretical physics and has also had independent ideas in this field.
>
> I should like to support the application of Dr. Infeld all the more since the conditions for his development in Warsaw, under the conditions prevailing there, are unfavorable.

In spite of the support of Natanson and Einstein, Infeld's application did not succeed. But, especially due to the help of Professor Stanisław Loria, he was able to obtain a senior lectureship at the University of Lwow, and then to habilitate there.

In 1932, he took a two-month leave from the University, in order to travel abroad. He went to Leipzig, no doubt attracted by the opportunity to learn more about quantum mechanics through contact with Heisenberg and his group. He found the political atmosphere extremely tense in the town, but:

> In this sea of hatred and fighting the physics department formed a small peaceful island free of anti-Semitism. Heisenberg's assistant was a Jew. Toward a foreigner from Poland the atmosphere was reserved but correct ... At [Heisenberg's] seminar I met a young professor of mathematics, Van der Waerden, who was also interested in theoretical physics ... I had with me the manuscript of a paper connected with a problem on which Professor Van der Waerden had worked [presumably EI Bibl. 18]. ... Two weeks after I had arrived in Leipzig Van der Waerden and I began to work together.

The result was the well-known Infeld–Van der Waerden paper on spinors, "Die Wellengleichung des Elektrons in der allgemeinen Relativitätstheorie," EI Bibl. 19. Apparently they sent the paper to Einstein, who suggested it be published in the *Sitzungsberichte* of the Prussian Academy of Sciences. On 15 November, after his return to Poland, Infeld wrote Einstein:

> I must thank you most heartily for the good will you have shown me, and for the proposal to submit the work I did together with v. d. Waerden to the Academy. Today I got Schrödinger's letter from v. d. Waerden. He will be glad to referee the paper for the Prussian Academy. He doesn't like the way it is presented, however, and he proposes many changes to us privately, which can easily be taken care of.
>
> It is a great pleasure and honor for me that you cite us in your work (the paper is called: The Wave Equation of the Electron in the General Theory of Relativity by L. Infeld and B. L. v. d. Waerden).
>
> I would be very grateful to you, highly honored Herr Professor, if you had the kindness to send me a reprint after your paper appears. I hope you wouldn't hold it against me if I permitted myself to ask you for a proof copy, if you have one to spare?

The paper mentioned is A. Einstein and W. Mayer, "Semi-Vektoren und Spinoren," *Sitzungsberichte der preussichen Akademie der Wissenschaften, phys.-math. Klasse*, 32, 1932. Infeld and van der Waerden are cited on p. 25.

In 1933, after the tragic death of his second wife Halina, Infeld again applied for a Rockefeller fellowship, and Einstein again supported him. On 1 July 1933 he wrote from exile in Le Coq-sur-mer, Belgium:

> Professor Loria of Lwow informs me that he is attempting to get a research stipend for the Assistant and Privat Docent at the University of Lwow, Dr. Leopold Infeld. I know a series of papers by Dr. Infeld in detail and can recommend him as a diligent young researcher in the fields of quantum and relativity theory. Infeld is a scientific writer, who possesses both the necessary originality and self-criticism, so that useful work can be expected from him.

This time, Infeld succeeded in getting a fellowship and, heeding the advice of Loria, went to Cambridge, England. Initially he hoped to work with Dirac, but communication proved difficult. After attending a course of lectures by Max Born on his new non-linear generalization of Maxwell's theory. Infeld became interested in it. He approached Born to explain a defect in the theory and how to remedy it. Born's reaction was not encouraging. He interrupted Infeld angrily to defend his position, finally leaving the lecture hall saying, "I shall think it over" (Q, p. 209).

> I was annoyed at Born's behavior as well as my own and was, for one afternoon, disgusted with Cambridge. I thought: 'here I met two great physicists. One of them does not talk. I could as easily read his papers in Poland as here. The other talks, but he is rude' (Q, p. 209).

But the next day Born admitted Infeld was right, and they started to collaborate. Soon he was writing to Einstein again, to describe his work with Born and ask another favor. On 12 February 1934, he wrote:

> I permit myself once again to request something from you. I wouldn't permit myself to do this, if I were not convinced that for you it means a trifle, and for me a quite essential thing.
>
> I am currently in Cambridge and am extraordinarily pleased with my stay here. Perhaps you saw in *Nature* the short notice written by Born and me. Together with Born, we wrote a large work (on the foundation of the new field theory ["Foundation of the New Field Theory," EI Bibl. 22]) that will appear in the *Proc. Roy. Soc.* Only now do I see what opportunities a center like that in Cambridge offers for scientific work. It means a lot to me to obtain the possibility of staying here longer. I don't have to give you the reasons why, because you know them very well. They lie not only in the scientific opportunities, but in the entire atmosphere of England, which I admire more and more. But I do not intend to look for any sort of position here. It would not be proper behavior towards my German colleagues, who find themselves in a much worse position. Luckily, there is another possibility for prolonging my stay here.
>
> I think I have a little talent for popular science. A half-year ago my popular-scientific book [*Nowe Drogi Nauki*] appeared in Poland, which has now been translated into English and will appear at the beginning of March. The publisher is Gollancz (the publisher of the Brown Book [on the Nazi terror]) and he believes it has very great possibilities on the English market. I have just received the proofs, and permit myself to send you a set at the same time.

> Now permit me to come to my request. Gollancz has told me that the possible sales of the book will be increased about tenfold, if you were inclined to write a few words about the book with permission to cite your judgement.
>
> It is very uncomfortable for me to presume so much on your kindness and friendliness. Therefore I have to be quite clear. I don't want you to do me the favor if you actually don't like the book. But I would be extraordinarily thankful if you would have the kindness to read through parts of the book and write a few lines or even a few words about it, if you like. Speed is very important for me, since the book is supposed to appear in March. (There are many errors of printing and translation in the proofs, that will be removed.)
>
> Let me once more apologize. I would never allow myself to burden you, if this were not a very important opportunity for me and if my further stay here did not depend on it.

Again, Einstein came through with a preface to the book, *The World in Modern Science* (London: Gollancz, 1934). He wrote Infeld on 1 March 1934: "I took great pleasure in your book and very much treasure its liveliness, clarity and simplicity. Enclosed I send you a short introduction, which I hope meets the need. If for any reason the wording has to be changed, please propose the relevant change to me."

But Infeld soon had to return to Poland, where the atmosphere at the University and in the country at large had not improved, nor had his prospects. He decided to try to go to the United States to work with Einstein. On 26 February 1936, he wrote:

> Again after a few years I have to turn to you with another request. I am doing so only as a last resort. Let me present the situation briefly. After another half-year's stay in England, during which I worked with Born in Cambridge, I returned to Poland. Unfortunately, for reasons well known to you, I have no possibility of advancement here. Working with Born was extraordinarily pleasant and instructive for me. I believe that I adopt myself well to collaboration. Now for many reasons I have to and I want to leave here, at least for a year. I have already made some efforts in this direction. My aim was to go to Princeton for a year, in order to be able to work in the atmosphere prevailing there. The financial side of my plan is not without hope. I will most probably get the fare and a little money from a newspaper, for which I write popular articles on physics.

As we learn from *Quest*, this newspaper—curiously enough—was a new paper, set up and subsidized by the government, *Gazeta Polska*, and Infeld had some hesitation about writing for it (Q, pp. 219–221).

> As far as I can see, a quite small aid from the side of the Institute in Princeton would make my stay possible. To speak concretely: a small stipend of $300 would allow a stay of half a year, one of $600 a whole year's stay. But even a small aid would also be essential for my plans. If that doesn't work out, I would still try to come to Princeton, since I really want to.
>
> I am still working on the new electrodynamics and have found some results lately. Born, with whom I am very friendly, will also write you about this matter. It will take time, however, because as you know he is in India. I speak good English, and intend to take my wife with me, who is an Englishwoman.

I would ask you strongly in any case to send me some sort of formal letter, which would facilitate the overcoming of formalities (visa, leave, etc.). I know very well that you will do what is possible. I would like to have the answer as soon as possible so that, in case help from the Institute is not possible, I still have time to think of something new. My goal is to come to Princeton for the next academic year (1936–37).

It is almost unpleasant for me to send you banal assurances of my gratitude. I have a bad conscience, that I am again imposing on your kindness.

Einstein replied on 18 March:

I spoke with Professor [Oswald] Veblen about your proposal; he is the most influential of the group of professors who are much occupied with administrative questions. He thought it would be possible that we invite you for a half or a whole year under the conditions proposed by you. But in any case, a decision on this is necessary at the next meeting of Professors, that will probably take place in a couple of weeks. Anyway, it would be justified that you soon assure your travel leave.

I would be pleased if you came here, especially since you are occupied with the same problem as I am. Reciprocal stimulation and criticism is always good.

On 4 May, Einstein was able to report:

The matter was somewhat delayed by an ebb in the Institute's assets. Now however I am happy to inform you that the $600 has been granted. My colleagues and I will be pleased to have you here during the next academic year (beginning of October).

The two soon started to collaborate, first on the problem of gravitational radiation—he and H. P. Robertson managed to find an error in Einstein's proof that it does not exist (Q, pp. 260–270); and then on the problem of motion in general relativity, the problem on which they continued to work for a decade and more. The story of their collaboration, as seen from Infeld's side, is told in *Quest*, and Peter Havas has provided additional material in his article "The Early History of the 'Problem of Motion' in General Relativity" (in D. Howard and J. Stachel, eds., *Einstein and the History of General Relativity*, Boston, Basel, Berlin: Birkhäuser, 1989, pp. 234–276); so I shall not discuss it here, except as it relates to the correspondence. By spring of the following year, it became clear that Einstein was not able to secure Institute support for Infeld for a second year. As he told Infeld: "My fame begins outside Princeton. My word counts for little in Fine Hall" [then the site of the Institute] (Q, p. 302).

Infeld proposed that he finance his further stay by writing another popular book—this time with Einstein.

I knew him well enough to understand that he would never lend his name to ghost-writing. A book with Einstein's name would really mean a book written together. ... I knew that if the book was to have any real historical value I must remain in the background and let Einstein express his views. It was important that the book should express Einstein's outlook on science. ... Next was the problem of the actual labor of writing, the tedious mechanism of popularization which would take a great deal of time. Here, I was sure, I could do better than Einstein and would be able to relieve him of most of the work (Q, p. 308).

After encouragement from Robertson and Melba Phillips, Infeld went to Einstein, but found himself almost tongue-tied.

> I wanted to explain my plan clearly and in logical fashion. ... But, apparently for no reason at all, I could not talk; my well-prepared speech went to pieces, and after a few meaningless phrases, 'It is difficult to explain ... I hope you won't misunderstand me ...' I gave it up.
>
> Einstein looked at me in utter astonishment. He had never heard me stutter or found me unable to express myself. ...
>
> 'For goodness' sake, shoot out what you have to say, I am beginning to be really interested in what it is.'
>
> I gathered my courage and began an incoherent explanation, finally making myself clear by repetition, by traversing the subject again and again. ...
>
> Einstein looked at me silently, stroked his moustache with his finger and then said quietly: 'This is not at all a stupid idea. Not stupid at all.' Then he got up, stretched out his hand to me and said: 'We shall do it.'

Thus was conceived the famous *Evolution of Physics*, published by Simon and Schuster in 1938 and still in print—and still used by many of us in popular courses on physics. The work, as described in *Quest*, proceeded rather speedily, and by the time Einstein left Princeton for his summer holiday on Long Island, it was essentially finished, it being left for Infeld to write up their discussions of the last two chapters.

> During the vacation months Princeton is one of the most uncomfortable places in the world. The town is dead. The few remaining inhabitants, moving languidly over the burning pavement, repeat over and over in pathetic voices that the humidity is to blame. I sweated and drank water, drank more water and sweated, got up at five to write my thousand words before the burning sun made a furnace of the town (Q, p. 315).

Finally the manuscript was finished and sent off for Einstein's imprimatur. By 27 August 1937, Einstein was able to write Infeld:

> I marvel at the energy, with which you have brought our worldly child into the world and outfitted it. The letter to the Dutch publisher [of a German translation] is going out with my signature by the same post. Cash the check and keep it as the first compensation for your pains.
>
> I am only returning to Princeton around the 19th of September and am already anticipating with pleasure the interesting work that lies ahead of us. I am now certain that we shall arrive at a decisive result.

The book was soon published, and a vigilant reader soon caught an embarassing error. Mr. J. T. Hayward, Vice President of the Barnsall Research Corporation, Petroleum Building, Tulsa, Oklahoma, wrote Einstein on June 15, 1938:

> In the second paragraph on Page 190 of your and Prof. Leopold Infeld's recent book ... the following statement occurs: "In viewing the setting sun, we note the event eight minutes after it has taken place." I have puzzled over this statement, and believe that it is possibly a slip of the pen. "The event" presumably refers to the setting of the sun, and the delay in perception would be equal to the velocity of light divided by the distance between the eye of the observer and the horizon [sic].

Again on holiday on Long Island, Einstein wrote a detailed reply to Mr. Hayward on June 20, sending a copy to Infeld with a note:

> A reader has informed me correctly, that a complete mistake about the time of sunset occurs on p. 190. It is vexing that this slip eluded us. Naturally, it must be removed from future editions.

In 1938, Infeld was offered a lectureship at the University of Toronto, where J. L. Synge was head of the department of applied mathematics.

> Einstein, Robertson and everyone else with whom I discussed it strongly urged me to accept the offer from the University of Toronto, although Einstein remarked many times how much he regretted that we should have to interrupt our collaboration (Q, p. 323).

It was during "the vacation period after the second year in Princeton" that Infeld "lived through the dramatic epilogue of a relationship which I have consistently omitted from my book" (Q, p. 323): he separated from his third wife, whom he had married in England. Einstein alludes to this in the next letter to Infeld, written 22 February 1939, when Infeld was already in Toronto.

> I am very happily looking forward to the prospect of seeing you again in April. As concerns the separation, this is a curious matter, in which neither the direct participants nor bystanders are sure of what sort of a face to put on it. Although it isn't the same, yet it calls to my mind again Schopenhauer's elegaic-sentimental dictum: The man who marries for a second time is not deserving of the death of his first wife. You are thus so to speak still taking the middle course.

Presumably Einstein's remarks have more to do with his own marriages than with Infeld's. They seem particularly inappropriate given the tragic circumstances surrounding the death of Infeld's second wife, Halina—but I don't know whether Einstein was aware of this.

Einstein went on to remark:

> Robertson told me that you are working on freeing the treatment of the problem of motion from coordinate conditions. That would be pretty. I see that this works for both integral conditions, but I still don't see how one can obtain the successive approximations in the calculation of the field in this way.

This apparently refers to the second paper on the problem of motion, in which the equations of motion are solved without distinguishing a special coordinate system: "Nothing is assumed in advance about the coordinate system except that it is galilean at infinity" (*The Gravitational Equations and the Problem of Motion II*, EI Bibl. 37, p. 455).

There follows an exchange of letters about this work, including: AE to LI, 2 March 1939; LI to AE, March 7, 1939; followed by a meeting in Princeton, described in *Quest* (p. 328). Apparently in the course of this trip to the United States he married Helen Infeld, a mathematician whom he had met seven months earlier at a meeting of the American Mathematical Society (Q, pp. 323–324, 326).

Perhaps he took her to meet Einstein; at any rate, he writes in his next letter: "I regret very much that we could speak to each other so little last time. I intend to come to Princeton again" (LI to AE, 19 April 1939). This letter was written while Helen and he were living in her New York apartment during his five-month summer vacation (Q, p. 326).

There follow four further letters about work on this paper: AE to LI, 22 April 1939; AE to LI, 25 April 1939; LI to AE, 27 April 1939; and 2 May 1939. By this time, the paper was nearing completion:

> I agree with your improvements, and believe that with these the train of thought has been brought to its most complete form. Somehow, this should be published in the Annals [of Mathematics], since the entire train of thought will thereby now really be made easily accessible. If you prepare it, we could then best do this together. It is only too bad that working together has been made so troublesome due to the spatial separation.

On 23 October Einstein wrote to Infeld—in English, for the first time. The shadow cast by the outbreak of war on Sept. 1, discussed so eloquently by Infeld in *Quest* (pp. 6–11), also falls on this letter:

> It is strange that the 'Annales' have not yet sent any proofsheets of our paper. I shall inquire occasionally about it at Dr. Lefschetz's office.
>
> I am very glad to learn that you are so happy with your wife. But I can imagine also how worried you are about your sisters in Poland. I hope that women are not so endangered in such situations. There is nothing one can do against those gangs of scoundrels. But it seems to me that destiny is en marche!

Infeld's and Einstein's worries were indeed justified:

> My younger sister [Bronia], who was also the best friend I ever had, vanished suddently in Cracow and no one knows how she died. [See also the essay "Bronia" in *Why I Left Canada*, pp. 123–129.] Her husband was shot on the street and their young son died of illness and hunger. My older sister [Fela] died in Bergen-Belsen . . . but I had better stop (Q, Epilogue, p. 353).

But in Leopold's home events followed a happier course, and on 18 January 1940 Helen Dukas, Einstein's secretary since 1928, wrote:

> Congratulations to the heir! [Eryk, of course] I can imagine your joy and happiness, particularly of the proud father. I hope everything went well and the family is quite allright again. Dear Ludwik, please write me soon and tell me all about it.

Einstein added a handwritten postscript: "Hearty congratulations! The youngster was earlier than the Annals of Mathematics."

The proofs finally arrived, and on 8 March Einstein sent them on to Infeld, with some suggestions for further changes. Then he went on:

> Your letter came today, and I am happy that all is going as you wish it, especially also with your little son. At Wigner's request, I spoke yesterday about the problem (in the colloquium). I only lectured about the thing in first approximation, so that the clouds of mathematical formalism don't hide the main idea. I succeeded in making the thing really understandable and I saw from the discussion that interest in it was exceedingly lively. We would have done well to have treated things this way in our first paper [*The Gravitational Equations and the Problem of Motion*, (EIH), EI Bibl.

34], because otherwise one can only see the wood for the trees with difficulty. But unfortunately this can no longer be changed.

Infeld sent back the proofs in an undated letter, probably from May 1940, agreeing to the suggested corrections, and adding some political comments.

> The political situation grows darker with each day. People who were very little concerned about the struggle in Spain [i.e., the Spanish Civil War of 1936–1939] are crying about the fate of the Finns [a reference to the Russo-Finnish War of 1940] and Mannerheim is a hero of democracy [a reference to the Finnish leader, who had presided over the White terror after the Soviet government gave Finland its independence in 1918]. The statements of the Communists exhibit a standard that is frightening [presumably a reference to what he later recalled as his 'revulsions against' the Party during the Molotov–Ribbentrop pact (Q, Epilogue, p. 351)]. Politically, one feels isolated and 'confused' just at the time when I would like to stand strongly on one side. All of this is covered up for me by the calm and good atmosphere in our home, but it emerges from time to time and even very strongly.

After mentioning some of his current work on the problem of motion in electrodynamics (EI Bibl. 36, with P. R. Wallace) and on Brownian motion (EI Bibl. 38), he goes on:

> Only now after being away from Princeton for two years, do I see how much I learned from you. I understand quite well, that you cannot report to me in detail on your work. Every day brought new attempts, new hopes, and also new disappointments. The endeavor and the work were so "exciting" that I even forgot to be concerned about if and when something would result from it. I have learned from you what it means to attack the most difficult problems and to have the courage for it. Although, like a de-excited atom, I have sunk back again to my normal level here, still I now know what it means to collaborate on great problems, and perhaps I will once again succeed in working near you.

Almost a year later, on 6 March 1941, Einstein wrote: "Our work on motion has met with considerably more interest than we expected at the time. ..." He added one of his now ever more-frequent pessimistic reports on the progress of the search for a unified field theory:

> Our attempts to set up a useful [unified] field theory have not led to any results. I am more and more inclining towards the viewpoint that one cannot make further progress with the theory of the continuum, because in it the Riemannian metric imposes itself almost necessarily as the only natural conceptual structure. Our attempts at a more general conceptual structure have had no success at all up to now.
>
> I have just received your book [*Quest*], and will look at it with pleasure.

Infeld replied the same month:

> I am extremely anxious to know whether you liked my book and especially my treatment of anti-Semitism. About you, as about myself, I wrote just as I thought and felt. My publisher thought that I made you too human which he regards as a small objection and I as a compliment to me.

Einstein replied, probably in April 1941:

> Your book is excellently written. I read through it completely with close attention. On the other hand, I must tell you that it is not right to publish remarks made in private by personal friends. Usually, one even asks for permission before one prints public statements. Just imagine how embittered, for example, Loria would be, if he happens to see such indiscretions and sees them put before his countrymen. Have you thought about how the professors here will behave, after your comments on Princeton and the Institute, if Infeld II wants to come to the Institute. If you had asked me ahead of time, I would have energetically advised you against publication. One should really not undertake anything that threatens the weak bridge of trust between human beings.
>
> Now since it has happened, don't have too many afterthoughts. It is meritorious to pitilessly expose wrongs and mendacity. And the grass grows quickly over what has already happened, especially in America.

Infeld must have reacted with alarm to the implications of this letter, for Einstein wrote him on 22 April:

> You mustn't take my criticism too seriously. There is no reason to decline an invitation to Princeton. Anyway, the book is so well written that in general, sympathetic interest will preponderate for most people. Also, in general the feeling of the inviolability of the personal sphere is not so strong here as it is for me.

On 2 May 1941 Einstein wrote to ask Infeld's help in finding a position for Leopold Halpern, "who, through an intrigue that is impenetrable for me, has lost his position at New York University." Aside from a short note of thanks for a photograph of young Eryk, appended to a letter by Helen Dukas (8 August 1941), the correspondence seems to have lapsed for several years, only resuming at the end of World War II.

On 30 October 1945, Infeld wrote to Einstein:

> First, I want to congratulate you on the fine statement on the atomic bomb. I read it with great relief, happy you spoke up so strongly and clearly for decency and world-cooperation. Perhaps with the exception of Prof. Urey's, all the other statements lacked clarity and definiteness.

This may be a reference to a letter to *The New York Times*, published October 10, and signed by Einstein and 19 other prominent persons (see *Einstein on Peace*, pp. 340–341).

After first explaining his long silence—"For the last few years I had a strong desire to ask your opinion on the things I was working on. But I did not want to take up your time"—Infeld went on to discuss his recent work on cosmology, done together with his student Alfred Schild (EI Bibl. 44–45; the discussion continued in the next few letters). The letter concluded on a personal note:

> As you see, I am still in Toronto and I like the place. We would be very happy here if not for the horrors of peace; the terrible news from my family. Of all the many people, my sisters, their children, cousins, friends, only one girl—my niece—was found alive in Belsen, and her father in Cracow!

Einstein replied on 29 November, 1945:

First of all my deepest sympathy on the dreadful news that you have also had about the fate of your relatives. The fate of the Jews is horrible and it is clear that the influence of National-Socialist propaganda still presents serious dangers to us for a long time.

Infeld replied on 19 December:

I was very much touched by your sympathetic letter. I find that what happened to my people in Poland haunts me days and nights. Out of my whole family there are only two people left alive, and I am doing my best to bring them to this continent. . . .

Perhaps the only good news that I have received from Poland was that my good friend Professor Loria is alive, and now reorganizing the University of Breslau [Wrocław]. However his letters are a tragic cry for help. There are no libraries, no books, no physical apparatus. He himself spent four years hiding with a farmer ["Despite his Jewish blood he had never denied his origin and, most exceptional among Christians with Jewish blood, he not only was not anti-Semitic but had the courage to fight anti-Semitism" (Q, p. 146).] He is yearning to come for a few months to this continent, and to organize some intellectual help for Poland. I wrote on his behalf to Professor [Karl K.] Darrow [perennial Secretary of the American Physical Society], who approached the Rockefeller Foundation, but they are not ready to intervene now. I wrote also to my friend, Professor Oscar Lange, the Polish Ambassador in Washington, suggesting that in the interests of Poland he would [should] do what he could to bring Professor Loria here for a few months. As you know, the present Polish Government is, on the whole, very progressive, and by far the best Poland ever had, but the situation in the country is still terrible, and the people are rotten and poisoned by Hitler's propaganda.

This is presumably a reference to the grave anti-Semitic incidents that took place in Poland after liberation. It has been estimated that there were 350 returning Jews murdered in 1945 alone, and pogroms took place in Rzeszow (July) and Cracow, Infeld's home town (11 August), culminating in the Kielce pogrom (4 July 1946), which left 42 dead and several dozen injured (see, e.g., Michal Borowicz, "Polish-Jewish Relations, 1944–1947," in C. Abramsky, M. Jachimczyk, and A. Polonsky, eds., *The Jews in Poland*, Oxford: Basil Blackwell 1986, pp. 190–198); and Iwona Irwin-Zaracka, *Neutralizing Memory/The Jew in Contemporary Poland*, New Brunswick, Oxford: Transaction Publishers, 1989, pp. 47–49 and Krystyna Kersten, "The Pogrom of Jews in Kielce on July 4, 1946," *Acta Poloniae Historica*, vol. 76 (1997), pp. 197–212).

I should be very grateful to you if you have any suggestions for help, or would you like to write a letter of a few lines to Professor Lange on Loria's behalf? It would have a great influence on the Polish Government. As you will know I don't like to bother you with such things. I do it only if I am absolutely convinced that you are very sympathetic to the cause as in this case.

Einstein replied on 25 December:

I can well empathize with your pain, especially since a number of members of my family were also killed by the Germans.

I am really shocked that the reaction in this country to these infamous facts is not as strong and spontaneous as one would have expected. . . .

I have sent some words of recommendation for Loria to the Polish Ambassador; from your book I know that he really deserves it.

On 21 April 1946, Einstein writes:

> I have read with great pleasure your excellent article on the atomic bomb [*Atomic Energy and World Government*, a pamphlet published by the Canadian Institute of International Affairs in 1946, based on a lecture Infeld gave "across the length and breadth of Canada, about fifty times" (*Why*. . ., p. 26)].

Einstein went on to discuss his latest attempt at a unified field theory, and apparently sent a copy of the paper, on which Infeld commented in his reply of 25 April:

> Your letter and your paper evoked all the wonderful memories of our collaboration, and I wish that I could have spent some hours in your study. I should like to send my best wishes for a great success in your work, which I believe with you may capture the truth for which you have been looking for such a long time.

Infeld's next letter of 13 July, 1946 is concerned with less pleasant matters. It is on the stationery of the Emergency Committee for Civil Rights, of which Infeld was an Executive Committee member. It is concerned with the aftermath of the defection of Igor Gouzenko, a code clerk in the Soviet Embassy in Canada, who claimed to know the real names of a number of Canadians in a Soviet spy ring, referred to in documents only by code names. Sixteen people were arrested under circumstances that Infeld and other prominent Canadians felt violated their rights under Canadian law. "Among those arrested I found the names of two people I knew well. I had no doubt they were innocent" (*Why*. . ., p. 29), and indeed both were subsequently cleared by the court. Infeld's letter appeals for Einstein's help by signing a proposed statement of American scientists protesting against such violations of civil rights. There is no indication of any action by Einstein on the matter.

The next letter, from Einstein to Infeld, is concerned with Infeld's contribution to the Schilpp volume, "General Relativity and the Structure of Our Universe" (EI Bibl. 53). The letter is dated 31 March 1947:

> At the beginning of your manuscript that you sent to Mr. Schilpp, there is an error, which is all the more disturbing, since you put special weight on this point.

There follows a discussion of Infeld's paper on relativistic cosmology, continued in a missing reply by Infeld, to which Einstein refers in his next letter of 12 April, continuing the discussion. Einstein ended with a political comment:

> I am very happy that you are so active in political matters. I quite agree with your comment about the U.S.A. Such a victory is a dangerous business.

In November of that year, Einstein wrote in support of a Guggenheim Fellowship for Infeld (dated Nov. 16, 1947 in Helen Dukas' hand), perhaps to work on his Galois book, published in 1948 (*Whom the Gods Love*):

> Leopold Infeld was one of the most remarkable men with whom I had the pleasure to work; he has since proved to be a physicist of creative faculty. I feel sure he will create some work of at least considerable educational value for students of science. Intellec-

> tually he is uncompromisingly honest and has a clear understanding for questions of principle. As a writer he is very gifted and his work will doubtlessly be easy reading (without avoidable technicalities) and not superficial.

In 1948, Einstein wrote Infeld in praise of the book; the letter is printed in *Why I Left Canada* (p. 23), so I shall not quote it.

The next sequence of letters is concerned with renewed work on the problem of the equations of motion, work that culminated in Einstein and Infeld's last joint published paper, "On the Motion of Particles in General Relativity Theory" (EI Bibl. 51), published in 1949 in the new *Canadian Journal of Mathematics* at Infeld's suggestion. Since they were not in physical contact, the correspondence about this paper documents their scientific collaboration in more detail than is possible for their two earlier papers. Consisting of some thirty-odd letters, it would itself merit a separate talk. Infeld discusses the collaboration, with citations from some of the letters, in a piece entitled "Einstein," (*Why*..., pp. 143–147), so I shall cite only a few points of human interest from them:

Infeld to Einstein, 18 October 1948:

> It is a great pleasure for me to work with you again and I don't think it can be bad for my ulcers because it makes me happy.

Infeld to Einstein, 25 October 1948:

> Your letter gave me great pleasure. The region of disagreement between us converges quickly to zero.

Infeld to Einstein, 1 November 1948:

> I hope this correspondence is not too great a strain on you. I enjoy it very much, though 550 miles make the progress slow. But, as you once remarked, the world waited for a good theory of motion for such a long time that a few months more matter very little.

Einstein to Infeld, 19 November 1948:

> Collaboration with you has given me extraordinary pleasure, and I believe that neither of us alone could have brought it to completion. For the material is downright insidious.

A number of Einstein's letters from 1949 are also quoted in "Einstein," (*Why*..., pp. 136–152), and I shall not cite these passages. This was the year in which Infeld made his first post-war visit to Poland, where he received an invitation to work for the following year. After his return to Canada, in June 1949, he visited Einstein for what was to be the last time, and told him about his plans to spend time in Poland (see "Einstein," *Why*..., pp. 147–148). Not long after, on June 20, Einstein wrote:

> I have often wondered whether, out of idealism, you might not get too deeply involved with the Polish problem. In spite of great sympathy with the present government in Poland I cannot help being very doubtful about the stability of conditions there. After some time the evil men may emerge from the mouse-holes in which they are now hiding—not much different from what happened in Germany in the twenties. These people will make life very difficult for you.

Infeld's death spared him from the full force with which this prophecy was realized in 1968.

> Even if conditions in the West are most disturbing today it is not to be assumed that the present hysteria will continue for a long time or even develop into an intolerable situation. People are too well off: they are not likely to go to extremes as long as they have enough to eat.

It is clear from the 1949 correspondence that joint work on a new paper was now underway:

> When it appeared that we had come to agreement on all details, again a difference appeared between us. I had already sent Einstein the finished manuscript of our joint paper when we once more failed to understand one another. I felt that we had to meet personally to discuss the new work, which still lies in my files. Unfortunately, something intervened making it impossible for me to see Einstein. Our last work together will never see the light of day. ("Einstein," *Why...*, p. 149).

What happened becomes clear from the correspondence. On 28 September 1949 Infeld wrote Einstein, commenting on the political situation, presumably in response to President Truman's announcement on 23 September that the Soviet Union had exploded an atomic bomb:

> We are coming through a crucial time and in the next few months the international situation must change. I do hope that it will change for the better. In my pamphlet that I sent you some time ago I have predicted that the Russians will have their atomic bomb by 48 and was attacked for this 'pessimistic' prediction by some military men.

In October 1949, Infeld received an invitation to spend the second term of the academic year at Princeton University. His request for a leave was turned down on the grounds that it was too late. Accordingly, Infeld applied in November for a one year leave for 1950–51, making it clear that he also intended to accept the invitation to spend time in Poland (see "Einstein," *Why...*, p. 147). His plan was approved by the university president (see Infeld to Dean Beatty, University of Toronto, 13 September 1950).

On 6 November 1949, Infeld wrote Einstein:

> I wanted to write to you this letter quite a long time ago. But then, unexpectedly, the invitation came from Princeton. I was surprised and happy about it. I looked forward (I can not tell you how much!) to seeing you and learning about the problems on which you are working. As I wrote to Wigner it seemed too good to be true. Briefly speaking I can not come! The dean (luckily you never needed to know what a dean means) said I am indispensible (!) here, that the notice was too short, that there would not have been any difficulty had he known in advance. So I am not coming this year. Whether they will extend the invitation for the next year is up to them and the best thing would be if Princeton would decide by itself without any outside pressure or suggestion. The only thing I can say is that it is almost sure that I would come if they would invite me.
>
> Now I would like to tell you about another thing. For a long time I wanted to come for a day or two, to discuss the problem of motion on which we got stuck. I want to tell you why I did not come and why I do not intend to come in the nearest future. This part of the letter it is *confidential and only for you.* (Of course it is all right for

> Miss Dukas and your household to know it.) I do not know whether you heard that a few people, certainly not communists, were lately not allowed to cross the unfortified frontier. This is rather a long story. My past is comparatively clean, although, as you well know, I am left of Louis XIVth. But as a foreign born, I have to, crossing the frontier carry a passport. In it is the Polish visa. Should have Princeton invited me, I would have needed a visa from Washington, and, so I think, the fact that Princeton invited me + perhaps some intervention from Princeton would have been sufficient to overcome this difficulty. But for a few days (up to 29 days to be exact) I do not need, as a Canadian, any visa. But they can send me back. I should like, therefore, to wait for a good opportunity, or for a time when the tension will ease up, especially as it is hardly possible to become worse than it is.

He then turned to the problem of motion, ending with the following comment:

> There are other interesting results concerning the gravitational radiation. I am sure now that you were right in 1938. If we adopt our approximation procedure, then gravitational radiation cannot exist. My paper (you remember, on the odd power expressions of λ) is wrong. These expressions can be created or wiped out purely by a choice of a coordinate system!

This is apparently a reference to a manuscript sent to Einstein. This letter seems to represent the first evidence of Infeld's dramatic reversal on the question of gravitational radiation. For Infeld's first paper arguing against its existence, see EI Bibl. 58.

A few more letters were exchanged before the well-known events of 1950, which led to Infeld's decision to resign from his Toronto post and stay in Poland (see "Why I Left Canada," in *Why...*, pp. 39–54). I shall not attempt to resume them here. On 5 October 1950 he wrote Einstein for the first time from Hoża 69:

> My family and I are very well off here. I regret very much indeed, that there is little chance that we will see each other again, but unfortunately this would probably have been true if I had remained in Toronto. I doubt very much that I would have been allowed to cross the border after daring to visit Poland. One of the pleasant things here is the absence of many of the worries which consume one's energy elsewhere.
>
> I'm sorry if the reporters bothered you because of me. It is always a great comfort to know that I can count on your understanding.

In his reply of 13 October, Einstein commented:

> In the past man was in the main only the plaything of blind forces—now he is additionally a plaything of bureaucracies, and learns to adjust. Do you know Lichtenberg's saying: 'Man learns little from experience because every new folly appears to him in a new light'?

He also commented on Infeld's application of the slow approximation method to Einstein's new field theory (EI Bibl. 55, "The New Einstein Theory and the Equations of Motion"):

> The attempt to transfer our theory of motion to the new theory would be completely unjustified. In fact, the localization of energy in the non-symmetric theory does not allow a quasi-stationary approximation. This finds expression, for example, in the fact that a homogeneous progressive wave in the antisymmetric part of the field does not

carry any energy at all. Energy transport must be based upon a substructure, to which there is no access based on successive weak field approximation.

Meanwhile, Infeld was in Holland in connection with his position as Vice-Chairman of the Polish peace organization, and wrote to Einstein on 11 October 1950 seeking his support for the efforts of the World Peace Council:

> I, myself, feel happier than I have ever felt in my life, and find the atmosphere here [i.e., Poland] very exhilarating. The people and the government are concentrating on reconstruction and peace. Of course, they make blunders too, but these are of an entirely different order of magnitude than those committed by the other side. But I do not intend to make propaganda. I will only add my voice to others which, I know, will reach you.

But in spite of a second letter asking for Einstein's support of the upcoming Peace Congress (LI to AE, 31 October 1950), Einstein demurred (AE to LI, 13 November 1950, cited in "Einstein," *Why*... p. 151). Perhaps Infeld did not know, or had forgotten about Einstein's unpleasant experience in connection with the 1948 Wrocław World Congress of Intellectuals, when a message he prepared criticising both sides in the Cold War was not read; instead, his letter agreeing to send a message was read as if it were the message itself (see *Einstein on Peace*, pp. 491–496).

The next few letters concern a mixup over the German-language rights to *The Evolution of Physics*. I shall only cite one characteristic sentence from Einstein to the Dutch publisher of the original German-language edition (26 December 1951):

> When Infeld asked me, on the occasion of a request by the American occupation authorities, if I would allow a German translation of the book to be published in Vienna, I gave my permission. I did so, although on grounds of principle I would never have allowed it, after the great crimes committed by the Germans against the Jewish people, if it had been a question of a book of which I was the sole author.

On 23 March 1952, Infeld wrote:

> The atmosphere in the Institute which the government built for me here is excellent. You would be interested to know that we have never been approached for anything for war. The only thing I miss is the opportunity to talk with you ...

On 9 April, referring to the Korean War, he wrote:

> I just came from Oslo where we have a meeting of the bureau of the Peace Council. I wish to tell you that from the very rich circumstantial evidence I am absolutely convinced that the bacteriological war is on though it is on a comparatively small and experimental scale.

Einstein replied only on 28 October, explaining that he had mislaid the letter:

> As concerns the Peace campaign, I cannot take part in it, because in my opinion it is more or less an act of propaganda, i.e., it concerns an action connected with the "cold war." The only thing that could really have practical success is an attempt at bringing together the main participants not a public blah-blah. The blah-blah only has the effect of unnecessarily exciting the opposing sides. I am always reminded of Heine's poem

The Disputation, which closes with the passage:

> Yet meseems it plain as ink
> That the Rabbi and the Monk
> That the both of them they stink.

On the side Einstein has added in his hand, "the poem is worth reading." He continued:

> The requested photograph I send you gladly, and only hope that the currently prevailing wind will not make it necessary for you to carefully hide it from time to time.

This seems to be an allusion to the anti-Einstein, anti-relativity campaign in the Soviet bloc.

On March 28, 1954 Infeld wrote from a Peace Council meeting in Vienna, asking Einstein whether he would accept an award:

> The peace council gives—every year—prizes for achievements in art and science. (The prizes have of course nothing to do with the Stalin prizes.) We are, as I tried to tell you many times a non-communist organization smeared by our enemies as such. (If you would only know how many lies are told about my country!) Now the jury would be greatly honored if you would accept the prize. You know that I am usually presenting to you a case without trying to influence you. Yet this time I cannot refrain from doing it. It may become an important factor for our battle for peace and for our battle of decency in scientifique critique. [This seems to be an allusion to Infeld's defense of the theory of relativity against attacks on it by Soviet philosophers as "idealist."] The others who are nominated for the prize are Chaplin and Shostakovich. Both agreed to accept it.

Again, Einstein demurred, writing on 3 April:

> Unfortunately, I cannot accept the honor intended for me in view of the experiences I had on the occasion of the Congress in Wrocław in 1948, when a message from me was suppressed or rather falsified.

On 2 December 1954 Infeld wrote:

> I am optimistic about the future; more so anyhow than a year ago ... I am feeling happy in my country. We have a lot of young very intelligent and capable men. It is nice to know that no one of them will be looking for a job. There is a lot to do here—too much, sometimes for my health.

Einstein replied on 8 December:

> I am happy about the good news about your life and activities. I share your optimistic outlook on the international situation, and one could hardly have hoped for such a favorable turn.

Both Infeld and Einstein seem to be alluding to the thaw inside the Soviet bloc and the international detente that followed the death of Stalin.

In 1955, a conference was organized in Berne to celebrate the fiftieth anniversary of the special theory of relativity, and Einstein was invited. On 3 January 1955 Infeld wrote Einstein at the behest of a committee preparing to celebrate the fiftieth anniversaries of quantum and relativity theory in Berlin, asking Einstein to attend. "Though I knew there was not much chance Einstein would come to Europe, I wanted to do as the organizing committee asked—for both East and West Berlin" ("Einstein," *Why*. . ., p. 152). Infeld was to lecture on relativity in Berlin, he informed Einstein.

Einstein answered on 17 January, in a letter that ends the correspondence:

> I am unfortunately (or should I say God be praised) not healthy enough any more to appear at such official occasions. I think it would be nice if you make clear in your sermon that the center of gravity of the theory lies in the *general* principle of relativity. Most contemporary physicists still haven't grasped this.

And in spite of the efforts of Infeld and many others, I suspect that this is still true.

Lanczos's Early Contributions to Relativity and His Relationship with Einstein

John Stachel

"... Wenn die Könige bauen, haben die Kärner zu tun." – Schiller
"When kings are at work, the carpenters are kept busy"
(Lanczos's epigraph to 1949b, celebrating Einstein's seventieth birthday!

This paper gives a selective survey of Lanczos's work before he left Germany (cited by reference to his Bibliography in this volume), and an account of his relationship with Einstein during the years 1919–1935, based on correspondence preserved in the Einstein Archive (cited by date and Control Number) and later reminiscences by Lanczos and others. The remarkable human story that emerges from these documents seems worth telling for its insights into two such remarkable yet different personalities.

Early Contacts with Einstein

The first contact between the two is an exchange of letters at the turn of 1919–1920, just when Einstein was becoming world famous as a result of the widespread publicity given the results of the English solar eclipse expeditions.

The 27-year old Kornel Lanczos, of Hungarian-Jewish professional background, was then an Assistant for Experimental Physics at the Budapest Technical University ([5], [7]). He wrote Einstein, enclosing a paper that "attempted to work out a new method in the theory of electrons." He employed quaternion techniques, which was not then quite so unusual as it might appear today. Silberstein's well-known textbook on special relativity [8], for example, made extensive use of quaternions. Apologizing for the fact that "the whole work based on the original [special] relativity principle could appear as an anachronism and the whole attempt as a priori unfortunate," Lanczos confides that he would not have sent it

Brown, J. David, Chu, Moody T., Ellison, Donald C., and Plemmons, Robert J., eds.
Proceedings of the Cornelius Lanczos International Centenary Conference

References to Lanczos's papers are based on the bibliography in this book, pp. lx–lxv in its present "incompleted stage" had he not felt "compelled by external circumstances," the urgent need "for intellectual reasons" to go abroad to study for "a doctorate in theoretical physics, and indeed, if my abilities are sufficient, at the University of Berlin."

Without minimizing his intellectual motives, it is possible that other reasons also motivated Lanczos's desire to go abroad. 1919 was a time of great turmoil in Hungary. Defeat in the first World War and collapse of the Austro-Hungarian Empire were followed by a red revolution that was bloodily crushed just a few months before Lanczos wrote; the Horthy dictatorship was consolidating itself to the accompaniment of a wave of anti-Semitism. His family had earlier "Magyarized" its name, but Kornel's signature, "Loewy-Lanczos," served to indicate the writer's Jewish origins to Einstein; and under the circumstances Lanczos might well have felt it prudent to go abroad.

Whatever his other reasons, Lanczos surely hoped that Einstein and Planck, to whom he also sent a copy of his paper (his letter also mentions Laue and Sommerfeld), would help him to gain admission to the charmed Berlin theoretical physics circle they headed. He asks Einstein "above all whether my feeling that I could do something independent in the field of theoretical physics is not based on a subjective illusion." In spite of these self-deprecatory words—perhaps intended to elicit reassurances—he begs leave to dedicate his proposed dissertation: "To the high standard bearers of constructive speculation: Albert Einstein and Max Planck in most respectful admiration" (Lanczos to Einstein, 2 December 1919, Doc. 15 206). Here we catch a first glimpse of the intensely personal and deeply philosophical motives that lie behind all Lanczos's work in physics, in particular his lifelong preoccupation with general relativity. He later described himself as having "the outlook of a person who took part in the great transformation that came about through the theory of relativity not as formalism but as inner experience" (1931a, p. 112, n. 1).

Einstein took almost two months to reply, praising the work: "I have studied your work insofar as my current overloading permitted me. I feel able to say this much, it is a diligent and original piece of intellectual work, on the basis of which the doctorate should be obtained." Yet he deftly if a little disingenuously sidestepped Lanczos's appeal for his help: "I am myself not at any university and so cannot help with the practical accomplishment of the matter." Einstein also excused himself from commenting on Lanczos's mathematics: "I myself have very little experience in the area of quaternions and currently take the view that one can get along very well with the simple tensor concept." In the light of their subsequent discussions on the equations of motion in general relativity (see below), one remark of Einstein's is particularly striking: "On the question of whether the electron is to be regarded as a singular point or as a singularly-free solution of the differential equations, I have not been able to form a definite opinion" (Einstein to Lanczos, 22 January 1920, Doc. 15 208).

His first attempt to leave Hungary having failed, Lanczos took his doctorate from the University of Szeged under Prof. Ortvay, a student of Sommerfeld. Open anti-Semitism precluded a post in a Hungarian university, and in 1921 he accepted a position as Assistant at the Institute for Theoretical Physics of the University of Freiburg im Breisgau in Germany, where he remained until 1924 ([7]).

He soon met Einstein.

> I had my first discussion with Einstein when I was in Germany in 1921. I had a chance to talk to him for a few minutes between two lectures. Of course, I already had a tremendous admiration for him in my student days, and so I was in a great palpitation of heart when this opportunity came. At that time, I was very enthusiastic, because I had found a way of successive approximations for his famous gravitational equations. And this is what I tried to communicate to him ... and I was, I must confess, rather proud of that. It was therefore quite a shock when he said, 'But why should anybody be interested in getting exact solutions of such an emphemeral set of equations?' I remember very well this word 'ephemeral.' It meant that he did not consider his gravitational equations as the last word. [Indeed, Einstein had already embarked on his thirty-odd year quest for a unified field theory. –JS] Although at the time it was quite a shock to me, later on I realized that it was a marvelous example of his complete lack of dogmatism, even in relation to his own intellectual work ([14], pp. 48–49).

Approximate Solutions and the Cauchy Problem

Lanczos is referring to 1923a, in which he used harmonic coordinate conditions (see below), to simplify the form of the gravitational field equations. As is well known, when these conditions are applied, the remaining terms containing second derivatives yield the (non-covariant) D'Alembertian of the components of the metric tensor. Lanczos takes this form of the field equations as the basis for his method of successive approximations, starting from the flat-space Minkowski metric. He proposes to solve the special-relativistic approximate equations that arise at each successive order by analogy with potential theory. That is, assuming the metric to be positive-definite (or "ellipsoidal" as he calls it) in signature for the moment, he sets up a boundary value problem, remarking that "the transition to the 'one-sheeted hyperbolic' character [i.e., hyperbolic normal or Minkowski signature –JS] of the real world can be carried out without difficulty, with only the corresponding modification of the boundary conditions." While this idea certainly works for analytic solutions, and leads him to some interesting results as we shall see in a moment, it also misleads him into thinking that he has proved that "A matter-free world that is everywhere regular, even at infinity [where it satisfies the boundary condition that the metric coefficients be constant], is necessarily Euclidean." He persisted in this error for a decade. As late as 1930b, he makes the even stronger statement: "As is well known, these equations [vanishing of the Ricci tensor –JS] have no everywhere regular solution, apart from the trivial Euclidean [i.e., Minkowski –JS] solution," and he made similar statements in 1931c. Only in 1932e, written after he had gone to Purdue, did he return to this question with more interesting results. He now realizes that the correct way to approach the question is to study the Cauchy problem (though he does not use the term) for the homogeneous Einstein equations. As initial data, he takes the 3-metric of the initial spacelike hypersurface and its derivative along the unit-normal to this hypersurface. He correctly formulates the constraint equations on the initial hypersurface, as well as the evolution equations, and understands the role of the contracted Bianchi identities in guaranteeing the validity of the constraint equations off the initial hypersurface. He does not seem to realize that Hilbert had worked on the Cauchy problem fifteen years earlier, nor that Georges Darmois had done all of the things mentioned five years earlier, as well as giving the geometrical interpretation of the normal derivatives of the 3-metric as the second fundamental form of the initial hypersurface. But Lanczos did make an important original contribution in this paper by formulating the concept of the number of degrees of freedom of the gravitational field. He gives an argument to show that there are two degrees of freedom per point of the initial hypersurface. While his argument is not without problems, the result is correct, and Lanczos became the first person to wrestle with what later was called the problem of the "true observables" by the Bergmann school. Anyone working on quantum gravity will realize the importance of this result.

Returning to 1923a, Lanczos's "ellipsoidal" method does prove suggestive when he finally turns to the hyperbolic case. He realizes that the boundary conditions for the metric on a hypersphere used in the elliptic case must be replaced by boundary conditions on a light cone in the hyperbolic case. He even considers the analogue of the hypersphere at infinity. "And indeed we obtain the conditions that are needed in physics if we prescribe the boundary conditions on a light cone whose vertex lies very far in the past and with its opening pointing towards the future." One can see here an anticipation of the method of prescribing initial data on a null cone for hyperbolic systems of equations, including the gravitational field equations, that was introduced by Roger Penrose four decades later [6]; and with a little good will and imagination, one can even recognize an anticipation of Penrose's introduction of scri-minus.

Cosmological Problems

Lanczos is well aware of the possibility of cosmological complications, making a comment indicative of the prevailing attitude:

> Such a boundary condition has little significance for the universe, since the 'connectivity relations in the large' [this term for the global topology of space-time comes straight out of Weyl's *Raum-Zeit-Materie* –JS] are unknown. ... It would not be very satisfactory, so to say, to postulate a temporal origin for the universe, while it should expand spatially into the infinite with increasing time. Here, probably, the 'cosmological equations' must enter into consideration.

Indeed, Lanczos was much occupied with cosmological questions at this time. As Ellis pointed out ([2], p. 373), in 1922b and 1923c Lanczos was the first one to show that the de Sitter solution could be written in a form that eliminated the so-called singularity at the "mass-horizon." De Sitter had originally proposed his solution as an anti-Machian metric, that is, to demonstrate that the field equations with the cosmological constant have an empty-space solution. But the static coordinate system in which de Sitter gave his metric has a (coordinate) singularity at what we would now call the event horizon of the central worldline. Einstein, unhappy with de Sitter's matter-free solution because it went against his Machian convictions, seized upon this "singularity" to maintain that the de Sitter universe was not really empty: "Until proof to the contrary, it is to be assumed that the de Sitter solution has a true singularity on the surface $r = \pi/2$ which lies at a finite distance" and corresponds to "the case of a world in which the matter is entirely concentrated on the surface $r = \pi/2$.

Lanczos supplied that "proof to the contrary" by casting the de Sitter metric into a non-static but globally singularity-free form, a form now characterized as a Friedman–Robertson–Walker metric with spatial sections of positive curvature and expansion factor of the "cosh t" form. Friedman's first article on expanding cosmological models appeared just shortly before Lanczos's, which was done quite independently, and at the time Lanczos attracted more notice. For example, in his early papers, Lemaitre acknowledges Lanczos rather than Friedman (see [4], p. 277). In 1923c, Lanczos calculated the red shift associated with his preferred family of non-static geodesic observers, finding the famous linear relation between red shift and distance. This work also contributed to the ultimate realization (also implicit in Weyl ([12], [13]) that a cosmological model does not consist of just a space-time, but of a space-time together with a congruence of timelike worldlines, representing the motions of a class of preferred observers.

In addition to its cosmological results, 1923c also contains an excellent discussion of the red shift in general relativity, including its invariant characterization and an argument why it should be considered kinematical in character (Doppler shift) rather than gravitational (see also [9], pp. 122–123). Lanczos later recalled corresponding with Einstein about the de Sitter solution, which would have been reasonable; but if he did, these letters are not in the Einstein Archive.

Curiously enough, Lanczos gave the non-static global form of the de Sitter metric as an aside in a footnote early in the first paper. Most of the paper is devoted to a discussion of the so-called "mass horizon," a term introduced by Weyl [11] to describe the apparent layer of matter in the original static form of the metric. To investigate this, Weyl removed a "shell" or "zone" of the empty (matter-free) space-time that included the "mass horizon" and replaced it with a fluid of constant density. Lanczos and Weyl now engaged in a controversy over what happened as the thickness of this layer went to zero. Did the resulting surface layer of matter have a finite mass limit, as Weyl claimed, or did it become infinite, as Lanczos maintained? The details of this controversy are rather tedious, but as a result, Lanczos finally realized that they were really dealing with an event horizon for the family of static observers in de Sitter space. He makes an analogy between the de Sitter horizon and the similar horizon in what we now call the Rindler form of the Minkowski metric, pointing out that "here we would also speak of an unreachable 'mass horizon'" (1923, p. 169), which is actually not unreachable since "it could be reached by an observer in a finite proper time" (ibid., p. 187). He makes the important observation that one could always cut out a piece of a perfectly regular space-time with enough symmetry, identifying the two edges in such a way that the metric on the edge is continuous but its first derivatives have a discontinuity in the direction normal to the edge, which corresponds to introducing a surface layer of matter there (see below).

In the course of his work, Lanczos was impressed by "how little one can infer an actual singularity of the field from the singular behavior of the functions $g_{\mu\nu}$ since it may be possible to remove the latter by a coordinate transformation" (translation from [1], p. 219). This remark occurs in his first paper, 1922a (which just precedes 1922b), in which Lanczos introduced harmonic coordinate conditions without realizing that they had been used by de Donder several years earlier. Applying this condition to Droste's form of the Schwarzschild metric, he became, as Eisenstaedt [1] puts it, "the first expert to cast doubt on the reality of the singular character of $r = 2m$." But rather than removing the coordinate singularity at the Schwarzschild radius by the transformation to harmonic coordinates, Lanczos had introduced a "new 'singularity' at a place where the [Schwarzschild] metric was completely regular before." It was this circumstance that provoked Lanczos's above-quoted comment about singularities. Considering his study of horizons in the de Sitter and Rindler metrics, one is struck by how easy it would have been for him to realize the similarity between these surfaces and the so-called Schwarzschild singularity. But hindsight vision is always 20–20, and the fact is that he did not recognize the regular nature of this surface.

Surface Layers of Matter

Returning to the papers on the de Sitter solution, the discussion of the "mass horizon" led Lanczos to carry out a detailed study of surface layers of matter in general relativity. He gives a brief account of the results in the first paper, and as noted above, applies them to a critical study of Weyl's [11] treatment of the de Sitter solution. As far as I know, this is the first treatment of the topic, which since then has been repeatedly discussed in the literature (notably by Papapetrou and Treder, and Israel), often without acknowledgement of Lanczos's work. This provides an example of a metatheorem I once proposed: Anything worth doing once in general relativity has been done (at least) twice by workers unaware of each other's work. I have not carried out an exhaustive proof; but insofar as inductive methods apply, I can supply many more examples, some involving Lanczos, such as his rediscovery of the harmonic coordinate condition already cited, and others we shall see later.

Lanczos wrote up detailed accounts of his study of surface layers of matter in general relativity twice: a manuscript entitled "Untersuchung über flächenhafte Verteilung der Materie in der Einsteinschen Gravitationstheorie" written in February 1922, and 1924b. The *Annalen* rejected his first paper as too "mathematically oriented" (Lanczos to Einstein, 12 October 1924, Doc. 15 209), so he confined himself to the brief account of the necessary results in his first paper on the de Sitter metric, 1922b. However, when another author used the *Annalen* to criticize this work, Lanczos seized the opportunity to submit a fuller account in the form of a rebuttal of his critic. I only mention these details, dating back seventy years and trivial in themselves, to emphasize the eternal nature of the struggle between authors on one side and editors and reviewers on the other, and the kind of ploys that an ingenious author can use to triumph over even a rejection.

The work itself is well done; but to those of us brought up to look at these matters in a more geometrical spirit, the absence of any reference to first or second fundamental forms in a discussion of matching conditions on the hypersurface is striking. Later we shall see further examples of Lanczos's ignorance of what we would now consider elementary points in Riemannian geometry. In spite of all later talk about general relativity having geometrised physics, talk to which Lanczos often contributed (although Einstein did not), it is sobering to realize how little the geometrical spirit had permeated the work of many relativists of that generation.

Tensorial Green Functions

In 1924 Lanczos moved to Frankfurt am Main as an Assistant in the University's Institute for Theoretical Physics (headed by Erwin Madelung) with the intention, which he soon carried out, to habilitate there. Habilititation is the process of becoming a Private Docent, which on the German academic ladder is one rung above an Assistant. It involves evidence of a substantial amount of research beyond the doctorate and the presentation of a *Habilitationschrift*. Apparently as part of this process, Lanczos sent Einstein his next work, 1925a. What he particularly recommends to Einstein's attention, however, is not its application to weak gravitational fields, but the method developed in the paper for solving tensorial differential equations.

> Herewith, permit me to send you a paper that I allow myself to recommend to your esteemed attention. In it I have developed a method for the solution of linear partial differential equations of second order in an arbitrary non-Euclidean manifold, which, if I evaluate it correctly, could be of great significance for the theory of relativity on account of its extremely great clarity and lucidity. As far as I know, this train of thought has not hitherto been carried out by anyone else.
>
> The impulse leading to the discovering of this method was a particular problem: the problem of infinitely weak [gravitational] fields when the background field is assumed to be *non*-Euclidean.

The method involves the introduction of tensorial Green functions. As Lanczos discovered in time to add a footnote in proof, he had been anticipated by Hilbert in 1912, who introduced what he called tensorial Green functions in flat space. But Hilbert did not recognize that, under general (non-linear) coordinate transformations, the two arguments of his functions transform independently of each other; that is that these functions are actually what were later called bitensors. Hence, Hilbert's "tensorial" Green functions are actually vectorial Green function, i.e., bitensors of rank 1 in each argument; and Lanczos could legimately claim credit for recognizing this, as well as for applying the method to general Riemannian space-times. Bitensors were later rediscovered and so named by

DeWitt and Brehme in the 1960's, and used extensively in studies of quantum gravity (providing another example of my metatheorem).

With Lanczos's letter to Einstein began another tragicomedy of journal publication. Einstein replied a few days later that the paper was too long and mathematical for the *Zeitschrift für Physik*, and asked whether Lanczos was agreeable to publication in the *Mathematische Zeitschrift* or the *Annalen der Physik* (Einstein to Lanczos, 3 November 1924, Doc. 15 210). Lanczos replied that he was withdrawing the paper from the *Zeitschrift*, and would send it to the *Mathematische Zeitschrift*. He demurred at the suggestion he send it to the *Annalen der Physik*, citing his earlier experiences with that journal (Lanczos to Einstein, 8 November 1924, Doc. 15 211). The next day he wrote to Einstein that he had changed his mind and decided to try the *Annalen*. He asked Einstein to intercede with Willy Wien, then the editor. "Printing would undoubtedly be assured if you would have the great kindness to recommend it in a few lines to Herr Geheimrat Wien." He ends, "Hoping that you do not take offense at this request ..." (Lanczos to Einstein, 9 November 1924, Doc. 15 212). Three days later Lanczos writes again: the editor of the *Zeitschrift für Physik*—the original destination—has written that Lanczos's manuscript was already in press! (It appeared in *Z. f. Phys.* 31 (1925), pp. 112–132.) Lanczos offers Einstein "most heart-felt thanks" for his aid in this matter, but now requests help in getting a more mathematically oriented version of his method published in the *Mathematische Zeitschrift* (1925b).

> In it I would treat the tensor-analytical introduction of the Green tensor, and in connection with this the setting up of tensor integral equations. As starting point, I could for example select the problem of solving the vector potential equation $\Delta\Phi_i = \rho_i$ in general relativity [i.e., in a curved space-time].

Lanczos must have thought Einstein might be irked by his repeated requests:

> I really would not be so pressing, but your esteemed support in these matters might be of great significance to me (perhaps also in my external advancement, which has finally come to the point of my habilitation) (Lanczos to Einstein, 12 November 1924, Doc. 15 213).

Over thirty and still on the bottom rung of the academic ladder, Lanczos aspired to move one step up. At forty-five, Einstein was at the height of his influence as an insider in the German scientific community. His word counted for a lot in decisions about publication, about promotion, about obtaining a new position in the academic world. He was already subject to anti-Semitic attacks, even within the academic community, but the Weimar establishment, both scientific and governmental, supported him vigorously—and gained much international prestige from his support of it. So Lanczos was far from alone in his role of supplicant for Einstein's support. The contrast with Einstein's situation in the United States in the 1930's, to which I allude later, is striking.

Equations of Motion

The next major topic in general relativity to which Lanczos turned his attention was the so-called equations of motion problem. At least since the triumph of Lorentz's interpretation of Maxwell's theory, the treatment of field and matter in electrodynamics had been dualistic. They were taken as conceptually independent entities. How matter acted as a source of fields is described by partial differential equations for the field variables (Maxwell's equations). How fields affected the motion of particles of matter is described by total differential equations of motion for the particles in the field (Lorentz force law). *A priori*, these two sets of equations are quite independent of each other. Many people were dissatisfied with this dualism, and hoped to find a unified treatment, explaining the fields entirely in terms of matter ("Mach's principle"), or matter entirely in terms of fields (unified field theory). But even if one accepted the dualism of matter and fields, soon after the general theory of relativity was developed, the question was raised whether in this case it might not be possible to deduce the equations of motion for the sources of the field from the field equations themselves. General relativity has two features that suggested this possibility: the field equations are non-linear, so that the superposition of two non-interacting solutions does not yield a new solution. But even more significant is the fact that the left-hand side of the Einstein equations obeys four identities, the contracted Bianchi identities, which impose an integrability condition on the right-hand side of the field equations: The covariant divergence of the stress-energy tensor must vanish. This condition is closely related to the conservation laws of energy and momentum for all sources of the gravitational field, which in turn are closely related to the equations of motion of these sources. So it is not surprising that, soon after general relativity was formulated, people like Eddington, Weyl, De Donder, etc., turned their attention to the problem of deriving the geodesic equation of motion for a particle in a gravitational field, originally postulated by Einstein in 1913, from the field equations themselves. (For a review of work on the problem of motion before 1950, see [3], on which I draw extensively.)

Einstein (with his collaborator Grommer) only turned to this problem in 1927, and it seems no accident that Lanczos did so later the same year (1927a). Both papers are marked by the complete absence of references to earlier work on the topic, Lanczos even citing the Einstein–Grommer paper as "the first time an attempt was made to transcend the dualism between field equations and dynamics" (p. 775), I don't know if anyone complained to Lanczos, but Einstein's lapse occasioned a strong letter of protest by Weyl. This letter elicited a curious reply by Einstein (see [3], pp. 246–248), and a reference to Weyl in Einstein's second paper on the subject, which also refers to Lanczos 1927a.

Both Einstein–Grommer's and Lanczos's work have this in common: They treat the particle as a singularity in the field, and attempt to obtain the equations of motion by imposing conditions on the exterior field in the neighborhood of the singularity. The paper by Lanczos (cited above and summarized in a talk published as 1927b) gives a derivation of the geodesic equation that is based on the form of the field equations he had established in 1922 using the harmonic coordinate conditions. In the words of his own summary in his second paper on the subject, he "tried to treat the dynamical problem ... by regarding it as a boundary value problem, which is solved by the interpolation of a weak external field by means of a transformation to accelerated coordinates" (1930b, citation from footnote on p. 531). But it seems to me that there is a problem here. Lanczos wants to set the boundary conditions on a small sphere around the particle whose equation of motion is being derived. What kind of conditions does he set? He prescribes the values of the metric tensor and their derivatives on the sphere, and says that the resulting matching conditions must hold only for "a very short time," but that "by a displacement of the origin ... the process can be infinitesimally continued." If I understand this correctly, this means that the boundary is a timelike world tube around the point particle; and the boundary conditions given are the metric and all its derivations, including those in directions normal to the tube. But this is not a well-posed problem for either hyperbolic or elliptic partial differential equations.

Lanczos ends up reversing his earlier Machian standpoint. "The matter tensor loses its primary significance for dynamics, into its place steps the moving field. The field is not carried along by the matter, but rather the matter by the field, where it is possible that 'matter' is to be understood as only the central portion of the field" (p. 789). It is not clear to me what a "moving field" means in general relativity, where the points of the manifold are only physically individuated by the metric tensor field; but at any rate Lanczos was now on the threshold of his adventures with unified field theory.

Assistant to Einstein

Apparently Einstein and Lanczos met in 1928, at a time when Einstein was looking for a successor to his long-time mathematical collaborator, Jakob Grommer, who was on the point of leaving for a university post in Minsk ([5], p. 488). Lanczos stated that "it was at the intervention of Szilard, the Hungarian physicist"—who was then working at a Kaiser-Wilhelm-Institut in Berlin and collaborating with Einstein on numerous inventions ([5], p. 489)—that he got this opportunity ([14], p. 49). Einstein was sufficiently impressed by Lanczos to write Lanczos's chief in Frankfurt, Erwin Madelung:

> When I recently discussed some problems in the theory of relativity with Mr. Lanczos I got the impression that joint work between him and me during the coming year might be profitable. My request is accordingly that you would bring it about that Mr. Lanczos is given the opportunity to take a year's leave to stay here in Berlin (Einstein to Madelung, 29 September 1928, Doc. 15 218).

Madelung replied good humoredly:

> I'm very cross with you because you want to unharness my dear colleague Lanczos. But it goes without saying that I support your project. Lanczos will have more working with you and also elsewhere in Berlin than we can offer him, and his progress is more important than my interests. Lanczos will thus be granted leave as Private Docent and Assistant here. His place will be kept for him and a younger man will substitute

for him. I have in mind the young Dr. Bethe, who has just gotten his degree with Sommerfeld ... (Madelung to Einstein, 6 October, 1928, Doc. 15 220).

Einstein must have also written to Lanczos, for on 1 October Lanczos replied:

> It goes without saying that I accept your kind offer with very great pleasure... Madelung is handling the matter from a completely generous standpoint. He repeatedly emphasized that in judging the matter only one thing, my personal interests, should enter into consideration (which is certainly not at all problematical in this case) and all other University and Institute interests, etc., are completely indifferent (Lanczos to Einstein, 1 October, 1928, Doc. 15 216).

Since Bethe's home was in Frankfurt, he was able to take over Lanczos's duties quite soon, and Lanczos left for Berlin at the beginning of November (Lanczos to Einstein, 6 October, 1928, Doc. 15 217).

Accordingly, Einstein applied to the *Notgemeinschaft Deutscher Wissenschaft*, an organization that gave financial support to German scholars, for a one-year research stipend to begin in November. According to the letter, Einstein wanted Lanczos to work on "the further extension of general relativity, in particular the problem of gravitation and electricity," that is, unified field theory. (Einstein to the *Notgemeinschaft Deutscher Wissenschaft*, 8 October, 1928, Doc. 15 221). Having heard nothing by January, Einstein complained to the *Notgemeinschaft*, noting that since November he had been paying Lanczos's salary from his personal funds (Einstein to NDW, 17(?) February, 1919, Doc. 15 223). The President of the *Notgemeinschaft* replied that its research stipends were not supposed to be given to foreigners, and since Lanczos was Private Docent at a Prussian University, the matter had been referred to the Prussian Minister of Science, Art and Education to decide whether an exception should be made in this case. A reply had just been received (NDW to Einstein, 19 January 1929, Doc. 15 225). A letter was sent by the NDW to Lanczos on the same day, with carbon copy to Einstein, informing him of the decision to grant him the stipend, which however was a little smaller than Einstein had asked (NDW to Lanczos, 19 January 1929, Doc. 15 226). Einstein's reply to the NDW is interesting for those who think Einstein didn't have a temper, or would like to see him on his high horse:

> ... Naturally I cannot judge whose fault the dragging out of the matter is. The fact of its dragging out remains indisputable. I put emphasis on this statement, because I am convinced that the large sums of money entrusted to the *Notgemeinschaft* can only bear adequate fruit if the distribution mechanism works not only reliably but rapidly. Otherwise people with initiative and vigor cannot base their scientific endeavors on the *Notgemeinschaft* , but must make desperate efforts to procure the means elsewhere. In this case the Notgemeinschaft would have forfeited in the main the justification for its existence (Einstein to the NDW, 23 January, 1929).

Returning to the calmer realms of research, it is clear in retrospect that, to some extent, Lanczos and Einstein were working at cross purposes during their year of collaboration. Many years later, in a letter to Einstein biographer Carl Seelig, Lanczos gave an account of their first meeting in Berlin:

We talked briefly about general things, about the technical side of my employment, and then he brought up his ideas concerning distant parallelism [Einstein's current candidate for "the" unified field theory –JS].

I already had a very specific program in mind when I came to Einstein, the so-called "problem of motion" of the general theory of relativity. He had already written a paper with Grommer on this, but the mathematical method used seemed to me insufficient and inconclusive, and I had a plan to illuminate the whole problem from a different side. In this I succeeded; I finally published the results in 1931 in a paper in the *Zeitschrift für Physik* [apparently referring to Lanczos 1930b]. But Einstein had the oddity (a peculiarity which showed itself especially in his later years) that he took almost no interest to thoughts that did not fit exactly into his frame of reference. At that time, I had attacked the dynamical problem using a different method, and there is little doubt that it was at least mathematically superior to Einstein's [probably a reference to 1927a] ... He was looking for somebody who would enter deeply into his train of thought and would not take any detours. But I was working intensely on the "problem of motion," which for the reasons mentioned above did not interest him. On the other hand, I could not become enthused with distant parallelism, since the idea appeared to me too artificial. So Einstein had gotten the impression that I did not have the necessary 'enthusiasm' for the cause, while actually the reasons were more basic and I really was extremely glad to be able to tackle fundamental problems without interference ([3], p. 250).

While there is undoubtedly a lot of truth in what Lanczos remembered over 25 years later, I think that some of his comments were colored by his later experiences with Einstein, one of which will be discussed in a moment. Work on the equations of motion problem and work on the distant-parallelism version of unified field theory is not really an either-or question. While Lanczos may have continued to work on the problem of motion in general relativity, with the primary aim of deducing the geodesic law, or some modified version of it, for the pure gravitational field, Einstein was interested in the problem of deducing the equations of motion for the sources of his unified field and showing that they yielded some version of the Lorentz-force law in the combined gravitational-electromagnetic field. Lanczos himself cites this potential confluence of their interests: "At the time I was very much interested in the equations of motion in general relativity. It was a sideline of his investigation, so he thought it might be a good occasion for us to work together" ([14], p. 49). Their correspondence during Lanczos's Berlin year shows that they did indeed work on this problem. Einstein wrote: "It has now become clear to me how the motoric behavior [i,e., the motion –JS] can be easily treated at those places that show singularities in the first approximation, but correspond however to a singularity-free solution. ... The question is if there are singularities in the first order solutions that one can interpret, as regards their behavior in the field, as ponderable and/or charged masses" (Einstein to Lanczos, 4 October, 1929, Doc. 15 233).

Another account of the first few collaborative meetings between Lanczos and Einstein was given by Lancelot L. Whyte, who was in Berlin in 1928 and based it on information given him by Lanczos at the time. It suggests that Lanczos threw himself into work on the unified field rather more enthusiastically than he later remembered—and suggests another reason for a possible early disillusionment.

> The scheme was that Lanczos would go to him [Einstein] once a week and Einstein would present him with some task which he would then take away and work on.
>
> Well, on the first occasion ... Einstein put before Lanczos a new type of wave equation or field equation and asked Lanczos to see if he could find a solution which should have certain properties. ... Lanczos understood the problem perfectly, was intensely proud of being given such a task by Einstein, and went away feeling very humble and unsure. But he studied the equation and after three or four days—flash!—there came into his mind the perfect solution. It had all the three properties that Einstein had asked for. Now, to Lanczos this was really something very extraordinary, because he was a very humble, religious type of personality and he felt that this inspiration had come from the heavens to him; and that he should be able to bring to Einstein in his first week in Berlin such a success moved him deeply. But he felt very humble about it, and when he reached Einstein ... he said, 'Yes, I have been able to find a solution.' He showed it to Einstein and demonstrated that it had the required three properties. Einstein looked at him and said, 'Yes, very interesting, quite remarkable.' There was a short silence, and then he exclaimed, rather impatiently, 'But don't you see, I gave you the wrong equation. It was quite wrong!' There was a silence. These two highly intelligent and sensitive men did not need to say anything, for they knew what a terrible thing had happened.
>
> What actually followed was that Einstein went and fetched his violin, Cornelius Lanczos went to the piano, and they played Bach together for the rest of the hour. This story I have heard from Lanczos within a weeek or two of it having happened ([14], pp. 56–57).

Residue of Bitterness

But his experiences with Einstein, some of which we have already discussed and some of which will be mentioned below, seem to have left an element of bitterness in Lanczos's attitute to Einstein that coexisted with the elements of undoubted reverence that we have already seen. This bitterness emerged in 1938, after the publication of the famous Einstein–Infeld–Hoffmann paper on the equations of motion problem. Lanczos wrote to H. P. Robertson, then at Princeton and working on the problem of motion in close contact with Einstein and Infeld:

> I cannot suppress some bitter remarks ... that I want to express to you, quite confidentially. I made a number of elaborate studies in the motion problem of relativity [he then lists his papers of 1927 and 1930 –JS] ..., which have some bearing on the subject, and still the authors have not found it worthwhile even to *mention* my name, not even in the bibliography of their paper. [He noted that in 1930 he] was able to show that the 'singularity method' [used by EIH –JS] is *completely equivalent* with the consideration of the inhomogeneous equations [that is, using a non-singular stress energy tensor on the right-hand side of the field equations –JS] avoiding any singularities. I have shown that the divergence-condition [for the stress-energy tensor –JS] actually permits to deduce a law of motion [again, Lanczos makes no mention of earlier work by others along these lines –JS] ... There was not the slightest reason to go back to the 'singularity method' again, applying a perturbation method which can

> hardly be surpassed in clumsiness and inadequateness [Lanczos apparently failed to grasp the significance of the slow-motion approximation scheme –JS]. Prof. Einstein did not like my method from the very beginning—merely by emotional reasons. But it would have been decent at least to *mention* it in this paper ([3], p. 261).

Robertson wrote a soothing reply, noting that "I know definitely that he [Einstein –JS] has a very high opinion of your work" ([3], p. 262). Lanczos replied that he "wanted to apologize for my foolish remarks in my letter before with regard to Prof. E. One should not take such personal matters so seriously and at present the whole cause seems to me rather unimportant" (Lanczos to Robertson, 19 May, 1938, cited from [3] 1989, p. 262). But the residue of bitterness remained.

Unified Field Theory

Returning to 1928–1929, it is clear from their correspondence during this period that Lanczos did work on Einstein's distant-parallelism theory during the year he was in Berlin. Although he did not publish any work in collaboration with Einstein, he did write a semi-popular account of "Einstein's neue Feldtheorie" for *Forschungen und Fortschritte*, a high-level journal presenting accounts of German accomplishments in science (including the social sciences) and technology. Einstein turned fifty in 1929, and the coincidence between this anniversary and his work on a new unified field theory led to widespread national and international publicity; and Lanczos helped to fulfill the demand for comprehensible explanations. His account is sympathetic, and at one point he even notes a contribution of his own (confirmed by a letter to Einstein) to the understanding of some of Einstein's results; but he states clearly that the new theory is far from having proved its validity as an explanation of electromagnetism.

As Lanczos explained, previous attempts at a unified field theory had been based on *abandoning* four-dimensional Riemannian geometry as the basic mathematical structure to define the unified field, either by increasing the number of dimensions (Kaluza–Klein), or taking the affine connection as basic (Weyl, Eddington, Einstein). The idea of distant parallelism is to *retain* four-dimensional Riemannian geometry, but *add* additional structure. In modern language, it adds a unique correspondence between the tangent spaces at all points of the manifold, so that vectors at different points can now be compared not only in length, as the Riemannian metric allows, but also in direction. This permits the statement that two vectors, each attached to a point distant from the other's point, are parallel (independently of any path between the two points). The hope was that the new structural element would allow a natural place for the electromagnetic field, while the original Riemannian structure still allowed a natural place for gravitation.

In 1931a, Lanczos gave a more detailed, technical account of unified field theories based on distant parallelism for the prestigious *Ergebnisse der Exakten Wissenschaften*, a series of volumes sponsored by the journal *Die Naturwissenschaften*, and published by Springer. Characteristically, he begins with a survey of "Solved and Unsolved Problems of the Earlier Relativity Theory," which delineates the motivation for the search for a unified field theory, and the sort of problems that such a theory might hope to solve. He places his comments on the theory in the context of the age-long struggle in science and philosophy between positivism and rationalism. He is aware that the interpretation of quantum mechanics then gaining ascendancy seems to strongly support a positivist viewpoint, but warns against a positivist interpretation of relativity theory or premature attempts to bury the rationalist approach. The editor of *Die Naturwissenschaften*, Arnold Berliner, thought so highly of this Introduction that he reprinted it in his journal "to make it available to a wider circle of readers." Lanczos's unusual talent for presenting philosophically informed accounts of highly technical theoretical issues was beginning to manifest itself and to be acknowledged.

Lanczos himself soon became intrigued by the possibility that the four-dimensional Riemannian structure itself, *without* any new elements, allowed a natural place for the electromagnetic field. This idea was actually propounded a few years earlier by G. Y. Rainich; in the 1960's Rainich's work enjoyed a renewed if brief vogue under the aegis of John Wheeler, who called it the "already unified theory." But Lanczos's first published work on the subject, 1931c, makes no mention of Rainich and is entirely based on Lanczos's own earlier ideas. Lanczos continued to work on this idea after he moved to the United States, but I shall say no more about it here.

Paradise Lost

Returning to his Berlin year, 1928–1929, Lanczos naturally took part in the very active scientific life of that city, including the famed Colloquium at the University and the meetings of the German Physical Society. He became close to Erwin Freundlich, head of the so-called Einstein Tower, an observatory set up largely to conduct astrophysical tests of general relativity. Freundlich invited Lanczos to join his staff, and shortly after his return to Frankfurt late in 1929, he wrote Einstein explaining his hopes:

> I would be really happy if this action initiated by Freundlich were to succeed; the year in Berlin meant a great deal to me, and I would experience the breaking off of all these connections as a very painful loss. Well—one cannot influence his fate, I base myself on only one thing, that Freundlich is a person who does not let himself be turned down so easily, and I am convinced of his good will toward me.
>
> Perhaps you could put in a good word for me with Laue [who was in effect heading the Physics Section of the Kaiser Wilhelm Institute for Sciences –JS]—possibly everything could depend on his advice (Lanczos to Einstein, 27 November 1929, Doc. 15 238).

Less than a month later, Lanczos again attempted to "influence his fate" with a most moving letter to Einstein:

> With heart-felt anxiety I send you these lines. For it is not easy to speak up in one's own cause, but often it cannot be avoided when so much is at stake.
>
> I have—although many weeks have passed since my departure—learned nothing about the fate of Prof. Freundlich's proposal, although I assume that a favorable answer would surely have been imparted to me immediately.

> When I consider that a single word from you at the right moment sufficed to create the scope for my activities in Berlin, then I am surely not mistaken when I rate your influence on the matter now pending as very great. And if I here once more put myself forward, I do so in the belief that perhaps you are not quite aware how much the success of Freundlich's project matters to me, how much such a position would suit my entire scientific and human nature. I can hardly think of anything more in accord with my nature, although the initiative for this plan in no way came from me, but rather from Prof. Freundlich and his collaborators.
>
> You know me well enough to know that I don't overvalue my abilities, that I attribute to myself only a quite average talent, which however is of a specific sort and utilized in the right place can evidence itself positively. And in its whole design, a pure research institute is much more salutory for me than the usual university business, where one is burdened with all sorts of petty tasks.
>
> Even if you are not completely convinced about the necessity of the reasons that can be objectively stated, I ask you to help me for humane reasons. You know how essential subjective factors are; in particular mutual trust, personal sympathy, often playing a codetermining role in the success of scientific work. And now there is the fortunate circumstance that from the beginning I made close contact with the entire Freundlich circle, which doesn't always happen so spontaneously, and if it is there at the beginning is to be valued as a strong positive feature.
>
> I wanted to submit this request to you, although I know that you are heavily overloaded with other matters. But just for that reason I wanted to briefly summarize what this means to me, so that you can clearly see and make your judgement accordingly (Lanczos to Einstein, 21 December 1929, Doc. 15 237).

A copy of Einstein's letter to Laue is not in the Einstein Archive, but Laue's reply makes it clear that Einstein did not write in the spirit Lanczos had hoped.

> Heartfelt thanks for your friendly letter of the 25 of this month. I am also not in favor of the project to bring Lanczos here, namely not in favor of using the funds of the K-W-I to pay him. ... I do not consider a position such as Lanczos is now trying for to be the best one for him. Since I heard him speak at the Physical Society about his development of Dirac's theory, I am of the opinion that one should make his excellent, really quite unusual teaching ability available quite soon to a university. On the other hand, I do not have such a high opinion of his research. If he could effect an exchange of gifts with Nernst so that Nernst gave up a part of his sense of reality in return for some mathematical insight, then perhaps he [Lanczos –JS] would become a really great physicist. But unfortunately he still lacks something for that (Laue to Einstein, 27 December 1929, Doc. 15 239).

Einstein broke the news to Lanczos:

> I have discussed your affairs with Laue. We have come to the conclusion that no funds are available for appointment at the Institute. On the other hand, we shall undertake that you obtain a teaching post at the university in the near future. In fact we both feel that clarity of thought and teaching ability are yours in uncommon measure (Einstein to Lanczos, 28 December 1929, Doc. 15 240).

In spite of this assurance, Lanczos did not receive an appointment as Associate Professor at Frankfurt until 1932, by which time he was at Purdue.

At the time, Lanczos must have experienced his inability to return to Berlin as an expulsion from Paradise after being allowed a brief glimpse of it. Ironically, within a few years Berlin was no paradise for any Jew—Einstein included—and in 1933 Lanczos might well have considered himself lucky to have gotten a head start on the exodus of Jewish and anti-fascist intellectuals from Germany. But, as his next move shows, around 1930 he seems to have been looking for a way out of his situation.

Move to America

In 1931, a friend of Lanczos named W. Maier (*not* the Walther Mayer who collaborated with Einstein), a professor at Purdue University, received a one-year Rockefeller Stipend, and recommended Lanczos to replace him. The Director of the Physical Laboratory at Purdue, K. Lark-Horovitz, was also interested in Lanczos's work and wrote Einstein for a recommendation, explaining that he and his colleagues were satisfied by what they knew of Lanczos's work but, "as you may know, in many American Universities and also here such proposals are not put to the faculty, but directly to the President and the Board of Trustees." A recommendation from Einstein would impress such dignitaries (Lark-Horovitz to Einstein, 28 April 1931, Doc. 15 241).

A few days later, Lanczos wrote Einstein explaining the circumstances, ending with some intensely personal comments:

> I have been living for some time already under a strong depression of spirit; getting out of my given situation, setting myself a completely new task appears to me as of fateful significance. Therefore I would greet the opportunity to go to America—even if only for a year, the University here will give me leave—with great happiness and inner enthusiasm.
>
> I planned to go to see you personally, but refrained from doing so in accord with your wishes. It is not easy for me—you will understand that—to request something concerning my own affairs, all the more since I have the feeling that deep in your heart you do not like me. But in the present case it is a question of a decision that is very important to me, in which a few words of recommendation from you will probably suffice to obtain a favorable solution ... Hoping that you will not take this importuning amiss, I remain with constant admiration and respect, etc. (Lanczos to Einstein, 8 May 1931, Doc. 15 242).

It is not impossible that Lanczos appealed to Elsa Einstein, Einstein's second wife, to whom he sends greetings in a number of his letters to Einstein. She appended a note at the bottom of Lanczos's letter (I know of no similar note): "You *absolutely* must do this, he deserves it. After all, he is an intelligent, capable person, even if he couldn't be of so much use to you. Please do this, he also has the feeling that you don't like him, he writes it too." Below this, Einstein noted: "Immediately written in support E[instein]." Lanczos got the job and went off to Purdue, where he remained until 1946. A year later, Einstein wrote supporting Lanczos's promotion at Frankfurt. But of course, when Hitler assumed power, rather than Lanczos returning to Germany, Einstein came to the United States.

Reconciliation

The two did not resume direct contact until 1935, when Einstein wrote Lanczos, commenting on the latest paper by the latter, which was published in the *Zeitschrift für Physik*. Einstein opened with a reproach:

> Dear Herr Lanczos: I was very interested in your latest publication, but I cannot understand how as a Jew you still publish in Germany. This is really a sort of betrayal. The German intellectuals have behaved disgracefully in connection with all the dreadful injustices and have richly merited being boycotted. If foreign non-Jews don't do it, that is sad enough (Einstein to Lanczos, 11 September 1935, Doc. 15 246).

Lanczos replied almost immediately:

> Dear Herr Professor! I was very pleased after a long time to hear again directly from you. ... May I perhaps say something in my defense, how it came about that, in spite of the very understandable ban on Germany, I allowed my work to appear there, without believing that thereby I disavowed myself as a Jew ... From various sources I have gotten the impression that a large part of the German physicists (and also mathematicians) are quite estranged from the present criminal mess and are in no way active in it, even if it is practically impossible to express oneself in any way actively *against* it. Since I consider the *Zeits. f. Phys.* as an organ of the German physicists and not as a German periodical, I felt no inhibitions about placing my work there ... The decisive element however was some practical considerations, which almost excluded any other solution. To wit, while I was well known almost everywhere in theoretical circles in Germany, here in America I am as good as unknown. The scientific periodicals here have a peculiar custom, of grading the intellectual value of colleagues and often rejecting them in an almost schoolmasterly fashion [presumably a reference to the refereeing system –JS]. My experiences in this department have been abundantly negative and in no way confined to mere touchiness. The ensuing difficulties by no means arose from my English, for I have precluded the well-known excuse of 'poor language' by undertaking a thorough revision of the text with good friends. I don't want to complain about America in any way, for I have very good relations at the university here and all colleagues and superiors are extremely well disposed towards me. But in general I have found very little good will on the scientific side and I hardly believe that I could publish a paper of such size anywhere in America without a great waste of time. So this was the motive that decided me for the *ZS f. Phys.*, where I didn't need any special recommendation to be accepted right away without any reservations (Lanczos to Einstein, 14 September 1935, Doc. 15 248).

Einstein sent a second letter to Lanczos on the same day that Lanczos wrote his reply to the first, adding some further scientific considerations. This letter elicted a second one from Lanczos. After replying to the scientific points Einstein had raised, Lanczos struck a more personal note:

> You will understand that I was very moved that you extended so much interest to my work and expressed yourself so approvingly about the contents. In the end we are bombarding more or less the same domain and it must appear really abundantly arrogant if I suddenly turn up announcing I have found "*the* solution" to a problem, which so many outstanding heads have tried in vain. But I believe you will understand my intentions. I have a fortunate disposition, in that I can derive much joy from mathematical illumination—even if it is only a deceptive illusion—and I am not among those

scientists who are plagued by feelings of inferiority. I know quite well of what I am capable, even if my colleagues, as most often happens, overlook my work. But perhaps just this inner security helps guard me from exaggeration in the other direction. I don't believe that I ever commit the stupidity of setting my own accomplishments on a par with your genius, which is quite secure for eternity. That would be quite senseless, even if, for example, the theory I worked out should happen to be a complete success. And yet I have the feeling that, perhaps as a consequence of mutual ["beiderseitig," written over "gegenseitig", probably to soften the word –JS] personal ambitions, that we have become deeply [innerlich] estranged in the course of the last years, while during the year in Berlin we spent together we were often so deeply close to each other. Perhaps also some sort of unintended misunderstandings on both sides have also played a role, as indeed are hardly to be avoided among natures that are basically "loners" [a play on Einstein's description of himself as a loner –JS].

And now all at once your noble and objective manner of reacting and the friendly and direct tone of your letter (even two letters). It has created a new situation for me and freed my heart from an oppressive feeling.

Hoping that all goes well with you, I remain with friendly greetings your devoted Lanczos

Einstein replied:

I was delighted by your friendly letter. There is no trace of an estrangement. I always felt that of all colleagues you are closest to me in your whole mode of thought. If I hadn't felt close to you, I wouldn't have reproached you over the publication in a German periodical. For one only speaks of such things *en famille* [within the family –JS]. So I reiterate to you, that according to my convictions you shouldn't have done so and I beg you not to do so again in the future. If people *here* don't show any understanding of your type of research, then I will do my bit so that at least your papers are published without difficulty. Naturally, all those who are hardly able to see a nose length beyond crude empiricism are opposed to me, and this means just about everyone (Einstein to Lanczos, 23 September 1935).

Among the ironies of this exchange is the fact that, a few years later, the *Physical Review* rejected one of Einstein's papers, provoking a furious reaction—but I must stop somewhere. So I shall end here, on this note of reconciliation between Einstein and Lanczos. His letter shows an Einstein who perhaps had gone as far as he was then able in expressing his emotions in such interpersonal matters. His experiences as a refugee forced to leave Germany and become an outsider in America had taught him some lessons; his experiences as nay-sayer to the chorus of hosannas to quantum mechanics by his fellow physicists had taught him some others.

Lanczos had reached an inner equilibrium and acceptance of his role in science that stood him in good stead for the rest of his years. As concerns his relationship with Einstein, one must bear in mind the strong element of ambivalence in all human relationships, which Freud taught us to recognize—and accept. If signs of inner resentment occasionally broke through Lanczos's placid facade, I do not believe they weaken the sincerity of his reverence for Einstein. As Lanczos well knew, it is a distinctive feature of the Jewish religious tradition always to question, and sometimes even to challenge, the ways of the Almighty.

ACKNOWLEDGMENTS. It is a pleasure to thank Mrs. Alice Lanczos for her kind permission to quote from her father-in-law's letters, and to Hebrew University of Jerusalem for permission to quote from the letters of Albert Einstein.

REFERENCES

[1] Eisenstaedt, Jean, "The Early Interpretations of the Schwarzschild Solution," in *Einstein and the History of General Relativity* (Einstein Studies, vol. 1), Boston, Basel, Berlin: Birkhäuser, 1989, pp. 213–233.

[2] Ellis, George F. R., "The Expanding Universe: A History of Cosmology from 1917 to 1960," in *Einstein and the History of General Relativity* (Einstein Studies, vol. 1), Boston, Basel, Berlin: Birkhäuser, 1989, pp. 367–431.

[3] Havas, Peter, "The Early History of the Problem of Motion in General Relativity," in *Einstein and the History of General Relativity* (Einstein Studies, vol. 1), Boston, Basel, Berlin: Birkhäuser, 1989, pp. 234–276.

[4] Kerszberg, Pierre, *The Invented Universe/The Einstein–de Sitter Controversy (1916–1917) and the Rise of Relativistic Cosmology*, Oxford: Clarendon Press, 1989.

[5] Pais, Abraham, *'Subtle is the Lord'... The Science and the Life of Albert Einstein*, Oxford, New York: Oxford University Press, 1982.

[6] Penrose, Roger, "Null Hypersurface Initial Data for Classical Fields of Arbitrary Spin and for General Relativity," preprint, 1963.

[7] *Cornelius Lanczos 1893–1974*, in *1993 Conference Program; Cornelius Lanczos International Centenary Conference*, [n.p.], Raleigh, 1993.

[8] Silberstein, Ludwik, *The Theory of Relativity*, London: MacMillan, 1914.

[9] Synge, John L., *Relativity: The General Theory*, Amsterdam: North-Holland, 1960.

[10] —— John L., *Relativity: The Special Theory*, 2nd ed., Amsterdam: North-Holland, 1964.

[11] Weyl, Hermann, "Über die statischen kugelsymmetrischen Lösungen von Einsteins 'kosmologischen' Gravitationsgleichungen," *Physik. Z.* **20** (1919), 31–34.

[12] —— "Entgegnungen auf die Bemerkungen von Herrn Lanczos über die de Sittersche Welt," *Physik. Z.* **24** (1923), 130–131.

[13] —— "Zur allgemeinen Relativitätstheorie," *Physik. Z.* **24** (1923), 230–232.

[14] Whitrow, G. J., *Einstein The Man and His Achievements*, British Broadcasting Corporation, London, 1967; New York: Dover, 1973.

Einstein and Bose

John Stachel

Introduction

Abraham Pais calls Bose's derivation of Planck's law "the fourth and last of the revolutionary papers of the old quantum theory (the other three being by, respectively Planck, Einstein and Bohr)" (Pais 1982, p. 425). This is high praise indeed, when one considers that: 1) The 1900 paper by Planck introduced the concept of the quantum of action and derived the law for the equilibrium spectrum of electromagnetic radiation that bears his name; 2) the 1905 paper by Einstein introduced the concept of the quantized structure of that radiation, and derived from it (among other things) the laws of the photoelectric effect; 3) the 1913 paper by Bohr introduced the concept of the quantized energy-level structure of the atom, and derived from it the atomic spectrum of hydrogen. Yet Pais's praise is surely deserved, for as he says "Bose's arguments divest Planck's law of all supererogatory elements of electromagnetic theory and base its derivation on the bare essentials" (*ibid*). In this paper, I shall review Bose's work and his contacts with Einstein, both epistolary and personal. But I will devote equal time to Einstein's reception of Bose's work. I shall discuss how and why Einstein was primed, so to speak, to immediately recognize the significance of Bose's approach and exploit its implications for a quantum theory of the ideal gas, as well as to recognize the limitations of that approach and the need for a deeper insight into the nature of things than the old quantum theory could provide. Finally I shall briefly indicate how the new quantum mechanics provides such a deeper insight, but at a price—quantum entanglement—that Einstein, for one, was unwilling to pay.

Talk at Bose Centenary Meeting, Calcutta, 1994
Max-Planck-Institut für Wissenschaftsgeschichte
Preprint No.4, 1994

Bose Derives Planck's Formula from the Photon Model

Bose's oral reminiscences, recorded by Jagdish Mehra (Mehra 1975, cited here from Mehra and Rechenberg 1982), together with the documentary record consisting of Bose's published papers and his correspondence with Einstein, allow reconstruction of at least the outlines of the path Bose took to his derivation of Planck's formula.

Satyendra Nath Bose must be seen as part of a generation of Bengali students, for whom patriotism and personal achievement were inseparable. He was one of the

> particularly brilliant lot—the famous 1909 batch of Presidency College [Calcutta, which] in all its hallowed history has not seen the likes of this batch since. These inspired youngsters were all destined to contribute to Indian Science, but perhaps their greatest contribution was a growing sense of national pride" (Chatterjee 1994).[1]

After getting his M.Sc. in 1915, Bose continued his study of mathematics and physics at University College, just founded in 1914 as part of the University of Calcutta, and began his teaching career there in 1917 as a lecturer in physics (Venkataraman 1992, pp. 3–9). While his close friend, fellow student, and then fellow lecturer Megh Nad Saha concentrated on thermodynamics, statistical mechanics and spectroscopy, Bose initially "decided upon electromagnetism and relativity." In the aftermath of the stir that followed the announcement in November 1919 of the results of the Eddington eclipse expeditions, Bose and Saha published the first English translation of a collection of original papers on relativity, Bose translating Einstein's 1916 review of the general theory. After an English edition of *The Principle of Relativity* was published, its publisher made an attempt to suppress the similar Indian edition, but as Bose recalled: "Einstein was very generous. He said that so long as the book remained in circulation inside India, he had no objection."

Though initially occupied with relativity, Bose did not neglect statistical mechanics. He recalled studying such books as Planck's *Vorlesungen über die Theorie der Wärmestrahlung* [*Lectures on the Theory of Thermal Radiation*] and *Vorlesungen über Thermodynamik* [*Lectures on Thermodynamics*], Boltzmann's *Vorlesungen über Gastheorie* [*Lectures on Gas Theory*], and Gibbs's *Elementary Principles in Statistical Mechanics*, as well as numerous papers on the topic. In 1918 he published his first paper, in collaboration with Saha, on a topic in statistical mechanics. He also read extensively in the literature of the then rather new quantum theory, to which we now refer as the *old* quantum theory. In addition to Planck's works, his known readings include papers by Einstein, Bohr, and Sommerfeld, and he soon published on that topic.

In 1921 Bose moved to the newly-established Dacca University, where he began to guide students in experimental work, as well as continuing his study and teaching of theoretical physics. He advised his students: "Never accept an idea as long as you are yourself not satisfied with its consistency and the logical structure upon which the concepts are based. Study the masters. These are the people who have made significant contributions to the subject. Lesser authorities cleverly bypass the difficult points" (Mehra and Rechenberg 1982, p. 564). One can hardly imagine better advice to give today to a student of theoretical physics. By following his own advice, Bose made himself a colleague of these masters of theoretical physics. Even though not in direct contact with them, his critical method of approach to their works made him privy to their difficulties, allowed him to share their perplexities, and ultimately enabled him to resolve one of the major contradictions at the heart of the old quantum theory.

The contradiction was in derivations of Planck's radiation law, which gives the energy distribution of electromagnetic radiation in equilibrium at a given temperature T (so-called black body radiation) as a function of its frequency ν. Planck found that:

$$\rho(\nu, T) = (8\pi\nu^2/c^3) \cdot (h\nu/[\exp(h\nu/kT) - 1]),$$

where ρ is the energy density per unit frequency interval at frequency ν and temperature T (c is the speed of light). I have written the right-hand side as the product of two factors, to which I shall refer as the first and second factor, since each plays a different role in derivations of the formula.

Now to the problem that Bose confronted. As he put it in his first 1924 paper, "In every case, the derivations do not appear to me to be sufficiently logically justified" (Bose 1924a, p. 179) because none was based on consistent use of quantum concepts. Each involved classically-derived results along with new quantum ideas. Late on, Bose described in more detail the matrix of theoretical difficulties within which his contribution was conceived:

> I had studied Planck's derivation of his radiation formula. I knew about phase space and Boltzmann statistics from Gibbs ... and Boltzmann. I knew about Einstein's [1916] deduction of Planck's formula and how he had never been satisfied with the derivation of this quantum law; he always kept on coming back to it. Planck's condition ran counter to the classical ideas. Planck was also aware of the difficulties, but he had never been able to resolve them. He had relations which were derived from Maxwell's electromagnetic theory and others in which discontinuities appeared (ibid., p. 564).

Bose's teaching duties continually confronted him with these issues:

> As a teacher who had to make these things clear to his students I was aware of the conflicts involved and had thought about them. *I wanted to know how to grapple with the difficulty in my own way.* It was not some teacher who asked me to go and solve this little problem. *I wanted to know.* And that led me to apply statistics (ibid., p. 565).

In the course of pondering this problem, he began to focus on Peter Debye's 1910 derivation of Planck's law. In the introductory paragraph of his second paper on the topic, Bose described this derivation and its relation to his own work:

> Debye has shown that Planck's law can be derived with the aid of statistical mechanics. However, his derivation is not completely independent of classical electrodynamics insofar as he makes use of the concept of proper vibrations of the ether [that is, normal modes of the field] and assumes that with respect to energy the spectral region between ν and $\nu + d\nu$ can be replaced by $(8\pi\nu^2/c^3)V\,d\nu$ oscillators, whose energy can only

> consist of multiples of $h\nu$. However, one can show that the derivation can be altered so that one does not have to borrow anything from classical theory (Bose 1924b, p. 384).

In other words, Debye derived the first factor in Planck's formula on the basis of the classical wave picture, by counting the number of normal modes of the electromagnetic field with frequencies in the range ν to $\nu + d\nu$ in a cavity of size V. Since the product of the number of such modes times the average energy per mode equals the energy density in that frequency range, the second factor is just the average energy per mode. Debye calculated it by:

1) assuming that, for every frequency ν, the total energy E is divided into energy elements $h\nu$, and that

2) the energy of a mode of frequency ν is an integral multiple of the corresponding energy element $h\nu$;

3) counting the number of ways W that such *indistinguishable* quanta of energy can be distributed among the *distinguishable* normal modes;

4) using Boltzmann's equation $S = k \ln W$ to find the entropy corresponding to each such distribution of quanta of energy; and finally

5) finding the equilibrium state by maximizing the entropy, subject to the constraint that the total energy E remain constant.

In short, Debye's derivation of the first factor is classical, while his derivation of the second factor is based on the hypothesis that there are *quanta of energy*, a hypothesis that from the classical standpoint is entirely unmotivated and even unnatural.

At this point, Bose took the decisive step: he based his derivation of both factors in Planck's law on the quite different *energy quantum* hypothesis that there are *particle-like entities* ("quanta") that can only have discrete energies, that is on what came to be known as the photon picture. To quote Pais again: "Bose's arguments divest Planck's law of all supererogatory elements of electromagnetic theory and base its derivation on the bare essentials. It is the thermal equilibrium law for particles with the following properties: they are massless, they have two states of polarization, the number of particles is not conserved, and the particles obey a new statistics" (Pais 1982, p. 425). In view of the currency of proposals to abandon strict energy conservation (notably the Bohr–Kramers–Slater theory of January 1924), it is worth emphasizing a fifth property: the total energy of the system is (strictly) conserved.

Bose's major contribution was to show that the first factor could be derived on the basis of the photon picture without any reference to classical wave concepts. I quote again from his summary:

> $8\pi\nu^2/c^3 V\, d\nu$ can be interpreted as the number of elementary cells of the six-dimensional phase space for the quanta. The rest of the calculation [i.e., of the second factor] remains essentially the same (Bose 1924b, p. 384).

To expand a little: Bose assumed that the six-dimensional phase space associated with a photon of energy $E = h\nu$ in the cavity is divided into cells of size h^3 (an idea that Planck had introduced earlier in discussing the energy of a massive oscillator) and just counted the number of such cells in the energy range between

E and $E+dE$, multiplying the result by two to take into account the photon's two possible independent states of polarization. "The rest of the calculation," far from being "essentially the same" as Debye's derivation, is based on the substitution of energy quanta for quanta of energy. This raises deep problems, to which I shall return in the third section; but first let us turn to Bose's contacts with Einstein.

Bose and Einstein

There is a conflict between two accounts, both based on discussions with Bose, of the sequence of events that led Bose to send the manuscript of his paper on Planck's law to Einstein. They differ by a half-year on when it was written, and on where it was first sent. According to Blanpied's account, based on an interview with Bose in 1971, "late in 1923 [Bose] submitted a paper on the subject to the *Philosophical Magazine*. Six months later the editors of the journal informed him that (regrettably) the referee's report on his paper was negative. Undeterred, he sent the rejected manuscript to Einstein. . ." (Blanpied 1972, p. 1215). According to Mehra and Rechenberg, it was only after a visit by Saha (who had moved to Allahabad shortly after Bose left for Dacca) in March 1924 that Bose finished his work: "In the beginning of June 1924 he had the manuscript of a paper ready, which he entitled 'Planck's Law and the Light Quantum Hypothesis.' Since he had the impression that his treatment very much followed Einstein's thought—indeed, it completed the proof which Einstein had been searching for many years—Bose decided to send his paper to Einstein. . ." (Mehra and Rechenberg 1982, p. 565).

I believe Blanpied's story is correct,[2] and that Mehra and Rechenberg's account actually concerns the circumstances in which Bose composed his *second* paper on the topic of thermal radiation, "Wärmegleichgewicht im Strahlungsfeld bei Anwesenheit von Materie" ["Thermal Equilibrium of the Radiation Field in the Presence of Matter"], also translated and sent to the *Zeitschrift für Physik* by Einstein. It refers to the first paper as "to appear in Phil. Mag." (Bose 1924b, p. 386, note 1), which is hard to square with the story Mehra and Rechenberg tell about the first paper, while its contents fit well with the assumption that the story applies to the second paper.

After finishing his first paper late in 1923 and sending it to *The Philosophical Magazine*, Bose continued to work on the radiation problem, turning his attention to the interaction between radiation and (ordinary) matter. In March 1924 Saha visited Bose in Dacca, and directed Bose's attention to recent papers by Pauli (1923) and by Einstein and Ehrenfest (1923), both of which attempted to generalize Einstein's famous 1917 derivation of Planck's law; and to Compton's 1923 explanation of the anomalous scattering of X-rays, which was based on the photon picture. Bose then prepared his second paper, which must have been completed before 4 June 1924, the date of his first letter to Einstein. By then he presumably had received notice of rejection by the *Philosophical Magazine* (if the second paper had not been finished before he received the rejection, it is inexplicable why it should refer to the first as forthcoming in that journal). The letter reads:

Respected Sir,

I have ventured to send you the accompanying article for your perusal and opinion. I am anxious to know what you think of it. You will see that I have tried to deduce the coefficient $8\pi\nu^2/c^3$ in Planck's Law independent of the classical electrodynamics, only assuming [*that finite* (crossed out in letter)] that the ultimate elementary regions in the Phase-space has the content h^3. I do not know sufficient German to translate the paper. If you think the paper worth publication, I shall be grateful if you arrange for its publication in Zeitschrift für Physik. Though a complete stranger to you, I do not feel any hesitation in making such a request. Because we are all your pupils through your writings. I do not know whether you still remember that somebody from Calcutta asked your permission to translate your papers on Relativity in English. You acceded to the request. The book has since been published. I was the one who translated your paper on Generalised Relativity.

Yours faithfully,
SN Bose
(Einstein Archive, Doc. 6-127).

Only eleven days later, on 15 June, Bose sent Einstein his second paper, with the following letter:

Respected Master,

I send herewith another paper of mine for your kind perusal and opinion. I hope my first paper has reached your hand. The result to which I have arrived seems rather important (to me at any rate). You will see that I have dealt with the problem of thermal equilibrium between Radiation and Matter in a different way, and have arrived at a different law for the probability for the elementary processes [presumably, different from the law proposed in Einstein and Ehrenfest 1923], which seems to have simplicity in its favor. I have ventured to send you the typewritten paper in English; it being beyond me to express myself in German (which will be intelligible to you). I shall be glad [*to* (crossed out in letter)] if its publication in Zeitschrift für Physik or any other German journal can be managed. I myself know not how to manage it. In any case, I shall be grateful if you express any opinion on the papers, and send it to me at the above address.

Yours truly
SN Bose
(Einstein Archive, Doc. 6-128).

On 2 July Einstein sent a postcard in response to Bose's first paper:

Dear Colleague:

I have translated your paper and given it to the *Zeitschrift für Physik* for publication. It signifies an important step forward and pleases me very much. However, I do not find your objection to my paper correct. For Wien's displacement law does not presuppose the wave theory and Bohr's correspondence principle is not used. But this is unimportant. You have derived the first factor quantum-theoretically even if not quite rigorously on account of the polarization factor 2. It is a beautiful step forward.

With friendly greetings your A. Einstein.
(copy in another hand, Dacca University Archives).[3]

Einstein said as much in a note he appended to his translation of the article: "Bose's derivation of Planck's law appears to me to be an important step forward.

The method used here also yields the quantum theory of the ideal gas, as I shall show elsewhere" (Bose 1924a, p. 181).

With the help of this postcard Bose was able to get quick approval of a study leave abroad, for which he had applied some time earlier.

> As soon as [the Senate Council] showed it to Hartog [the Vice-Chancellor], it solved all problems. As a student Hartog had spent some time at the University of Paris and he understood something of what such an experience could do for a young man. That little thing [the postcard from Einstein] gave me a sort of passport to the study leave. They gave me leave for two years and rather generous terms. ... Then I also got a visa from the German consulate just by showing them Einstein's card. They did not require me to pay the fee for the visa! (Mehra and Rechenberg 1982, p. 569).

Bose sailed for Europe in September 1924, going first to Paris, where he stayed for about a year. According to Mehra and Rechenberg, "Bose had not gone to Berlin straight away from Dacca because he felt that his knowledge of German was not good enough" (Mehra and Rechenberg 1982, p. 569). According to a physics student with whom he became friendly, he spoke excellent French by the time he left Paris; while there, he attended Paul Langevin's lectures at the Collège de France, but otherwise did not work too hard at physics (interview with Mme. Jacqueline Eisenmann, Paris, 21 May 1994).[4]

Soon after reaching Paris, he contacted Einstein:

> Dear Master,
>
> My heart-felt gratitude for taking the trouble of translating the paper yourself and publishing it. I just saw it in print before I left India. I have also sent you about the middle of June, a second paper entitled "Thermal Equilibrium in the Radiation field in the presence of matter".
>
> I am rather anxious to know your opinion about it [*before* (crossed out in letter)] as I think it to be rather important. I don't know whether it will be possible also to have this paper published in Zeit. für Physik. I have been granted study leave by my University for 2 years.
>
> I have [*just* (crossed out in letter)] arrived just a week ago in Paris. I don't know whether it will be possible for me to work with you in Germany. [*If* (crossed out in letter)] I shall be glad however if you grant me the permission to work under you, for it will mean for me the realisation of a long-cherished hope.
>
> I shall wait for your decision as well as your opinion of my second paper here in Paris.
>
> If the second paper has not reached you by any chance. Please let me know, I shall send you the copy I have with me.
>
> With respect
> yours sincerely
> SN Bose (Prof.)
> (26 October 1924, Einstein Archive, Doc. 6-129).

Einstein replied on 3 November, assuring Bose that his second paper was already in print:

> Most esteemed colleague!
>
> Friendly thanks for your letter of 26.10. I am happy to have the opportunity soon to meet you personally. Your papers have appeared already some time ago; unfortu-

nately the offprints were sent to me instead of you. You can have them at any time. I am not in agreement with your elementary law of probability for the interaction between radiation and matter, and have given the reasons in a note that appeared together with your article. To wit, your law is not compatible with the following two conditions:

1. The coefficient of absorption is independent of the density of the radiation.

2. The behavior of an oscillator in a radiation field must result from the statistical laws as a limiting case.

We can discuss this in more detail when you come here.

With friendly greetings

(retained carbon copy, Einstein Archive, Doc. 6-130).

On 27 January 1925 Bose replied from Paris:

Revered Master,

I received your kind note of 3rd Nov. in which you mentioned your objections against the elementary law of Probability. I have been thinking about your objections all along, and so did not answer immediately. It seems to me that there is a way out of the difficulty, and I have written my ideas in the form of a paper which I [*enclose* (crossed out in letter)] send under a separate cover.

Bose never published this third paper, and no copy of the manuscript has been found. As we shall see in a moment, Einstein was unmoved by Bose's new paper, and it is possible that when they met in Berlin he convinced Bose, but later remarks by Bose suggest that he still felt the paper contained important new results. Returning to the letter, it concludes:

I cannot exactly express how grateful I feel for your encouragement, and the interest you have taken in my papers. Your first postcard came at a critical moment, and it has more than any other made this sojourn to Europe possible for me. I am thinking of going to Berlin at the end of the winter, where I hope to have your inestimable help and guidance.

Yours sincerely

SN Bose

(27 January 1925, Einstein Archive, Doc. 6-131).

By the following October, Bose was in Berlin:

Dear Sir,

I have just arrived in Berlin, and am very anxious to see you. I shall be glad if you kindly write when it will be convenient for you.

Yours sincerely

S.N. Bose

(8 October, 1925, Einstein Archive, Doc. 6-132).

Here he met Einstein, and during the next year was plunged into one of the most exciting periods in physics. The following May, he wrote a friend:

Everybody (every physicist) seems to be quite excited in Berlin, about the way things have been going on with Physics, first on the 28th [of April] last Heisenberg spoke in the Colloquium about his theory, then, in the last Colloquium, there was a long lecture on the recent hypothesis of the spinning electron, (perhaps you have heard of it.) Everybody is quite bewildered, and then there is going to be very soon a discussion

> of Schrödinger's papers. Einstein seems quite excited about it; the other day coming from the Colloquium, we suddenly found him, jumping in the same compartment, where we were, and forthwith he began to talk excitedly about the things we have just heard. He has to admit that it seems a tremendous thing, considering the lot of things which these new theories correlate and explain, but he is very much troubled by the unreasonableness of it all. We were all silent, but he talked almost all the time; unconscious of the interest [*everybody is taking* (crossed out)] and wonder that he is exciting in the mind of the other passengers (Bose to Jacqueline Zadoc-Kahn, May 1926).

According to Mehra and Rechenberg: "Einstein introduced Bose to a number of people in Berlin, including Fritz Haber, Otto Hahn, Lise Meitner and Walter Gordon. He also proposed two problems to Bose to work on: first, the question of whether the new statistics implied a novel type of interaction between the light-quanta" (Mehra and Rechenberg 1982, p. 571). The second problem was described by Bose himself: "He wanted me to try to see what the statistics of light-quanta and the transition probabilities of radiation would look like in the new theory [i.e., quantum mechanics]" (cited from Venkataraman 1992, p. 120). As we shall see in the next section, these two questions go to the heart of Einstein's concerns about the interpretation of Bose's work.

Apparently Bose made no progress on either problem. In May 1926 he wrote:

> I have made an honest resolution of working hard during these months, but it is so hard to begin, when once you have given up the habit (letter cited above).

But the reasons for his failure probably lie deeper. Like many contemporaries who had mastered the old quantum theory, he seems never to have been comfortable with the new quantum mechanics.

A two-page note in Bose's hand, on "Fluctuations in density," in the Einstein Archive, which seems related to Einstein's second paper (see the next section), may be from Bose's stay in Berlin. But Bose published nothing during his stay in Europe, returning to Dacca in the late summer of 1926. A year later he was appointed Professor of Physics, supported by a letter from Einstein (Mehra and Rechenberg 1982, p. 571).

There seems to have been no further contact between the two until 1945, by which time Einstein had moved to Princeton and Bose to University College of Science in Calcutta, when Bose sent Einstein a paper "by an esteemed friend of mine Principal B. M. Sen" of Presidency College in Calcutta. He ended his letter on a personal note:

> As one who owes much to you and your guidance in life, I have always been anxious to have news of yourself, especially under the new conditions when you are in a new country. Some of my friends who are more lucky than me, had chances to visit the States recently, and had the good fortune of seeing you in your new house. I am glad to know that you are well, and still take a lot of interest in India and Indians (9 December 1945, Einstein Archive, Doc. 6-134).

During the 1950's Bose worked on Einstein's last unified field theory. A copy of a manuscript on this topic and drafts of two replies by Einstein are in the Einstein Archive.

Einstein Ponders Bose's Results

Now let us return to 1924 and look at matters from Einstein's point of view. As noted above, Bose's first letter to Einstein is dated 4 June 1924. Mehra and Rechenberg estimate that mail from Dacca to Berlin probably took two to three weeks at the time, so that Einstein received the letter in late June. By 2 July, when he sent his postcard to Bose, he had submitted his translation of Bose's paper to the *Zeitschrift für Physik*. Einstein's first paper based on Bose's work, "On the Quantum Theory of the Monatomic Ideal Gas" (Einstein 1924a), was presented to the Prussian Academy of Sciences at its session of 10 July. Thus, it took Einstein no more than a week or two to digest and translate Bose's work, and conceive and complete his own extension of it to the quantum ideal gas. Clearly, Bose's work triggered a creative outburst in a mind already well primed to respond to it. Let us try to understand why Einstein was so well prepared.

In his second paper on the subject, "On the Quantum Theory of the Monatomic Ideal Gas. Second Paper" (Einstein 1924b), Einstein notes that his theory "is based on the hypothesis of a far-reaching formal relationship between radiation and gas. According to this theory, the degenerate gas deviates from the gas of [classical-JS] statistical mechanics in a way that is analogous to that in which radiation obeying Planck's law deviates from radiation obeying Wien's law. If Bose's derivation of Planck's radiation formula is taken seriously, then this theory of the ideal gas is inevitable; for if it is justified to regard radiation as a quantum gas, then the analogy between quantum gas and gas of molecules must be complete" (Einstein 1924b, p. 3). In his third paper on the subject, "On the Theory of the Ideal Gas" (Einstein 1925), he reiterates this point: "This theory seems justified if one starts with the conviction that (apart from its property of being polarized) a light quantum differs essentially from a monatomic molecule only in that its rest mass is vanishingly small"—but hastens to add that this analogy is by no means sanctioned by all researchers (he may well have had Bohr—a well known opponent of the photon concept—in mind).

He also notes a second objection: "[T]he statistical method applied by Bose and myself is by no means indubitable, but on the contrary only appears to be justified *a posteriori* by its success in the case of radiation" (Einstein 1925, p. 18). Einstein thus highlighted two issues: 1) The analogy between radiation and matter (calculation of the first factor), and 2) the method of counting equiprobable states employed in deriving the probabilities W for use in Boltzmann's principle, $S = k \ln W$ (calculation of the second factor). By looking at the earlier history of these two issues in Einstein's work, we may hope to understand his ability to accept and apply Bose's method so quickly, as well as his abiding conviction that much still remained obscure about the reasons for its success.

Einstein's speculations on the possible analogy between matter and radiation can be documented as early as 1901, when he wrote: "Recently the idea came to me that in the production of light perhaps a direct transformation of motional energy into light takes place due to the parallelism kinetic energy of molecules—absolute temperature—spectrum (spatial radiant energy in the state of equilibrium" (Einstein to Marić, 30 April 1901, Doc. 101, Einstein 1987, p. 293). Since he was convinced early on of the correctness of the atomic model for matter, the analogy suggests the need for an atomic model of radiation.

In 1905 he developed this idea in his light quantum paper, "On a Heuristic Point of View Concerning the Production and Transformation of Light" (reproduced as Doc. 14, Einstein 1989, pp. 149–169). In it, he shows that:

> Monochromatic radiation of low density (within the range of validity of Wien's radiation formula) [high-frequency, low-temperature limit of Planck's: $\rho(\nu, T) = 8\pi\nu^2(h\nu)/\{c^3 \exp(-h\nu/kT)\}$], behaves thermodynamically as if it consisted of energy quanta that are independent of each other and of the magnitude, [$h\nu$, in modern notation] (ibid., p. 161).

He shows this by an ingenious inversion of Boltzmann's principle [which Einstein so named] $S = k \ln W$, a trick he often used thereafter. Instead of finding the entropy S from a known probability W, he read off the unknown probability from the entropy of radiation obeying Wien's law. This leads to the probability W for the total energy E of the radiation, at any given moment, to be concentrated in a subvolume v of the total volume V:

$$W = (v/V)^{E/h\nu}.$$

Now he introduces the analogy between radiation and matter. Consider a gas in equilibrium consisting of n particles in a volume V, each moving according to any law of motion, so long as it does not single out a preferred point or direction in V. Then the probability w that any one particle will be in the subregion v is clearly: $w = (v/V)$. If the particles are *statistically independent* of each other (as classical particles in an ideal gas in equilibrium always are), then the probability that all n particles are in v at a given moment is clearly the product of the probabilities of n statistically independent events:

$$W = w^n = (v/V)^n.$$

If we take the analogy between radiation and matter seriously, then comparison of the two formulae yields Einstein's conclusion: Radiation in the Wien region behaves like a gas of independent particles of energy $E = h\nu$.

Given these considerations, one might wonder why Einstein did not immediately apply a similar argument to Planck's law. Why did he not use the well-known form of the entropy of black body radiation to calculate the probability of a certain distribution of light quanta in phase space—in other words, why did Einstein not derive Bose statistics? I believe he probably did make some such attempt, but failed. Why publish what he calls "a *heuristic* point of view," based on the Wien limit of Planck's law, without first trying—and failing—to account for the entire spectrum covered by Planck's law? Such a failed attempt would explain an otherwise curious remark near the beginning of Einstein's next (1906) paper on radiation, "On the Theory of Light Production and Absorption" (reproduced as Doc. 34, Einstein 1989, pp. 350–358). After briefly summarizing his light quantum hypothesis in the first paragraph, Einstein goes on: "At the time [of his 1905 paper] it appeared to me as if Planck's theory of radiation in certain respects was counterposed to my work." That is, he originally believed that Planck's law did *not* support the light quantum hypothesis; now he was able to give an argument showing "that Planck's theory makes implicit use of the above-mentioned light quantum hypothesis" (ibid., p. 350).

If there was such a failed attempt to understand Planck's law in 1905, it could only have reinforced several quite independent reasons why Einstein did not pursue a "quantum gas" model of black-body radiation around 1905:

1) Aside from energy, he had no warrant to attribute other particle-like properties to light quanta. Only between 1909 and 1916 did Einstein develop arguments that finally led him to attribute momentum and spatial localization to them.

2) He was already well aware that, whatever they might be, light quanta could *not* be statistically independent of each other, as "normal" particles should be. Inverting the argument of his 1905 paper: if light quanta *were* independent, a light quantum gas would obey *Wien's* law and not Planck's.

He made both these points explicitly in 1909:

> Indeed, I am not at all of the opinion that one should think of light as composed of quanta localized in relatively small spaces and independent of each other. This would indeed be the most convenient explanation of the Wien end of the radiation formula. But, by itself, the division of a light ray at the surface of a refracting medium completely forbids this conception. A light ray can divide, but a light quantum cannot divide without change of frequency (Einstein to Lorentz, 23 May 1909, Doc. 163, Einstein 1993, p. 193).

3) If light quanta are *not* statistically independent, even in an ideal radiation gas, Einstein had no idea how to calculate the probabilities for an ensemble of light quanta to be distributed over various states. Einstein had strong views about the nature of probability: The time-ensemble definition of the probability of the state of a physical system was primary for him. Any *combinatorial* argument, based on counting the number of complexions of the elements composing the system that correspond to a given physical state, must be supplemented by a *dynamical* argument proving that these complexions are equally probable before such a count could be used to define the W needed in Boltzmann's principle.[5]

Planck's original derivation of the black-body spectrum applied such combinatorial considerations to the energy of an ensemble of material oscillators in thermal equilibrium with the radiation field, just assuming all complexions to be equiprobable; and a decade later Debye applied similar considerations to the radiation field itself, considered as an ensemble of normal modes in the cavity. In both cases, the problem was to divide the fixed total energy into quanta of magnitude $h\nu$ and distribute them among the material oscillators (Planck) or the normal modes of the radiation field (Debye). Complexion here is just the name for one such way of distributing the total energy. The counting of complexions is based on the assumption that, while the material oscillators or the normal modes of the radiation field are *distinguishable*, the quanta of energy of a given frequency, being just aliquot portions of the total energy at that frequency, are *indistinguishable*. If there are seven quanta of energy associated with one oscillator and five with another, it makes no sense to ask *which* seven and *which* five quanta of energy are involved.

Planck and Debye just *assumed* that the number of such complexions is a measure of the probability of a state, that is, that each complexion is equally probable. And here Einstein's criticism applies. They could offer no dynamical argument why such complexions should be equiprobable in *his* (time ensemble) sense of probability. Rather, their procedure was justified *a posteriori* by its success in producing Planck's formula. An attempt to turn this argument, based on quanta of energy, into an argument based on energy quanta, i.e., on a light-quantum gas, faces two problems. How to justify the count of the number of possible states of the gas corresponding to a given energy (or frequency)—which is the problem of the first factor; and how to justify the counting of equal probabilities—the problem of the second factor. If the first problem is solved, the second only becomes more acute. For, in addition to Einstein's *general* objection to the counting of complexions, a *new* objection arises when such a count of complexions in the Planck–Debye sense is interpreted as a count of complexions for *particles*. While elements of the total energy may be indistinguishable, in the sense that it makes no sense to ask *which* quanta of that energy are in a given state, how can one avoid this question for energy quanta, i.e., particles? Classically, one *cannot*, and it must have seemed by the nature of the particle concept that such a question makes sense. No wonder that even Einstein paused before entering such a statistical morass.

But Bose did so, with a courage perhaps born in part from an incomplete awareness of the perils[6]—the very reason that young men are supposed to make the best soldiers. (More seriously, it is often the case that knowing too much inhibits imaginative thinking.) As we have seen above, he solved the first problem—explaining the first factor in Planck's formula—fairly straightforwardly, by dividing the phase space into cells of size h^3 and counting them. To solve the second problem, he turned the problem of partitioning *quanta of energy* among normal modes of the field into one of distributing *energy quanta*—that is photons, considered as particles—among cells in phase space; but continued to count with the *old* method to solve the *new* problem.[7] Now the cells, like the normal modes, are certainly independent of each other; but, if the old counting method is used for the new problem, then the photons *cannot* be statistically independent. For if they were, as Einstein had known since 1905, Boltzmann's standard counting method for independent particles, accepted by everyone, would lead to Wien's radiation formula rather than Planck's.

As we have seen, Einstein's reaction was to accept the force of Bose's method applied to radiation—after all, it did lead directly to Planck's law—and combine it with his belief in the analogy between radiation and matter: If one wants the quantum theory of an ideal (monatomic) gas, the same method must be applicable. The only differences between the cases of radiation and matter are: 1) the relation between energy and momentum for a massive particle must be used; 2) as well as the total energy, the number of particles must be kept constant in maximizing the entropy.

It must have been clear from the start to Einstein that the use of the new counting method for distributions of particles among cells in phase space amounts to the assumption of a certain kind of statistical dependence between them: If they were counted as if they were independent—that is, in the Boltzmann way—this would have lead to the classical ideal gas law.

I say it must have been clear to Einstein from the outset, although he did not make the point in his first paper. But when it was pointed out to him, he responded in his second paper in a way that suggests he regarded the point as obvious:

> Ehrenfest and other colleagues have reproved Bose's theory of radiation and my analogous one of the ideal gas because in this theory the quanta and the molecules are not treated as statistically independent entities without any special indication of this circumstance in our papers. This is completely correct. If one treats the quanta as statistically independent of each other in their localisation [in cells in phase space–JS], one arrives at Wien's radiation law; if one treats the gas molecules analogously, one arrives at the classical equation of state of the ideal gas, even if one proceeds otherwise exactly in the same way as Bose and I have done (Einstein 1925a, p. 5).

Einstein proceeds to show this in detail, but rather than quote the paper, I shall quote a contemporaneous letter to Schroedinger, which makes the same point more graphically and less technically.

> In the Bose statistics employed by me, the quanta or molecules are not treated as being *independent of one other*. . . . A complexion is characterized through giving the number of molecules that are present in each individual cell. The number of complexions so defined should determine the entropy. According to this procedure, the molecules do not appear as being localized independently of one another, but rather they have a preference to sit together with another molecule in the same cell. One can easily picture this in the case of small numbers. [In particular] 2 quanta, 2 cells:

	Bose-statistics			independent molecules	
	1st cell	2nd cell		1st cell	2nd cell
1st case	o o	–	1st case	I II	–
2nd case	o	o	2nd case	I	II
3rd case	–	o o	3rd case	II	I
			4th case	–	I II

> According to Bose, the molecules stack together relatively more often [two out of three cases] than according to the hypothesis of the statistical independence of the molecules [two out of four cases] (Einstein Archives, Doc. 22 002, translation cited from Howard 1990, p. 67).

Indeed, this pecularity of Planck's counting was pointed out in Ehrenfest and Kammerlingh Onnes 1914 as an argument *against* light quanta (presumably known to Einstein, who was in close contact with Ehrenfest). Contrasting Planck's method of counting quanta of energy with Einstein's energy quantum hypothesis, they state:

> Planck does not deal with really mutually free quanta ε. The real *object which is counted* remains the number of all different distributions of N resonators [oscillators] over the energy grades [levels] 0, ε $2\varepsilon, \ldots$ with a given total energy $P\varepsilon$.

After contrasting the two counting methods for $P = 3$ and $N = 2$, setting up a table similar to Einstein's, they conclude:

> *Planck's formal device* (distribution of P elements of energy ε among N resonators) *cannot be interpreted in the sense of Einstein's light quanta* (Ehrenfest and Kammerlingh Onnes 1914, cited from Mehra and Rechenberg 1982, p. 560).

But apparently, until Einstein reacted to Bose's work, it never occurred to anyone to make a virtue of this "defect" by interpreting it as evidence for a new type of statistical *dependence* between light quanta, peculiar to quantum theory.

In short, what Bose called "essentially the same calculation" of the second factor in Planck's formula was actually a revolutionary act: in going from quanta of energy to energy quanta, he introduced a type of statistical dependence between the latter peculiar to quantum particles. This feature is often characterized as the *indistinguishability* or *identity* of particles. While these terms express certain features of the situation, as Howard 1990 points out there are other features that they do not. Using Einstein's example quoted above, while in the 1st and 3rd cases of Bose-statistics, the particles are indistinguishable, they are *not* identical (in at least one sense of the word); for if they were, there would be only *one* particle in the occupied cell. On the other hand, in the 2nd case they are certainly distinguishable (in at least one sense of the word): even if we cannot tell which is which (as we can in the classical 2nd and 3rd cases), one occupies a *different* cell of phase space from the other. What term, then, best describes this peculiar property of the Bose particles? Howard 1990 suggests *non-separability*, assimilating this property to that chacterizing any composite quantum system after interaction between its elements. Others, including myself, prefer the more general term *quantum entanglement* (but what's in a name?), a topic to which I shall return shortly.

From the beginning, Einstein saw the use of a new statistics in counting complexions as mysterious and disturbing. Two days after submitting his first paper on the quantum gas, he wrote to Ehrenfest that "the Indian Bose has given a beautiful derivation of Planck's law, including the constant ... but the essence remains obscure" (Einstein to Ehrenfest, 12 July 1924, Doc. 10 089). In his second paper, he notes that Bose's formula for the counting of complexions of equal probability

> thus expresses indirectly a certain hypothesis about how the molecules mutually influence each other that is temporarily of an enigmatical nature, which [hypothesis] determines just the equal statistical probability of the cases defined here as complexions (Einstein 1924b, p. 6).

In line with his earlier critique of the method of determining "probabilities in the sense of Planck" (as he almost always characterizes them), it was clear to Einstein that something more fundamental must lie behind Bose statistics.

This problem was still bothering Einstein when he met Bose and, as you may remember, "proposed two problems to Bose to work on: first, the question of whether the new statistics implied a novel type of interaction between the light-quanta; second, what the new statistics meant within the scheme of quantum mechanics" (Mehra and Rechenberg 1982, p. 571).

A word on the second problem, even though it is not strictly germane to my topic. If Bose or Einstein had been able to solve it, they would have discovered the need to totally symmetrize the wave function of N bosons discovered by Dirac in 1926. As Pais 1982 points out (pp. 424–425), the historical development of quantum statistical mechanics reversed the logical sequence. Bose, followed by Einstein, discovered how to do the course-grained counting for bosons before Dirac's fine-grained counting justified it. But there was considerable confusion over when to apply Bose–Einstein and Fermi–Dirac statistics, and for a time they were rivals until it became clear they peacefully divide the quantum kingdom on the basis of spin.

But, to return to the first problem, what Einstein had in mind is clear from a talk of September 1924 "On the Aether" (Einstein 1924), to which Howard 1990 drew my attention:

> Recently a paper of the Indian Bose on the derivation of Planck's formula has appeared, which is of particular importance for our theoretical outlook for the following reason: Previously all complete derivations of Planck's formula made use of the hypothesis of the undulatory structure of radiation. Thus, for example, the factor $8\pi h\nu^2/c^3$ in this formula was obtained in the well-known Ehrenfest–Debye derivation by counting the number of proper vibrations in the cavity in the frequency range $d\nu$. Bose replaced this counting based on the concepts of the wave theory by a gas-theoretical calculation, which he relates to the light quanta present in the cavity, the quanta being thought of as a kind of molecule. The question then arises, if the phenomena of diffraction and interference might not be connected with the quantum theory in the following way: the field-like concepts of the theory only represent expressions of the interactions between quanta, an independent physical reality no longer being ascribed to the field.[8]
>
> The important fact that, according to Bohr's theory, the frequency of the emitted radiation is not determined by electric charges that undergo periodic processes of the *same* frequency can only strengthen us in this doubt about the independent reality of the undulatory field" (Einstein 1924b, p. 93).

Einstein was looking for some way of assimilating *entanglement* of quantum systems, which, in retrospect, we can see first manifested itself in 1900 in Planck's radiation law—even though it was not understood as such until a quarter of a century later—to some more traditional type of *interaction* between particles. The circumstance that entanglement occurs even in the case of an ideal gas (of massive particles or photons), in which there are *no* interactions between particles except for collisions needed to randomize them, tended to militate against such a possibility from the start. The discovery of Fermi–Dirac statistics made the problem of quantum entanglement of otherwise-free particles of the same species more acute, by showing that it comes in two brands.

But there is more to entanglement than this. The work of Einstein, Podolsky and Rosen and subsequent tests of Bell-type inequalities show that, once two quantum systems of any type interact (in the ordinary sense of the word), they are quantum-entangled forever; they suggest that, even if quantum mechanics is modified or replaced, some such entanglement feature will almost surely characterize the new theory.

So far, I have discussed quantum entanglement of two or more particles. The Feynman approach suggests that, even in the one-particle case, entanglement is *the* feature distinguishing quantum theory from classical physics. This approach comes as close as can be to fulfilling Einstein's hope of eliminating the field and explaining interference effects in terms of "interactions" between particles. In the one-particle case, this can be done by introducing a new type of "interaction," or entanglement: one between all *possible* paths the particle could take between its initial preparation and final registration that are compatible with, and indistinguishable from each other within, the given physical arrangement. This peculiarly quantum entanglement between possible paths is implemented by assignment of a probability amplitude to each such path, and addition of all such amplitudes in order to get a total amplitude for the entire process.[9] This addition of amplitudes, rather than probabilities as for distinguishable paths, produces a non-local entanglement of the paths, entirely distinct from any classical interaction, but which accounts for all quantum interference effects. Perhaps it is not stretching things too much to assimilate this one-particle quantum entanglement to the many-particle entanglements discussed above, and say that entanglement is *the* distinctive feature of quantum mechanics.

As discussed in Stachel 1986b, even more than the probabilistic element it was entanglement, and the attendant non-locality that it introduced into the heart of physics, that most bothered Einstein about quantum mechanics. It is ironic that he played a crucial role in introducing both these troubling features of quantum mechanics.

Notes

[1] I am much indebted to Mrs. Enakshi Chatterjee for explaining to me the importance of situating Bose in this context.

[2] Here, I also differ with Bergia 1987, which follows Mehra and Rechenberg's account. Bergia's article contains a very interesting account of the prehistory of Bose–Einstein statistics, including the contributions of Natanson, Wolfke, Krutkow, and DeBroglie, which I shall not discuss.

[3] I am extremely grateful to Professor Parthe Ghose of the Bose Institute, Calcutta, for providing me with a photocopy of this letter.

[4] Interview with (May 21, 1994) and letter from (July 8, 1994) Mme. Jacqueline Eisenmann, Paris. I am grateful to Mme. Eisenmann (née Zadoc-Kahn) for her kindness in pro-

viding information on Bose's visit to Paris and copies of surviving letters from him. At the time of the visit, she was working in Aimé Cotton's laboratory.

[5] Einstein's 1909 criticism of Planck's derivation of his radiation formula shows how Einstein saw the problem:

If one adopts the standpoint that the irreversibility of natural processes is only *apparent*, and that irreversible processes consist of a transition to a more probable state, then one must first of all give a definition of the probability W of a state. The only such definition that can be considered, in my opinion would be the following:

Let $A_1, A_2, \ldots, A_l$ be all the states that an externally closed system with a definite energy is able to assume, or more precisely stated, that we are able to distinguish with certain instruments [i.e., coarse graining]. According to classical theory, after a definite time the system assumes one of these states (for example A_ν), and persists thereafter in this state (thermodynamical equilibrium). According to statistical theory, however, the system assumes all of the states $A_1 \ldots A_l$ in irregular sequence over and over again. If one observes the system for a very long time Θ, then there will be a certain part of this time τ such that during τ and only during τ the system is in the state A_ν. τ/Θ will have a definite limit which we call the probability W of the state in question A_ν.

Starting from this definition one can show that the following equation for the entropy S must hold

$$S = R/N \lg W + \text{const},$$

where the constant is the same for all states of equal energy.

Neither Boltzmann nor Planck have given a definition of W.

Proceeding quite formally, they set $W =$ the number of complexions of the state in question.

If one now requires that these complexions be equally probable, where the probability of a complexion is defined analogously to how we defined the probability of a state [above], then one arrives at exactly the definition of the probability of a state given [above]; except that the logically unnecessary element complexion has been used in the definition.

Although the relation given between S and W only holds when the probability of a complexion is defined in the indicated way or an equivalent one, neither Boltzmann nor Planck has defined the probability of a complexion. ... In the oscillator theory of radiation Planck was not free to choose his complexions. He could only postulate the pair of equations $S = R/N \lg W$ and $W =$ number of complexions if he added the condition that the complexions must be so chosen that, in his theoretical model, they are found to be equally probable on the basis of statistical considerations. In this way, he would have been led to the Jeans formula. (Cited from reproduction in Einstein 1989, Doc. 56, pp. 544–545).

[6] Bose is reported to have told Mehra: "I had no idea that what I had done was really novel ... I was not a statistician to the extent of really knowing that I was doing something which was really different from what Boltzmann would have done, from Boltzmann statistics" (Mehra and Rechenberg 1982). In addition, Bose's very isolation from the physics community in Europe shielded him from the skepticism there about the photon concept that prevailed before Compton's work.

[7] Since he was considering cells in phase space, this amounts to what is now called a coarse-grained counting problem.

[8] This formulation has a Machian ring to it, and indeed Einstein had originally applied what he called "Mach's principle" to the gravitational field. But in the final paragraph of this article, Einstein explicitly excepts general relativity, "because general relativity, to whose fundamental viewpoint physicists will probably always adhere, excludes an imme-

diate action-at-a-distance; but every local-action theory presupposes continuous field, and thus the existence of an 'aether' " (*ibid.*).

[9] The total amplitude must be squared in order to get the probability for the given process. For *distinguishable* alternative processes, the probability of each must be added in the usual (classical) way. For a discussion of the Feynman approach, with references to the original papers, see Stachel 1986a, pp. 314–315 and 331–333. [Note added for this edition: see Stachel 1997 for further elaboration of the relation between quantum entanglement and the Feynman approach.]

References

Bergia, Silvio (1987): "Who Discovered Bose–Einstein Statistics?," in Manuel G. Doncel et al., eds., *Symmetries in Physics (1600–1980)* (Barcelona; Siminari d'Historia de las Ciencies Universitat Autonoma de Barcelona), pp. 222–250.

Blanpied, William A. (1972): "Satyendranath Bose: Co-Founder of Quantum Statistics," *American Journal of Physics* 40: 1212–1220.

Bose, Satyendra Nath (1924a): "Plancks Gesetz und Lichtquantenhypothese," *Zeitschrift für Physik* 26: 178–181.

—— "Wärmegleichgewicht und Strahungsfeld bei Anwesenheit von Materie," *Zeitschrift für Physik* 27: 383–393.

Chatterjee, Enakshi (1994): "The Business of Freeing India," *Miscellany*, January 9, 1994, pp. 7, 15.

Dirac, Paul Adrian Maurice (1926): "On the Theory of Quantum Mechanics," *Proceedings of the Royal Society* A112: 661–667.

Ehrenfest, Paul and Kammerlingh Onnes, Heike (1914): "Vereinfachte Ableitung der kombinatorischen Formel, welche der Planckschen Strahlungstheorie zugrunde liegt," *Annalen der Physik* 46: 1021–1024.

Einstein, Albert (1924a): "Quantentheorie des einatomigen idealen Gases," *Sitzungsberichte der preussischen Akademie der Wissenschaften (physik.-math. Klasse) [n.v.]*: 261–267.

—— (1924b): "Quantentheorie des einatomigen idealen Gases. Zweite Abhandlung," *ibid.*, 3–14.

—— (1924c): "Ueber den Aether," *Schweizerische naturforschende Gesellschaft, Verhandlungen* 105: 85–93.

—— (1925): "Quantentheorie des idealen Gases," *Sitzungsberichte der preussischen Akademie der Wissenschaften (physik.-math. Klasse) [n.v.]*: 18–25.

—— (1987): *The Collected Papers of Albert Einstein*, vol. 1, *The Early Years, 1879–1902*, John Stachel et al., eds. (Princeton: Princeton University Press).

—— (1989): *ibid.*, vol. 2, *The Swiss Years: Writings, 1900–1909.*

—— (1993): *ibid.*, vol. 5, *The Swiss Years: Correspondence, 1902–1914*, Martin J. Klein et al., eds. (*ibid.*).

Einstein Albert and Ehrenfest, Paul (1923): "Zur Quantentheorie des Strahlungsgleichgewichts," *Zeitschrift für Physik* 19: 301–306.

Howard Don (1990): "'Nicht sein kann was nicht sein darf,' or the Prehistory of EPR, 1909–1935: Einstein's Early Worries About the Quantum Mechanics of Composite

Systems," *Sixty-Two Years of Uncertainty*, Arthur Miller, ed. (New York: Plenum), pp. 61–111.

Mehra, Jagdish (1975): "Satyendra Nath Bose," *Biographical Memoirs of Fellows of the Royal Society* 21: 117–154.

Mehra, Jagdish and Rechenberg, Helmut (1982): *The Historical Development of Quantum Theory*, vol. 1, part 2, *The Quantum Theory of Planck, Einstein, Bohr and Sommerfeld: Its Foundation and the Rise of its Difficulties 1900–1925* (New York/Heidelberg/Berlin: Springer-Verlag).

Pais, Abraham (1982): *'Subtle is the Lord ...' The Science and the Life of Albert Einstein* (Oxford: Clarendon Press/ New York: Oxford University Press).

Pauli, Wolfgang (1923): "Über das thermische Gleichgewicht zwischen Strahlung und freien Elektronen," *Zeitschrift für Physik* 18: 272–286.

Stachel, John (1986a): "Do Quanta Need a New Logic?," in *From Quarks to Quasars/ Philosophical Problems of Modern Physics*, Robert Colodny, ed. (Pittsburgh: University of Pittsburgh Press), pp. 229–347.

—— (1986b): "Einstein and the Quantum: Fifty Years of Struggle," in *From Quarks to Quasars/ Philosophical Problems of Modern Physics*, Robert Colodny, ed. (Pittsburgh: University of Pittsburgh Press), pp. 349–385.

—— (1997): "Feynman Paths and Quantum Entanglement: Is There Any More to the Mystery?", in *Potentiality, Entanglement and Passion-at-a-Distance*, Robert S. Cohen, Michael Horne and John Stachel, ets., (Dordrecht/Boston/London:Kluwer Academic), pp. 245–256.

Venkataraman, G. (1992): *Bose and His Statistics* (Hyderabad: Universities Press).

Einstein and "Zweistein"

John Stachel

Abstract: As the sobriquet "Zweistein" suggests, within the theoretical physics community Pauli came to be regarded as second in eminence only to Einstein. Over a period of 35 years, they interacted both intellectually and personally. This paper focuses on four interrelated topics of their discussions:

1) The theory of relativity: At the age of 20, Pauli prepared the definitive survey of relativity, a survey which he revised near the end of his life and which remains invaluable to physicists and historians of physics.

2) Unified field theories: Although initially sympathetic to Einstein's search for a unified theory of electromagnetism and gravitation and having made important contributions to the subject, Pauli came to regard such efforts as fruitless.

3) Foundations of quantum mechanics: Pauli's negative evaluation of Einstein's program grew out of their sharply differing evaluations of the role that quantum mechanics would play in the future development of theoretical physics.

4) Quantum gravity: Pauli recognized that, until the achievement of a successful reconciliation of quantum mechanics and general relativity—an achievement that still eludes us—Einstein's challenge to quantum mechanics could not be laid to rest.

Wolfgang Pauli's sobriquet "*Zwei*stein" well symbolizes his promotion to a place among the "augurs [Auguren]" of theoretical physics, as Einstein called them, second only to that of Einstein himself (according to an article in a Swiss newspaper, "the students at the ETH call him 'Einstein II' for short"). Indeed, during the final decades of both their lives (although Einstein was 21 years older, Pauli only outlived him by three years), Einstein's heretical views on quantum mechanics, the most rapidly developing field of physics, both theoretical and experimental, made him more and more into an icon, rather to be invoked on ceremonial occasions than heeded for his views on the development of physics; while Pauli assumed the role of the living "conscience of theoretical physics" (Weisskopf)—

Talk delivered at the Foundation F. Gonseth
Colloque 2000: Pensée et Science
100th Anniversaire de Wolfgang Pauli
25th Anniversaire mort de Ferdinand Gonseth

"No form of approval could be more precious to physicists, not excluding Bohr, than Pauli's benevolent nodding" (Rosenfeld)—or, in his less benevolent aspect (for, like a Hindu deity, Pauli had several aspects), the role of "the scourge of God" (Ehrenfest).

Indeed, there developed a sharp divergence of views between the two; but, as we shall see, some of the issues that are supposed to have divided Einstein and Pauli really did not, while those that really did remain challenges to contemporary physics.

The two may have first been introduced at the September 1920 Bad Nauheim meeting of the German Society of Scientists and Physicians, where both spoke, Einstein on relativity and Pauli on quantum theory (Pauli also commented on Weyl's paper on his unified field theory). Although only half Einstein's age, by this time Pauli had been at work on both the quantum theory and the theory of relativity for several years.

They are reported to have been introduced by Lise Meitner, who knew them both (von Meyenn–Pauli 1979, p. 536). However, Rudolph Peierls reports an anecdote ("which of course I know only from hearsay"), according to which there was an even earlier interaction between the two:

> [It] goes back to his student days under Sommerfeld. [Pauli] was still unknown but was writing his thesis on the hydrogen-molecule ion using the old quantum theory, while also preparing his big article on relativity... Einstein was visiting, and after a colloquium talk he made some comments, which moved young Pauli to get up at the back of the room and declare, "You know, what Mr. Einstein said is not so stupid!" (Peierls 1985, p. 46).

1) Now I shall turn to Pauli's work on relativity: At the age of 18, he published his first paper on general relativity, soon followed by two on Weyl's new, unified field theory of electromagnetism and gravitation. By early 1920, Pauli was at work on a review of relativity theory, both special and general, which included cosmology and unified field theory. The task was assigned to him by his teacher, Arnold Sommerfeld, editor of the physics portion of the prestigious *Enyclopedia of Mathematical Sciences*, in which it was published in 1921. The work drew the highest praise from the master himself. Einstein wrote:

> No one studying this mature, grandly conceived work would believe that the author is a man of twenty-one. One wonders what to admire most, the psychological understanding for the development of ideas, the sureness of mathematical deduction, the profound physical insight, the capacity for lucid, systematical presentation, the knowledge of the literature, the complete treatment of the subject matter [or] the sureness of critical appraisal (translation from Enz 1973, p. 768).

So invaluable—and imperishable—did this work prove that it was reissued unaltered in English translation in 1958, with twenty-five pages of Supplementary Notes added by Pauli in 1956. Suffice it to say that, when I had to prepare an article on the history of relativity for a *History of Twentieth Century Physics* a few years ago, I found Pauli's comments still to be the best starting point to begin working up the history of many topics (Stachel 1995).

In later years, Pauli did little work on what one may call the classical theory of gravitation. From the mid-twenties on, for about thirty years, general relativity entered upon a period that has been called "the traversal of the desert" (Eisenstaedt 1986, 1989), during which the attention of the physics community was by and large directed elsewhere. Of course, Pauli used the *special* theory of relativity as a tool in many of his greatest achievements, such as the development of the quantum theory of fields, the proof of the spin-statistics theorem, and of the CPT theorem; but: "For many reasons, the history of general relativity (from 1920 to 1960) has been much less spectacular" (Bargmann 1960, p. 195). However, during his stay at the Institute for Advanced Study (IAS) in the 1940s, Pauli and Einstein did collaborate on a paper, proving the non-existence of any stationary, non-singular, asympotically-flat solutions to the gravitational field equations (in four and five dimensions) except those falling off faster than $1/r$ at infinity (Einstein–Pauli 1943; later André Lichnerowicz showed that only Minkowski space-time satisfied these conditions). By the time of the second flowering of general relativity, which started in the sixties and continues to this day, both were dead.

2) From the beginning of his career, Pauli worked on problems of what we now refer to as the old quantum theory, a field to which Einstein had contributed so much. While Pauli's main inspiration in this area undoubtedly came from Sommerfeld and Bohr, here too Einstein's work proved invaluable, as Bohr himself later noted:

> Inspired by Einstein's general statistical considerations of energy and momentum exchange in radiative processes, [Pauli] showed that the scattering probability depends on the intensity of both radiation components involved in the process. The line followed in this work is indeed very closely related to the general dispersion theory, formulated by Kramers, which was to prove so important for the subsequent great developments (Bohr 1960, p. 2).

Indeed, it is not without significance that Pauli chose to write on "Einstein's Contributions to Quantum Theory" in the volume of the "Living Philosophers" series devoted to Einstein and published the year of his seventieth birthday (Pauli 1949). Elsewhere, Pauli emphasized the "fateful interplay of these two directions of research [i.e., relativity and quantum theory] in Einstein's life" (Pauli 1958a, p. 85).

But, with the advent of the new quantum mechanics in the mid-twenties, their views started to diverge. By time of the 1927 Solvay Congress, Pauli had become a firm adherent of the complementarity outlook developed by Bohr, while Einstein remained unconvinced of the completeness of the quantum-mechanical description of physical phenomena and committed to the search for such a complete description of physical reality. Thirty years later, Pauli summed up their differences:

> I myself am among those physicists, who see in the foundations of contemporary quantum mechanics, which postulates fundamental probabilities, a further development of the mode of thought created by Einstein. Specified experimental arrangements that under certain circumstances are complementary to each other here play the role that specified states of motion of the observer do in Einstein's theory of relativity; the finite size of the quantum of action, which sets a limit to the divisibility of

> phenomena in the atomic realm, here plays the role that the maximal signal velocity does in Einstein's special theory of relativity. The group of unitary transformations in quantum mechanics, which includes all possible specifications of the experimental arrangements, plays the role that the group of coordinate transformations plays in the general theory of relativity, which links all possible states of motion of observers and their assertions in conformity with the laws of nature. ... In spite of this, Einstein clings to the narrower conception of reality of classical physics, from which viewpoint a description of nature which admits individual events that are lawful yet not determinate had to appear to him as "incomplete" (Pauli 1956, p. 79).

Returning to the 1927 Solvay Congress, in a remark by Einstein and Pauli's response, one can see the germ of the later challenge in the well-know EPR paper, "Can Quantum-Mechanical Description of Physical Reality be Considered Complete?" (Solvay 1927, Einstein comments, pp. 253–256, Pauli comments, pp. 256–258). When the EPR paper was published in 1935, Pauli reacted with characteristic acerbity, but (after letting off some steam) equally characteristic acuity:

> Einstein has once again expressed himself publicly about quantum mechanics ... As is well known, this is a catastrophe every time it happens. ... Quite independently of Einstein, it seems to me that, in a systematic development of quantum mechanics, one should lay more stress from the beginning on the composition and separation of systems. ... This is indeed—as Einstein has *correctly* sensed—a very fundamental point in quantum mechanics ... (von Meyenn–Pauli 1985, pp. 402–404).

Although his views about quantum mechanics diverged so widely from Einstein's, Pauli understood the latter's views much better than most physicists:

> We have often discussed these questions with one another, and I always had the greatest profit from this, even if I could not agree with Einstein's views (Pauli 1958a, p. 89).

Indeed, in 1954 when Max Born engaged in a rather polemical exchange with Einstein—or rather an exchange in which Born became polemical, while Einstein became resigned to being misunderstood—Pauli, who was visiting the IAS undertook in two letters to Born to set him straight. Since Einstein's position is still often misunderstood, let me cite a few lines from these letters:

> Einstein gave me your manuscript to read; he was *not at all* annoyed with you, but only said you were a person who will not listen. This agrees with the impression I have formed myself insofar as I was unable to recognize Einstein whenever you talked about him in either your letter or your manuscript. It seemed to me as if you had erected some dummy Einstein for yourself, which you then knocked down with great pomp. In particular, Einstein does not consider the concept of 'determinism' to be as fundamental as it is frequently held to be (as he told me emphatically many times).... Einstein's point of departure is 'realistic' rather than 'deterministic,' which means that his philosophical prejudice is a different one. ... To *summarise* I should like to say this: ... your manuscript ... completely bypasses the problems which are of interest to Einstein. In particular, it seems to me misleading to bring the concept of determinism into the dispute with Einstein. ... I have tried in my last letter to explain Einstein's point of view to you. It *is exactly the same* in Einstein's printed work and

> in what he said to me (Pauli to Max Born, 31 March 1954, in Einstein–Born 1971, pp. 221, 224).

Indeed—if for somewhat different reasons—Einstein was as skeptical of deterministic, "hidden-variable explanations" of quantum mechanics as was Pauli (for Pauli, see Pauli 1953; for Einstein, see Stachel 1986, pp. 376–377). I emphasize this point because Einstein's attitude to hidden variables is still often misunderstood.

3) Initially, Pauli was attracted by the unified field theory program, that is the attempt to unite gravitation and electromagnetism into one, "naturally" (i.e., in a mathematically and physically unforced way) unified field theory, and his second paper was a contribution to the study of Weyl's proposed theory. However, from the beginning he was skeptical of Weyl's—and Einstein's—hope that such a unified field theory might contribute to the solution of the problem of the atomistic, quantum structure of matter and radiation:

> Perhaps we are on the wrong track in the first place with continuum theories for the field in the interior of the electron (Pauli 1919, translation modified from Enz 1973, p. 768).

In his 1920 review article he elaborated:

> ...[A] conceptual doubt should be mentioned. The continuum theories make direct use of the ordinary concept of electric field strength, even for fields in the interior of the electron. This field strength is however defined as the force acting on a test particle, and since there are no test particles smaller than an electron or a hydrogen nucleus, the field strength at a given point in the interior of such a particle would seem to be unobservable, by definition, and thus be ficticious and without physical meaning. ... [T]his much seems fairly certain: new elements which are foreign to the continuum concept of the field will have to be added to the basic structure of the theories developed so far, before one can arrive at a satisfactory solution of the problem of matter (Pauli 1921, translation from Pauli 1958b, p. 206).

Although it is doubtful whether Pauli was aware of it, even more radical doubts had been expressed privately by Einstein as early as 1916:

> If the molecular view of matter is the correct (appropriate) one; i.e., if a part of the universe is to be represented by a finite number of moving points, then the continuum of the present theory contains too great a manifold of possibilities. I also believe that this "too great" is responsible for the fact that our present means of description miscarry with the quantum theory. The problem seems to me [to be] how one can formulate statements about a discontinuum without calling upon a continuum (space-time) as an aid; the latter should be banned from the theory as a supplementary construction not justified by the essence of the problem, which corresponds to nothing "real." But we still lack the mathematical structure unfortunately. How much have I already plagued myself in this way! (Einstein letter to Walter Dällenbach, translation from Stachel 1993, p. 280).

Although Einstein is thought of as the field continuum advocate par excellence, there is another strand in Einstein's thought, which I have referred to as "The other Einstein: Einstein Contra Field Theory" (Stachel 1993, see also Part III of this volume), a strand that grew ever stronger in the last years of Einstein's life. So similar, even more strongly skeptical remarks could be cited from *both* Pauli's and Einstein's later writings. Yet, in the absence of viable non-continuum mathematical structures, on which to base theoretical physics, Einstein continued work on the search for a unified field theory to the end of his life, feeling that the continuum approach should be explored to its limits.

Pauli's satire could become savage in comments on Einstein's attempts. In 1931, Pauli wrote:

> [Einstein's] never-failing gift of invention together with his stubborn energy in pursuit of a definite goal has recently bestowed upon us on the average one such theory per year—in which connection it is psychologically interesting that the current theory is usually depicted by its author for a period of time as the "definitive solution." So, in a variant of the well-known historical saying, one might exclaim at the appearance of a new attempt on this subject: "The old field theory of Einstein is dead. Long live the new field theory of Einstein!" (Pauli 1932, translation from Stachel 1999).

Nevertheless, Pauli seems to have had a lingering fondness for unified field theories. He wrote two papers on the projective interpretation of the five-dimensional Kaluza–Klein unification of electromagnetism and gravitation, papers which in Valentine Bargmann's opinion give:

> by far the most satisfactory exposition of spinors in general relativity—quite independent of the problems of a unified field theory (Bargmann 1960, p. 197).

4) Pauli concerned himself early with the relations between quantum theory and general relativity. However, by the time of his fundamental work on the quantum theory of fields, undertaken with Heisenberg (1929), he seems to have felt that the methods of this theory would encounter no special difficulties when applied to gravitation. After stating that quantization of the gravitational field appears to be necessary for physical reasons, the two authors simply assert that it:

> may be carried out without any new difficulties by means of a formalism fully analogous to that applied here [to electromagnetism] (Heisenberg–Pauli 1929, p. 3, translation from Stachel 1999).

With one notable exception, almost the entire handful of physicists who worked on the problem of quantization of the gravitational field during the ensuing decade were associated with Pauli in one way or another: Leon Rosenfeld, Jacques Solomon and Markus Fierz. By the end of the decade, Pauli was ready to sum up and extend this work in one section of a report on "General Properties of the Elementary Particles," prepared for the never-to-be held 1939 Solvay Congress. At the end of this section on "Gravitational Waves and Gravitational Quanta (Spin 2)", he remarks:

> It is certainly a limitation of the quantum-theoretical aspects of this argument that one suffices with that approximation in which the general-relativistic field equations are linear. This limitation is connected most closely with the well-known divergence problems of [quantum] field theory (von Meyenn–Pauli 1993, p. 901).

Here Pauli references a 1938 work of Jacques Solomon. Although Pauli does not cite the following words, it seems likely that he read them:

> ...[I]n the case when the gravitational field is not weak, the very method of quantization based on the superposition principle fails. ... Such considerations are of a sort to put seriously in doubt the possibility of reconciling the *present* formalism of field quantization with the *non-linear* theory of gravitation (Solomon 1938, translation from Stachel 1999).

Solomon based his skeptical conclusions on the work of Matvei Petrovich Bronstein, a Soviet physicist who, constituting the notable exception alluded to above, worked on quantum gravity independently of Pauli. Unfortunately, both Bronstein and Solomon perished (thanks to Stalin and Hitler, respectively) well before work on quantum gravity resumed in the post-war period.

Only once, in the pre-World War II period, have I found a comment by Pauli indicating that he thought something deeper of relevance to elementary particle theory might be hidden within general relativity:

> [I]n contemporary classical field theories, including the relativistic theory of gravitation, the essentially *positive* character of κ [the gravitational coupling constant]—upon which rests the fact that gravitation always manifests itself as an attraction, and never in a repulsion of the gravitating masses—finds no satisfactory explanation. Such an explanation could only consist in reducing the constant κ to the *square* of another natural constant. This suggests looking for phenomena, in which the square root of κ plays a role (Pauli 1934, p. 75).

Pauli's speculation that this might be connected with the weak interactions proved fruitless, but an explanation of the essentially attractive nature of gravitation still eludes us.

By the end of his life, Pauli realized that quantum field theory had not been able to meet the challenge to it offered by general relativity:

> Had we been able to present Einstein a synthesis of his general relativity with quantum theory then the discussion with him would have been considerably easier. But the duality between the field and its means of measurement, although latently present in today's quantum theory of fields, is conceptually not clearly expressed. The relation of the applicability of the ordinary space-time concept in the small with the properties of the smallest physical objects, the so-called "elementary particles" is not disclosed. Einstein's life ended with a question [posed] to the science of physics and with a behest for a synthesis to us (Pauli 1958a, translation from Enz 1973, p. 791).

That challenge still remains before us, and in the course of struggling with it many surprises may still lie in store. Just as an indication of what turns the debate may take (and not because I necessarily agree with him), let me point out that no less an advocate of quantum field theory than Gerard 't Hooft, winner of the 1999 Nobel prize for his work on gauge field theories, wrote an article last year entitled "Quantum gravity as a dissipative deterministic system," from which I will quote just a sentence or two:

> In this paper, a theory [of quantum gravity]is developed that will *not* postulate the quantum states as being its central starting point, but rather classical, deterministic de-

grees of freedom. Quantum states, being mere mathematical devices enabling physicists to make statistical predictions, will turn out to be derived concepts, with a not strictly locally formulated definition ('t Hooft 1999, p. 3264).

Speaking more generally, no way has yet been found to fit together the strikingly different conceptual bases of quantum mechanics and general relativity, the two most successful framework-theories of modern physics. This problem is now usually referred to as the problem of quantum gravity. Two distinct communities of theoretical physicists have worked on this problem, the quantum field theorists and the general relativists. Those, like Pauli, who came to the problem from the side of the quantum theory of fields, which is based upon special relativity, have traditionally assumed that the quantum framework as it stands is sufficient for the solution of the problem, and that, if the methods of this theory do not work on general relativity, it may be necessary to modify the latter (e.g., supergravity). Einstein assumed that quantum mechanics did not provide a complete description of physical reality, and that the quantum effects might be explicable on the basis of some unified classical field theory, the requirement of non-singular solutions of which might provide the needed quantum conditions.

Both of these hopes have proved vain. Today, those approaching the problem of quantum gravity from the side of general relativity usually agree that some process of quantization is needed; but one that differs from the traditional ones of special-relativistic quantum field theory in that it must respect the fundamental invariance group of general relativity, the diffeomorphism group. Those who approach the problem from the side of quantum field theory now usually adopt an approach that starts from string theory in a background space-time of ten or more dimensions. So far, neither approach has been fully successful on its own terms, nor has either been able to incorporate the insights of the other. There is a growing minority, including 't Hooft and Roger Penrose, that believes that some fundamental new insights are needed. Meanwhile the problem of quantum gravity remains the outstanding challenge of the new millenium to theoretical physics.

In summary, we have seen that it was not so much their views about the significance of determinism, nor even about the role of the continuum in the future evolution of physics that separated Einstein and Pauli, but rather a fundamental difference about what constitutes a determinate, objective representation of the physical world. Until the challenge of quantum gravity is met, neither can be declared the victor on this issue.

But let me end, as did their lives, on a note of reconciliation. In spite of wide intellectual divergences, the personal bonds of respect and affection between the two men grew, particularly as a result of a lengthy period of day-to-day contact at the IAS (1935–1936, 1940–1946, and frequent later short visits by Pauli). The culmination of this relationship occurred during the dinner at the IAS to celebrate the award of the 1945 Nobel prize in physics to Pauli, an award for which Einstein among others had nominated him. In the middle of the after-dinner speeches, Einstein arose impromptu, looking (in the words of Franca Pauli) "like an ancient Greek" (Charles Enz, personal communication). On the occasion of Einstein's death, Pauli himself recalled this speech:

Einstein's death has also affected me personally. One who was such a well-meaning, fatherly friend to me is now no more. Never shall I forget the speech that he made in 1945 in Princeton about me and for me after I received the Nobel prize. It was like a king who abdicates and installs me as a sort of "chosen son" [Wahl-Sohn] as his successor (letter to Max Born, 24 April 1955, CERN Pauli Letter Collection. Pauli adds: "Unfortunately, no record of this speech of Einstein exists (it was improvised and a manuscript does not exist either)."

I thank Dr. Enz for drawing this letter to my attention, and for numerous other helpful remarks. Enz 2001 will undoubtedly constitute the definitive biography of Pauli for our time. I thank Mme. Anita Hollier, Section Leader, CERN Archives ETT-SI Division, for providing me with a copy of the German text, and of the newspaper article about Einstein II cited earlier.

References

Bargmann, Valentine (1960): "Relativity," in M. Fierz and V. F. Weisskopf, eds., *Theoretical Physics in the Twentieth Century: A Memorial Volume for Wolfgang Pauli* (New York/London: Interscience), pp. 187–198.

Bohr, Niels (1960): "Foreword" to M. Fierz and V. F. Weisskopf, eds., *Theoretical Physics in the Twentieth Century: A Memorial Volume for Wolfgang Pauli* (New York/London: Interscience), pp. 1–4.

Einstein, Albert and Born, Max (1971): Max Born, ed., *The Born–Einstein Letters: Correspondence between Albert Einstein and Max and Hedwig Born from 1916 to 1955* (New York: Walker and Company).

Einstein, Albert and Pauli, Wolfgang (1943): "Non-existence of regular stationary solutions of relativistic field equations," *Annals of Mathematics* **44**: 131–137.

Eisenstaedt, Jean (1986): "La relativité générale à l'étiage: 1925–1955," *Archive for History of Exact Sciences* **35**: 115–185.

—— (1989): "The Low Water Mark of General Relativity, 1925–1955," in Don Howard and John Stachel, eds., *Einstein and the History of General Relativity: Proceedings of the 1986 Osgood Hill Conference, Einstein Studies*, vol. 1 (Boston: Birkhäuser), pp. 277–292.

Enz, Charles P. (1973): "W. Pauli's Scientific Work," in J. Mehra, ed., *The Physicist's Concept of Nature* (Dordrecht: D. Reidel), pp. 766–799.

—— (2001): *Everything comes to him who knows how to wait: A biography of Wolfgang Pauli* (Oxford University Press).

Heisenberg, Werner and Pauli, Wolfgang (1929): "Zur Quantenelektrodynamik der Wellenfelder," *Zeitschrift für Physik* **56**: 1–61.

Pauli, Wolfgang (1919): "Zur Theorie der Gravitation und der Elektrizität von Hermann Weyl," *Physikalische Zeitschrift* **20**: 457–467.

—— (1921): "Relativitätstheorie," in Arnold Sommerfeld, ed., *Encyklopädie der Mathematischen Wissenschaften*, vol. 5, part 2 (Leipzig: Teubner), pp. 539–775.

—— (1932): Review of *Ergebnisse der exakten Naturwissenschaften, 10, Band, die Naturwissenschaften* **20**: 186–187.

—— (1934): "Raum, Zeit und Kausalität in der modernen Physik," cited from Pauli 1961, pp. 64–75.

—— (1953): "Remarques sur le problème des paramètres cachés dans la mécanique quantique et sur la théorie de l'onde pilote," in André George, ed., *Louis de Broglie: Physicien et Penseur* (Paris: Albin Michel), pp. 33–42.

—— (1956): "Relativitätstheorie und Wissenschaft," cited from Pauli 1961, pp. 76–80.

—— (1958a): "Albert Einstein in die Entwicklung der Physik," cited from Pauli 1961, pp. 85-90.

—— (1958b): *Theory of Relativity* (London/New York/Paris/Los Angeles: Pergamon) [translation of Pauli 1921, with Supplementary Notes]

—— (1961): *Aufsätze und Vorträge über Physik und Erkenntnistheorie* (Braunschweig: Vieweg).

Peierls, Rudolph (1985): *Bird of Passage: Recollections of a Physicist* (Princeton: Princeton University Press).

Solomon, Jacques (1938): "Gravitation et Quanta," *Journal de Physique et de Radium* **9**: 479–485.

Solvay (1927): *Electrons et Photons: Rapports et discussions du cinquième conseil de physique tenu à Bruxelles du 24 au 29 Octobre 1927 sous les auspices de l'Institut International de Physique Solvay* (Paris: Gauthier-Villars, 1928)

Stachel, John (1986): "Einstein and the Quantum: Fifty Years of Struggle," in Robert Colodny, ed., *From Quarks to Quasars: Philosophical Problems of Modern Physics* (Pittsburgh: University of Pittsburgh Press), pp. 349–385. [See this volume, pp. 367–402].

—— (1993): "The Other Einstein: Einstein Contra Field Theory," *Science in Context* **6**: 275–290. [See this volume, pp. 141–154].

—— (1995): "History of Relativity," in Laurie M. Brown, Abraham Pais, Brian Pippard, eds., *Twentieth Century Physics*, vol. 1 (Bristol: Institute of Physics Pub./Philadelphia: American Institute of Physics Press), pp. 249–356.

—— (1999): "Early History of Quantum Gravity (1916–1940)," Talk at the Fifth International Conference on the History of General Relativity, Notre Dame, July 9, 1999.

't Hooft, Gerard (1999): "Quantum gravity as a dissipative deterministic system," *Classical and Quantum Gravity* **16**: 3263–3279.

von Meyenn–Pauli (1979): A. Hermann, Karl von Meyenn, V. F. Weisskopf, eds., *Wolfgang Pauli: Wissenschaftlicher Breifwechsel mit Bohr, Einstein, Heisenberg u.a.*, vol. 1, *1919–1929* (New York/Heidelberg/Berlin: Springer).

von Meyenn–Pauli (1985): Karl von Meyenn, ed., *Wolfgang Pauli: Wissenschaftlicher Breifwechsel mit Bohr, Einstein, Heisenberg u.a.*, vol. 2, *1930–1939* (Berlin/Heidelberg/New York/Tokyo: Springer).

von Meyenn–Pauli (1993): Karl von Meyenn, ed., *Wolfgang Pauli: Wissenschaftlicher Breifwechsel mit Bohr, Einstein, Heisenberg u.a.*, vol. 3, *1940–1949* (New York/Heidelberg/Berlin: Springer).

Part VIII

Book Reviews

'Subtle is the Lord' . . .
The Science and Life of Albert Einstein

by Abraham Pais

Review by John Stachel

The literature on Einstein is enormous. Yet up to now someone trying to find out what Einstein actually did during his creative scientific lifetime (1900–1955) faced a choice between reading one or more popularizations of limited scope (and often even more limited depth) and trying to read and digest the almost 300 scientific papers he produced. Now at last one can enthusiastically recommend a third course: read this book. Here is a coherent account of almost everything of scientific significance that Einstein did, along with a great deal of needed historical background information and an eminent physicist's perspective on the significance of Einstein's achievements. This is not done through dry abstracts of the papers but with style and wit, by a man who knew Einstein during his last decade—if not well, perhaps as well as one could know a man like Einstein. Pais does not hestitate to leaven his text with personal reminiscences and opinions, even the occasional refreshing admission that there are questions about Einstein's *œuvre* that he cannot answer.

The book, moreover, is not merely an account of Einstein's scientific work but also includes a biography. Sometimes put into separate sections and sometimes mixed with the account of Einstein's scientific activities, and taking up perhaps a fifth of the book in all, the biographical material constitutes the most accurate account of Einstein's life yet written. Only someone who has consulted the existing biographies in an effort to establish a simple date can appreciate this achievement.

My recommendation of the book must be tempered by the statement that it cannot be fully understood except by someone trained as a physicist. Perhaps it is sufficient to note that the section of the book headed "The Quantum Theory" has already appeared in almost identical form in *Reviews of Modern Physics*, where it

Review of Abraham Pais, *'Subtle is the Lord' . . .*
The Science and Life of Albert Einstein
in *Science*, Vol. 218 ©1982, pp. 989–990

by no means seemed out of place. The introductory outline-summary of the book (pp. 5–51) provides a capsule account of Einstein's work that demonstrates that Pais could and should do the needed job of *haute vulgarisation* of his own book. Meanwhile, I urge everyone interested to read as much of this book as he or she can in order to learn something about the real Einstein.

After the introductory survey, Pais organizes his account of Einstein's scientific activities around five topics: Statistical Physics (pp. 55–107); Relativity, The Special Theory (pp. 111–174); Relativity, the General Theory (pp. 177–296); The Later Journey (on unified field theory, pp. 299–354); and The Quantum Theory (pp. 357–469). Within each section, the account is basically chronological. This method works quite well in the first four sections, since Einstein's progression from topic to topic was also roughly chronological. But it calls for extensive back-tracking in the section on quantum theory, which returns to 1905 and requires a good deal of cross-reference to the section on statistical physics. I fear this may cause some readers to lose sight of the threads that bound together all of Einstein's interests—statistical, relativistic, and quantum—in his *annus mirabilis* 1905, although Pais is well aware of the connections. The book concludes with Journey's End (an account of Einstein's last decade, pp. 473–479) and four valuable appendixes: accounts of Einstein's collaborators, his Nobel prize, his Nobel prize proposals for others and an accurate chronology. I cannot overemphasize the care that went into producing what seems a smooth and effortless narrative, studded with dates and facts hitherto unavailable in any biography. In this effort Pais was fortunate to have the full cooperation of the late Helen Dukas, who had worked since 1928 as Einstein's secretary and presided over the Einstein Archive after his death.

Pais presents the most convincing account of Einstein's complex personality I have read, based upon personal contacts as well as documentary material. The only issue on which I would seriously disagree is his effort to play down or even deny the rebellious element in Einstein's personality. Einstein described himself as "inwardly rebellious" as a student (letter to Walter Leich, 19 April 1950), and the rebelliousness certainly persisted for some time. Erik Erikson's paper on the young Einstein (in G. Holton and Y. Elkana, Eds., *Albert Einstein: Historical and Cultural Perspectives*, Princeton University Press, 1982) integrates this "defiant" element into the picture of Einstein's personality.

In general the translations of the many quotations from Einstein and others are excellent, although Pais notes the difficulties of finding the *mot juste* in rendering Einstein into English. I rather like Erikson's translation of the epigram from which the book draws its title: "Raffiniert ist der Herrgott aber boshaft ist er nicht": "The dear lord may be tricky but he is not mean."

The discussion of Einstein's political and social views is very brief indeed. His views on pacifism, internationalism, and Zionism are accurately summarized and his generally left-wing political orientation is noted, all without the prejudice and condescension that mark Ronald Clark's more extensive account. I assume it is only a concern for brevity that accounts for the absence of any reference to Einstein's socialist views in his later years or to his very interesting ideas on education.

Pais, unlike some other biographers, pays Einstein the compliment of taking all of his work and views seriously. He is not afraid to disagree with Einstein on quite fundamental matters, such as the interpretation of quantum mechanics; but he makes it clear that Einstein's views were not perverse or quixotic but were based on an alternative vision of the future development of physics that addressed itself to issues unresolved to this day. He also dispels the myth of Einstein the childlike *naïf* in his discussions of Einstein's social views.

Pais has utilized the work of Holton, Klein, Hirosige, McCormmach, and many other eminent Einstein scholars and historians of science, but the serious study of Einstein's life and work is still at too early a stage for anything approaching a definitive account to be written. New documents keep turning up and may be expected to do so for some time. I could cite several instances where factual details given in the book already stand in need of modification.

Even on the basis of the existing documentation, general agreement does not exist among Einstein scholars on such fundamental questions as the development of the special and general theories of relativity. It is not surprising, therefore, that some of Pais's views are open to question. I find it puzzling, for example (to use a favorite phrase of his), that he builds his account of the special theory around the Michelson–Morley experiment, devoting the first 8 1/2 pages of this account (pp. 111ff) to it, only to come to the conclusion that it played a minor role in Einstein's thinking; yet says next to nothing about stellar aberration or the Fizeau experiment, even though he characterizes Einstein's statement that these two results were sufficient for him as "*the most crucial statement Einstein ever made on the origins of the special theory of relativity*" (p. 117, his italics). More exasperating than puzzling is Pais's repetition of the old chestnut that the explanation of the "twin paradox" requires the general theory of relativity—especially since he misattributes this claim to Einstein (p. 145). I think his discussion of the two postulates of the special theory is also somewhat off the mark, largely owing to neglect of evidence that Einstein seriously considered an emission theory of light before adopting the second postulate.

Pais's discussion of the development of the general theory—in particular of the three-year delay between Einstein's adoption of the metric tensor in late 1912 and his discovery of the correct field equations in late 1915—is a careful reworking of what I may call the received version, filled in with many new details based on a careful reading of the relevant papers and correspondence in the Einstein Archive. This version attributes the delay (insofar as it was not due to external circumstances) to Einstein's failure to grasp certain elementary mathematical features of generally covariant equations. Unfortunately, Pais does not seem to have studied Einstein's notebooks from that period, also preserved in the Archive. One of these makes the received version simply untenable, as John Norton will soon show in a forthcoming paper. A number of minor features of Pais's account of the general theory and Einstein's ensuing quest for a unified field theory are slightly awkward, perhaps understandably given that he himself has not worked extensively in this area. In addition, he never explains why Einstein was so taken with the mathematical concept of displacement field or affine connection, on which he based most of his efforts at constructing a unified field theory after 1925. Einstein put it succinctly in 1945: "The essence of the general theory of relativity is to go beyond the inertial system ... it [is] the displacement field which allows one to be freed from the obstacle of the inertial system." (The important letter to Besso in which this statement appears is translated in full in A. P. French, Ed., *Einstein: A Centenary Volume*, Harvard University Press, 1979, pp. 267–269.)

For me, the sections on statistical physics and quantum theory are the most successful in the book. Not only does Pais give an excellent presentation of Einstein's contributions to the development of quantum theory, he explains why Einstein felt that it never became a fundamental theory in his sense, even after the development of quantum mechanics. He makes clear the relationship between Einstein's critique of quantum mechanics and his search for a unified field theory: the non-singular solutions to the unified field equations were somehow finally to provide a fundamental explanation of the quantum riddle. His negative evaluation (p. 456) of the paper by Einstein, Rosen, and Pololsky, however, totally neglects the current discussion of nonseparability in quantum mechanics. There are some smaller weak points. For example, I do not feel the account of Einstein's work on "Gespensterfelder" and its influence on Born is adequate, nor does Pais explore the role of Einstein's challenge to reproduce his blackbody fluctuation formula on Jordan's work on the quantum theory of fields.

Other fascinating topics on which Pais expresses novel views that I don't have space to discuss include the reason for Einstein's fame (pp. 311–312), the interpretation of the concept of scientific revolution (pp. 29–30), and the concept of going beyond "the edge of history" in the analysis of individual creative acts (pp. 163–164).

This book is both unique and indispensable for any serious Einstein scholar, not least for its extensive documentation. May it serve not only as a source of profound insight and pleasure to many readers but as a further spur to the current renaissance of Einstein studies.

Albert Einstein: A Biography

by Albert Fölsing

Review by John Stachel

Einstein scholarship took a great leap forward with the start of work on his *Collected Papers* (Princeton University Press, 1987, John Stachel *et al.*, Morton Klein *et al.*, eds). It then became possible to supplement or replace anecdote, conjecture and myth with documented accounts of important events in Albert Einstein's life, many of them (such as the birth of a daughter out of wedlock and the existence of a notebook recording crucial early steps on his road to general relativity) hitherto unknown. In recent years, numerous biographers have attempted to use the available material to shape a new picture, less hagiographic and more realistic, of Einstein the man and the scientist.

As his early years started to emerge from the twilight, it became tempting to replace the familiar icon of ancient sage and seer with the portrait of a more youthful, impudent, rebellious and altogether more appealing figure. If the stick is sometimes bent too far and an attempt made to turn a plaster saint into an inveterate sinner, we may hope that sooner or later an equilibrium state will be reached: an understanding that Einstein was a human being, although a most extraordinary one, whose genius constantly challenges our ability to comprehend its unique features—one whose life we are better advised to attempt to understand rather than to judge.

Albrecht Fölsing's *Albert Einstein* is certainly the fullest and probably the best of the recent attempts to bring the "new," more human Einstein to a wider public. It is surely no accident that he is most successful in covering the first 40 years of Einstein's life-—just the years the editors of the Einstein papers have already investigated. Although Fölsing has done some admirable research of his own, he is still most heavily indebted to these editors, who made available to him much then-unpublished material. Readers of even the best earlier biographies will be impressed by the wealth of new material about the young Einstein that Fölsing presents in a highly readable and usually accurate form. If he, too, occasionally

Review of Albrecht Fölsing, *Albert Einstein: A Biography*, in *Physics Today*, January 1998 bends the stick a bit too far, I am inclined to pardon him in view of the book's overall biographical merits.

One could wish, however, that Fölsing had been content to end his book two-thirds of the way through it, with Einstein's apotheosis in 1919, saving the last 35 years for a future second volume, after more material on this period becomes available. As it is, his treatment of topics concerning Einstein's later years is often glib and superficial.

To give just one example: His treatment of Einstein's relations with the Jewish community in Germany and abroad, probably the nonscientific topic closest to Einstein's heart, is based almost exclusively on Einstein's letters and writings on a topic—particularly as concerns the history of the world Zionist movement—that calls for much more background study. The treatment is further marred by constant references to Einstein's "tribal companions"—a bad mistranslation of *Stammesgenossen*, roughly "fellow members of the tribe." Fölsing finds Einstein's use of this term singular ("a phrase that sounded odd even to Jewish ears"); but most Jews will find in it the familiar mixture of affection and irony that is as common among Jews in the US today as it was in pre-Hitler Germany.

In spite of the lack of any indication by the publisher, this book constitutes a substantial abridgment of the German original (*Albert Einstein Eine Biographie*, Suhrkamp, 1993). I estimate that almost 20% of the original text remains untranslated, and cuts have not always been made smoothly; cryptic allusions become clear only when one consults the uncut German text. Fölsing has also not always been well served by his translator. For example, well-known scientific terms appear under sometimes bizarre pseudonyms: What should be "non-commutativity" appears as "non-interchangeability,""angular momentum" becomes "torque," "contradictions" becomes "conflicts," "quantum of action" becomes "quantum of effect"—to take only a few examples from chapter 30.

In spite of his physics background and career as a radio and TV science journalist, Fölsing himself "shoots quite a few goats" (the German idiom *einen Bock schiessen* was Einstein's favorite expression for committing a blunder) in scientific matters. Between misstatements and mistranslations, many of the scientific explanations are quite unreliable—and sometimes even unintelligible. Short of an exhaustive list of blunders, one can only advise extreme caution here. It is curious that the author found it necessary to include, and the English editors to preserve (when they cut so much else) a 23-page list of Einstein's scientific publications. A list of his nonscientific publications would have been more useful—especially since many that are not included in earlier lists have been discovered in recent years.

In spite of these caveats, readers who want an up-to-date account of Einstein's life and extrascientific activities, particularly up to about 1920, will find here the best currently available biography. Readers who want a superb and accurate account of Einstein's scientific work and its impact on 20th-century physics are still advised to turn to Abraham Pais's *Subtle is the Lord . . . The Science and the Life of Albert Einstein* (Oxford, 1982).